High-Energy Radiation
from Magnetized Neutron Stars

Theoretical Astrophysics

David N. Schramm, *series editor*

PETER MÉSZÁROS

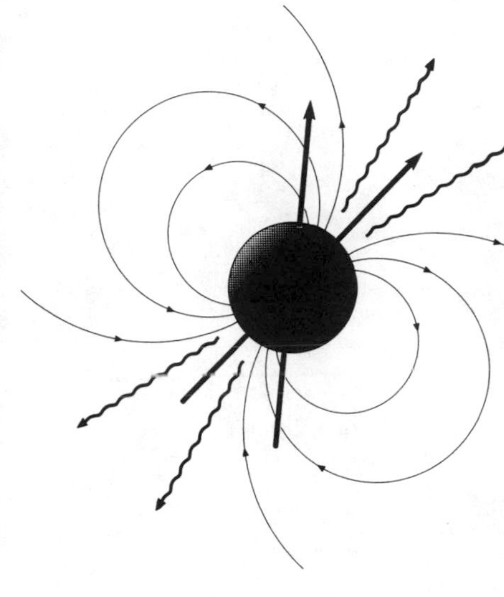

# High-Energy Radiation from Magnetized Neutron Stars

THE UNIVERSITY OF CHICAGO PRESS

*Chicago & London*

Peter Mészáros is professor of astronomy and astrophysics at the Pennsylvania State University.

The University of Chicago Press, Chicago 60637
The University of Chicago Press, Ltd., London


Printed in the United States of America
00 99 98 97 96 95 94 93 92 54321

ISBN 0-226-52093-5 (cloth)
ISBN 0-226-52094-3 (paper)

Library of Congress Cataloging-in-Publication Data

Mészáros, Peter.
High-energy radiation from magnetized neutron stars / Peter Mészáros.
p. cm. — (Theoretical astrophysics)
Includes bibliographical references and index.
1. Neutron stars. 2. Pulsars. 3. Astrophysics. I. Title. II. Series.
QB843.N4M47 1992
523.8′874—dc20

The paper used in this publication meets the minimum requirements of the American National Standard for Information Sciences—Permanence of Paper for Printed Library Materials, ANSI Z39.48-1984.

Szüleimnek
Deborahnak, Andornak

# Contents

# Preface

The beauty of studying neutron stars is that, at its foundations, the subject involves a wide range of challenging physics, while at its forefront ramifications it presents a number of fascinating theoretical and experimental puzzles. The subject of this book covers a significant fraction of what we currently know as high-energy astrophysics. While in principle concentrating on the high-energy radiation from magnetized neutron stars, a proper treatment of these topics necessarily entails the discussion of a large amount of material relating to neutron stars in general and to radiative phenomena of broader applicability. To put the subject of magnetized neutron stars in its context, I have tried where possible to point out their relationships with other types of compact objects and their general role in the overall picture of stellar evolution.

The general approach in this book is a theoretical one, supplemented with a condensed discussion of the most relevant observational material. I have tried to present the material in such a way as to progress from the basic physics and the more general, better established aspects of the subject to the latest developments, which necessarily involve a considerable degree of uncertainty. While new experiments will no doubt eventually lead to a need for revising some of the views presented in the latter chapters, the material of the first half of the book is intended to provide a framework onto which such an evolving picture can be built. In this spirit, after a general introduction the first half of the book discusses the basic physics of magnetized neutron stars, the plasma physics and the radiative processes, as well as some techniques and results of radiative transfer in magnetized atmospheres. The second half of the book covers more specifically the phenomenology and a sample of the models that have been developed, covering in particular accreting X-ray pulsars, rotation-powered pulsars, gamma-ray bursters, and super-high-energy sources. The last chapter considers the evolution of magnetized neutron stars. The choice of the specific material discussed at greater length has naturally been influenced by my own involvement in the field or by my reading and my scientific interactions. While I have attempted to at least refer to work that has not been included here, the area covered is sufficiently large that any quest for completeness would be unrealistic. In this manner the author can communicate information most efficiently on those aspects where he has had experience. The purpose of the book is to provide a basic discussion of the phenomena and the theoretical concepts involved, and to serve as a general guide and introduction to the subject leading into more specific topics. It is intended for researchers in the field as well as for graduate students and advanced undergraduates in physics and astronomy with an interest in the astrophysics of neutron stars. With current experiments from space and ground continuing to provide new results, with a number of well-advanced projects dedicated to such studies under way, and with vigorous theoretical investigations being pursued by a number of groups, this field is growing fast and will no doubt contribute significant scientific excitement in the future.

# Acknowledgments

I list here with special thanks those people who most directly contributed to shaping this book, through either their published work, their collaboration, or their feedback on the manuscript: S. Alexander, J. Arons, R. Blandford, R. Bussard, V. Canuto, A. Harding, H. Herold, E. van den Heuvel, C. Ho, I. Iben, J. Kirk, D. Lamb, F. Lamb, W. Nagel, G. Pavlov, B. Paczyński, D. Pines, H. Riffert, M. Ruderman, J. Taylor, V. Trimble, J. Ventura, I. Wasserman, and T. Weekes. On a more general level, I am indebted to G. Field, M. Rees, J. Ostriker, L. Spitzer, R. Ramaty, J. Silk, A. Lightman, R. Kippenhahn, J. Truemper, G. Garmire, G. Clark, and F. Nagase. The initial push for starting to write this book was provided by D. Schramm, and during its preparation I have benefited from the congenial editorial advice and management of S. Abrams and J. Lightner. The research was supported in part by the NSF and by NASA.

# 1. Neutron Stars: An Overview

## 1.1 Formation

A few times per century, in a typical galaxy like ours, a massive star approaches the end of its normal luminous life, consuming at an accelerating rate a nuclear fuel of increasing weight and complexity which provides a steadily decreasing energy yield. This process culminates when the internal chemical composition of the stellar core has evolved to such a degree that energy can no longer be extracted from further nuclear reactions. The exhausted core is thus deprived of its previous thermal pressure support so that, unable to continue holding itself up against gravity, it collapses. Within a free-fall time scale the collapse liberates a huge amount of gravitational energy, reverses its motion, and gives rise to one of the most energetic events in nature, a supernova explosion which for a brief period outshines the whole galaxy. This massive outburst blows away the entire outer envelope of the star, leaving behind as a glowing ember the remnant of its collapsed, ultracompact core. In a significant fraction of supernova explosions, the direct outcome of the collapse of the stellar core is a neutron star. In other cases a white dwarf is the primary remnant, which in a binary system may accrete enough additional matter to provoke a second, somewhat less intense catastrophic event, which according to some calculations may also result in a neutron star.

The supernova explosion is a consequence of the thermonuclear evolution of the progenitor star, in which the lighter chemical elements have gradually undergone fusion into heavier elements up to the iron group, elements heavier than this being synthesized during the collapse. The direct collapse to a neutron star is possible for stars whose original mass is greater than about eight to ten solar masses, while for progenitor masses below about six to eight solar masses the result is a white dwarf, which is left over after extensive mass loss in the form of a planetary nebula.

The sequence of reactions leading to the collapse may be summarized as follows (e.g. Schwarzschild, 1965; Clayton, 1983). During the main-sequence life of a star, $^{1}$H is fused into $^{4}$He at temperatures in excess of about $10^7$ K in the central portion of the star (von Weizsäcker, 1937; Bethe and Critchfield, 1938). This transformation occurs either via direct proton-proton reactions or via the CNO cycle. Once $^{1}$H burning has proceeded to completion in the core, the latter starts to shrink owing to the lack of pressure support previously provided by the hydrogen-burning energy input, even though hydrogen burning continues in a shell around the now thermonuclearly inactive $^{4}$He core. This relatively rapid gravitational contraction leads to a compact $^{4}$He core whose size is roughly that of a planet, but whose mass is of the order of a solar mass, so that it resembles more closely a white dwarf. The outer layers of the star have meanwhile expanded to the size of a red giant, which is how the star now appears.

Once the core has contracted sufficiently to raise the temperature to about $10^8$ K, $^{4}$He burning produces $^{12}$C via a resonant reaction with the highly unstable $^{8}$Be, which itself was formed from the fusion of two $^{4}$He (Salpeter, 1952; Hoyle *et al.*, 1953). The existence of this resonance is one of a number of interesting coincidences without which neutron stars, and indeed humans to study them, would not be possible (e.g. Barrow and Tipler, 1986). Once a $^{12}$C core is formed, a second core contraction follows, which for stars less massive than about four solar masses fails to produce high enough temperatures to ignite the next set of nuclear reactions, following which the star ejects its envelope and the core settles down to become a cooling white dwarf star. However, in the more massive stars gravity is able to compress the core sufficiently for it to reach a central temperature in excess of about $6 \times 10^8$ K (e.g. Iben and Renzini, 1984). At this temperature the collision of two $^{12}$C, accompanied by the ejection of an alpha particle to carry off excess energy, can lead to the formation of $^{20}$Ne; to the formation of $^{23}$Na plus a proton, and $^{24}$Mg, with the excess energy carried off by gamma-rays; and, occasionally, with the ejection of two alpha particles, to the formation of $^{16}$O. Upon exhaustion of the $^{12}$C supply in the core, another contraction ensues, which can ignite the next set of reactions provided the star is massive enough; otherwise the remnant is again a white dwarf. If the star was originally more massive than about eight to ten times the mass of the sun, its next core contraction proceeds far enough to reach the next threshold temperature at about $10^9$ K, at which point $^{20}$Ne burning sets in, which by absorption of a $^4$He produces a $^{24}$Mg nucleus plus gamma-rays, or by absorption of a gamma-ray leads to a $^{16}$O nucleus plus an alpha particle.

A star's subsequent evolution depends on how much larger than $\sim 9\ M_\odot$ its original mass is (e.g. Woosley and Weaver, 1986). For stars which start out only slightly above this value, the CO core becomes more massive than the Chandrasekhar mass, $M_{Ch} \simeq 1.4\ M_\odot$ (eq. 1.2.6), which is the maximum mass for a stable white dwarf, so that the core eventually goes into a final collapse at this stage. For the more massive stars, further nuclear reactions are still possible. At temperatures above $\sim 1.5 \times 10^9$ K, $^{16}$O burns into various isotopes of Si, P, and Mg, and above $\sim 3 \times 10^9$ K, Si gets involved in a series of nuclear reactions leading ultimately to $^{56}$Fe. This set of reactions is what is expected for stars of total mass $M \gtrsim 20\ M_\odot$. No further exothermic nuclear reactions are possible beyond $^{56}$Fe, since that is the most tightly bound atomic nucleus, and thereafter the core goes into a final gravitational collapse. This final free fall of the core, accompanied by photodisintegration of the nuclei, leads to rapid neutronization of the imploding core via inverse $\beta$-decay reactions.

The gravitational collapse of the core can stop when the neutrons become degenerate, causing a sudden stiffening of the core, which leads to a bounce and to a reversal of the collapse. For this to occur the core mass must not exceed a value $M_{max} \sim 4\ M_\odot$ (see eq. 1.3.10); otherwise a general relativistic gravitational instability sets in whose only possible outcome is a black hole. When the bounce does occur, the reversed implosion steepens into a shock wave traveling outward at speeds in excess of $10^{-1}\ c$, which is further accelerated by the momentum absorbed from the copious emission of neutrinos produced in the core. Although calculations of this bounce are as yet uncertain, there is no doubt that this is observed to occur in nature. The shock wave accelerating the outer layers of the star leads to the disruption and dispersal of the stellar envelope in the enormously energetic type II supernova event. Other types of supernova explosions, e.g. type I explosions, may proceed via thermonuclear runaway instabilities and disruption, or possibly via the further collapse of a white dwarf core (see Chap. 11-1). For type II supernovae, the cores left over from stars of masses between about 9 and 25 $M_\odot$ are shown by calculations to be neutron stars, and this is also the probable outcome of a fraction of the type I supernova events. The kinetic energy imparted to the envelope by the explosion is of the order of $10^{51}$ ergs. This is only about 1% of the gravitational binding energy liberated by the core when it collapses to neutron star size, the other $\sim 99\%$ going mainly into neutrino emission. Such a burst of neutrinos has been observed in the case of the most recent nearby supernova, SN 1987a, which occurred in the Large Magellanic Cloud.

The net result after such a cataclysmic event as a supernova explosion is an expanding, cooling shell of material made up of the disrupted envelope of the progenitor star, which slowly disperses within less than about a hundred thousand years, and inside it an extremely compact, degenerate remnant. For the conditions described above, this remnant is a cooling neutron star rotating at high speed because of its share of the original angular momentum left over from the contracted core; its magnetic field is determined by the conditions prior to and during the explosion, enhanced by the compression resulting from the collapse. Often the neutron star is solitary, its progenitor having been either a single star or a binary system which was disrupted by the explosion. In cases in which the progenitor star was in a binary system which was not disrupted by the explosion, the result is a neutron star with a less evolved binary companion or, in rare cases, with an evolved companion. These neutron stars are much later detected as intense sources of gamma-rays, X-rays, and radio waves, as funeral beacons indicating the locations where at some distant epoch a dazzlingly powerful supernova detonation marked the demise of a massive star.

## 1.2 Neutron Star Physical Parameters

The collapse of the core in a supernova cannot be halted by the thermal gas pressure, which is insufficient. This can be achieved only when the degeneracy pressure of the particles in the core has grown to the point where it can balance gravity. In fact the precollapse core is already dominated by the degeneracy pressure of the electrons, which coexist with the high-Z nuclei, but this electron degeneracy pressure is insufficient for stabilizing core masses that grow beyond a critical value. The core is essentially a white dwarf, and this critical mass for a white dwarf, or for a core where nuclear reactions have stopped, has been calculated by Chandrasekhar (1935). In neutron stars, a similar situation exists, with the difference that in these most of the electrons have been absorbed by the protons, and it is the neutrons that are degenerate and provide the necessary opposition to the gravity forces (Landau, 1932; Baade and Zwicky, 1934a, b; Oppenheimer and Volkoff, 1939).

The critical mass and size of stars dominated by degenerate electron and neutron pressure can be estimated fairly simply. The degeneracy pressure arises from Pauli's exclusion principle, which states that particles cannot be brought closer together once they have filled all their lowest-energy quantum levels. The momenta $p_x$ of the particles along an arbitrary direction, $x$, are related to their position by the uncertainty principle, $p_x \sim \Delta p_x \sim \hbar/\Delta x$. The limiting density is $n \sim (\Delta x)^3$, so we have $p \sim \hbar n^{1/3}$, or more exactly (see Landau and Lifshitz, 1977)

$$p_F = (6\pi^2/g)^{1/3} \hbar n^{1/3}, \tag{1.2.1}$$

which is called the Fermi or limiting momentum for a degenerate gas. The factor $g$ here is 2 for electrons or neutrons. The pressure is given by $P \sim n v_x p_x$, and for nonrelativistic particles $v_x = p_x/m$ so that the nonrelativistic degenerate pressure is $P \sim \hbar^2 m^{-1} n^{5/3}$, or more exactly

$$P = (1/5)(6\pi^2/g)^{2/3} \hbar^2 m^{-1} n^{5/3}, \tag{1.2.2a}$$

which is valid for temperatures below the Fermi temperature,

$$kT \ll kT_F = m^{-1} \hbar^2 n^{2/3}. \tag{1.2.2b}$$

When the particles are relativistic, $p_F$ remains the same but $v_x \sim c$ so that the relativistic degeneracy pressure is $P \sim \hbar c n^{4/3}$, or

$$P = (1/4)(6\pi^2/g)^{1/3} \hbar c n^{4/3}. \tag{1.2.3}$$

This degeneracy pressure must balance the weight of the overlying layers of the star according to the hydrostatic equilibrium condition, which is $\nabla P = GM\rho/r^2$. Dimensionally, the central pressure $P_c$ must be $P_c \sim Gm^2/R^4$, where $R$ is the outer radius of the star, which for a self-gravitating sphere of nonrelativistic gas with $P \propto \rho^{5/3}$ (e.g. Schwarzschild, 1965) is

$$P_c = 0.77 G \frac{M^2}{R^4}, \tag{1.2.4}$$

with the central density given by $\rho_c = 5.99\langle\rho\rangle = 1.43(M/R^3)$. For a self-gravitating sphere of relativistic degenerate gas with $P \propto \rho^{4/3}$, the central pressure is

$$P_c = 11.0 G \frac{M^2}{R^4} \tag{1.2.5}$$

and the central density is $\rho_c = 54.2\langle\rho\rangle = 54.2(3/4\pi)(M/R^3)$.

The critical mass for a white dwarf, the Chandrasekhar mass $M_{Ch}$, determines not only the maximum mass of a stable white dwarf but also the mass value of the core of an evolved star above which collapse to a neutron star occurs. Such a core, made up of heavy elements with average atomic weight and charge $A$ and $Z$, will have an electron density $n_e = (Z/A)(\rho/m_p)$, where $m_p$ is the mass of the proton. The pressure, however, is provided by the degenerate electrons so that in equations (1.2.2)–(1.2.3) the density and particle mass are $n_e$ and $m_e$. Near the upper mass limit of white dwarfs the degenerate electrons are typically relativistic so that for the central pressure we must equate (1.2.2) with (1.2.4) and use the corresponding definition of $\rho_c$ together with the previous definition of $n_e$. In the resulting expression the radius $R$ cancels out and the limiting mass is

$$M_{Ch} \simeq \pi \left(\frac{Z}{A}\right)^2 \left(\frac{\hbar c}{G m_p^2}\right)^{3/2} m_p, \tag{1.2.6}$$

which for a typical value $(Z/A) = 0.5$ is $1.45\ M_\odot$. The typical number of nucleons in such a configuration is $N_N \sim M_{Ch}/m_p \sim 10^{57}$. The limiting mass obtained from a more sophisticated numerical calculation for a nonrotating star (e.g. Shapiro and Teukolsky, 1983) comes out essentially the same, although for rotating configurations the value can be somewhat higher. A degenerate star, or stellar core, whose mass exceeds the value $M_{Ch}$ must inevitably collapse upon itself in search of a new equilibrium, which for some range of masses can be provided by a neutron star configuration.

For a neutron star, the same principles apply, with the difference that here the electrons have mostly been incorporated into the neutrons and it is the neutrons, of mass $m_N \simeq m_p$, that are degenerate and provide the pressure opposing gravity. Here the relevant particle density is $n_N = \rho/m_N$ and the particle mass is $m_N \simeq m_p$ in equations (1.2.2)–(1.2.3). Using the relativistic degenerate neutron pressure at the center (1.2.2), equating it to the relativistic central pressure (1.2.4) for a spherical configuration with $P \propto \rho^{4/3}$, and using the previously defined $n_N$, the radius $R$ again cancels out and one obtains for the limiting mass of a neutron star the estimate

$$M_{lim} \sim \pi \left(\frac{\hbar c}{G m_N^2}\right)^{3/2} m_N. \tag{1.2.7}$$

This would give $\sim 5\ M_\odot$ as the limiting mass, but this value is an overestimate, since a number of effects were neglected, such as general relativity, as were the details of the nuclear equation of state. These details are not yet too well understood. However, under very general conditions, a numerical calculation including general relativistic effects (Rhoades and Ruffini, 1974; Hartle and Sabbadini, 1977) leads to a maximum mass estimate for a nonrotating neutron star of $M_{max} \lesssim 3\text{–}5\ M_\odot$ (eq. 1.3.10), which is not much below (1.2.7). The quantity $\alpha_G = (Gm_N^2/\hbar c) = 6.2 \times 10^{39}$, where $m_N \simeq m_p$, is called the gravitational coupling constant by analogy with the quantum electrodynamic fine structure constant; it also appears in the Chandrasekhar mass (1.2.6). The number of nucleons is again $N_N \sim \alpha_G^{-3/2} \sim 10^{57}$.

The existence of a maximum stable mass for a degenerate stellar configuration can be understood in terms of a simple physical argument (Schwarzschild, 1965). The average density is $\langle\rho\rangle \propto M/R^3$, so the gravitational force is

$$F_G \propto \frac{\langle\rho\rangle GM}{R^2} \propto \frac{M^2}{R^5}. \tag{1.2.8}$$

For low stellar masses the pressure force will be due to nonrelativistic degenerate effects,

$$\left(\frac{dP}{dR}\right)_{nr,\,d} \propto \frac{\langle\rho\rangle^{5/3}}{R} \propto \frac{M^{5/3}}{R^6}, \tag{1.2.9}$$

whose dependence on the radius is different from that of the gravitational force so that the star can always adjust its radius until it finds a unique equilibrium. However, above a certain stellar mass one expects the degenerate particles to become relativistic, so the pressure force becomes

$$\left(\frac{dP}{dR}\right)_{r,\,d} \propto \frac{\langle\rho\rangle^{4/3}}{R} \propto \frac{M^{4/3}}{R^5}. \tag{1.2.10}$$

One sees that in this case both forces depend on the same power of the radius, so the star in general cannot bring the two forces into equilibrium by changing its radius, as the more conventional stars can. Also, the two forces depend on different powers of the mass, so there is a particular (limiting) mass value for which the two forces can be in balance. Above that mass, gravity will always exceed pressure and the star will collapse. Much below it, gravity is less than the pressure and the star will expand until it becomes nonrelativistic and finds its equilibrium.

The size of a neutron star can be estimated by using the nonrelativistic degenerate pressure, which is approximately appropriate for the lower-mass neutron stars. Using the pressure (1.2.1) with $n_N$ and $m_N \simeq m_p$, and setting it equal at the center to (1.2.4) with the corresponding relation for $\rho_c$, one obtains the mass-radius relation for a neutron star,

$$R \simeq 4.50 \frac{\hbar^2}{G m_p^{8/3}} M^{-1/3}. \tag{1.2.11}$$

This equation shows that as the neutron star mass increases, the radius decreases. By replacing in (1.2.11) the characteristic value $M \simeq 1.4\, M_\odot$ of a neutron star arising from the collapse of a white dwarf or stellar core pushed beyond the Chandrasekhar limit, one obtains a typical neutron star radius $R \sim 1.2 \times 10^6$ cm. This value is a factor $(A/Z)^{5/3} m_e/m_p$ smaller than the typical radius of a white dwarf, as can be seen by deriving the analogue of (1.2.11) but using $m_e$ for the degenerate electron pressure in (1.2.1).

As the mass, or equivalently the central density, increases, the neutrons will become relativistic. One can see how this affects the mass-radius dependence (1.2.11) using an approximate scaling argument (Salpeter, 1967). The stellar equilibrium is given by the virial theorem as $3\langle P/\rho\rangle \sim M/R \propto M^{2/3}\langle n^{1/3}\rangle$, or

$$M \propto \left(\frac{P}{\rho}\right)^{3/2} \langle n^{1/3}\rangle^{-3/2}. \tag{1.2.12}$$

For a nonrelativistic degenerate gas, $P \propto \langle n\rangle^{5/3}$ and $\langle\rho\rangle \sim m_N\langle n\rangle$, while for a relativistic degenerate gas, $P \propto \langle n\rangle^{4/3}$ and $\langle\rho\rangle \sim (\varepsilon_F/c^2)\langle n\rangle \propto \langle n\rangle^{4/3}$, where the rest mass has been neglected, the relativistic Fermi energy $\varepsilon_F = p_F c$, and the Fermi momentum $p_F$ is given by (1.2.1). Replacing these expressions in (1.2.12) and using $\langle n\rangle \sim MR^{-3}$, one sees that

$$M \propto \begin{cases} R^{-3} & \text{nonrelativistic;} \\ R & \text{relativistic (unstable).} \end{cases} \tag{1.2.13}$$

Thus one sees that for an increasing mass the radius decreases, in agreement with 1.2.11, until a maximum mass $M_{\max}$ is reached.

Alternatively, the mass initially increases with increasing central density and then starts to decrease. The exact behavior is more subtle due to general relativistic effects and changes in the equation of state; in particular, configurations more compact than the turnover point, corresponding to the second line in (1.2.13), are subject to gravitational instability. However, the qualitative behavior of the first line of equation (1.2.13) at masses below $M_{max}$ agrees with that obtained from more accurate numerical calculations with standard nuclear equations of state (see Fig. 1.3.2).

The magnetic field of neutron stars can depend on a number of processes (see Chap. 11.4), but a simple estimate of its magnitude can be obtained by calculating the field component which is primeval in origin. This primeval field should be determined largely by the original magnetic field of the progenitor star, where it was maintained by a dynamo type of mechanism. The currents that sustained such a field in the progenitor star can no longer flow in the rigid crust or in the interior of the neutron star, but the whole star remains ionized to such a degree that it is an almost perfect conductor. In such a high-conductivity medium, the magnetic flux is essentially trapped, or frozen in (see Chap. 2.5), and the flux conservation condition is approximately valid:

$$\Phi = \int \vec{B} \cdot d\vec{l} = \iint \vec{B} \cdot d\vec{A} \sim B \cdot A = \text{constant}, \qquad (1.2.14)$$

where $d\vec{l}$ is a differential of length along a closed path enclosing a bundle of magnetic field lines $\vec{B}$, and $d\vec{A}$ is a differential of area directed normal to a surface that is bounded by the integration circuit. Roughly, one can envisage $A = \pi R^2$ to be an equatorial section of the star, with normal parallel to the magnetic field through it. From equation (1.2.14) one sees that, if a star of characteristic dimension $R \sim 10^{11}$ cm and magnetic field $B \sim 10^2$ G (gauss) collapses to $R \sim R_{NS} \sim 10^6$ cm, the neutron star is expected to have a field $B_{NS} \sim 10^{12}$ G, close to the average value observed in many of the strongly magnetized neutron stars such as radio pulsars or X-ray pulsars. Actually the collapse involves the core, which is typically smaller, $R_{core} \sim 0.2\ R$, where the magnetic field is expected to be correspondingly higher, leading to an approximately similar result. By the same arguments, or from observations, one can find that the magnetic field of a white dwarf ($R \sim 10^9$ cm) is of order $B_{WD} \lesssim 10^6$–$10^7$ G so that a white dwarf collapsing from this radius to $R_{NS} \sim 10^6$ cm can again lead to a value $\sim 10^{12}$ G. The approximate orders of magnitude of the neutron star mass, radius, and magnetic field are therefore

$$M_{NS} \sim 1.4\ M_\odot, \quad R_{NS} \sim 10^6 \text{ cm}, \quad B_{NS} \lesssim 10^{12}\text{–}10^{13} \text{ G}. \tag{1.2.15}$$

The first two equations of (1.2.15) show that general relativistic effects can be of some importance, since the ratio of the stellar radius $R$ to the gravitational radius $R_S = 2GM/c^2 = 3 \times 10^5(M/M_\odot)$ cm is about a factor 2–4 depending on the radius, getting smaller for more massive stars or softer equations of state, which imply a smaller radius (see §1.3). The magnetic field value in (1.2.15) is an approximate upper end of the range observed thus far. Thermomagnetic effects, which may cause field growth under some circumstances (Chap. 11), are unlikely to produce any fields higher than those of (1.2.15). Fields lower than those of (1.2.15), however, can be encountered both because of the possibility of a lower initial field in the progenitor and because the neutron star conductivity is not strictly infinite so that the magnetic field may actually dissipate over a long period of time, as discussed in Chapter 11.2. Indeed, the observations indicate that while many neutron stars have fields as large as (1.2.15), a large number have much smaller or undetectable fields.

## 1.3 Structure of the Envelope and the Interior

The gravitational binding energy per unit mass on the surface of a neutron star is $E_G/m_p = GM/R \simeq 0.15c^2$ ergs/g, where $M$ and $R$ are the mass and radius of the star, and the corresponding surface gravity $g = GM/R^2 \simeq 0.15c^2/R \simeq 10^{14}$ cm s$^{-2}$ is enormous. An isolated neutron star, with initial formation temperature $T_i \gtrsim 10^{11}$ K, is rapidly cooled by neutrino losses, dropping to $10^{10}$–$10^9$ K within a day and to $T_i \sim 10^8$ K within $10^4$–$10^5$ yr (e.g. Shapiro and Teukolsky, 1983). The corresponding surface temperatures at this stage are of order $10^{-2}$ of the interior values, $T_s \sim 10^6$ K (Gudmundsson, Pethick, and Epstein, 1983; Nomoto and Tsuruta, 1987). The outermost layer of a neutron star will therefore be a thin, nondegenerate atmosphere of typical scale height

$$h \simeq \frac{P}{g\rho} \simeq \frac{kT}{\mu g m_p} \sim 6R_6 T_7 \mu^{-1} \text{ cm}, \tag{1.3.1}$$

where $\mu = A/(Z+1)$ is the mean molecular weight, $A$ and $Z$ being the atomic weight and atomic number of the predominant nuclei. The nuclear composition of the envelope and the crust is typically assumed to be that for cold catalyzed matter. The sequence of nuclei as a function of density has been calculated by Baym, Pethick, and Sutherland (1971). At densities $\lesssim 6.6 \times 10^6$ g cm$^{-3}$ the dominant nucleus is $^{56}$Fe, with increasingly heavier nuclei at higher densities (provided the temperature is low with respect to the Fermi energy and any intrusion of new matter, e.g. from accretion, is neglected).

As the density increases inward, electron degeneracy sets in for (see eq. 1.2.2b)

$$\rho_F \gtrsim 1.36 \times 10^2 (\mu_e/2) T_7^{3/2} \text{ g cm}^{-3}, \tag{1.3.2}$$

where the molecular weight per electron $\mu_e = A/Z \sim 2$. The pressure can be written as

$$P_e = 10^{23} \frac{(\rho_6/\mu_e)^{5/3}}{1 + \left[1 + (\rho_6/\mu_e)^{2/3}\right]^{1/2}} \text{ dynes cm}^{-2}, \tag{1.3.3}$$

which contains a factor allowing for the switchover from nonrelativistic to relativistic degenerate pressure at a density near $\rho_6 \sim 1$ (e.g. eq. 1.3.5). The above expressions neglect any magnetic effects. In the presence of a strong magnetic field, however, the phase space available to electrons is reduced in the transverse direction, the transverse momentum being quantized with typical transverse dimension of the electron Landau orbit $\lambda_\perp = (\hbar/m_e\omega_c)^{1/2} \simeq 2.6 \times 10^{-10} B_{12}^{1/2}$ cm, while the longitudinal momentum is unconstrained as in the nonmagnetic case. This leads to a degeneracy that sets in at densities higher by a factor $\sim \hbar\omega_c/kT$ than in the absence of magnetic field effects (Ventura, 1989), where $\omega_c = eB/m_e c$ is the ground cyclotron energy. In this case the Fermi temperature and pressure scale as $T_F \sim \rho^2$ and $P_e \sim \rho^3$ rather than as in the nonmagnetic case $T_F \sim \rho^{2/3}$ and $P_e \sim \rho^3$ (Hernquist, 1984b).

The strong magnetic field alters the atomic structure of the nuclei in the envelope (Ruderman, 1971). This is caused by the fact that the typical transverse dimension of the electron Landau orbit $\lambda_\perp$ becomes smaller than the Bohr radius already for $B \gtrsim 10^9$ G so that the atomic binding energy along the magnetic field becomes stronger. Qualitatively, this opens the interesting possibility that atoms may form long chains along the magnetic field, a sort of magnetic polymer. In such a one-dimensional magnetic lattice, the outer envelope would be solid and would end abruptly at a density typical of this solid, $\rho_{\rm ml} \sim 10^4$ g

$\mathrm{cm}^{-3}$. However, according to improved binding energy calculations (Flowers *et al.*, 1977; Müller, 1984), such a magnetic solid is not expected to form if the composition is mainly iron group elements (Chap. 2.5), although some caution is required, since the binding energies are obtained as the difference between two very large quantities. In this discussion, therefore, the atmosphere and envelope are assumed to be in the gas or liquid phase until the much larger crystallization density (1.3.6) is reached, as discussed below.

Deeper in the envelope a density $\rho_B$ is reached above which magnetic effects lose their importance for the thermodynamic and transport properties. This density can be estimated by equating the nonmagnetic Fermi energy to the energy of the first Landau level, which gives (Alpar, 1989)

$$\rho_B \simeq 10^4 \mu_e B_{12}^{3/2} \text{ g cm}^{-3}. \tag{1.3.4}$$

Significantly above this density, the electrons populate the higher Landau levels, whose increasing number eventually restores the three-dimensionality of the nonmagnetic phase space.

At even higher densities, the degenerate electrons become relativistic when the momentum of an electron on the Fermi surface $p_F$ becomes comparable to $m_e c$, which occurs at

$$\rho_r = \mu_e n_e m_N \simeq \frac{2 m_N}{3\pi^2 (mc/\hbar)^3} \simeq 3 \times 10^6 \text{ g cm}^{-3}. \tag{1.3.5}$$

At densities $\sim 10\rho_r$ the electrons become extremely relativistic, $\varepsilon_F \simeq p_F c$, and the electrons are essentially free, providing very little screening of the ions by the clouds of electrons around them. Under these circumstances, the nucleons will feel the bare Coulomb potential of their neighbors, and a nuclear Coulomb lattice or crystal will form. This is the depth at which the outer crust begins. More accurately, one has to compare the thermal energy of the ions with their Coulomb energy through the dimensionless parameter $\Gamma = \langle Z\rangle^2 e^2 / r_i kT$, where $r_i = [(4\pi/3)n_i]^{-1/3}$ is the radius of the Wigner-Seitz sphere. The value of $\Gamma$ increases with density, and at a critical value $\Gamma_m \simeq 158$ a liquid-solid phase transition occurs (Pollock and Hansen, 1973). Essentially this is saying that when the temperature $kT$ becomes about $10^{-2}$ of the typical lattice Coulomb energy $\langle Z\rangle^2 e^2 / r_i$, the lattice starts to fall apart. The corresponding melting density of the solid lattice can be written as

$$\rho_m \simeq 2.7 \times 10^7 T_8^3 \mu_e (\langle Z \rangle/26)^{-5} (\Gamma_m/158)^3 \text{ g cm}^{-3}. \qquad (1.3.6)$$

Notice that the melting density $\rho_m$ depends on the temperature. As the neutron star cools, this melting surface moves outward. Above it, the envelope is liquid; below it, it is a Coulomb lattice, whose onset represents the beginning of the solid outer crust of the neutron star. Stresses exerted anywhere locally quickly extend along the lattice due to the strong coupling. The lattice responds elastically, with a simple proportionality between stress and strain given by Hooke's law. This is valid up to a critical dimensionless strain angle $\theta_c$, which if exceeded leads to a cracking or breaking of the lattice, which then rearranges itself. In terrestrial crystals, which have effective screening and weak coupling forces, the critical strain angles are $\theta_c \sim 10^{-4}$–$10^{-5}$, but in a neutron star crust with an unscreened Coulomb lattice the critical strain angle could be much larger. No definitive calculations exist, but estimates range as high as $\theta_c \sim 10^{-2}$–$10^{-1}$. The corresponding yield stress, or bulk modulus, near the surface (Dyson, 1969; Smolukowski and Welch, 1970) is $Y \sim \theta_c \langle Z^2 \rangle e^2 a_N^{-4} \sim 10^{12} \rho^{4/3}$ dynes cm$^{-2}$, where $\theta_c \sim 10^{-2}$ and $a_N$ is the internucleon separation.

Deeper down, a qualitatively new region, the inner crust, is reached. This sets in when the density is high enough that the increasing energy of the degenerate electrons leads to large rates of electron capture by the nuclei, which become increasingly neutron rich until eventually a density is reached at which the lattice nuclei can no longer accommodate the growing number of neutrons. The density at which the inner crust begins is taken to coincide with this neutron drip density (e.g. Shapiro and Teukolsky, 1983)

$$\rho_{\rm nd} \sim 4 \times 10^{11} \text{ g cm}^{-3}. \qquad (1.3.7)$$

In the inner crust, some of the neutrons are localized in bound states in nuclei, while an increasing number of neutrons occupy extended Bloch states of the crystal. The neutrons in these continuum states are superfluid at temperatures below $10^{10}$–$10^9$ K, since at the inner-crust density the interparticle spacing is larger than the range of the repulsive forces ($\lesssim 1$ fm), leading to Cooper pairs in the $^1S_0$ state, with a gap energy $\sim 1$ MeV (e.g. Pines and Alpar, 1985). The inner crust carries most of the crystal mass and moment of inertia, and extends down to a density $\sim 2.8 \times 10^{14}$ g cm$^{-3}$.

The core region of the neutron star begins at the density where neutronization has become so extensive that the neutron density in the interstitial spaces between the lattice sites is comparable to that inside the nuclei. This nuclear density is

$$\rho_N \simeq 2.8 \times 10^{14} \text{ g cm}^{-3}. \tag{1.3.8}$$

As this density is reached, the decreasing lattice spacing leads to a melting of the inner-crust lower boundary into a homogeneous medium made up predominantly of neutrons, with approximately a 5% admixture of protons and electrons. The neutrons in the core are superfluid, the combination of the repulsive force and spin-orbit coupling leading to a condensate of $^3P_2$ Cooper pairs, while the core protons, being much more dilute, are expected to be in a $^1S_0$ superconducting state (Pines, 1987). Because of the large difference in their Fermi energies, no pairing between protons and neutrons is expected. The liquid core, representing most of the total mass and moment of inertia of the star, may extend down to the center of the star at densities $\gtrsim 10^{15}$ g cm$^{-3}$. More speculatively, and depending on the equation of state (e.g. Baym and Pethick, 1979), there could also be an inner core dominated by a pion condensate, or possibly by quark matter (Alcock, Farhi, and Olinto, 1986). The various regions of a neutron star interior are shown in Figure 1.3.1.

The equation of state (e.o.s.) of the core is crucial in determining the central density of the star, whether an inner core will develop, the maximum mass of the neutron star, and the size of the outer radius of the neutron star. The maximum mass is of interest because above that value a compact remnant should be a black hole, while the outer radius is of interest mainly in order to know the surface gravity and gravitational redshift, which are needed to model the surface emission and the transport. Calculations of the equation of state of nuclear matter and neutron star cores are hampered by a lack of detailed knowledge concerning the interactions at densities $\rho \gtrsim \rho_N = 2.8 \times 10^{14}$ g cm$^{-3}$ (e.g. Baym and Pethick, 1979; Shapiro and Teukolsky, 1983). Even assuming a given interaction potential, the calculation of the e.o.s. is a complicated many-body problem. It is nonetheless possible to construct a variety of models of the nuclear interactions consistent with laboratory constraints (Pandharipande, Pines, and Smith, 1976) which range all the way from very stiff (e.g. the TI, or tensor interaction, model) to very soft (e.g. the R, or phenomenological Reid, potential). In the stiff e.o.s. the average system interaction energy becomes repulsive at subnuclear densities, while in the soft e.o.s. the average system interaction energy is attractive. Neutron star

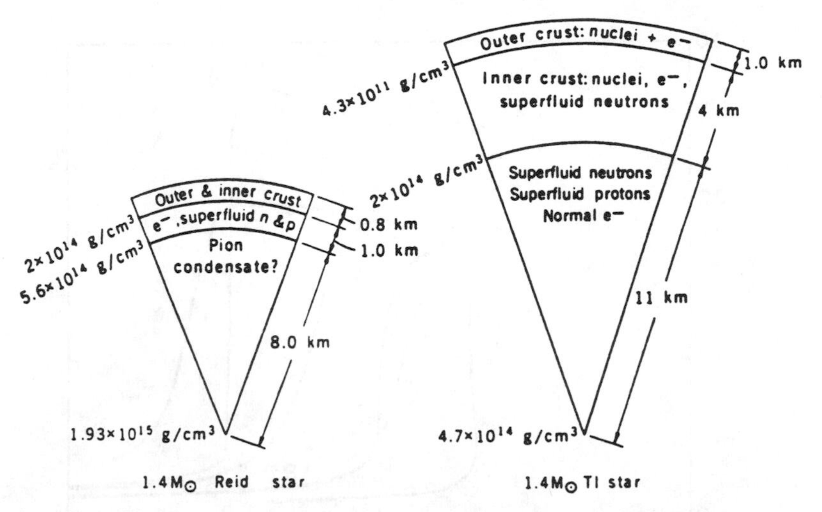


FIGURE 1.3.1 Cross sectional view of neutron stars for two different equations of state. From David Pines, 1980, "Accreting Neutron Stars, Black Holes and Degenerate Dwarf Stars," *Science*, 207 (February 8), 597. Copyright 1980 by the American Association for the Advancement of Science.

models are calculated by integrating the general relativistic equation of hydrostatic equilibrium (Oppenheimer and Volkoff, 1939),

$$-\frac{dP}{dr} = \frac{G[\rho(r) + P(r)/c^2][m(r) + 4\pi r^3 P(r)/c^2]}{r^2[1 - 2Gm(r)/rc^2]}, \quad (1.3.9)$$

where $m(r)$ is the mass within radius $r$ and $P(r)$, $\rho(r)$ are the pressure and density at $r$. It is evident that, in the general relativistic case, pressure contributes to the effective mass density so that the pressure gradient is greater than in the nonrelativistic case. This fact also emphasizes the importance of the e.o.s., which is a relationship between $P(r)$ and $\rho(r)$. Neutron star structure models calculated for a large variety of equations of state by Arnett and Bowers (1977) show that, for the same baryon mass, the softer equations of state lead to a higher central density and a smaller outer radius, while the stiffer equations of state lead to a lower central density and a larger outer radius. The gravitational mass versus outer radius is shown for several examples of nonrotating stars in Figure 1.3.2. In this figure, MF is the mean field approximation, TI the tensor interaction, BJ the Bethe-Johnson interaction, R the Reid potential, and $\pi$, $\pi'$ the Reid potential

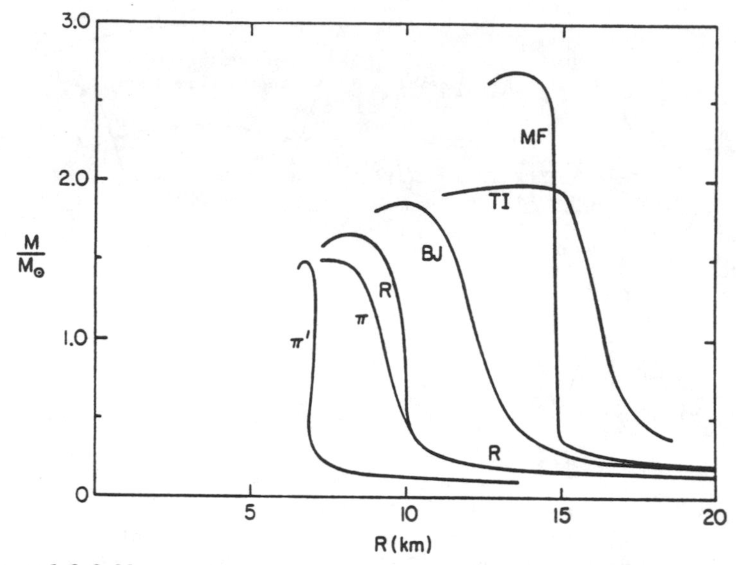


FIGURE 1.3.2 Nonrotating neutron star mass-radius relation. The labels refer to the various equations of state mentioned in the text. From Baym and Pethick (1979). Reproduced, with permission, from the *Annual Review of Astronomy and Astrophysics*, vol. 17, © 1979 by Annual Reviews Inc.

modified by pion condensation of increasing strength, after Baym and Pethick (1979), who list the corresponding references. If the stellar interior consisted of quarks with a "strange matter" e.o.s. (Alcock, 1987), the mass-radius relation would have the opposite tendency. Whereas for nuclear matter the radius decreases with increasing mass up to a point, for strange matter the radius increases with increasing mass, a behavior connected to the asymptotic freedom of quarks at high energies. The uncertainty concerning the latter type of star is considerable, as might be expected.

The maximum mass of nonrotating neutron stars for a given e.o.s. can be read off from the mass-radius curves of Figure 1.3.2. Because of the uncertainties concerning the choice of e.o.s., however, attempts have been made to determine the maximum theoretical mass under the most general assumptions possible. In general, the e.o.s. is better known at densities $< \rho_N$ than above it. Assuming the accuracy of the e.o.s. up to density $\rho_f$ and assuming the e.o.s. above that obeys $\rho > 0$, $\partial P/\partial \rho > 0$, and the causality relation $\partial P/\partial \rho > c^2$, one can show (Hartle and Sabbadini, 1977; Baym and Pethick, 1979) that the maximum nonrotating mass should be

$$M_{\max} = 4.0\ M_{\odot}\left(\frac{\rho_N}{\rho_f}\right)^{1/2}. \tag{1.3.10}$$

For rotating configurations, the maximum mass values are increased because centrifugal force can contribute to stabilizing the heavier stars. Calculations for rotating configurations have been made, e.g., by Ray and Datta (1984) and Friedman, Ipser, and Parker (1986).

## 1.4 Production of High-Energy Radiation from Magnetized Neutron Stars

Neutron stars do not have large internal energy sources other than the heat content left over from their collapse or that acquired by interaction with their environment. The typical temperatures in the initial few seconds of the collapse are characterized by the free-fall or binding energy of a proton in the gravitational potential well of the neutron star,

$$kT_{\mathrm{ff}} \sim \frac{GMm_p}{R} \sim 0.15 m_p c^2 \sim 140\ \mathrm{MeV}, \tag{1.4.1}$$

which is about $T \sim 1.5 \times 10^{11}$ K. This temperature drops several orders of magnitude in a matter of seconds due to copious neutrino emission, and the surface temperature drops to nearly $10^6$ K in a time of the order $10^4$ yr. The thermal emission of an isolated cooling neutron star, in the absence of other energy sources, is therefore a modest

$$L \sim 4\pi R^2 \sigma T^4 \sim 6 \times 10^{32} R_6^2 T_6^4\ \mathrm{ergs\ s^{-1}} \tag{1.4.2}$$

and is concentrated at soft X-ray photon energies $\hbar\omega \lesssim 0.1$–$1$ keV.

There are, however, two important energy sources that can be tapped by neutron stars to produce radiation. One of these is the stellar rotation energy in the presence of a strong magnetic field. The spin rotation rate of a neutron star is determined initially by the spin rate of the progenitor star and by any angular momentum dissipation mechanisms. Neglecting the latter, if the initial configuration has $R_* \sim 10^{10}$ cm and an initial rotation period above a few weeks $P_* = 2\pi/\Omega_* \gtrsim 10^6$ s, one would expect for a neutron star $P_{\mathrm{NS}} \sim$

$P_*(R_{NS}/R_*)^2 \gtrsim 10^{-4}$ s. But dissipation is bound to be present to an unknown extent, and there could also be other effects which would spin the core up, so that a more sensible lower limit to the rotation period is obtained from the breakup rotational velocity, above which the neutron star would fly apart from centrifugal forces. Neglecting distortions and relativistic effects, the latter condition is

$$P \gtrsim 4 \times 10^{-4} R_6^{3/2} (M/M_\odot)^{-1/2} \text{ s}. \tag{1.4.3}$$

If the neutron star in addition has a magnetic field $\vec{B}$, the rotation will produce an induced electric field which in vacuum is of order $\vec{E} = c^{-1}\vec{v} \times \vec{B}$, where $v/c \sim (R/c)\sin\alpha(2\pi/P) \sim 10^{-4}P^{-1}$, with $\alpha$ the angle between the rotation and magnetic axes, and $P$ the rotation period in seconds. For magnetic fields $B \sim 10^{12}B_{12}$ G, and for an acceleration distance of order of the polar cap radius $d \sim R_p = R\theta_p \sim 10^4 d_4$ cm given in equation (1.4.8), the maximum particle energies attainable in the induced electric field are

$$\mathscr{E} \sim eEd \lesssim 3 \times 10^{14} B_{12} P^{-1} d_4 \text{ eV}, \tag{1.4.4}$$

provided the plasma environment is tenuous enough. The primary accelerated particles will give rise to $e^+e^-$ cascades, producing high-energy radiation in the strong magnetic field via the synchrotron or inverse Compton mechanisms. The particles will be relativistic, and being constrained to move along the field, the radiation will be strongly beamed, which coupled with the rotation gives a pulsed emission pattern. The radiation will be strongly nonthermal and will extend in energy from the radio range into the ultrahigh gamma-ray energy range, depending on the period and field strength.

The other major source of energy that is available is the potential energy of matter gravitationally captured by the star and accreted onto its surface. Such matter is amply available if the neutron star is in binary orbit about a stellar companion with an envelope that is not too tightly bound. The gravitational capture of material at a rate $\dot{M}$, falling at the free-fall velocity onto the neutron star, where half of the energy is converted into heat and subsequently into radiation, leads to an accretion luminosity of order

$$L = \frac{GM\dot{M}}{R} \simeq 1.26 \times 10^{38} R_6^{-1} \left(\frac{M}{M_\odot}\right) \times \left(\frac{\dot{M}}{9.3 \times 10^{17} \text{ g s}^{-1}}\right) \text{ ergs s}^{-1}. \tag{1.4.5}$$

The accretion rate $\dot{M} = \dot{M}_{\rm Ed} \simeq 9.33 \times 10^{17}$ g s$^{-1}$, corresponding to about $10^{-8}$ solar masses per year, is the critical accretion rate that produces a luminosity equal to the Eddington luminosity

$$L_{\rm Ed} = \frac{4\pi G M m_p c}{\sigma_T} = 1.26 \times 10^{38} \left( \frac{M}{M_\odot} \right) \text{ ergs s}^{-1}. \tag{1.4.6}$$

Above this luminosity, if it comes from the entire surface, the radiation pressure exceeds the force of gravity and prevents further accretion. However, if the accretion occurs over a smaller fraction of the area $A < 4\pi R^2$, the Eddington luminosity is correspondingly smaller. If accretion is assumed to occur over the area $A$ where the infall energy is thermalized, the typical photon energy will be

$$\hbar\omega \sim kT_{\rm BB} = k\left(L/A\sigma\right)^{1/4} \simeq 10 L_{38}^{1/4} A_{10}^{-1/4} \text{ keV}, \tag{1.4.7}$$

where the area $A$ is normalized to $10^{10}$ cm$^2$ corresponding to a typical polar cap area (e.g. eq. 1.4.8). For an unmagnetized neutron star, however, the area is $4\pi R^2$ and the corresponding temperature is $\sim 3$ keV.

The emission of radiation from neutron stars is obviously strongly influenced by the presence of a magnetic field, which, if sufficiently strong, can dominate the radiation mechanism, control the motion of the charged particles, and determine the geometry of the emission region. The geometric effect plays a major role in rotation-powered and accretion-powered magnetized neutron stars, which rely for their luminosity on particles accelerated to speeds much higher than the thermal velocity of the particles gravitationally bound in the atmosphere. The strong magnetic field allows particles to move freely only along the field lines so that any acceleration of particles toward or away from the star over a substantial path length can be achieved only along field lines that reach far from the star. The distant field of any magnetic configuration anchored in a conducting body such as a star will always be dominated by the dipole contribution to the field. The field strength of a dipole varies as $B \propto r^{-3}$, where $r$ is radial distance. However, the magnetic field lines can be in approximate corotation with the star only out to a limiting radius $r_0$, which is either the light cylinder in a low-density plasma environment or the Alfvén radius in a high-density plasma environment. The light cylinder is the distance at which the tangential velocity of the corotating field lines would reach the speed of light, while an approximate definition of the Alfvén distance is the radius at which the material stresses $\sim \rho v^2$ (where $\rho$ is the ambient plasma density and $v$ is the relevant kinematic or thermal velocity) equal the magnetic dipole stresses $\sim B^2/8\pi$. The light cylinder radius depends only on the rotation period, while the Alfvén radius depends on the magnetic field strength, the accretion rate $\dot{M}$, and the mass, typical values for both radii being in the range $r_0 \sim 10^7$–$10^9$ cm. The field lines of a dipole satisfy the relation $\sin^2\theta/r =$ constant, where $\theta$ is the polar angle with respect to the magnetic axis, and the limiting field line which still closes within $r_0$ delimits on the neutron star surface a polar cap of radius

$$R_p = R\sin\theta_p \sim R\left(\frac{R}{r_0}\right)^{1/2}, \tag{1.4.8}$$

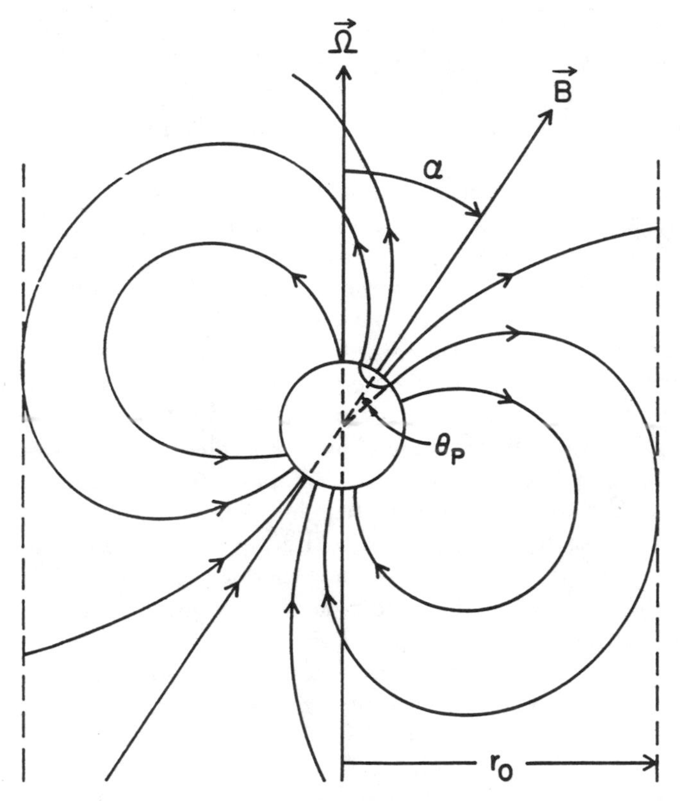


FIGURE 1.4.1 Schematic view of a magnetized neutron star with a dipole magnetic moment whose axis makes an angle $\alpha$ with the rotation axis $\Omega$, the closed magnetic field lines $\vec{B}$ extending out to a radius $r_0$ representing either the light cylinder (isolated neutron star) or the Alfvén surface (in the accretion flow of a companion).

which is typically $R_p \sim 10^{-1}$–$10^{-2}R$. In an isolated neutron star the lines inside $\theta_p$ will be open and reach outside the light cylinder, while in an accreting neutron star the accreting matter can reach the neutron star only along the field lines inside $\theta_p$ (see Fig. 1.4.1). Since charged particles reach high accelerations or crash onto the surface only for those lines inside $\theta_p$, the emission is concentrated in beams or hot spots centered on the polar caps rather than on the whole surface. If the rotation axis of the neutron star does not coincide with the magnetic dipole axis ($\alpha \neq 0$), the rotation gives rise to the appearance of radiation pulses. In an accreting neutron star this requires fields $B \gtrsim 10^9$ G, since otherwise the field is too weak to channel the incoming matter and the latter falls in more isotropically, heating all of the surface so that no pulsations are expected. In an isolated neutron star, it is a combination of the magnetic field strength and the rotation rate that must be sufficiently high for significant particle acceleration and radiation to occur.

## 1.5 Observations of High-Energy Radiation from Neutron Stars

Although the existence of neutron stars was predicted theoretically in the 1930s, the first observational discovery of one class of such objects by Bell and Hewish is relatively recent (Hewish *et al.*, 1968). These objects are now called pulsars, and we shall refer to them here more specifically as rotation-powered radio pulsars (RPPs), underlining the fact that the radiation produced is derived at the expense of the rotational energy and that most objects in this class have been detected only in the radio band. However, some of the best known young objects of this class produce most of their emission at higher energies, in the form of gamma-rays, X-rays, and energetic particles, as well as weaker optical emission. The particle emission is suspected of being the crucial element in producing the emission at all energies. The fact that the radiation is strongly beamed, is polarized, and has a nonthermal spectrum indicates that these objects are strongly magnetized, spinning neutron stars.

A different class of astrophysical objects involving a magnetized neutron star are the binary X-ray pulsars (Giacconi *et al.*, 1971), which derive their X-ray luminosity from conversion of the gravitational energy of gas accreted from a binary stellar companion and are here referred to as accreting X-ray pulsars (AXPs). Some of the RPPs are also in binary systems, but these do not accrete or emit X-rays, whereas the AXPs are exclusively accreting systems and do not show any radio emission, which would be absorbed in the dense plasma of the accretion flow. The radiation is again beamed, and a strong magnetic field has been directly measured via cyclotron lines in a number of objects. This magnetic field is responsible for the temporal modulation of the emission with the spin and orbital rotation period.

A third class of objects are the very-high-energy (VHE) and ultra-high-energy (UHE) sources (Fazio *et al.*, 1972; Samorski and Stamm, 1983). This class appears associated with, or includes as members, a small subset of the RPPs and of the AXPs, as well as a few other not easily classified X-ray sources. The radiation may be pulsed with a period similar to that at lower energies, and at very high energies it appears to consist of $\gamma$-rays in the $10^{12}$ eV range, while at ultrahigh energies it may be a mixture of $\gamma$-rays and neutral particles in the $10^{15}$ eV energy range. The VHE-UHE emission is sporadic, but when present it appears to be substantial. Because of the positional identification with known pulsating X-ray sources in the majority of cases and the similar time modulation in some of them, these sources are believed to be magnetized spinning neutron stars.

A fourth class of objects associated with neutron stars are the gamma-ray bursters (GRBs), discovered as intense flashes of $\gamma$-radiation in the $10^{-1}$–$10^{2}$ MeV range, with very short rise times and overall durations of tens of seconds (Klebesadel, Strong, and Olson, 1973). These bursts contain less than a few percent contribution in X-rays, and there is only scanty or no evidence for weak emission at other wavelengths. The short rise times and the spectrum and luminosities are evidence for a neutron star origin, and the presence of lines interpreted as due to the cyclotron mechanism indicates a strong magnetic field. This evidence is not as strong as for RPP and AXP sources, and with one exception there is no definite indication that these objects are spinning or in a binary system, nor are there identifications with other steady sources identified at other wavelengths.

Other classes of neutron stars appear to have a very low magnetic field. One of these classes is that of the quasi-periodic oscillator (QPO) X-ray sources (van der Klis *et al.*, 1985), which are accreting binaries with a luminosity which is time modulated with a variable and rather broadened period distribution whose central frequency depends on luminosity. The luminosity and the partial overlap in membership with the X-ray bursters described below indicates that QPOs are neutron star sources whose quasi-periodic luminosity modulation can be understood in terms of accretion by a spinning neutron star with a very weak magnetic field $B \lesssim 10^{9}$–$10^{10}$ G, much lower than those of the previously discussed neutron star classes. Another class of very weakly magnetized sources are the X-ray bursters (XRBs), which produce bursts of X-rays with durations of 10–30 seconds and luminosities close to the Eddington value (Grindlay *et al.*, 1976; Belian, Conner, and Evans, 1976), indicating emission from the entire surface of an object of neutron star size. The widely accepted interpretation of this occasionally repetitive behavior invokes a thermonuclear origin for these bursts on the surface of a nonmagnetized or weakly magnetized ($\lesssim 10^9$ G) neutron star. These sources are not central to our discussion of magnetized neutron stars, but because they seem to differ only in the strength of their field, their origin and relationship to the former class of sources are of great interest.

# 2. Physics in a Strong Magnetic Field

## 2.1 Classical Motion of Charged Particles

### *a) Hamiltonian Mechanics in an Electromagnetic Field*

The classical motion of charged particles in electromagnetic fields is generally best described in the Hamiltonian formalism. This is also useful for the quantum treatment of the electron interaction with the radiation field. Since the transition from classical to quantum dynamics is most natural in the Hamiltonian formulation of mechanics, we set down here some of the basic equations, without going into much detail (see e.g. Goldstein, 1950; Landau and Lifshitz, 1965a).

Hamilton's principle of least action states that the laws of physics are such that the time integral of the Lagrangian function $L(q, \dot{q}, t)$ of the system under consideration assumes an extremum value, which in most cases is a minimum,

$$J = \int_{t_1}^{t_2} L(q_i, \dot{q}_i, t)\, dt = \text{extremum}. \tag{2.1.1}$$

For a system with $f$ degrees of freedom, there are $f$ generalized coordinates $q_1, \ldots, q_f$. For a point particle, there are three, the space coordinates $x$, $y$, $z$ of the particle. The Lagrangian also depends on the time $t$ and on $\dot{q}_i$, where

$$\dot{q}_i = dq_i / dt, \tag{2.1.2}$$

the $\dot{q}_i$ being the generalized velocities. The integral (2.1.1) is taken between two fixed end points of the trajectory $q_i(t)$, and the minimum is obtained by varying the trajectory between these two fixed points. This leads to a set of differential equations, called the Lagrange equations of motion,

$$\frac{d}{dt}\left(\frac{\partial L}{\partial \dot{q}_i}\right) - \frac{\partial L}{\partial q_i} = 0 \tag{2.1.3}$$

In order to make the transition to quantum mechanics, it is convenient to transform these equations of motion to the canonical, or Hamiltonian, form. This is done by introducing canonically conjugate variables to the $q_i$, which are the generalized momenta

$$p_i = \partial L / \partial \dot{q}_i . \tag{2.1.4}$$

For a free point particle, these are just the usual components of the momentum $p_x$, $p_y$, $p_z$. The canonical equations of motion use a new function of the canonical coordinates, generalized momenta and time, called the Hamiltonian, which is related to the Lagrangian by

$$H(q_i, p_i, t) = \left\{ \sum_{i=1}^{f} p_i \dot{q}_i \right\} - L(q_i, \dot{q}_i, t). \tag{2.1.5}$$

In the Hamiltonian, the variables $\dot{q}_i$ are expressed as functions of the $p_i$ with the help of equations (2.1.4). The new equations of motion can then be obtained from equations (2.1.3). These are the Hamiltonian equations of motion,

$$\dot{q}_i = \partial H / \partial p_i , \quad \dot{p}_i = -\partial H / \partial q_i . \tag{2.1.6}$$

The advantage of a Hamiltonian formulation is that $H$ has a direct interpretation as the total energy of the system.

### *b) The Point Particle in a Scalar Potential Field*

As an example, for a particle in a field of force $\vec{F}$ derivable from a potential $V$, the Hamiltonian is

$$H = T + V, \tag{2.1.7}$$

where the kinetic energy $T$ is

$$T = \frac{1}{2}m\left(p_x^2 + p_y^2 + p_z^2\right). \tag{2.1.8}$$

From the canonical equations (2.1.5)–(2.1.6) we get

$$\dot{x} = p_x / m , \quad \dot{y} = p_y / m , \quad \dot{z} = p_z / m \tag{2.1.9}$$

and

$$\dot{p}_x = -\partial V / \partial x , \quad \dot{p}_y = -\partial V / \partial y , \quad \dot{p}_z = -\partial V / \partial z . \tag{2.1.10}$$

Eliminating $\dot{p}_i$ in equation (2.1.10) by means of equation (2.1.9) leads to the usual Newtonian equations of motion

$$\vec{F} = m\ddot{\vec{r}} = -\vec{\nabla}V, \tag{2.1.11}$$

where $\vec{F}$ is the force on the particle.

### *c) The Charged Point Particle in an Electromagnetic Field*

It is well known that a point particle in an electric field $\vec{E}$ and magnetic field $\vec{B}$ is subject to the Lorentz force $\vec{F}_L$ defined by

$$m\ddot{\vec{r}} = \vec{F}_L = e\left(\vec{E} + \frac{\dot{\vec{r}}}{c} \times \vec{B}\right), \tag{2.1.12}$$

where $m$ is the mass and $e$, the electromagnetic charge of the particle. This Lorentz equation of motion can also be derived from the Hamiltonian equations of motion. In order to do this, however, it is necessary to define the canonical momentum of the particle, $p_i$, by

$$p_i \to p_i - \frac{e}{c}A_i, \tag{2.1.13}$$

where $p_i$ is the physical (or usual) momentum, $e \equiv -|e|$ for electrons, and $A_i$ is the $i$th component of the vector potential of the electromagnetic field. The physical fields $\vec{E}$ and $\vec{B}$ in vacuum can be derived (Jackson, 1975) from the vector potential $\vec{A}$ and the scalar potential $U$ by means of

$$E_i = -\partial U/\partial x_i - \partial A_i/\partial t, \tag{2.1.14}$$

where the vector potential is related to the magnetic field strength $B$ through

$$B_i = \left(\vec{\nabla} \times \vec{A}\right)_i. \tag{2.1.15}$$

Using equation (2.1.13), the Hamiltonian is

$$H = 1/2m\sum_i \left(p_i - \frac{e}{c}A_i\right)^2 + eU, \tag{2.1.16}$$

where $V = eU$. The canonical equations of motion (2.1.6) lead to

$$\dot{r}_i = \left(p_i - \frac{e}{c}A_i\right)\Big/m \tag{2.1.17a}$$

and

$$\dot{p}_i = (e/m)\left[\sum_{j=1}^{3}\left(p_j - \frac{e}{c}A_j\right)\partial A_j/\partial x_i\right] - e\,\partial U/\partial x_i. \quad (2.1.17b)$$

Differentiating equation (2.1.17b) with respect to time and substituting equation (2.1.17b) for $p_i$, we get

$$\begin{aligned} m\dot{r}_i &= \dot{p}_i - (e/c)\dot{A}_i \\ &= (e/m)\left[\sum_{j=1}^{3}\left(p_j - \frac{e}{c}A_j\right)\partial A_j/\partial r_i\right] - \frac{e}{c}\dot{A}_i - e\,\partial U/\partial r_i, \end{aligned} \quad (2.1.18)$$

which can be simplified using equation (2.1.17a) to

$$m\ddot{r}_i = \frac{e}{c}\left[\sum_{j=1}^{3}\left(\dot{r}_j\,\partial A_j/\partial r_i\right)\right] - \frac{e}{c}\dot{A}_i - e\,\partial U/\partial r_i. \quad (2.1.19)$$

We can write the components explicitly as

$$\dot{A}_x = \frac{\partial A_x}{\partial t} + \frac{\partial A_x}{\partial x}\dot{x} + \frac{\partial A_x}{\partial y}\dot{y} + \frac{\partial A_x}{\partial z}\dot{z}, \quad (2.1.20)$$

and similarly for the components $y$ and $z$. Introducing equation (2.1.20) into equation (2.1.19), we obtain for the $x$-component

$$\begin{aligned} m\ddot{x} &= (e/c)\left[\dot{y}\left(\frac{\partial A_y}{\partial x} - \frac{\partial A_x}{\partial y}\right) - \dot{z}\left(\frac{\partial A_x}{\partial z} - \frac{\partial A_z}{\partial x}\right)\right] \\ &\quad - \frac{e}{c}\frac{\partial A_x}{\partial t} - e\frac{\partial U}{\partial x}. \end{aligned} \quad (2.1.21)$$

We can rewrite this equation in terms of the electric and magnetic fields using equations (2.1.14) and (2.1.16) to get

$$m\ddot{x} = e\left[E_x + \left(\frac{\dot{\vec{r}}}{c} \times \vec{B}\right)_x\right], \quad (2.1.22)$$

and similarly for $y$, $z$. These are the components of the Lorentz equation of motion (2.1.12). Thus, in a Hamiltonian formulation, the correct canonical momentum to be used is $(\vec{p} - \vec{A}e/c)$ rather than the physical momentum of the particle $\vec{p}$.

*d) Classical Electron Motion in a Uniform Magnetic Field*

The nonrelativistic equation of motion of an electron in arbitrary electric and magnetic fields was derived above. A relativistic equation of motion can be similarly derived from a covariant formulation of mechanics and electrodynamics (e.g. Bekefi, 1966; Rybicki and Lightman, 1979). This is done by equating the electron four-momentum $p^\mu = (\mathscr{E}/c, \vec{p})$, where Greek indices range from 1 to 4 and $\mathscr{E}$ is energy, to the Lorentz four-force $F^\mu = (e/c)F^\mu_\nu U^\nu$, where $U^\nu$ is the particle four-velocity and $F^\mu_\nu$ is the electromagnetic field tensor defined in equation (2.5.1). For a uniform, static magnetic field, one obtains then

$$\frac{d}{dt}(\gamma m\vec{v}) = (q/c)\vec{v} \times \vec{B}, \tag{2.1.23}$$

$$\frac{d}{dt}(\gamma mc^2) = q\vec{v} \cdot \vec{E}, \tag{2.1.24}$$

which express, respectively, the conservation of relativistic momentum $\gamma m\vec{v}$, and total energy $\gamma mc^2$, where $\gamma = (1 - \beta^2)^{-1/2}$ is the Lorentz factor and $\vec{\beta} = \vec{v}/c$. In the energy equation, radiative losses have been left out, which is correct over time scales much shorter than $t_l \sim \gamma/(d\gamma/dt)$. From the second equation, we see then that $\gamma$ = constant, or $|\vec{v}|$ = constant, so that

$$m\gamma\frac{d\vec{v}}{dt} = \frac{q}{c}\vec{v} \times \vec{B}. \tag{2.1.25}$$

The velocity can be separated into components along the field and perpendicular to it, $\vec{v}_\parallel$ and $\vec{v}_\perp$, satisfying

$$\frac{dv_\parallel}{dt} = 0, \quad \frac{d\vec{v}_\perp}{dt} = \frac{q}{\gamma mc}\vec{v}_\perp \times \vec{B}. \tag{2.1.26}$$

Therefore $\vec{v}_\parallel$ = constant, and since $|\vec{v}|$ = constant, we have also $|\vec{v}_\perp|$ = constant. The particle trajectory, when projected on the plane perpendicular to $\vec{B}$, represents uniform circular motion, since $|\vec{v}_\perp|$ is constant and the acceleration is perpendicular to the velocity. Combined with the longitudinal uniform motion along $\vec{B}$, the total motion corresponds to a helicoid (see Fig. 2.1.1). The frequency of

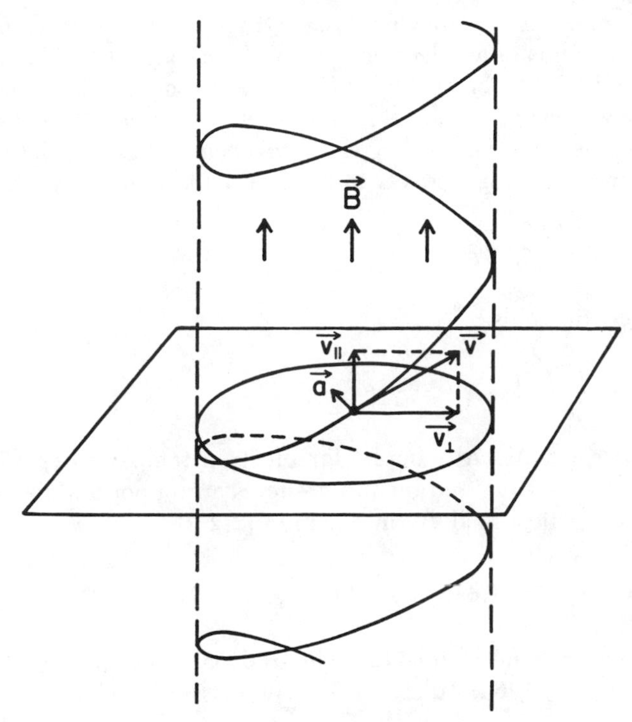


FIGURE 2.1.1 Classical helicoidal motion of a charged particle in a magnetic field $\vec{B}$.

rotation, or gyrofrequency, is seen to be

$$\omega_g = \frac{qB}{\gamma m_e c}, \tag{2.1.27}$$

while the acceleration, which is normal to the velocity, is $a_\perp = \omega_g v_\perp$. The classical radius of gyration in the plane perpendicular to $\vec{B}$ is

$$r_L = \frac{v_\perp}{\omega_g} = \gamma v_\perp \frac{m_e c}{eB}, \tag{2.1.28}$$

which is called the Larmor radius of the electron.

## 2.2 The Onset of Quantum Effects in a Strong Magnetic Field

In a strong magnetic field $B \sim 10^{12}$–$10^{13}$ G such as that encountered in neutron stars, a quantum mechanical treatment of the plasma and radiation processes becomes necessary. To see this, we may consider the Larmor radius of an electron gyrating in a magnetic field (see eq. 2.1.28). When $r_L$ becomes comparable to or smaller than the de Broglie wavelength of the electron, $\hbar/p = \hbar/(\gamma m v_\perp)$, the localization of the electron can no longer be given in a classical description and we expect quantum effects to become important. This sets in for

$$\frac{\hbar}{\gamma m v} \gtrsim \frac{\gamma m c v_\perp}{eB} \tag{2.2.1}$$

or for magnetic fields

$$B \gtrsim \frac{m^2 c^3 \gamma^2}{e\hbar}\left(\frac{v_\perp}{c}\right)^2 = \gamma^2 \beta_\perp^2 B_Q, \tag{2.2.2}$$

where $\beta \equiv v_e/c$. As an example, for electron temperatures $kT = 10$ keV, $\beta^2 = 5.6 \times 10^{-2}$ and quantum effects are important above $B \simeq 10^{12}$ G. The critical field $B_Q$ in equation (2.2.2),

$$B_Q = \frac{m^2 c^3}{e\hbar} = 4.413 \times 10^{13} \text{ G}, \tag{2.2.3}$$

plays a crucial role in the quantum electrodynamics of strong magnetic fields (see Erber, 1966; Adler, 1971). The classical cyclotron energy

$$\hbar\omega_c = \hbar\frac{eB}{mc} = mc^2(B/B_Q) \tag{2.2.4}$$

becomes equal to the electron rest mass energy at $B = B_Q$. This is similar to the situation encountered in a strong electric field, where there is a critical field $E_Q = m^2c^3/e\hbar$ for which an electron, if accelerated through a distance equal to the Compton wavelength $\lambda\!\!\!^{-}_c$ by the field $E_Q$, acquires an energy $\mathscr{E} = mc^2$; that is, $eE_Q\hbar/mc = mc^2$. For $E > E_Q$, the vacuum becomes unstable against spontaneous breakdown into electron-positron pairs, the well-known "Klein catastrophe." However, in the case of a very strong magnetic field $B > B_Q$, a similar breakdown does not occur. Because of the inherent asymmetry of electric and magnetic fields, manifested through the absence (so far) of magnetic monopoles, magnetic fields of arbitrarily large strength $B > B_Q$ remain stable against spontaneous vacuum breakdown. The quantity $B_Q$ nonetheless remains of crucial significance, being the natural quantum mechanical measure of field strength, which parametrizes all radiative and plasma processes occurring in a strong magnetic field.

For $v$ approaching $c$, the condition of equation (2.2.2) is equivalent to stating that, for $B = B_Q$, the Larmor radius becomes comparable to or less than the electron Compton wavelength $\lambdabar_c = \hbar / mc$. This is clearly the smallest spatial region within which the electron can be localized. Setting the classical Larmor circumference of the electron orbit $2\pi r_L$ given by equation (2.1.28) equal to the de Broglie wavelength of the electron $h/p = h/\gamma m_e v$, we get a characteristic magnetic quantum length scale for the electron

$$\lambda \equiv \left(\frac{\hbar c}{eB}\right)^{1/2} = 2.6 \times 10^{-10} B_{12}^{1/2} \text{ cm}, \tag{2.2.5}$$

where $B_{12}$ is the field in units of $10^{12}$ G. The characteristic length can also be obtained formally (e.g. Ruderman, 1974) by considering the classical equation of motion of the electron,

$$\frac{m_e v^2}{r} = \frac{ev}{c} B, \tag{2.2.6}$$

where $r$ is the cylindrical radius of gyration perpendicular to $\vec{B}$ and $v$ is perpendicular velocity, and by imposing the quantization condition

$$\vec{p} \times \vec{r} = n\hbar \left(\frac{\vec{B}}{B}\right). \tag{2.2.7}$$

According to (2.1.13), the momentum must be understood to be

$$\vec{p} = m_e \vec{v} - (e/c)\vec{A}, \tag{2.2.8}$$

and using the Landau gauge for the vector potential

$$\vec{A} = \frac{1}{2} B r \hat{\varphi}, \tag{2.2.9}$$

where $\hat{\varphi}$ is a unit vector along the azimuthal angle coordinate, one obtains the quantized transverse energy values

$$E_n = n\hbar\omega_c, \qquad n = 1, 2, \ldots, \tag{2.2.10}$$

where $\omega_c = eB/m_e c$ is the usual cyclotron energy. The circular orbits of the electrons have quantized radii

$$r_n = (2n)^{1/2} \lambda, \tag{2.2.11}$$

where $\lambda$ is the characteristic length given by equation (2.2.5). In a more detailed quantum treatment, one must solve the Schrödinger equation, which modifies the energy values by a constant and reveals a degeneracy involving the freedom of translation of the center of the orbit perpendicular to the field.

## 2.3 Quantum Treatment of the Electron in a Magnetic Field

### *a) Nonrelativistic Hamiltonian Operator*

The Hamiltonian operator of an isolated nonrelativistic, spinless particle of charge $q$ and mass $m$ is $H = p^2/2m$, where $\vec{p}$ is the canonical momentum. In an external magnetic field $\vec{B}$, of vector potential $\vec{A}$ given by

$$\vec{A} = \frac{1}{2}\vec{B} \times \vec{r}, \tag{2.3.1}$$

the canonical momentum $\vec{p}$ must be replaced (eq. 2.1.13) by $\vec{p} \to \vec{p} - (q/c)\vec{A}$, where $q$ is the charge. The spinless nonrelativistic Hamiltonian is therefore

$$H = \frac{1}{2m}\left(\vec{p} - \frac{q}{c}\vec{A}\right)^2. \tag{2.3.2}$$

The spin contribution to the Hamiltonian, if we forgo for the moment a full treatment via the Dirac equation, can be included by adding the term

$$H_{\text{spin}} = -\frac{q\hbar}{2mc}\vec{\sigma}\cdot\vec{B}, \tag{2.3.3}$$

where $\vec{\sigma}$ is a Pauli spin matrix operator of eigenvalues $\sigma = \pm 1$. The difference in energy between the spin-up and spin-down electrons is given by

$$\Delta\mathscr{E}_{\text{sp}} = \frac{|q|\hbar}{mc}B = mc^2\left(\frac{B}{B_Q}\right). \tag{2.3.4}$$

The value of $B_Q$, given by equation (2.3.3), again appears as a natural unit of field strength for which the separation of spin levels equals the electron rest mass energy.

The total Hamiltonian in the nonrelativistic case is therefore, from (2.3.2) and (2.3.3),

$$H_{\text{nr}} = \frac{1}{2m}\left(\vec{p} - \frac{q}{c}\vec{A}\right)^2 - \frac{q\hbar}{2mc}\vec{\sigma}\cdot\vec{B}. \tag{2.3.5}$$

This equation ignores the effect of any admixture of negative energy states—i.e. the effects of the Dirac positron sea—present in a relativistic treatment. This is justifiable, provided that the particle states expressed in equation (2.3.5) represent wave functions uncontaminated by the vacuum. A simple criterion for this to be satisfied is that the particle (in this case an electron) be localizable over a region not smaller than the Compton wavelength $\lambdabar_c$. An approximate spatial localization for the particle is given by the Larmor gyroradius, given by equation (2.1.28). The particle Hamiltonian (2.3.5) is therefore valid for nonrelativistic energies and fields $B \lesssim \beta B_Q$. For $kT \sim 10$ keV, this means $B \lesssim 0.2 B_Q$. Under these circumstances, one can develop a complete formulation (Sokolov and Ternov, 1968; Canuto and Ventura, 1977) of the quantum behavior of an electron in a superstrong magnetic field.

### *b) Commutation Relations and Operator Algebra*

For simplicity, let us neglect temporarily the spin part of the Hamiltonian and assume that $\vec{B}$ is in the $z$-direction. Using $q = -e$, the physical (as opposed to the canonical) momentum of the electron is given by

$$\vec{\pi} = m\vec{v} \equiv \vec{p} + (e/c)\vec{A}, \tag{2.3.6}$$

and therefore

$$H = \frac{1}{2m}\left(\pi_x^2 + \pi_y^2 = \pi_z^2\right), \tag{2.3.7}$$

where, using equation (2.3.1),

$$\pi_x = p_x - \frac{eB}{2c}y, \quad \pi_y = p_y + \frac{eB}{2c}x, \quad \pi_z = p_z. \tag{2.3.8}$$

The components of the $\vec{\pi}$ operator do not all commute, since, using the usual commutation properties $[p_x, x] = [p_y, y] = -i\hbar$, $[x, y] = [x, p_y] = 0$, we obtain

$$\begin{aligned} [\pi_x, \pi_y] &= -i\hbar eB/c = -im^2c^2(B/B_Q), \\ [\pi_x, \pi_z] &= [\pi_y, \pi_z] = 0. \end{aligned} \tag{2.3.9}$$

Equation (2.3.9) gives a quantum indeterminacy in the transverse nonrelativistic velocity

$$\Delta v_x\, \Delta v_y \gtrsim c^2(B/B_Q). \tag{2.3.10}$$

In the classical limit, the circular motion in the plane transverse to $\vec{B}$ is represented by a harmonic motion. By analogy, we may define here a quantity

$$Q_x \equiv (c/eB)\pi_y \tag{2.3.11}$$

and rewrite equation (2.3.7) as

$$H = \frac{1}{2m}\left[\pi_x^2 + \omega_c^2 Q_x^2\right] + \frac{1}{2m}p_z^2 \tag{2.3.12}$$

with

$$[\pi_x, Q_x] = -i\hbar, \tag{2.3.13}$$

where $\omega_c = eB/mc$ is the cyclotron frequency. Equation (2.3.12) represents the superposition of a free motion along $z$ and a harmonic motion of frequency $\omega_c$ in the $x$, $y$ plane. The energy eigenvalues of such a system are given in elementary quantum mechanics texts (e.g. Gasiorowicz, 1974). They are, for the zero spin case of equation (2.3.12),

$$\begin{aligned} E &= (n + 1/2)\hbar\omega_c + (1/2m)p_z^2 \\ &= (n + 1/2)mc^2(B/B_Q) + (1/2m)p_z^2, \quad n = 0, 1, \ldots \end{aligned} \tag{2.3.14}$$

The spectrum is a superposition of a continuum along $z$ and a discrete set of levels in the transverse direction, called Landau levels in the magnetic case. This differs from the energy levels derived from a simple quantization of the classical motion (eq. 2.2.10) by an additive factor of 1/2, which is due to the zero-point energy of the ground state $n = 0$, which is $\frac{1}{2}mc^2(B/B_Q)$. As in the case of the harmonic oscillator, the various eigenfunctions and eigenvalues can be obtained from each other via the application of up and down ladder operators. In terms of $v_\pm = v_x \pm iv_y$, the up and down operators can be defined as

$$a^\dagger = (m/\hbar\omega_c)^{1/2}v_+ \text{ (up)}; \quad a = (m/\hbar\omega_c)^{1/2}v_- \text{ (down)} \tag{2.3.15}$$

with the effect

$$\begin{aligned} a^\dagger|n\rangle &= \text{const}\left(n + \frac{1}{2} + \frac{1}{2}\right)^{1/2}|n+1\rangle, \\ a|n\rangle &= \text{const}\left(n + \frac{1}{2} - \frac{1}{2}\right)^{1/2}|n+1\rangle. \end{aligned} \tag{2.3.16}$$

One can therefore write

$$H_\perp = \frac{1}{2m}\left(\pi_x^2 + \pi_y^2\right) = \hbar\omega_c\left(a^\dagger a + \frac{1}{2}\right) \tag{2.3.17}$$

for the transverse Hamiltonian.

*c) Degeneracy and Constants of the Motion*

The ground Landau level $n = 0$ is defined by $a|0\rangle = 0$, and it is infinitely degenerate, as are all higher levels. Clearly, a state of a given $n$, $p_z$ must be infinitely degenerate, since the energy cannot depend on the position of the guiding center. The coordinates of the guiding center are, from the solution of equations (2.1.26),

$$x_0 = x - \frac{v_+}{\omega}\sin\omega_c t = x - v_y/\omega_c,$$

$$y_0 = y + \frac{v_+}{\omega}\cos\omega_c t = y + v_x/\omega_c, \tag{2.3.18}$$

where in the nonrelativistic limit the gyrofrequency $\omega_g = \omega_c$ and $v_i = \pi_i/m$. One can verify that $x_0$ and $y_0$ each commute with $\pi_\perp^2 = \pi_x^2 + \pi_y^2$, which means that they are quantum mechanical constants of motion, and therefore the energy does not depend on the guiding center. At the same time $x_0$ does not commute with $y_0$, since

$$[x_0, y_0] = i\hbar/m\omega_c = i\lambda^2, \tag{2.3.19}$$

where $\lambda$ is the characteristic magnetic transverse length scale of equation (2.2.5). This equation shows again that the transverse position of the guiding center cannot be defined closer than within a length given by $\lambda$. One can define, by means of $r_\pm = 2^{-1/2}(x_0 \pm iy_0)$, operators $b$ and $b^\dagger$ similar to those of equations (2.3.15), which lead to transitions between the various degenerate states, i.e. a translation of the guiding center,

$$b = (m\omega_c/\hbar)^{1/2}(r_+ + iv_+/\omega),$$

$$b^\dagger = (m\omega_c/\hbar)^{1/2}(r_- - iv_-/\omega), \tag{2.3.20}$$

with an associated quantum number $s$ (not to be confused with spin). A complete set of eigenfunctions can be defined as the set which are simultaneous eigenstates of $H_\perp$, of $p_z$, and of any function of $x_0$, $y_0$ (Canuto and Ventura, 1977). If the latter function is taken to be, say,

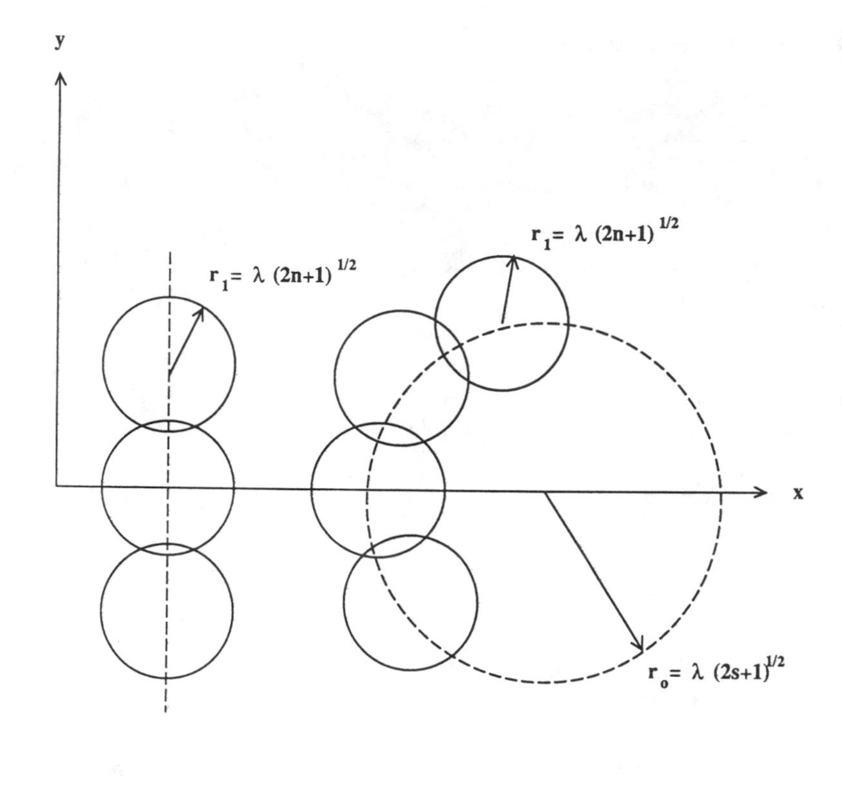


FIGURE 2.3.1 Degeneracy of the orbits corresponding to the eigenfunctions of (a) the operator $x_0$ and (b) the operator $r_0^2$. After V. Canuto and J. Ventura, 1977, *Fund. Cosmic Physics*, 2, 203. Reprinted with permission of Gordon and Breach Science Publishers.

$x_0$, equation (2.3.19) shows that $y_0$ is totally indeterminate, with the electron orbit localized along the line $x = x_0$, as shown in Figure 2.3.1a. Alternatively, one can choose for the arbitrary function of $x_0$, $y_0$ the more convenient form

$$r_0^2 = x_0^2 + y_0^2 = b^\dagger b. \tag{2.3.21}$$

This, together with (2.3.19), defines an algebra of oscillators with eigenvalues

$$r_0^2 = (2s + 1)\lambda^2, \quad s = 0, 1, 2, \ldots \tag{2.3.22}$$

In this case, the electron orbits are localized on a cylindrical shell of radius given by $r_0$, as shown in Figure 2.3.1b. The radius $r_1$ of an individual orbit, which is located somewhere along the shell of radius $r_0$, is a constant of the motion which can be evaluated for a given Landau level,

$$r_1^2 \equiv (x - x_0)^2 + (y - y_0)^2 = \frac{1}{\omega_c}(v_x^2 + v_y^2) = (2n + 1)\lambda^2. \tag{2.3.23}$$

This equation also gives the orbit radius of the operator $x_0$ shown in Figure 2.3.1a. An additional constant of the motion is the $z$-component of the angular momentum $L_z = xp_y - yp_x$, where

$$l_z = (n - s)\hbar \tag{2.3.24}$$

are the eigenvalues.

*d) Eigenfunctions and Eigenstates*

From the previous considerations, the complete set of transverse eigenfunctions can be taken to be

$$|n, s\rangle = (n!s!)^{-1/2}(a^\dagger)^n(b^\dagger)^s|0, 0\rangle. \tag{2.3.25}$$

The number of nodes of the transverse wave function is $n_r = \min(n, s)$. As for the longitudinal part of the wave function, corresponding to a free particle, this is $e^{ip_z z}$ so that the total nonrelativistic, spinless wave function is the product of the transverse and longitudinal parts, and is given by (Canuto and Ventura, 1977):

$$\Psi = L^{-1/2}\phi_{ns}(r, \varphi)\exp(ip_z z), \tag{2.3.26}$$

with the definitions

$$\begin{aligned} \phi_{ns}(r, \varphi) &= (\gamma/\pi)^{1/2}\exp[i(n - s)\varphi]\, I_{ns}(\gamma r^2), \\ \gamma &= (1/2)\lambda_c^{-2}(B/B_Q), \qquad \lambda_c = \hbar/mc, \\ I_{ns}(x) &= (-1)^s(n!s!)^{-1/2}\exp(-x/2)\,x^{(n-s)/2}Q_n^{n-s}(x), \end{aligned} \tag{2.3.27}$$

where $r$ and $\varphi$ are cylindrical coordinates, and the $Q_n^{n-s}$ are Laguerre polynomials. The total energy, without spin contributions, is given by equation (2.3.28). The associated system of levels is shown in Figure 2.3.2, where the values of $E$ and $l_z$ and the corresponding Landau

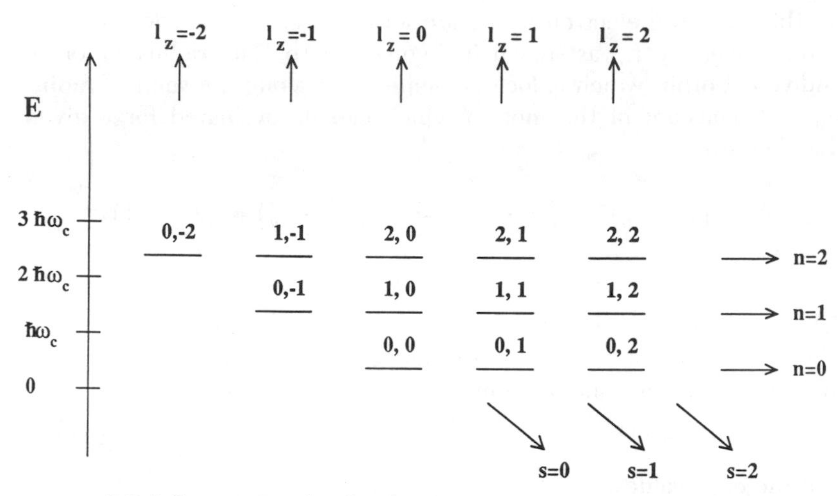


FIGURE 2.3.2 Lowest Landau levels of nonrelativistic spinless electron in a magnetic field. The terms are identified by the values of the nodal number $n_r$ of the radial wave function and by the $z$-component of the angular momentum $l_z$. The quantum numbers $n$ and $s$ are given on the ordinate and abscissa, respectively, where $n$ is the principal (Landau) quantum number and $s$ is the guiding center quantum number, the state being labeled by $|n, s\rangle$. The angular momentum along the magnetic field is $l_z = (n - s)\hbar$. After Nagel (1982).

quantum number $n$ and guiding center quantum number $s$ are indicated. This level structure is the same as for the standard harmonic oscillator, except that the energies are displaced by a constant value. This analogy leads to useful simplifications, since the harmonic oscillator is one of the few systems for which quantum and classical calculations agree (this is no longer true for the relativistic case).

Including now the spin contribution to the nonrelativistic energy of the free electron in a magnetic field, the eigenvalues (2.3.14) become

$$E = \frac{1}{2}\hbar\omega_c(2n + 1 + \zeta) + \frac{1}{2m}p_z^2,$$
$$n = 0, 1, 2, \ldots, \qquad \zeta = \pm 1, \tag{2.3.28}$$

where $\zeta$ is the spin eigenvalue. Clearly the level structure is the same as in Figure 2.3.2, except that each level is doubly degenerate due to the spin, the same energy corresponding to $2n + \zeta =$ constant. The solution of Dirac's equation, as discussed in Chapter 5, gives

$$E = \left[m^2c^4 + c^2p_z^2 + (2n + \zeta + 1)mc^2\hbar\omega_c\right]^{1/2}$$

$$= \left[m^2c^4 + c^2p_z^2 + (2n + \zeta + 1)m^2c^4(B/B_Q)\right]^{1/2}, \qquad (2.3.29)$$

which is the relativistic generalization of equation (2.3.28). The level structure, when relativistic effects are important, departs somewhat from that of Figure 2.3.2 in that the levels are no longer equally spaced, i.e. an anharmonicity is introduced. These and other relativistic effects are discussed in Chapter 5.

## 2.4 Atomic Structure in a Strong Magnetic Field

### *a) Qualitative Behavior*

In the absence of a magnetic field, the size of a hydrogen atom is characterized by the Bohr radius $a_0$ of the lowest quantized orbit. This radius can be obtained by equating the centrifugal force on an electron orbiting at radius $r$ with the electrostatic attraction force that binds it to the nucleus, $m_e v^2/r = Ze^2/r^2$, and then setting the circumference of the orbit equal to the de Broglie wavelength of the electron, $2\pi r = h/p = h/m_e v$, which yields

$$a_0 = \frac{\hbar^2}{m_e e^2} = 0.529 \times 10^{-8}\ \text{cm}. \qquad (2.4.1)$$

This atomic radius scales $\propto Z^{-1}$ with the charge of the nucleus, which was taken here to be $Z = 1$ for the hydrogen atom. In the presence of a magnetic field, however, besides the electrostatic force there is also a magnetic force $\vec{f}_L = (e/c)\vec{v} \times \vec{B}$ acting on the electron, and it is the combination of both the electrostatic and the magnetic force that will control the motion of the electron.

The effect of a magnetic field on the motion of a free electron was shown in the last section to consist of a free particle motion along the field, plus a transverse part consisting of a gyromotion with orbit of radius $r_1 = (2n + 1)^{1/2}\lambda$ (eq. 2.3.23) whose guiding center is undetermined along a circular shell of radius $r_0 = (2s + 1)^{1/2}\lambda$ (Fig. 2.3.1) for the particular choice of translation operator $r_0$. The transverse degeneracy of the guiding center will be removed in the presence of a weak electric field $\vec{E}$, because the latter forces the particle to drift with the velocity

$$v_d = \frac{\vec{E} \times \vec{B}}{B^2} c, \tag{2.4.2}$$

which causes the gyrating electron to drift sideways in a direction perpendicular to both the electric and the magnetic field. Thus, for the case of the choice of translation operator $x_0$ in Figure 2.3.1, where $\vec{B}$ is out of the page, if $\vec{E}$ is along $y$ the electron orbit will actually drift sideways along $x$ at the same time that it gyrates. A radial electric field originating at $r = 0$ (e.g. a nucleus) presents a similar situation. In this case, when the translation operator $r_0$ is chosen (eq. 2.3.21), the degeneracy with respect to translation along the guiding center circle $r_s$ is now lifted, and the electron actually starts to drift in the $x, y$ plane transverse to $\vec{B}$ along the guiding center circle of radius $r_0$ shown in Figure 2.3.1. The orbital motion of radius $r_1$ combined with the drift along the guiding center circle $r_0$ results in a cycloidal motion similar to that of the moon orbiting the earth and moving with the earth's guiding center around the sun.

In the case of a very strong magnetic field, we expect $n = 0$ or at any rate $n$ not very large, so the shell $r_0 = (2s + 1)^{1/2}\lambda$ has a larger radius than the orbit $r_1 = (2n + 1)^{1/2}\lambda$. The characteristic magnetic transverse length given in equation (2.2.5) can also be written

$$\lambda = \left(\frac{\hbar}{m_e \omega_c}\right)^{1/2} = \lambda\!\!\!^{-}_c \left(\frac{B_Q}{B}\right)^{1/2}, \tag{2.4.3}$$

where $\lambda\!\!\!^{-}_c = \hbar / m_e c = 3.862 \times 10^{-11}$ cm is the Compton wavelength. The order of magnitude of $\lambda \simeq 2.6 \times 10^{-10} B_{12}^{-1/2}$ cm is much smaller than the characteristic Bohr radius $a_0$ of an unmagnetized hydrogen atom, given by equation (2.4.1). The magnetic fields satisfying $\lambda \ll a_0$ are strong as far as the atomic structure of matter is concerned, even though they are of modest strength by neutron star standards,

$$B \gg B_s \left(\frac{e^2}{\hbar c}\right)^2 \frac{m_e^2 c^3 Z^2}{\hbar e} = \alpha^2 Z^2 B_Q \simeq 2.35 \times 10^9 Z^2 \text{ G}. \tag{2.4.4}$$

Here we have included the $Z^{-1}$ scaling of $a_0$. When condition (2.4.4) is satisfied, we can expect the transverse dimension of the atom to be much smaller than the longitudinal, being constricted by the magnetic field to have a size $\sim \lambda$ in the transverse direction.

Under these circumstances, most of the transverse force on the electron is magnetic, so the Coulomb force is now essentially one-

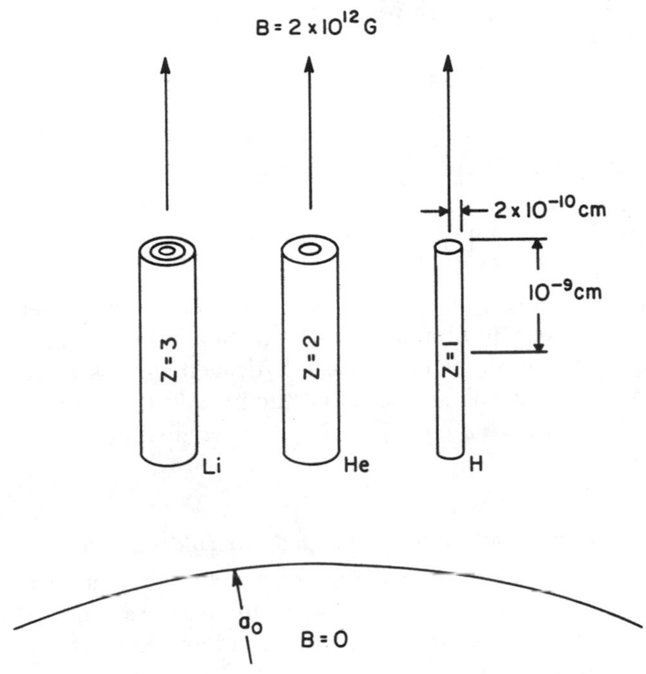


FIGURE 2.4.1 Schematic shape of a hydrogen atom in a strong magnetic field $B = 2 \times 10^{12}$ G also showing the relative sizes of the next few light atoms. The bottom part shows for comparison the zero field size ($a_0$) of the hydrogen atom. After M. Ruderman, in *IAU Symp. 53*, *Physics of Dense Matter*, ed. C. Hansen (Reidel, Dordrecht), p. 117. Reprinted by permission of Kluwer Academic Publishers.

dimensional, being effective only along $z$. It will be stronger along that one dimension than what it was in the three-dimensional field-free case. The magnetic atom is therefore expected to be cylindrical, and the electron in the $n = 0$ lowest-energy level has a typical transverse radius $\lambda$ and a much larger length $l$ along the magnetic field, as shown in Figure 2.4.1. The length $l$ can be obtained by considering the longitudinal component of the energy of the electron, $p_z^2/2m_e + V(z)$, where $V(z)$ is the one-dimensional Coulomb potential (e.g. Ruderman, 1974; Haines and Roberts, 1969). Again using the de Broglie wavelength for $p_z$, one has roughly

$$E \sim \frac{\hbar^2}{m_e l^2} - \frac{e^2}{l} \ln\left(\frac{l}{\lambda}\right). \tag{2.4.5}$$

Minimizing this result with respect to $l$ yields

$$l \sim \frac{a_0}{\ln\left(\dfrac{a_0}{l}\right)}, \tag{2.4.6}$$

and the ground state energy is approximately

$$E \sim -\frac{\hbar^2}{m_e a_0^2} \ln^2\left(\frac{a_0}{\lambda}\right). \tag{2.4.7}$$

Thus, in a very strong field, $\lambda \ll l \ll a_0$ and the binding energy of the ground state grows in proportion to $\ln^2 B$. Since the binding energy of the ground state of the unmagnetized hydrogen atom is $I_H = e^2/(2a_0) = \hbar^2/m_e a_0^2 = 13.6$ eV, one sees that the ground state of the magnetic atom with $B$ satisfying equation (2.4.4) is comparatively much more tightly bound.

*b) Energy Level Structure of the Magnetized Hydrogen Atom*

The detailed energy level structure of a hydrogenic atom requires a knowledge of the corresponding wave functions. In a magnetic field, the Hamiltonian of an electron in the Coulomb field of a nucleus is given by

$$H = H_\perp + \frac{p_z^2}{2m_e} - \frac{e^2}{\left(r_\perp^2 + z^2\right)^{1/2}}, \tag{2.4.8}$$

where $r_\perp^2 = x^2 + y^2$ and $z$ is along $\vec{B}$. When the magnetic field is very strong, the wave functions will be approximately separable (Landau and Lifshitz, 1965b; Canuto and Ventura, 1977), so they may be written as

$$\Psi_{0,s,\nu} = \phi_{0,s}(r_\perp, \varphi) f_\nu(z), \tag{2.4.9}$$

where $\phi_{0,s}$ is the transverse wave function of the free electron defined in equation (2.3.27) corresponding to an electron with $n = 0$ and a guiding center at $r_s = (2s+1)^{1/2}\lambda$. The $f_\nu(z)$ in (2.4.9) is the longitudinal wave function, which for the bound atom is of course no longer given by the free particle plane wave, as was the case in (2.3.26). The form of $f_\nu(z)$ can be found by replacing (2.4.6) into Schrödinger's equation and integrating the transverse variables. This procedure gives the longitudinal wave equation

$$\left[-\frac{\hbar^2}{2m_e}\frac{d^2}{dz^2} + V_{0s}(z)\right] f_{s\nu}(z) = E f_{s\nu}(z), \tag{2.4.10}$$

with

$$V_{ns}(z) = -\langle ns | \frac{-e^2}{r} | ns \rangle = -e^2 \gamma^{1/2} \int_0^\infty \frac{I_{ns}^2(t)\, dt}{\sqrt{t + \gamma z^2}}, \tag{2.4.11}$$

where use is made of the definitions (2.3.28). For $n = 0$ and $s \to \infty$,

$$V_{0s}(z) = \langle 0s | \frac{-e^2}{\left(r_\perp^2 + z^2\right)^{1/2}} | 0s \rangle \sim \frac{-e^2}{\left(\lambda_s^2 + z^2\right)^{1/2}}, \tag{2.4.12}$$

where $\lambda_s = (2s + 1)^{1/2}\lambda$. The energy of the ground state with $s = 0$, which has the highest binding energy, can also be roughly estimated as

$$|V_{00}| \sim \frac{e^2}{\left(\lambda^2 - z^2\right)^{1/2}} < \frac{e^2}{\lambda} \sim 20 B_{12}^{1/2} \text{ Ry}, \tag{2.4.13}$$

where $\text{Ry} = I_H = e^2/2a_0 = 13.6$ eV is the Rydberg energy unit.

There are two kinds of excited states for the magnetic hydrogen atom, neither of which involves excitation out of the ground Landau level $n = 0$. One type involves exciting the $s$ quantum number; that is, instead of having the smallest radius cylinders with $s = 0$, these atoms consist of larger cylinders of radius $\lambda_s = (2s + 1)^{1/2}\lambda$, within which the electron is localized (Canuto and Kelly, 1972). From equation (2.4.7) one sees that the binding energy depends only logarithmically on the radius of the cylinder $\lambda$ so that excited states with $s \neq 0$ have almost the same large binding energy as the ground state. A second type is obtained by exciting the quantum number $\nu$ of the longitudinal wave function $f_{s\nu}$ given by the solution of equation (2.4.10). This longitudinal wave function has $\nu$ nodes along $z$, so the ground level is nodeless, while the successive excited states have $\nu = 1, 2, \ldots$ nodes or zeros. All of the excited states with $\nu > 0$ are relatively weakly bound, having a $z$ extent of the order of $a_0$ or greater rather than the extent $l$ of the $\nu = 0$ ground state (eq. 2.4.6). Thus, the $n = 0$, $\nu = 0$, $s$ states are all much more tightly bound than the nonmagnetic atom, while the $n = 0, \nu, s$ for $\nu = 1, 2, \ldots$ states have a binding energy not too different from that of the nonmagnetized atom. This fact becomes more apparent in the limit $B \to \infty$ or $\lambda_s \to \infty$. In this case (Loudon, 1959) the level structure is just

$$E_N = -\left(\frac{1}{N^2}\right) \text{Ry}, \qquad N = 0, 1, 2, \ldots, \tag{2.4.14}$$

which, aside from the level $N = 0$, is just the structure of the principal quantum number of the hydrogen atom. The level $N = 0$ corresponds to an infinitely tightly bound ground state, while the $N = 1, 2, \ldots$ levels correspond to the usual Balmer levels. This is because in the limit $B \to \infty$ the longitudinal wave equation (2.4.10) is identical to the radial wave equation for the field-free hydrogen atom. In the finite $B$ case, the level structure of (2.4.14) is modified to

$$E_\nu = -\frac{1}{\left(\nu + \delta\nu\right)^2}\ \mathrm{Ry}, \qquad \nu = 0, 1, 2, \ldots \quad \left(0 < \delta\nu < 1\right), \tag{2.4.15}$$

where $\delta\nu$ is a quantum correction depending on $\nu$ and $B$. The ground state wave function is given by a $\nu = 0$ nodeless (i.e. even parity), nonsingular wave function with a finite but large binding energy, while the other excited states corresponding to $\nu \geq 1$ have $\nu$ nodes and consist of closely spaced doublets (for even and odd parity) with binding energies very close to the nonmagnetic atom (Canuto and Kelly, 1972; Simola and Virtamo, 1978). For this reason, these $\nu \geq 1$ levels are called the "hydrogen-like" levels, while the $\nu = 0$ are the "tightly bound" levels. Both the hydrogenic and the tightly bound levels are subdivided not only into parity doublets but also into the $s = 0, 1, 2, \ldots, \infty$ transverse sublevels (see Fig. 2.4.2). Unlike the nonmagnetic atom, the total angular momentum here is not conserved, but the longitudinal component of the angular momentum $L_z$ is, the eigenvalues being $L_z = (n - s)\hbar = -s\hbar$ (the latter equality is for the Landau magnetic quantum number $n = 0$). Notice that the electron spin in the ground Landau level is antiparallel to the magnetic field, so there is no spin degeneracy for the levels being considered here. Also, unlike the case of the nonmagnetic atom, the $s$ transverse excitation quantum number does not have an upper limit for a given $\nu$.

The transition from nonmagnetic behavior to magnetic behavior in Figure 2.4.2 occurs for the hydrogen atom at about the field strength given by equation (2.4.4). As the field is gradually increased, the usual $n$, $l$, $m$ levels of the hydrogen atom split into first the usual linear and later the quadratic Zeeman sublevels, and as the field increases beyond (2.4.4) the levels rearrange themselves into the completely different pattern given by the $n$, $\nu$, $s$ magnetic quantum numbers. The dipole transitions arise for $\Delta s = 0, \pm 1$, with the $\Delta s = 0$ being the strongest, since the $\Delta s = \pm 1$ involve the transverse dipole moment

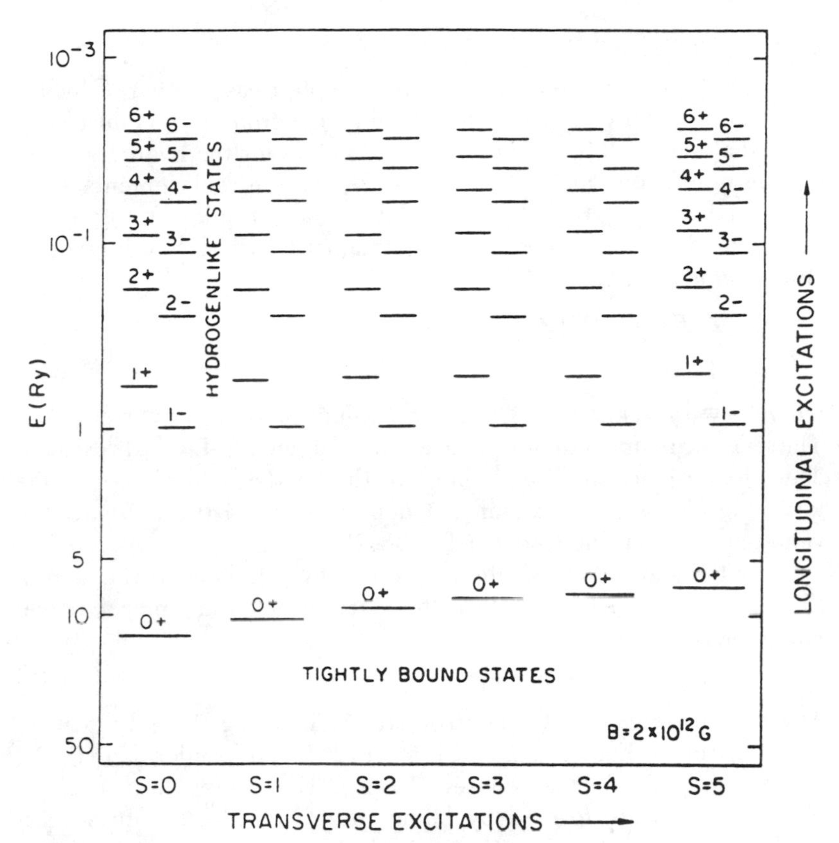


FIGURE 2.4.2 The hydrogen atom energy level structure in a strong magnetic field $B = 2 \times 10^{12}$ G. From V. Canuto and J. Ventura, *Ap. Space Sci.*, 17, 277. Reprinted by permission of Kluwer Academic Publishers.

which is suppressed by the magnetic field (Wunner and Ruder, 1980). The strengths of the transitions are different from the nonmagnetic case, but in the hydrogen-like levels, not by much. These hydrogen-like transitions look like blue- and redshifted Lyman, Balmer, etc., states, with intensity ratios depending on the magnetic field strength (Rösner *et al.*, 1984). The $\Delta s = 0$ transitions are linearly polarized and have a directionality $\propto \sin^2\theta$, while the $\Delta s = \pm 1$ transitions are circularly polarized with directionality $\propto (1 + \cos^2\theta)$, as in the Zeeman effect. The hydrogen-like transitions are in the eV range, while the tightly bound transitions (involving $\nu = 0$) are in the keV range for $B_{12} \gtrsim 1$.

*c) Heavier Elements*

For high-Z atoms in neutron stars, the simplest case is that of hydrogenic atoms, with $Z \geq 1$ but only one electron, i.e. almost fully ionized atoms. In this case, the previous values for the hydrogen atom can be used if one makes in the $Z \geq 1$ case the replacements $a_0 \to a_0/Z$, and $\mathrm{Ry} \to Z^2\,\mathrm{Ry}$. The energy levels and transition strengths can be obtained from the corresponding magnetic hydrogen values using (Ruder *et al.*, 1981)

$$E_s(Z, B) = Z^2 E_s(Z = 1, B/Z^2),$$
$$f_{\tau,\tau'}(Z, B) = f_{\tau,\tau'}(Z = 1, B/Z^2), \tag{2.4.16}$$

where $\tau \equiv [n, s, \nu]$ and $f_{\tau,\tau'}$ is the $f$-value for the $\tau \to \tau'$ transition. The level structure is then similar to that of Figure 2.4.2, but rescaled. Thus, for instance, for $Fe^{+25}$ most of the low-lying levels are in the keV range for $B_{12} \sim 1$, significant departures from the nonmagnetic values already starting to occur for $B \gtrsim 10^{11}$ G.

For the many-electron atoms, the energy is in general the sum of the kinetic, the electron-nuclear, the electron-electron, and the correlation energy terms

$$E = E_{\mathrm{kin}} + E_{eN} + E_{ee} + E_{\mathrm{corr}}. \tag{2.4.17}$$

The first investigations for neutron star fields along these lines were those of Cohen, Lodenquai, and Ruderman (1970), Ruderman (1971), and Kadomtsev and Kudryavtsev (1971a). The lightest $Z \geq 1$ multi-electron atoms are formed by filling the $n = 0, \nu = 0$ state successively with $s = 0, 1, 2, \ldots, Z$ single electrons of spin antiparallel to $\vec{B}$. From Figure 2.4.2, it is apparent that this is the energetically favorable way to fill levels rather than filling the larger energy gap $\nu \geq 1$ levels. This can go on, however, only as long as $\lambda_s \lesssim a_0 Z$. The length $l$ remains approximately the same, while the radius of the orbitals increases approximately as

$$\lambda_Z = (2Z + 1)^{1/2}\lambda. \tag{2.4.18}$$

The relative sizes of several of the lightest atoms with respect to the Bohr radius $a_0$ are shown in Figure 2.4.1. The ionization energy of the last electron of light atoms in a field $B = 2 \times 10^{12}$ G can be estimated as (Cohen, Lodenquai, and Ruderman, 1970)

$$E_I \sim 160 + 140 \ln\sqrt{Z}\ \mathrm{eV}, \tag{2.4.19}$$

where the $\ln\sqrt{Z}$ term comes from the exchange energy term. This is about two orders of magnitude larger than the single ionization energy of nonmagnetic atoms. A rough estimate of the total binding energy of the atom can be obtained by summing over all the terms like equation (2.4.7), which gives (Kadomtsev and Kudryavtsev, 1971a)

$$E_a \sim -\frac{9}{8}\frac{Z^3\hbar^2}{m_e a_0^2}\ln^2\eta, \tag{2.4.20}$$

where the dimensionless parameter

$$\eta \equiv \frac{a_0}{Z\lambda_Z} = \left(\frac{B}{Z^3 4.6 \times 10^9 \text{ G}}\right)^{1/2} \tag{2.4.21}$$

characterizes the filling pattern of the atom. For a strong field, $\eta \gg 1$, there is one electron per orbital while for marginally strong fields, $\eta \sim 1$, some of the $\nu = 0$ outer orbitals are left empty in favor of putting those electrons in the inner orbitals of the next lowest excited $\nu$ levels. An interesting property of the strongly magnetized atoms is that, since $s$ does not have a maximum upper value, one does not have the equivalent of the nonmagnetic atom closed-shell effects, which give the usual periodic table of elements its periodic recurrence of chemical properties.

*d) Condensed Matter Effects*

The elongated shape of atoms in a strong field causes them to have an enormous electrostatic quadrupole moment, $Q_{zz} \sim \langle 3z^2 - r^2\rangle e \sim l^2 e$, which will result in a strong interatomic attraction at some orientations. In a three-dimensional orthorhombic body-centered cubic lattice, nearest neighbors repel and next nearest attract each other. Since the quadrupole force falls off very fast, $\propto r^{-6}$, the attraction forces dominate (Ruderman, 1971; Canuto and Ventura, 1977). This can reduce the electrostatic energy by $l/\lambda$ and results in successive layers where the ions are displaced relative to the layers below and above by $l$. The density of such a lattice is $a_0^2/\lambda^2 \sim 0.4 \times 10^3 B_{12}$ times larger than that of ordinary matter.

Another effect is the magnetic equivalent of the conventional quantum mechanical covalent bond, which may lead to the formation of polymer-like chains, i.e. long molecules consisting of many atoms. This effect arises from the sharing of electrons by several atoms. However, this effect can be much stronger in the magnetic atoms described above. In normal atoms with $\eta \ll 1$, the binding is determined by the least bound (valence) electrons. The shared electrons have a binding energy which is much less than the total binding energy of the atoms. Also, the wavelength of all the electrons except the valence ones is much less than the interatomic spacing so that most electrons are insensitive to their environment. However, in a strong field with $\eta \gg 1$, this is no longer true. All of the atomic electrons can participate in the covalent binding process, and the covalent binding energy can greatly exceed the total binding energy of individual atoms as $B \to \infty$ (Kadomtsev and Kudryavtsev, 1971b).

The qualitative difference between nonmagnetic and magnetic covalent binding can be understood from a consideration of the differences in the energy level structures (Ruderman, 1974). In normal hydrogen atoms, the binding of two atoms with parallel electron spins does not occur easily because the Pauli exclusion principle requires one of the $1s$ atoms to change to $2s$, $2p$, etc. (Fig. 2.4.3a). This jump in energy is comparable to the energy of the $1s$ ground state itself, which is energetically demanding. By contrast, two magnetic atoms approaching along the magnetic axis can satisfy the exclusion principle relatively easily, since all that is needed is for one of the $s = 0$ atoms to be promoted to $s = 1$, which is a small amount of energy relative to the binding energy of the $\nu = 0$ state. The radius of the electron changes only by $\delta r_s \sim (2s)^{-1}\lambda$, which enters only logarithmically into the binding energy (Fig. 2.4.3b). A similar argument holds for $Z > 1$ atoms. It is energetically relatively easy to accommodate further shared electrons, since all that is needed is a step of the quantum number $s$, involving an energy change small by comparison to the energy of the level.

The covalent binding can lead to the formation of long magnetic molecular chains, or polymers, by the addition of further atoms at the ends of a diatom. The interatomic spacing in such a polymer is $l_p$, surrounded by a sheath of electrons of radial extent $r_p$, which can be estimated by minimizing the total energy per atom with respect to these quantities (Ruderman, 1974),

$$E_{a,p} \simeq -\frac{(Ze^2)}{l}\left[\ln\left(\frac{2l}{r}\right) - \left(\gamma - \frac{5}{8}\right)\right] + \frac{2Z^3\pi^2\hbar^2}{3m_e l^2}\left(\frac{\lambda}{r}\right)^4, \tag{2.4.22}$$

where $\gamma$ is Euler's constant. In the limit $\eta \to \infty$, this gives

$$r_p \sim \frac{1.3a_0}{Z\eta^{4/5}}, \quad l_p \sim \frac{2.4a_0}{Z\eta^{4/5}}, \tag{2.4.23}$$

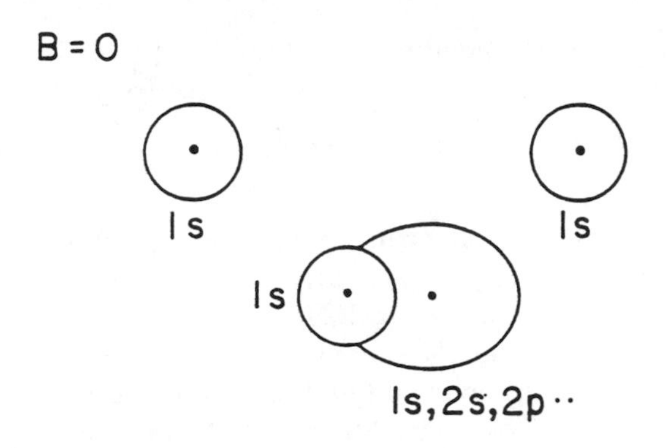


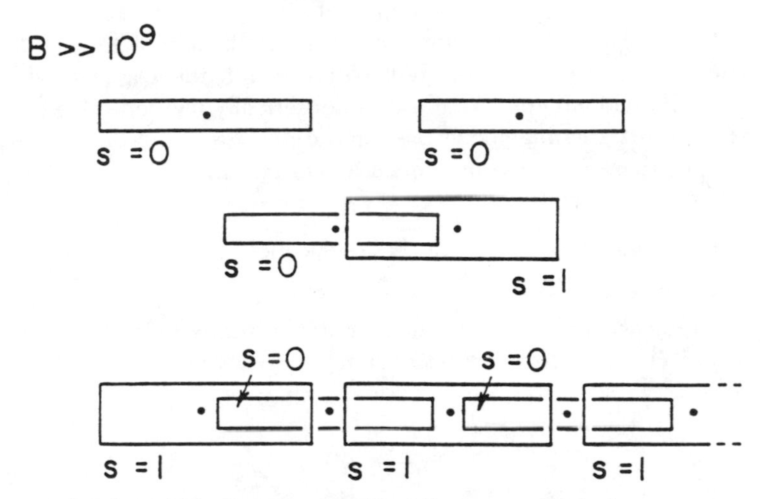


FIGURE 2.4.3 (a) *Top*: Covalent binding of nonmagnetic hydrogen atoms with parallel spins. (b) *Bottom*: Covalent binding of magnetic atoms with spins antiparallel to *B*, leading to formation of molecules and polymer chains. After M. Ruderman, in *IAU Symp. 53, Physics of Dense Matter*, ed. C. Hansen (Reidel, Dordrecht), p. 117. Reprinted by permission of Kluwer Academic Publishers.

and a total binding energy per atom

$$E_{a,\,p} \sim -\frac{1}{2}\frac{e^2}{a_0}Z^3\eta^{4/5}. \tag{2.4.24}$$

From $\eta \gg 1$, this molecular chain cohesion energy per atom can be larger than the total binding energy of an individual atom $E_a$ as estimated in (2.4.20). The corresponding density is

$$\rho_p \sim \frac{Z^4 m_N}{6 a_0^3} \eta^{12/5} \sim 10^6 \left( \frac{Z}{26} \right)^4 \eta^{12/5} \text{ g cm}^{-3}, \tag{2.4.25}$$

which is much greater than that of ordinary molecular matter. The numerical value of $\eta \sim 14.7 B_{12}^{1/2} Z^{-3/2}$ is, at $B_{12} \sim 1$, about 5.2 for He but 0.11 for Fe, so that the estimate (4.2.24) may be valid for the former but not for the latter. More detailed calculations have been performed for the ground state atomic and molecular binding energies using a variational approach (Flowers *et al.*, 1977; Müller, 1984), the Hartree-Fock approach (Simola and Virtamo, 1978; Pröschl *et al.*, 1982), and a density functional approach (Jones, 1986, 1988; Kössl *et al.*, 1988). These calculations appear to confirm that for He the molecular chains would be favored over the individual atoms, while for Fe the individual atoms would be energetically favored. However, since the results are obtained as the small difference between two very large terms, these conclusions remain as yet tentative.

## 2.5 Classical Electrodynamics in the Weak-Field Limit

### *a) The Microscopic Maxwell Equations*

Maxwell's equations for the classical electromagnetic field are written most concisely in terms of the electromagnetic field tensor

$$F_{\mu\nu} = \begin{pmatrix} 0 & -E_1 & -E_2 & -E_3 \\ E_1 & 0 & B_3 & -B_2 \\ E_2 & -B_3 & 0 & B_1 \\ E_3 & B_2 & -B_1 & 0 \end{pmatrix} \tag{2.5.1}$$

and the four-current $j_\mu$ defined as

$$j_\mu = \begin{pmatrix} \rho c \\ \vec{j} \end{pmatrix}, \tag{2.5.2}$$

e.g. Jackson (1975). The two microscopic Maxwell equations containing sources

$$\vec{\nabla} \cdot \vec{E} = 4\pi\rho, \quad \vec{\nabla} \times \vec{B} - \frac{1}{c}\frac{\partial \vec{E}}{\partial t} = \frac{4\pi}{c}\vec{j} \tag{2.5.3}$$

can be expressed in tensor notation as

$$\partial F_{\mu\nu}^{,\nu} = \frac{4\pi}{c} j_{\mu}, \tag{2.5.4}$$

where the subscript (superscript) ",$\nu$" denotes differentiation with respect to the component $\nu$ of the four-coordinate and $\rho, \vec{j}$ represent total charges and currents. The corresponding sourceless microscopic Maxwell equations

$$\vec{\nabla} \cdot \vec{B} = 0, \quad \vec{\nabla} \times \vec{E} + \frac{1}{c}\frac{\partial \vec{B}}{\partial t} = 0, \tag{2.5.5}$$

in turn, are written as

$$F_{\mu\nu,\sigma} + F_{\sigma\mu,\nu} + F_{\nu\sigma,\mu} = 0, \tag{2.5.6}$$

while charge conservation,

$$\frac{\partial \rho}{\partial t} = \vec{\nabla} \cdot \vec{j} = 0, \tag{2.5.7}$$

is expressed as

$$j^{\mu}_{,\mu} = 0. \tag{2.5.8}$$

*b) Derivation from the Variational Principle*

Maxwell's equations can be derived from the variational principle, given a Lagrangian or a Hamiltonian for the electromagnetic field. The only true scalar density that can be constructed from the field tensor that is invariant under space inversion (which leaves out $\vec{B} \cdot \vec{E}$) is

$$F_{\mu\nu} F^{\mu\nu} = 2\left(|B|^2 - |E|^2\right). \tag{2.5.9}$$

Introducing the four-potential $A_{\mu} = (\phi, \vec{A})$ through

$$A_{\nu,\mu} - A_{\mu,\nu} \equiv F_{\mu\nu}, \tag{2.5.10}$$

the Lagrangian density for the field plus charge system is (e.g. Jackson, 1975)

$$\begin{aligned} \mathscr{L} &= -\frac{1}{16\pi} F_{\mu\nu} F^{\mu\nu} + \frac{1}{c} j_{\mu} A^{\mu} \\ &= \frac{1}{8\pi} + \left(|\vec{E}|^2 - |\vec{B}|^2\right) + \frac{1}{c}\vec{j} \cdot \vec{A}. \end{aligned} \tag{2.5.11}$$

The first term represents the classical contribution of the electromagnetic (EM) field, while the second term represents the interaction between the particles $\vec{j}$ and the EM field described through its vector potential $\vec{A}$. As seen from the first line of (2.5.11) $\mathscr{L} = \mathscr{L}(A_\mu, \partial A_\mu / \partial x_\nu)$, where $x_\nu$ is the four-vector $(\vec{x}, ict)$. The variational principle, which leads to the Euler-Lagrange equations, states that

$$\frac{\partial}{\partial x_\nu} \frac{\partial \mathscr{L}}{\partial(\partial A_\mu / \partial x_\nu)} - \frac{\partial \mathscr{L}}{\partial A_\mu} = 0. \tag{2.5.12}$$

Using the definition (2.5.1), this leads to the two inhomogeneous Maxwell equations given by equations (2.5.4) (e.g. Sakurai, 1967; Bjorken and Drell, 1964), while the two homogeneous ones given by equations (2.5.6) are automatically satisfied, given the definition (2.5.10).

*c) The Macroscopic Maxwell Equations*

Maxwell's equations (2.5.3), (2.5.5), and (2.5.7) or, equivalently, (2.5.4), (2.5.6), and (2.5.8) are written in their microscopic form, that is, $\rho$ and $\vec{j}$ represent all the charges and currents, whether bound or free (e.g. Jackson, 1975). In a medium such as a plasma, which can respond to an applied field by producing its own induced fields, it is useful to consider Maxwell's equations in macroscopic form, written in terms of a second set of fields $\vec{D}$ and $\vec{H}$. The constitutive relations

$$\vec{D} = \vec{D}(\vec{E}, \vec{B}), \quad \vec{H} = \vec{H}(\vec{E}, \vec{B}) \tag{2.5.13}$$

can be obtained from the Lagrangian density via the prescription

$$D_\mu = 4\pi \frac{\partial \mathscr{L}}{\partial E_\mu}, \quad H_\mu = -4\pi \frac{\partial \mathscr{L}}{\partial B_\mu}. \tag{2.5.14}$$

For weak enough fields, the constitutive relationships lead to a linear dependence between the applied and induced fields

$$\begin{aligned} D_i &= \sum_j \epsilon_{ij} E_j, \\ H_i &= \sum_j \mu_{ij}^{-1} B_j, \end{aligned} \tag{2.5.15}$$

where $\epsilon_{i,j}$ and $\mu_{i,j}^{-1}$ are the dielectric and inverse magnetic permeability tensors. The response of the medium is then limited to an induced electric dipole density $\vec{P}$ and an induced magnetic moment density $\vec{M}$ related to $\vec{E}$ and $\vec{B}$ via

$$\vec{D} = \vec{E} + 4\pi\vec{P},$$
$$\vec{H} = \vec{B} - 4\pi\vec{M}. \tag{2.5.16}$$

In strong enough fields, however, the basically linear relationships expressed in equations (2.5.15) are replaced by a more involved dependence on the applied fields. Maxwell's microscopic equations (2.5.3), (2.5.5) involve only $\vec{E}$, $\vec{B}$ and the total charge $\rho$ and total current $\vec{j}$. These are made up, in a medium, of

$$\rho = \rho_0 + \rho_i,$$
$$\vec{j} = \vec{j}_0 + \vec{j}_i, \tag{2.5.17}$$

where $\rho_0$, $j_0$ are the free (or external) charges and currents, and $\rho_i$, $\vec{j}_i$ are the induced charges and currents which arise in the medium as a response to the fields $\vec{E}$ and $\vec{B}$. Maxwell's macroscopic equations are obtained by a suitable averaging of the microscopic equations (e.g. Jackson, 1975) and by incorporation of the induced charges and currents into the new fields, $\vec{D}$, $\vec{H}$. The resulting *macroscopic* Maxwell equations are

$$\vec{\nabla}\cdot\vec{D} = 4\pi\rho_0, \quad \vec{\nabla}\times\vec{H} - \frac{1}{c}\frac{\partial\vec{D}}{\partial t} = \frac{4\pi}{c}\vec{j}_0, \tag{2.5.18a}$$

$$\vec{\nabla}\times\vec{E} + \frac{1}{c}\frac{\partial\vec{B}}{\partial t} = 0, \quad \vec{\nabla}\cdot\vec{B} = 0. \tag{2.5.18b}$$

This passage from the microscopic to the macroscopic equations is easily achieved by analyzing what is meant by the induced charges and currents in equation (2.5.17). The total charge $\rho$ used in the microscopic equations can be written as

$$\rho = \rho_0 + \vec{\nabla}\cdot\vec{P}, \tag{2.5.19}$$

where the second term contains the induced charges due to the volume polarization of the plasma $\vec{P}$. The total current $\vec{j}$ of the microscopic equations is similarly

$$\vec{j} = \vec{j}_0 + c\vec{\nabla}\times\vec{M} + \frac{\partial\vec{P}}{\partial t}, \tag{2.5.20}$$

where the second and third terms are the induced currents due to the volume magnetic moment and the change of the electric dipole. Placing equations (2.5.19), (2.5.20) into (2.5.3), (2.5.5) leads to the microscopic equations (2.5.8). An additional equation which will be used is Ohm's generalized law,

$$j_k = \sigma_{kl} E_l, \tag{2.5.21}$$

where $\sigma_{kl}$ is the conductivity tensor, which is also related to the dielectric tensor $\epsilon_{kl}$ (see e.g. Chap. 3).

*d) Magnetohydrodynamic Equations*

In the presence of plasma motions under forces other than electromagnetic, the microscopic Maxwell equations (2.5.3)–(2.5.5) must be supplemented with the matter continuity equation

$$\frac{d\rho}{dt} + \vec{\nabla} \cdot \left(\rho \vec{v}\right) = 0 \tag{2.5.22}$$

and the momentum conservation (Euler) equation

$$\rho \frac{d\vec{v}}{dt} = -\vec{\nabla} p + \frac{1}{c}\left(\vec{j} \times \vec{B}\right) + \rho \vec{g}. \tag{2.5.23}$$

Here $\rho$ is the mass density, $p$ is the pressure (assumed here to be isotropic; otherwise a tensor would be needed), and $\rho\vec{g}$ is the body force per unit volume (e.g. gravity). The electric force term $\rho_e \vec{E}$ has been neglected, since most cosmic plasmas are charge neutral. If viscosity is to be considered, another term $\eta' \nabla^2 \vec{v}$ must be added to (2.5.23), where $\eta'$ is the coefficient of viscosity. The derivatives in (2.5.21) and (2.5.22) are the usual convective derivatives, $d/dt \equiv \partial/\partial t + \vec{v} \cdot \vec{\nabla}$. The electromagnetic fields are determined by the second of equations (2.5.5) and the second of equations (2.5.3), where in the latter one can drop the displacement current $(1/c)(\partial \vec{E}/\partial t)$, which is appropriate provided the plasma velocities are $v \ll c$,

$$\vec{\nabla} \times \vec{E} + \frac{1}{c}\frac{\partial \vec{B}}{\partial t} = 0, \quad \vec{\nabla} \times \vec{B} = \frac{4\pi}{c}\vec{j}. \tag{2.5.24}$$

These two are enough in this limit, since $\vec{\nabla} \cdot \vec{B} = 0$ follows from the first of (2.5.24), and $\vec{\nabla} \cdot \vec{E} = 0$ is consistent with neglecting the displacement current, and $\rho_e \simeq 0$. Using the second of (2.5.24) and cross-multiplying with $\vec{B}$, one can rewrite the magnetic force density

$\vec{j} \times \vec{B}$ using the identity

$$\frac{1}{2}\vec{\nabla}(\vec{B} \cdot \vec{B}) = (\vec{B} \cdot \vec{\nabla})\vec{B} - (\vec{\nabla} \times \vec{B}) \times \vec{B}, \tag{2.5.25}$$

so that Euler's equation (5.2.23) becomes

$$\rho \frac{d\vec{v}}{dt} = -\vec{\nabla}p - \rho\vec{g} - \frac{1}{8\pi}\vec{\nabla}B^2 + \frac{1}{4\pi}(\vec{B} \cdot \vec{\nabla})\vec{B}. \tag{2.5.26}$$

In this equation the term $B^2/8\pi$ is the magnetic pressure, while the last term gives the magnetic tension force per unit volume.

An additional relation which is needed is the equivalent of Ohm's law (2.5.21), $\vec{j} = \sigma\vec{E}$, which is valid in the comoving frame of the plasma. In the laboratory frame, where the plasma velocity is $\vec{v}$, Ohm's law becomes

$$\vec{j} = \sigma\left(\vec{E} + \frac{\vec{v}}{c} \times \vec{B}\right). \tag{2.5.27}$$

Under some circumstances it is possible to approximate the plasma conductivity to be infinite, in which case

$$\vec{E} + \frac{1}{c}(\vec{v} \times \vec{B}) = 0, \tag{2.5.28}$$

which is often referred to as the flux-freezing condition.

The time evolution of the magnetic field in the plasma can be obtained from equations (2.5.24) with the use of (2.5.25) and (2.5.27), which gives

$$\frac{\partial\vec{B}}{\partial t} = \vec{\nabla} \times (\vec{v} \times \vec{B}) + \frac{c^2}{4\pi\sigma}\vec{\nabla}^2\vec{B}. \tag{2.5.29}$$

In the limit of infinite conductivity, or for time scales much shorter than the flux diffusion time scale implied by (5.2.29) (which is typically very long compared to most time scales of interest),

$$\frac{\partial\vec{B}}{\partial t} = \vec{\nabla} \times (\vec{v} \times \vec{B}). \tag{2.5.30}$$

In this case one can show (e.g. Spitzer, 1962) that the magnetic flux $\Phi_M = \int_S \vec{B} \cdot d\vec{S}$ remains constant in time through any closed surface $S$ moving with the fluid, so that the magnetic field lines are trapped and forced to comove with the same fluid element. This is referred to as the flux-freezing condition. Thus, for a circular surface of area $\pi R^2$ traversed by a bundle of field lines of density $B$, one has $\Phi_M = BA = B\pi R^2$ and the flux-freezing condition implies

$$\Phi_M \propto BR^2 = \text{constant}, \tag{2.5.31}$$

where $R$ is the typical radius of the loop enclosing the bundle of field lines.

## 2.6 Quantum Electrodynamics in Strong Fields

### *a) The Quantum Electrodynamic Vacuum*

A quantum electrodynamic (QED) formulation of the radiation field reduces in the classical limit to the results obtained in § 2.5. This limit is obtained for a density of photons large enough, and for fields weak enough, that one can neglect higher-order corrections (e.g. Bjorken and Drell, 1964). On the other hand, when we are dealing with the quantum regime and we are considering sufficiently strong fields, the Lagrangian density of the EM field is no longer given by equation (2.5.11), since higher-order interactions have to be taken into account. While the Lagrangian (2.5.11) leads to Maxwell's equations, which are strictly linear, the modifications introduced by QED in the strong-field limit destroy this linearity. The linearity of the classical Lagrangian (2.5.11) implies that two EM waves cannot interact with each other (superposition of the solutions of Maxwell's equations), whereas the nonlinearities introduced by QED imply that there is a small but finite probability of light scattering by light, which represents a polarizability of the vacuum. A Lagrangian including these effects was first derived by Euler and Kockel (1934). Expanded to the second order in powers of the fine structure constant $\alpha$, the nonlinear correction term to be added to equation (2.5.11) in the limit $\hbar\omega/mc^2 < 1$, $(B/B_Q) \ll 1$, due to graphs like Figure 2.6.1b, is

$$\mathscr{L}_v = \frac{2\alpha^2}{45(4\pi)^2 m^4}\left[\left(\vec{E}^2 - \vec{B}^2\right)^2 + 7\left(\vec{E}\cdot\vec{B}\right)^2 + \cdots\right], \tag{2.6.1}$$

where the index $v$ refers to the quantum electrodynamic vacuum. Since in QED a light quantum is treated like an interaction with a quantum of the electrostatic field, these effects are easiest to observe when one of the quanta involved is due to the strong Coulomb field of nuclei. Alternatively, one of the two quanta may be due to a strong external magnetic field. Examples of processes due to the strongly magnetized vacuum are photo-pair creation, vacuum polarization,

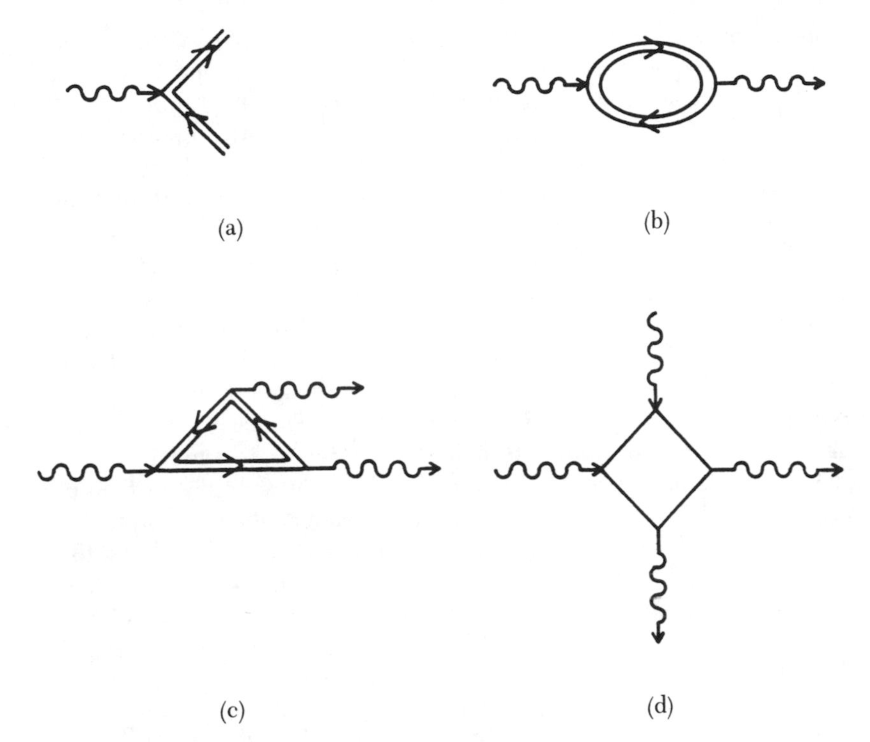


FIGURE 2.6.1 Lowest-order Feynman graphs for processes in an external magnetic field. *Wavy lines*: photon; *single full lines*: virtual $e^-$ or $e^+$; *double full lines*: $e^-$ or $e^+$, where the effect of the magnetic field can be taken into account exactly. (a) Photo-pair creation, possible for $\omega > 2m$. (b) Vacuum polarization in a magnetic field. (c) Photon splitting at $\omega < 2m$ in a magnetic field. (d) Photon-photon scattering.

photon splitting, and other higher-order processes. Some of the lowest-order ones are shown in Figure 2.6.1. Photo-pair creation (Toll, 1952) can occur from the interaction of one photon $\omega > 2m$ with the magnetic field (we set $\hbar = c = 1$ for now). Unlike the nonmagnetic case, where momentum conservation requires the presence of two initial photons, $\gamma + \gamma \rightarrow e^+ + e^-$, the magnetic field can take up the recoil, and the one-photon process $\gamma + B \rightarrow e^+ + e^-$ is possible (Fig. 2.6.1a). At frequencies $\omega < 2m$, photo-pair creation is energetically not allowed; however, virtual pairs can be produced, which interact with the photon, so that the vacuum acquires a polarizability (Fig. 2.6.1b). A higher-order process which can also occur at $\omega < 2m$, involving three vertices, is that of photon splitting (Adler, 1971; Adler *et al.*, 1970), $\gamma_1 + B \rightarrow \gamma_2 + \gamma_3$. Photon-photon scattering is a four-vertex process, as shown in Figure 2.6.1d. This process differs from that of Figure 2.6.1b only in that, in Figure 2.6.1d, two photon lines have been replaced by two field interactions so that both processes can be treated similarly. The scattering of the photon from the left in Figure 2.6.1d can be considered to be due to the virtual $e^+e^-$ produced by the photon from the top, which has polarized the vacuum (Euler and Kockel, 1934). Delbrück scattering (Jauch and Rohrlich, 1976), in which photons are scattered by virtual pairs created by the strong Coulomb field of a nucleus, is described by a diagram similar to the right-hand side of Figure 2.6.1b, with the magnetic field interactions replaced by Coulomb field interactions. Similarly, one-photon pair creation in the Coulomb field of a nucleus is similar to the right-hand side of Figure 2.6.1a, with Coulomb interactions instead of magnetic field interactions. A number of related effects in a strong Coulomb field have been measured in the laboratory, such as the Lamb shift and the anomalous magnetic moment of the electron. While the formalism is the same for quantum electrodynamic effects induced by the magnetic field, laboratory measurements do not yet exist for them. Experiments have been proposed by Erber (1964) and Mészáros and Ventura (1979). The natural setting for this kind of experiment, however, is provided by neutron stars, where fields of the order of $10^{12}$–$10^{13}$ G have been measured (see Chaps. 7 and 8), which approach $B_Q$.

The various effects associated with the magnetic vacuum operate in conjunction with the more usual radiative processes due to the plasma component, such as electron scattering and bremsstrahlung. In general, the latter are proportional to the first or second power of the plasma density times a cross section of order of or less than the Thomson cross section, while the vacuum effects are proportional to various powers of $B/B_Q$. Thus, in general, vacuum processes may predominate for high field strengths and low plasma densities, e.g. in nonaccreting radio pulsars (some of which are most luminous in X-rays or gamma-rays). However, in accreting X-ray pulsars, where plasma densities are high, there are selected frequency regions where plasma effects are outweighed by vacuum effects, in particular as concerns the polarization properties. This topic will be discussed in more detail in the next chapter. Before doing that, however, it should be useful to show explicitly what is meant by the polarization of a vacuum in a strong magnetic field.

*b) Wave Propagation in the QED Vacuum*

Consider a pure vacuum (no electrons), permeated by a strong external magnetic field $\vec{B}^0$, and consider a plane electromagnetic wave of wave vector $\vec{k}$ and wave field components $\vec{E}^w$, $\vec{B}^w$ (Adler, 1971). The approximate nonlinear Lagrangian for the field plus radiation is

$$\mathscr{L} = \frac{1}{8\pi}\left(\vec{E}^2 - \vec{B}^2\right) + \frac{2\alpha^2}{45(4\pi)^2 m^4}\left[\left(\vec{E}^2 - \vec{B}^2\right)^2 + 7\left(\vec{E}\cdot\vec{B}\right)^2\right], \tag{2.6.2}$$

where $\hbar = c = 1$. Defining

$$E_i = E_i^w, \quad B_i = B_i^w + B_i^0, \tag{2.6.3}$$

where the superscripts $w$ and 0 denote the wave and external field contributions, the response of the medium is given by the constitutive equations (2.5.15),

$$D_i = 4\pi\frac{\partial\mathscr{L}}{\partial E_i}, \quad H_i = -4\pi\frac{\partial\mathscr{L}}{\partial B_i}. \tag{2.6.4}$$

We are interested in the high-frequency response, so we linearize the constitutive equations with respect to the wave fields $\vec{E}^w$, $\vec{B}^w$. This gives the $\vec{D}^w$ and $\vec{H}^w$ fields as functions of $\vec{E}^w$, $\vec{B}^w$:

$$D_i^w = \epsilon_{ij}E_j^w, \quad H_i^w = \mu_{ij}^{-1}B_j^w. \tag{2.6.5}$$

These define a dielectric tensor $\epsilon_{ij}$ and an inverse magnetic permeability $\mu_{ij}^{-1}$ for the vacuum,

$$\begin{aligned}\epsilon_{ij} &= \delta_{ij}(1 - 2\delta) + 7\delta b_i b_j,\\ \mu_{ij}^{-1} &= \delta_{ij}(1 - 2\delta) - 4\delta b_i b_j,\end{aligned} \tag{2.6.6}$$

where

$$\delta = \alpha^2(45\pi)^{-1}\left(B/B_Q\right)^2 = 0.5\times 10^{-4}\left(B/B_Q\right)^2 \tag{2.6.7}$$

is a parameter characterizing the strength of the magnetic vacuum, $\alpha$ is the fine structure constant, and $\hat{b}$ is a unit vector along the external large-scale magnetic field $\vec{B}^0$. Maxwell's equations for the sourceless radiation field in macroscopic form are

$$\begin{aligned}&\vec{\nabla}\cdot\vec{D}^w = 0, \quad \vec{\nabla}\cdot\vec{B}^w = 0,\\ &\vec{\nabla}\times\vec{E}^w = -\frac{1}{c}\frac{\partial\vec{B}^w}{\partial t}, \quad \vec{\nabla}\times\vec{H}^w = \frac{1}{c}\frac{\partial\vec{D}^w}{\partial t}.\end{aligned} \tag{2.6.8}$$

For a plane wave we have

$$\vec{E}^w = \vec{e}^w e^{i(\vec{k}\cdot\vec{r}-\omega t)}, \quad \vec{H}^w = \vec{h}^w e^{i(\vec{k}\cdot\vec{r}-\omega t)}, \tag{2.6.9}$$

where $\vec{e}^w, \vec{h}^w$ are unit polarization vectors. Substituting equations (2.6.9) into (2.6.8), we find two propagation eigenmodes, depending on whether $\vec{e}^w$ is parallel to $\vec{B}^0$ or perpendicular to it. These are

$$\vec{e}^w_\perp = \hat{k} \times \hat{b},$$
$$\vec{h}^w_\perp = -n_\perp^{-1} a\hat{b} + n_\perp a\hat{k}\cdot\hat{b}\hat{k} \tag{2.6.10}$$

and

$$\vec{e}^w_\parallel = n_\parallel^{-1} a^{-1}\hat{b} - n_\parallel a^{-1}\hat{k}\cdot\hat{b}\hat{k},$$
$$\vec{h}^w_\parallel = \hat{k} \times \hat{b}. \tag{2.6.11}$$

Here we use the definitions

$$a = 1 - 2\delta, \quad \cos\theta = \hat{k}\cdot\hat{b}, \tag{2.6.12}$$

and the (real) refractive indices $n = k/\omega$ are given by (Adler, 1971):

$$n_\perp = 1 + 2\delta\sin^2\theta,$$
$$n_\parallel = 1 + (7/2)\delta\sin^2\theta. \tag{2.6.13}$$

Note that in these equations the $\parallel, \perp$ convention is based on the usual wave electric vectors, instead of on the wave magnetic vectors (i.e. the opposite of Adler, 1971).

Several striking facts emerge from equations (2.6.6) to (2.6.13). First of all, looking at the dielectric and permeability tensors (eqs. 2.6.6), one sees that in the limit $B/B_Q \to 0$ one regains the "usual" vacuum: $\epsilon_{ij} = \mu_{ij} = 1$. However, for $B/B_Q \sim 10^{-1}$, as in neutron stars, the departure from unity is of order $10^{-6}$. Typical departures from unity due to purely plasma effects (discussed in Chap. 3) are either of order

$$(\omega_p/\omega)^2 = 4\pi e^2 n_e m^{-1}\omega^{-2} \sim 3\times 10^{-7} n_{20}\omega_{18}^{-2} \tag{2.6.14}$$

or of order

$$(\omega_p/\omega_c)^2 = 4\pi n_e mc^2 B^{-2} \sim 10^{-9} n_{20} B_{12}^{-2}. \tag{2.6.15}$$

Hence, in typical pulsar exteriors ($n_e \lesssim 10^{20}$ cm$^{-3}$, $B \gtrsim 10^{12}$ G, and X-ray frequencies $\omega \gtrsim 10^{18}$ Hz), vacuum effects easily dominate plasma effects. Also, we notice that the vacuum magnetic permeability tensor is of the same order of magnitude as the dielectric tensor. This departs radically from the normal plasma case, laboratory or cosmic, where $\mu_{ij}$ is unity for all practical purposes (exceptions are ferromagnets and similar substances in solids). We see that the eigenmodes for waves in a strong field (eqs. 2.6.10–2.6.11) are linearly polarized. However, unlike the case in an unmagnetized vacuum, where the two polarization modes can be chosen to be oriented arbitrarily so long as they are perpendicular to each other, in a magnetic vacuum the eigenmodes have well-determined orientations either parallel or perpendicular to the external field. It also emerges, from equations (2.6.13), that the magnetic vacuum is birefringent, unlike the usual vacuum, since $n_{\parallel} \neq n_{\perp}$ except for propagation along the field $\vec{B}$. Of course, in the presence of a plasma in a magnetic field, birefringence is well known (e.g. Jackson, 1975). Here, however, birefringence occurs even in the absence of any plasma. For waves of frequency $\omega$, an optical path length difference amounting to one wavelength is achieved over a distance $l_v$ given by

$$l_v = \frac{2\pi c}{\omega \,|\, n_{\perp} - n_{\parallel} \,|} = \frac{4\pi c}{3\omega\delta \sin^2\theta} \sim 2.4 \times 10^{-3} \omega_{18}^{-1} \left(\frac{B}{B_Q}\right)^{-2} \frac{1}{\sin^2\theta}. \tag{2.6.16}$$

This path length, in the pure magnetic vacuum case (no plasma), tends to infinity as $\sin\theta \to 0$, since in that limit the two waves travel at the same speed. At any other value of $\sin\theta$, however, for $B \gtrsim 10^{-1} B_Q$ and frequencies in the X-ray range, the plane of polarization can rotate by 90° over distances less than a neutron star radius, $R_{\rm NS} \sim 10^6$ cm.

# 3. Magnetized Plasma Response Properties

## 3.1 Classical Wave Propagation in a Magnetized Plasma

In investigating high-energy radiation from magnetized neutron stars, we are typically concerned with plasmas in the high-frequency regime, in the sense that the wave frequency $\omega$ satisfies

$$\omega \gg \left(\omega_{pe}, \omega_{ci}\right), \tag{3.1.1}$$

where

$$\omega_{pe} = \left(4\pi n_e e^2/m_e\right)^{1/2} \tag{3.1.2}$$

is the plasma frequency of the electrons and

$$\omega_{ci} = \frac{m_e}{m_i}\omega_c = \frac{eB}{m_i c} \tag{3.1.3}$$

is the cyclotron frequency of the ions, $m_i$ being the ion (proton) mass. The first part of restriction (3.1.1) allows us to neglect collective effects, while the second allows us to neglect the response motions of the ions, so we need deal only with the plasma electrons. This limit is sometimes called the magnetoionic regime.

To analyze the response properties, we start out with Maxwell's macroscopic equations (2.5.18), together with the constitutive equations (2.5.13) and Ohm's law (eq. 2.5.21). We are interested in the response of a charge-neutral plasma to the passage of an electromagnetic wave. This means that, in equations (2.5.18), the free or externally maintained charges and currents are identically zero:

$$\rho_0 = \vec{j}_0 = 0. \tag{3.1.4}$$

The only charges and currents in the medium are those due to the polarizability and magnetizability of the plasma. Furthermore, in plasmas one can usually neglect the volume magnetic moment, i.e.

$$\vec{M}_{(\mathrm{pl})} \equiv 0, \quad \mu^{-1}_{ik(\mathrm{pl})} \equiv 1. \tag{3.1.5}$$

We add the subscript pl to indicate the plasma component, since, as we saw in the last chapter, the magnetic vacuum component (appreciable for $B$ not much less than $B_Q$) can be significant. We concentrate for the moment on the plasma component only, and return to the vacuum effects below. In this case, the total current is just (see eqs. 2.5.16 and 2.5.20)

$$\vec{j} \equiv \frac{\partial \vec{P}}{\partial t} = \frac{1}{4\pi}\frac{\partial}{\partial t}\left(\vec{D} - \vec{E}\right). \tag{3.1.6}$$

We consider the propagation of a plane wave, or perturbation, of the form

$$e^{i(\vec{k}\cdot\vec{r}-\omega t)}, \tag{3.1.7}$$

which is equivalent to performing a Fourier transform of all our quantities. We have then from equations (3.1.6) and (2.5.15)

$$j_i = -i\frac{\omega}{4\pi}\left(D_i - E_i\right) = -i\frac{\omega}{4\pi}\left(\epsilon_{ik}\left(\vec{k}, \omega\right) - \delta_{ik}\right)E_k, \tag{3.1.8}$$

where $i$, $k$ are Cartesian components. Since the current $j_i$ is related to the electric field via Ohm's law (eq. 2.5.21), we obtain the relationship, valid for harmonic perturbations, between the dielectric and conductivity tensors

$$\epsilon_{ik}\left(\vec{k}, \omega\right) = \delta_{ik} + i\frac{4\pi}{\omega}\sigma_{ik}\left(\vec{k}, \omega\right), \tag{3.1.9}$$

where both tensors are functions of $\vec{k}$ and $\omega$, since we are now dealing with the Fourier transforms of the field quantities.

We consider now the propagation of a wave given by equation (3.1.7) in a medium characterized by the dielectric tensor of equation (3.1.9). We have to solve the wave equation, for which we may use either the microscopic or the macroscopic form of Maxwell's equations. However, it is clear that a wave equation derived from the microscopic equations (2.5.3) and (2.5.5) will involve the source terms $\rho = \vec{\nabla}\cdot\vec{P}$ and $\vec{j} = \partial\vec{P}/\partial t$, leading to an inhomogeneous wave equation. It is therefore preferable to deal with the formally equivalent macroscopic equations (2.5.18), since in these the nonvanishing source terms have been incorporated into the fields $D, H$ (and $\rho_0, j_0$ are zero). We have then the sourceless equations

$$\vec{\nabla} \cdot \vec{D} = 0, \quad \vec{\nabla} \times \vec{H} - \frac{1}{c}\frac{\partial \vec{D}}{\partial t} = 0,$$

$$\vec{\nabla} \times \vec{E} + \frac{1}{c}\frac{\partial \vec{B}}{\partial t} = 0, \quad \vec{\nabla} \cdot \vec{B} = 0. \tag{3.1.10}$$

With the wave dependence given by equation (3.1.7), the two curl equations for the Fourier transforms (using $\partial/\partial t \to -i\omega$, $\nabla \to i\vec{k}$) give

$$\vec{k} \times \vec{B} = -\frac{\omega}{c}\vec{D}, \quad \vec{k} \times \vec{E} = \frac{\omega}{c}\vec{B}. \tag{3.1.10a}$$

The two divergence equations, on the other hand, imply that $\vec{B}$ and $\vec{D}$ must be orthogonal to $\vec{k}$. Eliminating $\vec{B}$ from equations (3.1.10a), we have

$$\vec{k} \times \left(\vec{k} \times \vec{E}\right) + \frac{\omega^2}{c^2}\vec{D} = k^2\vec{E} - \vec{k}\left(\vec{k} \cdot \vec{E}\right) - \frac{\omega^2}{c}\vec{D} = 0, \tag{3.1.11}$$

or, since $D_i = \epsilon_{ij}E_{ij}$, we can write in Cartesian components

$$\left(k_i k_j - k^2\delta_{ij} + \frac{\omega^2}{c^2}\epsilon_{ij}\right)E_j = 0. \tag{3.1.12}$$

The susceptibility tensor $\vec{\vec{\alpha}}$, defined through $\vec{P} = \vec{\vec{\alpha}} \cdot \vec{E}$, is related to $\vec{\vec{\epsilon}}$ via

$$4\pi\vec{\vec{\alpha}} = \left(\vec{\vec{\epsilon}} - \vec{\vec{1}}\right), \tag{3.1.13}$$

where $\vec{\vec{1}}$ is the unit tensor. An alternative susceptibility often used is $\vec{\vec{\chi}} = 4\pi\vec{\vec{\alpha}}$. Equation (3.1.12) can then be rewritten as

$$\left\{\left[\delta_{ij} - \frac{k^2c^2}{\omega^2}\left(\delta_{ij} - k_i k_j\right)\right] + 4\pi\alpha_{ij}\right\}E_j = 0. \tag{3.1.14}$$

Depending on the type of problem considered, one may consider in equation (3.1.7) either the wave vector $\vec{k}$, or the frequency $\omega$, to be a complex number. For problems with spatial boundary conditions, where one investigates the stability of a spatial eigenmode of given wavelength, it is customary to treat $\omega$ as complex, $\omega = \omega_r + i\omega_i$, and the growth or decay in time is given by the imaginary part of $\omega$. For problems of wave propagation, where an external source produces a wave of given fixed frequency, it is customary to treat the wave vector $\vec{k}$ as complex,

$$k = k_1(\omega) + ik_2(\omega). \tag{3.1.15}$$

In this case, the refractive index represented by $kc/\omega$ is also a complex number. The complex refractive index $N$ is given by

$$k = \frac{\omega}{c}N = \frac{\omega}{c}(n + i\kappa), \tag{3.1.16}$$

where $n$ is the usual (real) refractive index and $\kappa$ is the (real) absorption coefficient, implying spatial growth or decay of the wave. The wave equation is therefore

$$\left\{\left[\vec{\vec{1}} - N^2\left(\vec{\vec{1}} - \hat{k}\hat{k}\right)\right] + 4\pi\vec{\vec{\alpha}}\right\} \cdot \vec{E} = 0, \tag{3.1.17}$$

and all the physical information required to solve it is in the plasma susceptibility tensor $\vec{\vec{\alpha}}$ or the related dielectric tensor $\vec{\vec{\epsilon}} = \vec{\vec{1}} + 4\pi\vec{\vec{\alpha}} = \vec{\vec{1}} + \vec{\vec{\chi}}$.

We restrict ourselves now to the cold-plasma limit $kT \to 0$, which means that thermal electron motions are neglected compared to those induced by the wave. In this case, the dielectric tensor takes a simple form. The nonrelativistic equation of motion for charges $q$ is

$$m\frac{d\vec{v}}{dt} = q\vec{E}^w + \frac{q}{c}\vec{v} \times \left(\vec{B}^w + \vec{B}_0\right) + m\nu\vec{v}, \tag{3.1.18}$$

where $\vec{B}^w$, $\vec{E}^w$ correspond to the wave field, $\vec{B}_0$ is the external field, and we have included a phenomenological damping term $m\nu\vec{v}$ proportional to the mass and velocity of the charge and to an effective damping frequency $\nu$. Under the action of the wave of equation (3.1.7), the equation of motion becomes

$$-i\omega\left(1 + i\frac{\nu}{\omega}\right)\vec{v} = \vec{v} \times \vec{\omega}_c + \frac{q}{m}\vec{E}, \tag{3.1.19}$$

where now $\vec{E} \equiv \vec{E}^w$, we neglected $\vec{B}^w$ with respect to $\vec{B}_0$, and we defined

$$\vec{\omega}_c = q\vec{B}_0/mc. \tag{3.1.20}$$

Here $q$, $m$ are $-e$, $m_e$ for electrons, $+e$, $m_i$ for ions, and similarly for other species. Taking vector and scalar products of equation (3.1.19) with $\omega_c$, we find

$$\vec{v} = -i\frac{q}{m}\frac{\omega\lambda}{\left(\omega_c^2 - \omega^2\lambda^2\right)}\left[\vec{E} - \frac{\vec{\omega}_c\left(\vec{E}\cdot\vec{\omega}_c\right)}{\omega^2\lambda^2}\right] + \frac{q}{m}\frac{\vec{E}\times\vec{\omega}_c}{\left(\omega_c^2 - \omega^2\lambda^2\right)}, \tag{3.1.21}$$

where

$$\lambda = \left(1 + i\nu/\omega\right). \tag{3.1.22}$$

Now, from equation (3.1.6) we have

$$\vec{j} = -\frac{i\omega}{4\pi}\left(\vec{D} - \vec{E}\right), \tag{3.1.23}$$

and, since $\vec{j}$ is the total current,

$$\vec{j} = \sum qn\vec{v}, \tag{3.1.24}$$

where the sum is over the different types of particles. Hence, from equations (3.1.23), (3.1.24), and (3.1.21) we obtain

$$\vec{D} = \left(1 - \sum\frac{\omega_p^2}{\omega^2\lambda^2 - \omega_c^2}\right)\vec{E} + \sum\frac{\omega_p^2\,\vec{\omega}_c\left(\vec{\omega}_c\cdot\vec{E}\right)}{\left(\omega^2\lambda^2 - \omega_c^2\right)\lambda\omega}$$

$$+ i\sum\frac{\omega_p^2\left(\vec{\omega}_c\times\vec{E}\right)}{\left(\omega^2\lambda^2 - \omega_c^2\right)\omega}, \tag{3.1.25}$$

where $\omega_p = 4\pi nq^2/m$ for the individual species. Since $D_i = \epsilon_{ij}E_j$, we can identify the components of the tensor $\epsilon_{ij}$. In a coordinate system $x_0$, $y_0$, $z_0$ with $\vec{B}_0$ along $z_0$, the cold-plasma dielectric tensor is then (e.g. Shafranov, 1967)

$$\epsilon_{ij}^0 = \begin{pmatrix} \varepsilon & ig & 0 \\ -ig & \varepsilon & 0 \\ 0 & 0 & \eta \end{pmatrix}, \tag{3.1.26}$$

where

$$\varepsilon = 1 - \sum\omega_p^2\lambda\left(\omega^2\lambda^2 - \omega_c^2\right)^{-1},$$

$$g = -\sum\omega_p^2\omega_c\omega^{-1}\left(\omega^2\lambda^2 - \omega_c^2\right)^{-1},$$

$$\eta = 1 - \sum\omega_p^2\omega^{-2}\lambda^{-1}. \tag{3.1.26a}$$

The sums in the diagonal terms have the same sign for all particles, but in the off-diagonal terms $g$, the factor $\omega_c = qB/mc$ changes sign depending on sign($q$), e.g. $\omega_c = \mp eB/mc$ for electrons (protons). If we neglect the ion component $\omega \gg \omega_{ci}$ and damping, the cold-electron dielectric tensor is

$$\overleftrightarrow{\epsilon} = \begin{pmatrix} 1 - v/(1-u) & -ivu^{1/2}/(1-u) & 0 \\ ivu^{1/2}/(1-u) & 1 - v/(1-u) & 0 \\ 0 & 0 & 1 - v \end{pmatrix}. \qquad (3.1.27)$$

The quantities $u$ and $v$ are the normalized squares of the electron cyclotron and plasma frequencies, respectively:

$$u = (\omega_c/\omega)^2, \quad v = (\omega_p/\omega)^2, \qquad (3.1.28)$$

where the dimensionless parameter $v$ is not to be confused with the velocity.

## 3.2 Normal Modes of the Cold Magnetized Plasma

The wave equation (3.1.17) can be written compactly as

$$\left(\overleftrightarrow{\Lambda} + 4\pi\overleftrightarrow{\alpha}\right) \cdot \vec{E} = 0, \qquad (3.2.1)$$

where the Maxwell tensor $\overleftrightarrow{\Lambda}$, given by

$$\overleftrightarrow{\Lambda} = \overleftrightarrow{1} - N^2\left(\overleftrightarrow{1} - \hat{k}\hat{k}\right), \qquad (3.2.2)$$

describes the propagation in the absence of a medium, while the susceptibility $\overleftrightarrow{\chi} = \overleftrightarrow{1} - \overleftrightarrow{\epsilon}$ includes the plasma effects. In order to have a solution, we need

$$\det\left(\overleftrightarrow{\Lambda} + 4\pi\overleftrightarrow{\alpha}\right) = 0, \qquad (3.2.3)$$

which gives a dispersion relation between $k$ and $\omega$, or, equivalently, this condition determines the allowed values of the complex refractive index $N$, which are the eigenvalues of equation (3.2.3). The corresponding eigenvectors will be the polarization eigenmodes, i.e. unit vectors along the electric field $\vec{E}$ which solve equation (3.2.1). In terms of the quantities $\varepsilon$, $g$, $\eta$ which characterize the cold-plasma dielectric tensor with $\vec{B}_0$ along $z$ (eqs. 3.1.26, 3.1.27), the wave equation (3.2.1) (assuming $\vec{k}$ in the $x$, $z$ plane) is

$$\begin{pmatrix} \varepsilon - N^2\cos^2\theta & ig & N^2\sin\theta\cos\theta \\ -ig & \varepsilon - N^2 & 0 \\ N^2\sin\theta\cos\theta & 0 & \eta - N^2\sin^2\theta \end{pmatrix} \begin{pmatrix} E_x \\ E_y \\ E_z \end{pmatrix} = 0. \tag{3.2.4}$$

If we define

$$\varepsilon_r = \varepsilon + g, \quad \varepsilon_l = \varepsilon - g, \tag{3.2.5}$$

the dispersion relation (3.2.3) can be written as

$$AN^4 - BN^2 + C = 0, \tag{3.2.6}$$

where

$$\begin{aligned} A &= \varepsilon\sin^2\theta + \eta\cos^2\theta, \\ B &= \varepsilon_r\varepsilon_l\sin^2\theta + \varepsilon\eta(1 + \cos^2\theta), \\ C &= \varepsilon_r\varepsilon_l\eta. \end{aligned} \tag{3.2.7}$$

Thus the eigenvalues are given by

$$N^2 = (B \pm D)/A \tag{3.2.8}$$

with

$$D = \left[(\varepsilon_r\varepsilon_l - \varepsilon\eta)^2\sin^4\theta + 4\eta^2 g^2\cos^2\theta\right]^{1/2}. \tag{3.2.9}$$

The dispersion relation can also be cast in the form (e.g. Krall and Trivelpiece, 1973; Stix, 1962)

$$\tan^2\theta = \frac{-\eta(N^2 - \varepsilon_r)(N^2 - \varepsilon_l)}{(\eta N^2 - \varepsilon_r\varepsilon_l)(N^2 - \eta)}, \tag{3.2.10}$$

which is particularly illustrative for the limiting cases of $\theta = 0, \pi/2$ (propagation parallel and perpendicular to $\vec{B}_0$).

Neglecting the damping $\lambda = 1$ and restricting ourselves to $\omega \gg \omega_{ci}$, so only the electron component is important in equation (3.1.27), we have for parallel propagation ($\theta = 0$)

$$\begin{aligned} N_1^2 &= \varepsilon_r = 1 - \frac{\omega_p^2}{\omega^2}\frac{\omega}{\omega - \omega_c}, \\ N_2^2 &= \varepsilon_l = 1 - \frac{\omega_p^2}{\omega^2}\frac{\omega}{\omega + \omega_c}, \\ \eta &= 0. \end{aligned} \tag{3.2.11}$$

The polarization eigenvectors are given by ($\theta = 0$)

$$\begin{pmatrix} \varepsilon - N^2 & ig & 0 \\ -ig & \varepsilon - N^2 & 0 \\ 0 & 0 & \eta \end{pmatrix} \begin{pmatrix} E_x \\ E_y \\ E_z \end{pmatrix} = 0. \tag{3.2.12}$$

The mode $\eta = 0 = 1 - \omega_p^2/\omega^2$ corresponds to a wave where $E_z \neq 0$, i.e. a longitudinal oscillation ($\vec{E} \parallel \vec{k} \parallel \hat{z}$) of frequency $\omega = \omega_p$, the electron plasma frequency. This is a purely electrostatic oscillation, which does not propagate. The other two modes are transverse, since they involve $E_x$, $E_y$ only, while $\vec{k}$ is along $z$. These are, therefore, two different types of possible transverse electromagnetic waves, which propagate with different speeds $v_{1,2} = c/N_{1,2}$, where the refractive indices $N_{1,2}$ (real, since we neglected damping) are given by equation (3.2.11). Since the first two lines of equation (3.2.12) give $(\varepsilon - N^2) = \pm g$, the transverse polarization eigenvectors satisfy

$$\left(\varepsilon - N^2\right)\begin{pmatrix} 1 & \pm i \\ \mp i & 1 \end{pmatrix} \begin{pmatrix} E_x \\ E_y \end{pmatrix} = 0, \tag{3.2.13}$$

and therefore their components are related by

$$E_x = \pm iE_y. \tag{3.2.14}$$

This equation represents a transverse, circularly polarized wave, since $|E_x| = |E_y|$ and they are 90° out of phase. The wave with $N_1$ is right-hand circularly polarized (R), i.e. in the same sense as the electron gyration. Clearly a wave which rotates in the same sense as an electron can resonate with it if $\omega \to \omega_c$, and this is seen to be the case from the first of equations (3.2.11). At the resonance itself, however, equation (3.2.11) is no longer valid, since the energy transfer between wave and electron would heat the latter up, violating the cold-plasma assumption. Damping effects would also become important in this case. The other transverse wave, with index $N_2$, is left-hand circularly polarized (L), and in the cold-plasma limit does not show a resonance. If we normalize to unity the transverse part of the polarization vector, it is presented by

$$\vec{e}^{\,t}_{1,2} = 2^{-1/2}\left(1, \pm i, 0\right). \tag{3.2.15}$$

The R (index 1) wave, which can resonate, is often called the extraordinary wave (X), and the L (index 2) wave, which does not, the ordinary wave (O). The (real) refractive indices for the ordinary and extraordinary wave at $\theta = 0°$ are shown in Figure 3.2.1.

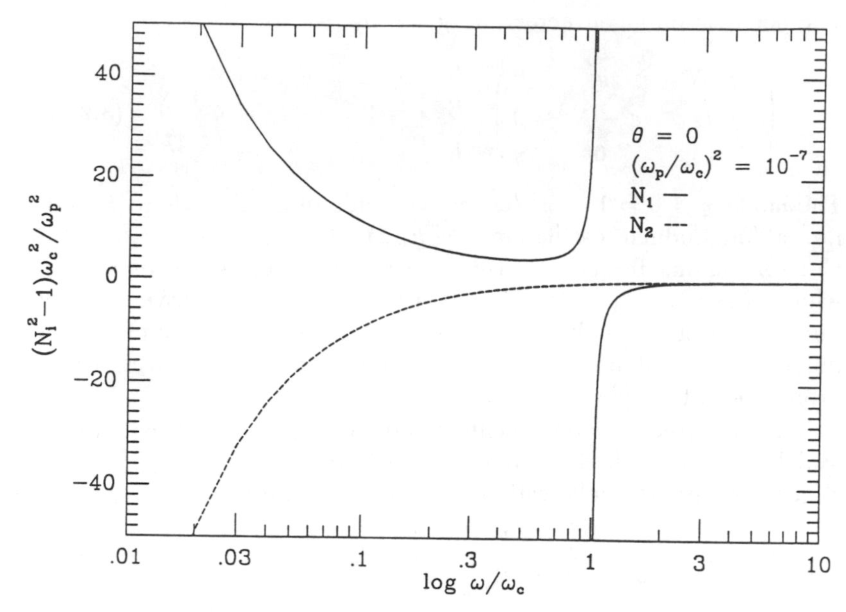


FIGURE 3.2.1 Square of the index of refraction of a cold magnetized plasma for the extraordinary (1) and ordinary (2) waves traveling parallel to the field $\theta = 0°$ for a particular choice of the ratio $\omega_c/\omega_p$.

The case of transverse propagation, $\theta = \pi/2$, illustrates the anisotropy introduced by the magnetic field. For this case, the dispersion relation is

$$N_1^2 = \eta = 1 - \frac{\omega_p^2}{\omega^2},$$

$$N_2^2 = \frac{\varepsilon_r \varepsilon_l}{\varepsilon} = \frac{\left(\omega^2 - \omega_p^2\right)^2 - \omega^2\omega_c^2}{\left(\omega^2 - \omega_c^2 - \omega_p^2\right)\omega^2}, \tag{3.2.16}$$

which is shown in Figure 3.2.2. The corresponding eigenvectors at $\theta = \pi/2$ satisfy

$$\begin{pmatrix} \varepsilon & ig & 0 \\ -ig & \varepsilon - N^2 & 0 \\ 0 & 0 & \eta - N^2 \end{pmatrix} \begin{pmatrix} E_x \\ E_y \\ E_z \end{pmatrix} = 0. \tag{3.2.17}$$

The mode $N_1$ of equation (3.2.16) represents a wave of electric vector along $z \parallel \vec{B}_0$, and it is a transverse (electromagnetic) wave, $\vec{E} \perp \vec{k}$.

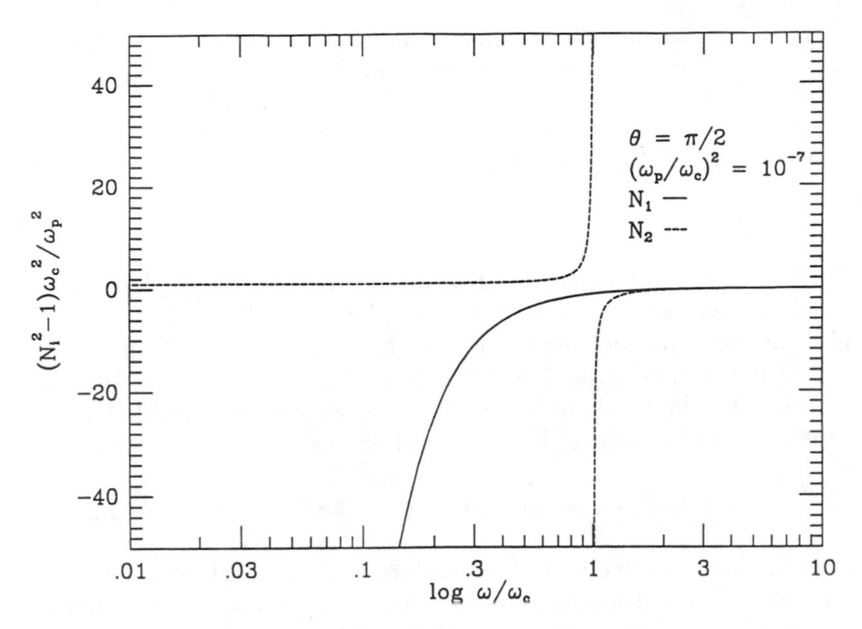


FIGURE 3.2.2 Square of the index of refraction of the cold magnetized plasma for the ordinary and extraordinary waves at $\theta = \pi/2$.

Since the electric field is along $\vec{B}_0$, no magnetic quantities enter the dispersion relation (the first of eqs. 3.2.16), which is not surprising since the electron is free along the $\vec{B}_0$ direction. This mode corresponds to the ordinary (or L) mode discussed above. The corresponding eigenvector is

$$\vec{e}_2 = (0, 0, 1). \tag{3.2.18}$$

The second line of equation (3.2.16) represents a mixture of transverse and longitudinal modes. From the first line of equation (3.2.17) we get the ratio of the longitudinal to transverse components,

$$\left|\frac{E_x}{E_y}\right| = \left|\frac{g}{\varepsilon}\right| = \left|\frac{\omega_p^2 \omega_c}{\omega(\omega^2 - \omega_c^2 - \omega_p^2)}\right|. \tag{3.2.19}$$

For low densities $\omega_p \to 0$, or for $\omega \gg \omega_c, \omega_p$, the wave is essentially transverse ($E_y \neq 0$, $E_x \to 0$). From $\omega$, $\omega_c \gg \omega_p$ the longitudinal component is generally small. On the other hand, at the frequency

$$\omega = \left(\omega_c^2 + \omega_p^2\right)^{1/2} \tag{3.2.20}$$

there is a resonance, which implies that $|E_x/E_y| \to \infty$ or $E_y \to 0$, i.e. the wave becomes purely longitudinal (in the cold-plasma limit). For other frequencies $\omega, \omega_c \gg \omega_p$, where the longitudinal component is small, one can normalize this extraordinary eigenvector by using equation (3.2.19) and requiring $E_x^2 + E_y^2 = 1$ so that

$$\vec{e}_1 \cong (0, 1, 0). \tag{3.2.21}$$

Thus, for propagation at $\theta = \pi/2$ in the high frequency limit, the ordinary and extraordinary waves are linearly polarized, with the electric vector parallel and perpendicular to the magnetic field.

In the general case of $0 < \theta < \pi/2$, the normal modes are, not surprisingly, elliptically polarized, but they retain the general characteristics described above. Using the definitions

$$u = (\omega_c/\omega)^2, \quad v = (\omega_p/\omega)^2, \quad \gamma = \nu/\omega, \tag{3.2.22}$$

the complex refractive index for an arbitrary angle, neglecting the ions but including the damping, is given by equation (3.2.8) (e.g. Ginzburg, 1970):

$$N_{1,2}^2 = 1 - \frac{2\nu(1+i\gamma-v)}{2(1+i\gamma)(1+i\gamma-v) - u\sin^2\theta \pm \left[u^2\sin^4\theta + 4u(1+i\gamma-v)^2\cos^2\theta\right]^{1/2}}, \tag{3.2.23}$$

where the upper/lower sign of the root gives the ordinary/extraordinary wave index. In the absence of damping, this is

$$N_{1,2}^2 = 1 - \frac{2\omega_p^2(\omega^2-\omega_p^2)}{2(\omega^2-\omega_p^2)\omega^2 - \omega_c^2\omega^2\sin^2\theta \pm \left[\omega^4\omega_c^4\sin^4\theta + 4\omega_c^2\omega^2(\omega^2-\omega_p^2)^2\cos^2\theta\right]^{1/2}}. \tag{3.2.24}$$

The polarization eigenmodes are found more easily in a frame of reference with $\vec{k}$ along $z$ (and $\vec{B}_0$ in the $x$, $z$ plane; see Fig. 3.2.3). The components of the dielectric tensor in this new frame $x$, $y$, $z$ (as opposed to $x_0$, $y_0$, $z_0$, where $\vec{B}_0$ was along $z$), and which were used

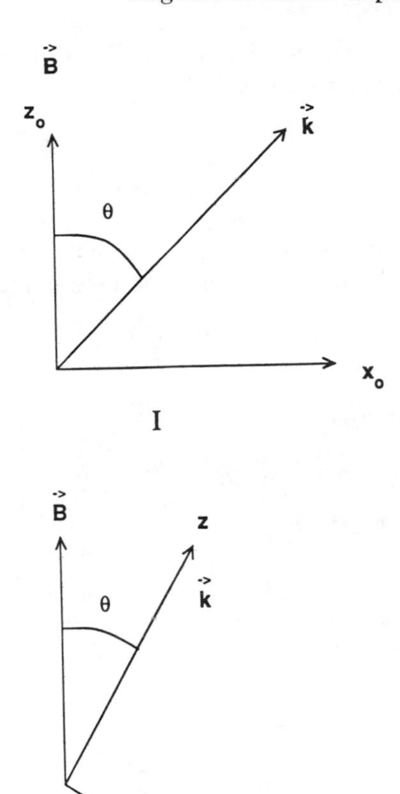


FIGURE 3.2.3 The two systems of reference I and II used for defining the polarization tensor.

for equation 3.2.4, are $(\vec{k} \parallel z)$:

$$
\begin{aligned}
\epsilon_{xx} &= \varepsilon \cos^2\theta + \eta \sin^2\theta, \\
\epsilon_{xy} &= -\epsilon_{yx} = ig \cos\theta, \\
\epsilon_{xz} &= \epsilon_{zx} = (\varepsilon - \eta)\sin\theta \cos\theta, \\
\epsilon_{yy} &= \varepsilon, \\
\epsilon_{yz} &= -\epsilon_{zy} = -ig \sin\theta, \\
e_{zz} &= \varepsilon \sin^2\theta + \eta \cos^2\theta.
\end{aligned}
\qquad (3.2.25)
$$

The advantage of using this frame can be seen by examining wave equation (3.1.17), which can be written

$$\left[N^2\left(\vec{\vec{1}} - \hat{k}\hat{k}\right) + \vec{\vec{\epsilon}}\right] \cdot \vec{E} = 0. \tag{3.2.26}$$

The tensor multiplying $N^2$ is perpendicular to $\vec{k}$. In a system with $\vec{k}$ along $z$, the third line of the matrix equation (3.2.26) is independent of $N^2$. The $z$-component of the electric field is therefore dependent on the other two,

$$E_z = -\epsilon_{zz}^{-1}\left(\epsilon_{zx}e_x + \epsilon_{zy}E_y\right). \tag{3.2.27}$$

The remaining two equations are

$$\begin{pmatrix} \eta_{xx} - N^2 & \eta_{xy} \\ \eta_{yx} & \eta_{yy} - N^2 \end{pmatrix} \begin{pmatrix} E_x \\ E_y \end{pmatrix} = 0, \tag{3.2.28}$$

where

$$\begin{aligned} \eta_{xx} &= \left(\epsilon_{xx} - \epsilon_{xz}\epsilon_{zx}\epsilon_{zz}^{-1}\right), \\ \eta_{xy} &= -\eta_{yx} = \left(\epsilon_{xy} - \epsilon_{xz}\epsilon_{zy}\epsilon_{zz}^{-1}\right), \\ \eta_{yy} &= \left(\epsilon_{yy} - \epsilon_{yz}\epsilon_{zy}\epsilon_{zz}^{-1}\right). \end{aligned} \tag{3.2.29}$$

The ratio of the components of the polarization eigenvectors satisfy

$$\begin{aligned} K_x &= E_x/E_y = \left(N^2 - \eta_{xx}\right)\eta_{xy}^{-1}, \\ K_z &= E_z/E_y = \left(K_z - \varepsilon_{zy}\right)\varepsilon_{zz}^{-1}. \end{aligned} \tag{3.2.30}$$

Neglecting the damping, the transverse, normalized polarization eigenvectors are

$$\vec{e}^{\,t}_{1,2} = C\left(K_{x1,2}, 1\right), \tag{3.2.31}$$

where $C = (1 + K_{x1,2}^2)^{-1/2}$, and

$$K_{x1,2} = b\left[1 \pm \zeta\left(1 + b^{-2}\right)\right]^{-1/2}, \quad \zeta = \operatorname{sign}\left(\eta_{xx} - \eta_{yy}\right), \tag{3.2.32}$$

and the ellipticity parameter $b$ is given by

$$b = u^{1/2}\sin^2\theta\left[2\cos\theta\left(1 - v\right)\right]^{-1}. \tag{3.2.33}$$

The ratio $E_x/E_y$ gives the degree of the ellipticity of the normal modes. In terms of the quantities (3.2.22), and with damping included, this is

$$\frac{E_{x1,2}}{E_{y1,2}} = K_{x1,2}$$

$$= i\frac{2u^{1/2}(1 + i\gamma - v)\cos\theta}{u\sin^2\theta \pm \left[u^2\sin^4\theta + 4u(1 + i\gamma - v)^2\cos^2\theta\right]^{1/2}}, \tag{3.2.34}$$

where again $\pm$ corresponds to modes (2, 1). Waves of either type are elliptically polarized, and in the absence of damping, the two ellipses described by the tip of the $\vec{E}$ vector are in the $x$, $y$ plane, being of equal shape but perpendicular to each other (see Fig. 3.2.4). The quantity $K_{x1,2}$ is the ratio of semiaxes of the ellipses, and satisfies

$$K_{x1}K_{x2} = 1. \tag{3.2.35}$$

It is seen that, for $\theta = 0°$, $K_{x1} = i$, $K_{x2} = -i$, so that the two waves are circularly polarized and rotating in opposite senses. For $\theta = \pi/2$, $K_{x1} = 0$, $K_{x2} = i\infty$, that is, $E_{x1} = 0$, $E_{y2} = 0$, which represents linear polarization along the $x$ axis (1) and $y$ axis (2). The longitudinal field

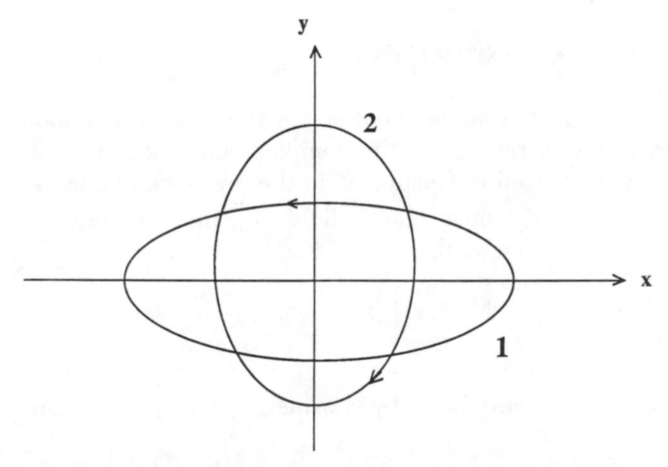


FIGURE 3.2.4 Polarization ellipses for the two normal modes in a magnetized plasma.

is given by equation (3.2.27), which in the absence of absorption is

$$E_z = i\frac{u^{1/2} v \sin\theta}{u - (1 - v) - uv\cos^2\theta} E_y + \frac{uv\cos\theta\sin\theta}{u - (1 - v) - uv\cos^2\theta} E_x. \tag{3.2.26}$$

Thus, $E_z$ is in phase with $E_x$ and is $\pi/2$ out of phase with $E_y$, and the vector $\vec{E}$ describes an ellipse in a plane parallel to the $y$ axis.

## 3.3 Quantum Mechanical Derivation of the Dielectric Tensor

The dielectric tensor derived in § 3.2 by classical means can also be obtained from a quantum calculation. The quantum description is important because in a strong magnetic field one expects quantum effects to become important whenever the de Broglie wavelength is comparable to the Larmor radius, and one approaches or surpasses that regime in neutron stars. In general, it is sufficient to treat the electromagnetic field as a classical entity while dealing with the particles quantum mechanically (e.g. Svetozarova and Tsytovich, 1962; Pavlov, Shibanov, and Yakovlev, 1980; Kirk and Cramer, 1985). We follow in this section the nonrelativistic formulation of Canuto and Ventura (1972). The quantity whose quantum equivalent we need is the current density $\vec{j}$, since as we saw from equations (3.1.23)–(3.1.25), that determines the dielectric tensor. The current $\vec{j}$ for a free particle is defined quantum mechanically as

$$\vec{j}(\vec{r}, t) = -\frac{e}{2m}\Psi^*(\vec{r}, t)\vec{p}\Psi(\vec{r}, t) + c.c., \tag{3.3.1}$$

where $\Psi(\vec{r}, t)$ is the wave function, $\vec{p} \equiv -i\hbar\nabla$ is the momentum operator, and $c.c.$ represents the complex conjugate of the previous quantity. As discussed in Chapter 2, in the presence of a wave field of vector potential $\vec{A}_w$ and magnetic field $\vec{A}_B$, the (canonical) momentum must be modified so that

$$\vec{j}(\vec{r}, t) = -\frac{e}{2m}\Psi^*(\vec{r}, t)\left[\vec{p} + \frac{e}{c}\vec{A}_B + \frac{e}{c}\vec{A}_w\right]\Psi(\vec{r}, t) + c.c. \tag{3.3.2}$$

The wave functions are found by considering the Hamiltonian

$$H = \frac{1}{2m}\left(\vec{p} + \frac{e}{c}\vec{A}_B + \frac{e}{c}\vec{A}_w\right)^2 \equiv \frac{1}{2m}\left(\vec{\pi} + \frac{e}{c}\vec{A}_w\right)^2, \tag{3.3.3}$$

where

$$\vec{A}_B = \frac{1}{2}\vec{r} \times \vec{B}, \quad \vec{A}_w = \vec{A}_0 \exp\left(i\vec{k}\cdot\vec{r} - i\omega t\right) + c.c. \tag{3.3.4}$$

Neglecting terms in $\vec{A}_w^2$, equation (3.3.3) becomes

$$H = \frac{\pi^2}{2m} + \frac{e}{mc}\vec{\pi}\cdot\vec{A}_w = H_0 + H', \tag{3.3.5}$$

where $H'$ is treated as a perturbation due to the wave on the basic $H_0$, of the form

$$H' = V\exp(-i\omega t). \tag{3.3.6}$$

Calling $\Psi$ the total wave function and $\phi$ that of the unperturbed system, we have for the individual eigenfunctions

$$H\Psi_\alpha = \eta_\alpha \Psi_\alpha, \quad H_0\phi_\alpha = E_\alpha \phi_\alpha. \tag{3.3.7}$$

The correction to $\phi$ due to the time-dependent perturbation $V\exp(-i\omega t)$ is given by the first-order perturbation theory (e.g. Sakurai, 1967) as

$$\Psi_\alpha(r,t) = \phi_\alpha(r,t) + \sum_\beta c_{\beta\alpha}(t)\phi_\beta(r,t), \tag{3.3.8}$$

where $\alpha$, $\beta$ are the set of quantum numbers representing an electron state and

$$c_{\beta\alpha}(t) = \frac{\langle\beta|V|\alpha\rangle}{\hbar(\omega-\omega_{\beta\alpha})}\exp\left[i(\omega_{\beta\alpha}-\omega)t\right] - \frac{\langle\alpha|V|\beta\rangle}{\hbar(\omega+\omega_{\beta\alpha})}\exp\left[i(\omega_{\beta\alpha}+\omega)t\right], \tag{3.3.9}$$

$$V = \frac{e}{2mc}\left(\vec{\pi}\cdot\vec{A}_0\exp\left(i\vec{k}\cdot\vec{r}\right) + \exp\left(i\vec{k}\cdot\vec{r}\right)\vec{\pi}\cdot\vec{A}_0\right), \tag{3.3.10}$$

$$\hbar\omega_{\beta\alpha} = E_\beta - E_\alpha. \tag{3.3.11}$$

The time dependence of $\phi_\alpha(r,t)$ is of the form

$$\phi_\alpha(r,t) = \phi_\alpha(r)\exp(-iE_\alpha t/\hbar) \tag{3.3.12}$$

so that one gets

$$\Psi_\alpha(r,t) = \left[\phi_\alpha(r) + \phi_\alpha^{(1)}(r,t) + \phi_\alpha^{(2)}(r,t)\right]\exp(-iE_\alpha t/\hbar), \tag{3.3.13}$$

where

$$\phi_\alpha^{(1)}(r,t) = \sum_\beta A_{\beta\alpha} \exp(-i\omega t)\phi_\beta(r),$$

$$\phi_\alpha^{(2)}(r,t) = -\sum_\beta B_{\beta\alpha} \exp(i\omega t)\phi_\beta(r),$$

$$A_{\beta\alpha} = \frac{\langle \beta | V | \alpha \rangle}{\hbar(\omega - \omega_{\beta\alpha})}, \quad B_{\beta\alpha} = \frac{\langle \alpha | V | \beta \rangle}{\hbar(\omega + \omega_{\beta\alpha})}. \tag{3.3.14}$$

Notice here (from eq. 3.3.12) the denominators $\omega \pm \omega_{\beta\alpha}$, which are the quantum equivalents of the classical $\omega \pm \omega_c$. One can now calculate the current (3.3.2), or rather its Fourier transform, since that is the classical quantity needed,

$$\begin{aligned} \vec{j}(\vec{k},\omega) &= \int d^3r\,dt\,\vec{j}(\vec{r},t)\exp(i\vec{k}\cdot\vec{r} + i\omega t) \\ &= -\frac{e}{2m}\int d^3r\,dt \\ &\quad \cdot\left\{\left[\phi_\alpha^*(r) + \phi_\alpha^{(1)*}(r,t) + \phi_\alpha^{(2)*}(r,t)\right]\vec{\pi} + \frac{e}{c}\vec{A}_w\right\} \\ &\quad \cdot\left[\phi_\alpha(r) + \phi_\alpha^{(1)}(r,t) + \phi_\alpha^{(2)}(r,t)\right] \\ &\quad \times \exp(-i\vec{k}\cdot\vec{r} + i\omega t) + c.c. \end{aligned} \tag{3.3.15}$$

This can be written as

$$\vec{j}(\vec{k},\omega) = \sum_{i=1}^{3} \vec{j}_i(\vec{k},\omega), \tag{3.3.16}$$

where

$$\begin{aligned} \vec{j}_1(\vec{k},\omega) &= -\frac{e}{2m}\int d^3r\,dt\,\phi_\alpha^*(r)\left(\vec{\pi} + \frac{e}{c}\vec{A}_w\right) \\ &\quad \times \phi_\alpha(r)\exp(-i\vec{k}\cdot\vec{r} + i\omega t) + c.c., \\ j_2(\vec{k},\omega) &= -\frac{e}{2m}\int d^3r\,dt\left[\phi_\alpha^*(r)\vec{\pi}\phi_\alpha^{*(1)}(r,t)\right. \\ &\quad \left. + \phi_\alpha^{*(1)}(r,t)\vec{\pi}\phi_\alpha(r)\right] \\ &\quad \cdot \exp(-i\vec{k}\cdot\vec{r} + i\omega t) + c.c., \\ j_3(\vec{k},\omega) &= +\frac{e}{2m}\int d^3r\,dt\left[\phi_\alpha^*(r)\vec{\pi}\phi_\alpha^{*(2)}(r,t)\right. \\ &\quad \left. + \phi_\alpha^{*(2)}(r,t)\vec{\pi}\phi_\alpha(r)\right] \\ &\quad \cdot \exp(-i\vec{k}\cdot\vec{r} + i\omega t) + c.c. \end{aligned} \tag{3.3.17}$$

In these expressions, quantities in $\vec{A}_w^2$ or higher have been neglected, because $\phi^{(1)}, \phi^{(2)}$ are themselves proportional to $\vec{A}^w$. Substituting in $\vec{j}_1, \vec{j}_2$ the values of $\phi_\alpha^{(1)}, \phi_\alpha^{(2)}$ from equation (3.3.14) and in $\vec{j}_1$ the expression for $\vec{A}_w$ of equation (3.3.4), we get

$$
\begin{aligned}
\vec{j}_1(\vec{k}, \omega) &= -\frac{e}{2m} \int d^3r\,dt \Big[ \phi_\alpha^*(r) \vec{\pi} \phi(r) \exp(-i\vec{k} \cdot \vec{r} + i\omega t) \\
&\quad + \phi_\alpha^*(r) \vec{A}_0 \phi_\alpha(r) \Big] + c.c., \\
\vec{j}_2(\vec{k}, \omega) &= -\frac{e}{2m} \sum_\beta \int d^3r\,dt\, A_{\beta\alpha} \Big[ \phi_\alpha^*(r) \vec{\pi} \exp(-i\vec{k} \cdot \vec{r}) \phi_\beta(r) \\
&\quad + \phi_\beta^*(r) \vec{\pi} \exp(-i\vec{k} \cdot \vec{r}) \phi_\alpha(r) \exp(2i\omega t) \Big] + c.c., \\
\vec{j}_3(\vec{k}, \omega) &= +\frac{e}{2m} \sum_\beta \int d^3r\,dt\, B_{\beta\alpha} \Big[ \phi_\alpha^* \vec{\pi} \exp(-i\vec{k} \cdot \vec{r}) \\
&\quad \times \phi_\beta(r) \exp(2i\omega t) \\
&\quad + \phi_\beta^*(r) \vec{\pi} \exp(-i\vec{k} \cdot \vec{r}) \phi_\alpha(r) \Big] + c.c.
\end{aligned}
\tag{3.3.18}
$$

However, the time integrals of the terms proportional to $\exp(i\omega t)$, $\exp(2i\omega t)$ vanish for $\omega \neq 0$, and using the definition (3.3.9) of $V$ and the normalization of the $\phi_\alpha$, we finally have

$$
\begin{aligned}
\vec{j}_1(\vec{k}, \omega) &= -\frac{e^2}{mc} \vec{A}_0, \\
\vec{j}_2(\vec{k}, \omega) &= -\frac{e^2}{m^2 c} \\
&\quad \times \sum_\beta \frac{\langle \alpha | \exp(-i\vec{k} \cdot \vec{r}) | \beta \rangle \langle \beta | \vec{\pi} \cdot \vec{A} \exp(-i\vec{k} \cdot \vec{r}) | \alpha \rangle}{\hbar(\omega - \omega_{\beta\alpha})}, \\
\vec{j}_3(\vec{k}, \omega) &= \frac{e^2}{m^2 c} \\
&\quad \times \sum_\beta \frac{\langle \alpha | \vec{A}_0 \cdot \vec{\pi} \exp(i\vec{k} \cdot \vec{r}) | \beta \rangle \langle \beta | \vec{\pi} \exp(-i\vec{k} \cdot \vec{r}) | \alpha \rangle}{\hbar(\omega + \omega_{\beta\alpha})}.
\end{aligned}
\tag{3.3.19}
$$

Instead of $\vec{A}_w$, one can use $\vec{E}_w$, which in the radiation gauge (zero scalar potential) is

$$
\vec{E}_w = -\frac{1}{c} \frac{\partial \vec{A}_\omega}{\partial t} = \frac{i\omega}{c} \vec{A}_w. \tag{3.3.20}
$$

Now, Ohm's law is

$$j_i = \sigma_{ij} E_j. \tag{3.3.21}$$

However, the $\vec{j}$ of equations (3.3.19) is for a single particle in quantum state $\alpha$. The $\vec{j}$ in a medium is given by a statistical average, obtained by multiplying equations (3.3.19) by the probability $f_\alpha$ of having that state populated, and summing over $\alpha$, with $f_\alpha$ normalized to give the electron density $N_e$,

$$\sum_\alpha f_\alpha = N_e. \tag{3.3.22}$$

Doing this and comparing the resultant equation with equation (3.3.21), one obtains

$$\sigma_{ij} = i\frac{e^2}{m\omega} N_e \delta_{ij} + \frac{ie^2}{m^2\omega} \sum_{\beta,\alpha} f_\alpha \left[ \frac{\langle \alpha | \pi_i e^{-i\vec{k}\cdot\vec{r}} | \beta \rangle \langle \beta | \pi_j e^{i\vec{k}\cdot\vec{r}} | \alpha \rangle}{\hbar(\omega - \omega_{\beta\alpha})} - \frac{\langle \alpha | \pi_j e^{i\vec{k}\cdot\vec{r}} | \beta \rangle \langle \beta | \pi_i e^{-i\vec{k}\cdot\vec{r}} | \alpha \rangle}{\hbar(\omega + \omega_{\beta\alpha})} \right], \tag{3.3.23}$$

where the $\pi_i \exp(i\vec{k} \cdot \vec{r})$ are symmetrized products, since $V$ is Hermitian. Writing the dielectric tensor as

$$\epsilon_{ij} = (1 - v)\delta_{ij} - v\tau_{ij}, \tag{3.3.24}$$

where $v = (\omega_p/\omega)^2 = (4\pi N_e e^2/m\omega^2)$, one gets

$$\tau_{ij} = \frac{1}{N_e m\hbar} \sum_{\alpha,\beta} \left[ \frac{\langle \alpha | \pi_i e^{-i\vec{k}\cdot\vec{r}} | \beta \rangle \langle \beta | \pi_j e^{i\vec{k}\cdot\vec{r}} | \alpha \rangle}{(\omega - \omega_{\beta\alpha})} - \frac{\langle \alpha | \pi_i e^{i\vec{k}\cdot\vec{r}} | \beta \rangle \langle \beta | \pi_j e^{-i\vec{k}\cdot\vec{r}} | \alpha \rangle}{(\omega + \omega_{\beta\alpha})} \right] f_\alpha (1 - f_\beta). \tag{3.3.25}$$

By interchanging the dummy indices in equation (3.3.25), this equation can be recast as

$$\begin{aligned} \tau_{ij} &= \frac{1}{N_e m\hbar} \sum_{\alpha\beta} \frac{\langle \alpha | \pi_i e^{-i\vec{k}\cdot\vec{r}} | \beta \rangle \langle \beta | \pi_i e^{i\vec{k}\cdot\vec{r}} | \alpha \rangle}{(\omega - \omega_{\beta\alpha})} (f_\alpha - f_\beta) \\ &= \frac{1}{N_e m\hbar} \sum_{\alpha\beta} \left[ \frac{\langle \alpha | \pi_i e^{-i\vec{k}\cdot\vec{r}} | \beta \rangle \langle \beta | \pi_i e^{i\vec{k}\cdot\vec{r}} | \alpha \rangle}{(\omega - \omega_{\beta\alpha})} - \frac{\langle \alpha | \pi_i e^{i\vec{k}\cdot\vec{r}} | \beta \rangle \langle \beta | \pi_i e^{-i\vec{k}\cdot\vec{r}} | \alpha \rangle}{(\omega + \omega_{\beta\alpha})} \right] f_\alpha. \end{aligned} \tag{3.3.26}$$

Either of the two latter forms can be used, as convenient. In these expressions one has a formal canceling of the factor $(1 - f_\beta)$, expressing the Pauli exclusion principle. Defining

$$A_{ij}^{\alpha\beta} \equiv \frac{\langle\alpha\,|\,\pi_i e^{-i\vec{k}\cdot\vec{r}}\,|\,\beta\rangle\langle\beta\,|\,\pi_j e^{i\vec{k}\cdot\vec{r}}\,|\,\alpha\rangle}{m\hbar(\omega - \omega_{\beta\alpha})}, \tag{3.3.27a}$$

$$B_{ij}^{\alpha\beta} \equiv \frac{\langle\alpha\,|\,\pi_j e^{i\vec{k}\cdot\vec{r}}\,|\,\beta\rangle\langle\beta\,|\,\pi_i e^{-i\vec{k}\cdot\vec{r}}\,|\,\alpha\rangle}{m\hbar(\omega + \omega_{\beta\alpha})}, \tag{3.3.27b}$$

which satisfy

$$B_{ij}^{\alpha\beta} = A_{ij}^{\beta\alpha}, \tag{3.3.28}$$

equation (3.3.26) becomes

$$\tau_{ij} = N_e^{-1}\sum_{\alpha,\,\beta}\left[A_{ij}^{\alpha\beta} - B_{ij}^{\alpha\beta}\right]f_\alpha. \tag{3.3.29}$$

This expression is correct whether one is dealing with classical statistics, where $f_\alpha$ is the Maxwell-Boltzmann distribution, or quantum statistics, in which case $f_\alpha$ is the Fermi-Dirac distribution. An explicit expression for $\tau_{ij}$ can be obtained using the wave function $\Psi = |\,\alpha\rangle$ of equations (2.3.26)–(2.3.27), where $\alpha \equiv (n, p_z, s)$, namely

$$\begin{aligned} |\,\alpha\rangle &= L^{-1/2}(\gamma/\pi)^{1/2} I_{ns}(\gamma r^2)\exp(ip_z z)\exp[(n-s)\phi],\\ \gamma &= (1/2)\lambda\!\!\!^{-}_c{}^{-2}(B/B_{cr}), \quad \lambda\!\!\!^{-}_c = \hbar/mc,\\ I_{ns}(x^2) &= (-1)^s(n!s!)^{-1/2}\exp(-x^2/2)\,x^{(n-s)}Q_n^{n-s}(x^2). \end{aligned} \tag{3.3.30}$$

The effect of the $\pi_i$ on $|\,\alpha\rangle$ is as follows:

$$\begin{aligned} \pi_x\,|\,n\rangle &= i\left(\frac{1}{2}m\hbar\omega\right)^{1/2}\left[\sqrt{n+1}\,|\,n+1\rangle - \sqrt{n}\,|\,n-1\rangle\right],\\ \pi_y\,|\,n\rangle &= \left(\frac{1}{2}m\hbar\omega\right)^{1/2}\left[\sqrt{n-1}\,|\,n+1\rangle + \sqrt{n}\,|\,n-1\rangle\right],\\ \pi_z\,|\,n\rangle &= p_z\,|\,n\rangle. \end{aligned} \tag{3.3.31}$$

Thus, one only needs to use the integral (e.g. Canuto and Ventura, 1977)

$$\langle\alpha'\,|\exp(i\vec{q}\cdot\vec{r})\,|\,\alpha\rangle = CI_{nn'}(x)\,I_{ss'}(x), \tag{3.3.32}$$

with $x^2 \equiv q_\perp^2/4\gamma$, $C = \delta_{p'_z,\, p_z+q_z} i^{n'+n+s'+s}$, $q_\perp^2 = q_x^2 + q_y^2$. The matrix elements involved are

$$(\hbar m)^{-1/2} \langle \alpha' \mid \pi_i \exp(i\vec{k}\cdot\vec{r}) \mid \alpha \rangle = \delta_{p',\, p+k} (F_i)_{nn'} I_{ss'}, \tag{3.3.33}$$

with

$$(F_1)_{nn'} = i\left(\frac{\hbar\omega_c}{2}\right)^{1/2} \left(\sqrt{n+1}\, I_{n+1,\, n'} - \sqrt{n}\, I_{n-1,\, n'}\right),$$

$$(F_2)_{nn'} = \left(\frac{\hbar\omega_c}{2}\right)^{1/2} \left(\sqrt{n+1}\, I_{n+1,\, n'} + \sqrt{n}\, I_{n-1,\, n'} + \frac{1}{2}\gamma^{-1/2} k_\perp I_{nn'}\right),$$

$$(F_3)_{nn'} = (\hbar m)^{-1/2} \left(p_z + \frac{1}{2}\hbar k_\parallel\right) I_{n,\, n'}, \tag{3.3.34}$$

and the argument of the $I_{mm'}$ is $x^2 = k_\perp^2/4\gamma$. In equations (3.3.34), it has been assumed that $\vec{k}$ lies in the $y$, $z$ plane (with the opposite $x$, $z$ choice, $F_1$ and $F_2$ are interchanged). The tensor $\tau_{ij}$ becomes

$$\tau_{ij} = N_e^{-1} \sum_{\alpha\alpha'} \left[\frac{F_i^* F_j}{D_1} - \frac{\tilde{F}_i^* F_j}{D_2}\right] I_{ss'}^2 f_\alpha, \tag{3.3.35}$$

with $D_{1,2}$ defined below and $(\tilde{F}_i)_{nn'} = (F_i)_{n'n}$, the transpose matrix. However, the $F_l$ and $f_\alpha$ depend only on the energy $n$, and $\alpha \equiv (n, p_z, s)$, so the sum over $\alpha$ in equation (3.3.35) acts only on the $I_{ss'}^2$. Using the relationship (Shafranov, 1967)

$$\sum_{s''} I_{ss''} I_{s''s'} = \delta_{ss'}, \tag{3.3.36}$$

one can eliminate $I_{ss'}^2$ from $\tau_{ij}$ and have the final nonrelativistic expression

$$\tau_{ij} = N_e^{-1} \sum_{\alpha\alpha'} \left[\frac{F_i^* F_j}{D_1} - \frac{\tilde{F}_i^* \tilde{F}_j}{D_2}\right] f_\alpha, \tag{3.3.37}$$

where

$$D_{1,2} = \omega \mp (n' - n)\omega_c \mp \frac{\hbar k_\parallel^2}{2m} - \frac{\hbar k_\parallel p_z}{m}, \tag{3.3.38}$$

the $F_i$ being given by equations (3.3.34). Placing this into the definition (3.3.24), one obtains as the dielectric tensor (Canuto and Ventura, 1972)

$$\epsilon_{ij}^{(p)} = (1 - v)\delta_{ij} - vN_e^{-1}\sum_{\alpha\alpha'}\left[\frac{F_i^*F_j}{D_1} - \frac{\tilde{F}_i^*\tilde{F}_j}{D_2}\right]f_\alpha. \tag{3.3.39}$$

This equation contains nonrelativistic quantum effects, the discreteness effects associated with quantization of the transverse levels, and allows for a study of temperature effects, since electrons in higher Landau levels are included. One can verify that equation (3.3.39) reduces to the correct cold plasma in the limit where $p_z, k \to 0$ (see also § 3.5). One sees from the definition (3.3.31) of $I_{nn'}$ that, for vanishing arguments $x \propto k$,

$$\lim_{x\to 0} I_{nn'}(x) \to \delta_{n,n'}, \tag{3.3.40}$$

so in this limit

$$(F_1)_{nn'} = i\left(\frac{\omega_c}{2}\right)^{1/2}\left[\sqrt{n+1}\,\delta_{n',n+1} - \sqrt{n}\,\delta_{n',n-1}\right],$$

$$(F_2)_{nn'} = \left(\frac{\omega_c}{2}\right)^{1/2}\left[\sqrt{n+1}\,\delta_{n',n+1} + \sqrt{n}\,\delta_{n',n-1}\right],$$

$$(F_3)_{nn'} = (\hbar m)^{-1/2}\left(p_z + \frac{1}{2}\hbar k_\parallel\right) \simeq 0, \tag{3.3.41}$$

where for $F_3$ we used $k \to 0$ and $p_z \to 0$ (since $T \to 0$). We have then

$$|F_1|^2 = |F_2|^2 = \frac{\omega_c}{2}\left[(n+1)\delta_{n',n+1} + n\delta_{n',n-1}\right],$$

$$F_1^*F_2 = -F_2^*F_1 = -i\frac{\omega_c}{2}\left[(n+1)\delta_{n',n+1} - n\delta_{n',n-1}\right],$$

$$F_1^*F_3 = F_2^*F_3 = F_3^*F_3 = 0,$$

$$|\tilde{F}_1| = |\tilde{F}_2| = \frac{\omega_c}{2}\left[n\delta_{n',n+1} + (n+1)\delta_{n',n-1}\right],$$

$$\tilde{F}_1^*\tilde{F}_2 = \tilde{F}_2^*\tilde{F}_1 = -i\frac{\omega_c}{2}\left[n\delta_{n',n-1} - (n+1)\delta_{n',n+1}\right],$$

$$\tilde{F}_1^*\tilde{F}_2 = \tilde{F}_2^*\tilde{F}_3 = \tilde{F}_3^*\tilde{F}_3 = 0. \tag{3.3.42}$$

Using the real quantities

$$\tau_{\pm} = \tau_{xx} \pm i\tau_{xy}, \tag{3.3.43}$$

we can write this as

$$\tau_{\pm} = \pm \frac{\omega_c}{N_e} \sum_{\alpha} \left[ \frac{n}{D\pm} - \frac{n+1}{D\pm} \right] f_{\alpha},$$

$$\tau_{zz} = \tau_{xz} = \tau_{zx} = \tau_{yz} = \tau_{zy} = 0,$$

$$D\pm = \omega \pm \omega_c, \tag{3.3.44}$$

and from equation (3.3.44) we get

$$\tau_{\pm} = \mp \frac{\omega_c}{\omega \pm \omega_c} = \mp \frac{u^{1/2}}{1 \pm u^{1/2}}, \tag{3.3.45}$$

or

$$\tau_{xx} = \tau_{yy} = \frac{u}{1-u}, \quad \tau_{xy} = -\tau_{yx} = i\frac{u^{1/2}}{1-u}. \tag{3.3.46}$$

From equation (3.3.24), we finally obtain

$$\epsilon_{ij} = \begin{pmatrix} 1 - v(1-u)^{-1} & -ivu^{1/2}(1-u)^{-1} & 0 \\ ivu^{1/2}(1-u)^{-1} & 1 - v(1-u)^{-1} & 0 \\ 0 & 0 & 1-v \end{pmatrix}, \tag{3.3.47}$$

which agrees with the classical cold-plasma tensor of equation (3.1.27). In the limit $k \to 0$ it is of course irrelevant whether $\vec{k}$ was in the $x$, $y$ or $y$, $z$ plane, and in fact such considerations did not arise in evaluating equation (3.1.27). In practice, the range of validity of the classical cold tensor (3.3.47) is confined to the regions

$$\hbar\omega \ll m_e c^2, \quad kT_e \ll m_e c^2,$$

$$|\omega - n\omega_c| \gg \omega\left(kT_e/m_e c^2\right)^{1/2} |\cos\theta|, \qquad n = 1, 2, \ldots, \tag{3.3.48}$$

i.e. for photons away from the resonances, and, say, $\hbar\omega$, $kT \lesssim 50$ keV. However, when temperature effects are important, e.g. near the resonances, or for $kT \gtrsim \hbar\omega_c$, the quantum mechanical tensor (3.3.37) has to be used, with an appropriate statistical averaging over the distribution function $f_\alpha$ (see § 3.5).

## 3.4 Vacuum Polarizability Effects in Strongly Magnetized Plasmas

In a very strong magnetic field, the cyclotron energy of the ground Landau level, $\hbar\omega_c = 11.6$ keV $(B/10^{12}$ G), may be rather larger than the plasma temperature, in which case the cold-plasma approximation may be valid far from the resonances (see eqs. 3.3.48). However, in the same regime, the vacuum polarization effects due to the virtual $e^+e^-$ pairs of the magnetic field start becoming significant compared to the plasma effects (Mészáros and Ventura, 1978, 1979; Gnedin, Pavlov, and Shibanov, 1978a, b). As seen from equations (2.6.6), the vacuum itself contributes a departure from unity of the dielectric tensor $\epsilon_{ij}$. In addition, the magnetic permeability tensor $\mu_{ij}^{-1}$, which would be essentially unity if only plasma effects were considered, acquires a departure from unity due to the vacuum. If we are interested in relatively low densities, the magnitude of the plasma dielectric susceptibility (or polarizability) is small:

$$|4\pi\alpha_{ij}^{(p)}| = |\varepsilon_{ij} - \delta_{ij}| \ll 1. \tag{3.4.1}$$

The vacuum susceptibility is proportional to the vacuum factor $\delta = 0.5 \times 10^{-4}(B/B_Q)^2$, which is also small unless $B \gg B_Q$. Under these circumstances, the plasma and vacuum susceptibilities and magnetizabilities can be added linearly, so that the total dielectric tensor and total inverse magnetic permeability are

$$\epsilon_{ij} = \delta_{ij} + 4\pi\left[\alpha_{ij}^{(p)} + \alpha_{ij}^{(v)}\right], \quad \mu_{ij}^{-1} = \mu_{ij}^{-1(v)}. \tag{3.4.2}$$

In a system where $\vec{B}_0$ is along $z$ and $B \ll B_Q$, we have according to equations (2.6.6) and (3.1.26)

$$4\pi\alpha_{ij}^{(v)} = \begin{pmatrix} -2\delta & 0 & 0 \\ 0 & -2\delta & 0 \\ 0 & 0 & 5\delta \end{pmatrix}, \tag{3.4.3}$$

$$4\pi\alpha_{ij}^{(p)} = \begin{pmatrix} S & iD & 0 \\ -iD & S & 0 \\ 0 & 0 & P \end{pmatrix}, \tag{3.4.4}$$

$$\mu_{ij}^{-1(v)} = \begin{pmatrix} 1-2\delta & 0 & 0 \\ 0 & 1-2\delta & 0 \\ 0 & 0 & 1-6\delta \end{pmatrix}, \tag{3.4.5}$$

where

$$\delta = \left(e^2/\hbar c\right)^2 (45\pi)^{-1} \left(B/B_Q\right)^2 \ll 1, \tag{3.4.6a}$$

$$S = -v\lambda\left(\lambda^2 - u^2\right)^{-1}, \quad D = vu^{1/2}\left(\lambda^2 - u\right)^{-1}, \quad P = -v\lambda^{-1}, \tag{3.4.6b}$$

$$v = \left(\omega_p/\omega\right)^2, \quad u = \left(\omega_c/\omega\right)^2, \quad \lambda = 1 + i\gamma = 1 + i\Gamma/2\omega,$$
$$\gamma \ll 1. \tag{3.4.6c}$$

In arbitrary Cartesian coordinates,

$$\begin{aligned} \overleftrightarrow{\epsilon} &= a\overleftrightarrow{1} + q\hat{b}\hat{b} + 4\pi\overleftrightarrow{\alpha}^{(p)}, \\ \overleftrightarrow{\mu}^{-1} &= a\overleftrightarrow{1} - h\hat{b}\hat{b}, \\ a = 1 - 2\delta, &\quad q = 7\delta, \quad h = 4\delta, \end{aligned} \tag{3.4.7}$$

where $\hat{b}$ is a unit vector along $\vec{B}$. The macroscopic Maxwell equations (2.5.18), given the constitutive relations (2.5.15) and arbitrary $\epsilon_{ij}$, $\mu_{ij}$, lead in the presence of plane wave perturbations $\exp(i\vec{k}\cdot\vec{r} - i\omega t)$ to the wave equation (Pavlov and Shibanov, 1979)

$$\{\epsilon_{ij} + N^2 e_{iqs} k_q \mu_{sk}^{-1} e_{kpj} k_p\} E_j = 0, \tag{3.4.8}$$

where $N$ is the complex refractive index, $\hat{k}$ is a unit vector along $\vec{k}$, and $e_{iqs}$ is the unit antisymmetric tensor. One can also directly replace the specific $\epsilon_{ij}$ and $\mu_{ij}$ of equations (3.4.7) into Maxwell's equations (2.5.3) and derive the vacuum-corrected wave equation (Mészáros and Ventura, 1978, 1979)

$$\{N^2[k_\alpha k_\beta - \delta_{\alpha\beta} + 4\delta a^{-1}(\hat{k}\times\hat{b})_\alpha(\hat{k}\times\hat{b})_\beta] + a^{-1}\varepsilon_{\beta\alpha}\}E_\beta = 0. \tag{3.4.9}$$

In a more compressed form (Kirk, 1980), this can be expressed as

$$\left(\overleftrightarrow{\Lambda} + \overleftrightarrow{\chi} + \overleftrightarrow{\Psi}\right)\cdot\vec{E} = 0, \tag{3.4.10}$$

where

$$\overleftrightarrow{\Lambda} = \overleftrightarrow{1} - N^2\left(\overleftrightarrow{1} - \hat{k}\hat{k}\right) \tag{3.4.11}$$

is the empty-space Maxwell wave tensor,

$$\overleftrightarrow{\chi} = 4\pi\alpha^{(p)} \tag{3.4.12}$$

is the renormalized plasma dielectric susceptibility ($\vec{\vec{\chi}} = \vec{\vec{\epsilon}} - \vec{\vec{1}}$), and $\vec{\vec{\Psi}}$ is the vacuum correction tensor for $B \ll B_Q$,

$$\vec{\vec{\Psi}} = -2\delta\vec{\vec{\Lambda}} + N^2 h\hat{k} \times \hat{b}\hat{k} \times \hat{b} + q\hat{b}\hat{b}, \tag{3.4.13}$$

in terms of the quantities defined in equation (3.4.11). The wave equation (3.4.11) reduces to the normal magnetized plasma relation (eq. 3.1.17) in the limit $\vec{\vec{\Psi}} = 0$. The dispersion relation is given by

$$\det\left(\vec{\vec{\Lambda}} + \vec{\vec{\chi}} + \vec{\vec{\Psi}}\right) = 0. \tag{3.4.14}$$

In order to solve for the value of the complex refractive index and the polarization eigenvectors we use again the coordinate frame II, with $z$ along $\vec{k}$ and $\vec{B}$ in the $x$, $z$ plane. The treatment follows that in Mészáros and Ventura (1979). The strategy is similar to that used in equations (3.2.25) and the following, except that the $\varepsilon$, $\eta$, $g$ appearing in $\epsilon_{ij}$ in (3.2.25) must now be extended to include the vacuum contribution, as in equation (3.4.7). Also, the wave equation (3.2.26) must be replaced by (3.4.9). Looking at the new wave equation (3.4.9), one sees that the tensor multiplying $N^2$ is still orthogonal to $\vec{k}$. Using the frame with $\vec{k}$ along $z$ will then cause $N^2$ to appear only in the $x$, $y$ components of the wave equations, and again one can solve for $E_z$ as a function of $E_x$, $E_y$ and the dielectric tensor elements (see eq. 3.2.27). The remaining two components of the plasma plus vacuum wave equation read

$$\begin{pmatrix} \eta_{xx} - N^2 & \eta_{xy} \\ \eta_{yx} & \eta_{yy} - N^2\rho \end{pmatrix} \begin{pmatrix} E_x \\ E_y \end{pmatrix} = 0, \tag{3.4.15}$$

where we have defined a vacuum correction factor

$$\rho \equiv 1 - 4\delta a^{-1}\sin^2\theta, \tag{3.4.16}$$

and

$$\begin{aligned} \eta_{xx} &= a^{-1}\left(\epsilon_{xx} - \epsilon_{xz}\epsilon_{zx}\epsilon_{zz}^{-1}\right), \\ \eta_{xy} &= -\eta_{yx} = a^{-1}\left(\epsilon_{xy} - \epsilon_{xz}\epsilon_{zy}\epsilon_{zz}^{-1}\right), \\ \eta_{yy} &= a^{-1}\left(\epsilon_{yy} - \epsilon_{yz}\epsilon_{zy}\epsilon_{zz}^{-1}\right), \end{aligned} \tag{3.4.17}$$

in terms of $\theta = \cos^{-1}(\vec{k} \cdot \vec{B})$ and the total dielectric tensor elements of equations (3.4.2). One can see that, in the absence of the plasma component ($\alpha_{ij}^{(p)} \equiv 0$, $\mu_{ij}^{-1} \equiv \delta_{ij}$), the off-diagonal terms in equation (3.4.15) vanish. For a pure magnetic vacuum the modes are therefore plane-polarized, with $\vec{E}^{w}$ either parallel or perpendicular to the $\vec{B}, \vec{k}$ plane. The corresponding vacuum refractive index is

$$N_{1,2}^2 = \begin{cases} \eta_{xx} = 1 + 7\delta \sin^2\theta, \\ \rho^{-1}\eta_{yy} = 1 + 4\delta \sin\theta. \end{cases} \tag{3.4.18}$$

For $N_{1,2}$ we may expand the square root to first order in $\delta \ll 1$, which leads to the same expressions as equations (2.6.13). In the combined case of vacuum plus cold plasma, the refractive index is then given by

$$N^4 - (B/A)N^2 + C/A = 0, \tag{3.4.19}$$

where

$$B/A = \eta_{xx} + \rho^{-1}\eta_{yy}, \quad C/A = \rho^{-1}\left(\eta_{xx}\eta_{yy} - |\eta_{xy}|^2\right). \tag{3.4.20}$$

The solution of this equation is

$$\begin{aligned} N_{1,2}^2 &= (B/2A) \pm (F/2A) \\ &= \frac{1}{2}\left(\eta_{xx} + \rho^{-1}\eta_{yy}\right) \\ &\quad \pm \frac{1}{2}\left[\left(\eta_{xx} - \rho^{-1}\eta_{yy}\right)^2 + 4\rho^{-1}|\eta_{xy}|^2\right]^{1/2} \\ &= \left[2a^2\left(\sin^2\theta\,\varepsilon + \cos^2\theta\,\eta\right)\right]^{-1} \\ &\quad \times\Big\{\left[\varepsilon\eta + \rho^{-1}\left(\cos^2\theta\,\varepsilon\eta + \sin^2\theta\left(\varepsilon^2 - g^2\right)\right)\right] \\ &\quad \pm\Big[\varepsilon\eta - \rho^{-1}\left(\cos^2\theta\,\varepsilon\eta + \sin^2\theta\left(\varepsilon^2 - g^2\right)\right)^2 \\ &\quad +4\rho^{-1}g^2\eta^2\cos^2\theta\Big]^{1/2}\Big\}, \end{aligned} \tag{3.4.21}$$

where

$$\begin{aligned} &\varepsilon = 1 - va^{-1}\lambda\left(\lambda^2 - u\right)^{-1}, \quad \eta = 1 + a^{-1}\lambda^{-1}(q - v), \\ &g = vu^{1/2}a^{-1}\left(\lambda^2 - u\right)^{-1}. \end{aligned} \tag{3.4.22}$$

We have assumed $\omega, \omega_c \gg \omega_p$ and neglected the proton component, and as before $v = (\omega_p/\omega)^2$, $u = (\omega_c/\omega)^2$, $\lambda = 1 + i\gamma$, while $a, q$ are related to the vacuum factor $\delta$. From equation (3.4.15) we obtain the ratios of the components of the electric field in the wave, defining the polarization eigenmodes

$$i\alpha_x = K_x \equiv E_x/E_y = \left(N^2\rho - \eta_{yy}\right)\eta_{yx}^{-1} = \eta_{xy}\left(N^2 - \eta_{xx}\right)^{-1},$$
$$i\alpha_z = K_z = E_z/E_y = \varepsilon_{zz}^{-1}\left(i\alpha_x - \varepsilon_{zy}\right). \qquad (3.4.23)$$

For the polarization vectors, it is permissible to neglect the imaginary part of the plasma dielectric component; i.e. one can set $\lambda = 1$. From equation (3.4.23), the transverse components of the polarization vector are again

$$\vec{e}^{\,t}_{1,2} = C\left(i\alpha_{x1,2}, 1\right), \qquad (3.4.24)$$

where

$$\alpha_{x1,2} = \rho^{1/2} b\left[1 \pm \zeta\left(1 + b^{-2}\right)^{1/2}\right],$$
$$\zeta = \mathrm{sign}\left(\rho\eta_{xx} - \eta_{yy}\right),$$
$$C = \left(1 + \alpha_{x1,2}^2\right)^{-1/2}. \qquad (3.4.25)$$

To lowest order in $v$ and $\delta$, the ellipticity parameter is now

$$b \equiv \left(\rho\eta_{xx} - \eta_{yy}\right)\left(2\rho^{1/2} i\eta_{yx}\right)^{-1}$$
$$\simeq u^{1/2}\sin^2\theta\left[2\cos\theta\left(1 - v\right)\right]^{-1}\left[1 + \delta 3\left(uv\right)^{-1}\left(1 - u\right)\right]. \qquad (3.4.26)$$

Thus, the effect of the vacuum polarization is to modify the plasma ellipticity parameter $b_0 = u^{1/2}\sin^2\theta[2\cos\theta(1 - v)]^{-1}$ by the vacuum correction factor $V$, in the form

$$b = b_0\left(1 + V\right). \qquad (3.4.27)$$

The numerical value of this factor is

$$V = \delta 3\left(uv\right)^{-1}\left(1 - u\right) = 0.77u^{-2}\left(1 - u\right)n_{20}^{-1}B_{12}^4, \qquad (3.4.28)$$

normalized to a (real) electron density

$$n_{20} = \left(n_e/10^{20} cm^{-3}\right) \qquad (3.4.29)$$

and magnetic field $B_{12} = (B/10^{12}\ \mathrm{G})$. Notice that, depending on whether $u = (\omega_c/\omega)^2$ is greater or smaller than unity, the correction $V$ can be of either sign.

Defining another ellipticity-related quantity $\alpha(\theta)$ by means of

$$\alpha = -b^{-1}\left[1 + \left(1 + b^{-2}\right)^{1/2}\right]^{-1}, \tag{3.4.30}$$

whose modulus is bounded between 0 and 1 (unlike $\alpha_{x1,2}$), we may rewrite the transverse polarization vectors as

$$\vec{e}_1^{\,t} = \left(\rho^2 + \alpha^2\right)^{-1}\left(\rho^{1/2}, i\alpha, 0\right),$$
$$\vec{e}_2^{\,t} = \left(1 + \rho^2\alpha^2\right)^{-1}\left(\rho^{1/2} i\alpha, 1, 0\right) \tag{3.4.31}$$

for $\zeta > 0$, and the same with interchanged indices for $\zeta < 0$. The longitudinal component of the polarization vectors can be found from equation (3.2.27) as

$$\vec{e}^{\,l} = -\varepsilon_{zz}^{-1}\left(\hat{k} \cdot \vec{\vec{\varepsilon}} \cdot \vec{e}_t\right), \tag{3.4.32}$$

so that for $\zeta > 0$ we obtain

$$\vec{e}_1^{\,l} \equiv \lambda_1 = v u^{1/2} \sin\theta \left(1 - u\right)^{-1}$$
$$\times\, \epsilon_{zz}^{-1}\left\{\left[1 + 7\delta\left(1 - u\right)\left(uv\right)^{-1}\right]\alpha u^{1/2} \cos\theta + 1\right\},$$
$$\vec{e}_2^{\,l} \equiv \lambda_2 = i v u^{1/2} \sin\theta \left(1 - u\right)^{-1}$$
$$\times\, \epsilon_{zz}^{-1}\left\{\left[1 + 7\delta\left(1 - u\right)\left(uv\right)^{-1}\right] u^{1/2} \cos\theta - \alpha\right\}, \tag{3.4.33}$$

while for $\zeta < 0$ the indices are again swapped. The complete polarization eigenmodes can therefore be written in rectangular Cartesian coordinates, for $\zeta > 0$, as

$$\vec{e}_1 = \left(\rho^2 + \alpha^2\right)^{-1}\left(\rho^{1/2}, i\alpha, \lambda_1\right),$$
$$\vec{e}_2 = \left(1 + \rho^2\alpha^2\right)^{-1}\left(\rho^{1/2} i\alpha, 1, i\lambda_2\right), \tag{3.4.34a}$$

and, for $\zeta < 0$, as

$$\vec{e}_1 = \left(1 + \rho^2\alpha^2\right)^{-1}\left(\rho^{1/2} i\alpha, 1, i\lambda_2\right),$$
$$\vec{e}_2 = \left(\rho^2 + \alpha^2\right)^{-1}\left(\rho^{1/2}, i\alpha, \lambda_1\right). \tag{3.4.34b}$$

The quantities $\lambda_1$, $\lambda_2$, defined in equations (3.4.33) can be expressed as

$$\begin{pmatrix} \lambda_1 \\ \lambda_2 \end{pmatrix} \cong v u^{1/2} \sin\theta (1-u)^{-1} \times \left[1 - v(1 - u\cos^2\theta)(1-u)^{-1} + q\cos^2\theta\right]^{-1} \times \begin{pmatrix} \left[1 + 7\delta(1-u)(uv)^{-1}\right]\alpha u^{1/2}\cos\theta + 1 \\ \left[1 + 7\delta(1-u)(uv)^{-1}\right]u^{1/2}\cos\theta - \alpha \end{pmatrix}, \quad (3.4.35)$$

where we took

$$\epsilon_{zz} = \hat{k}_\alpha \epsilon_{\alpha\beta} \hat{k}_\beta = 1 - v(1 - u\cos^2\theta)(1-u)^{-1} + q\cos^2\theta,$$

where $q$ is defined in equations (3.4.7). The parameter $\alpha(\theta)$ is the eccentricity of the normal mode ellipse, and is related to $K_x = i\alpha_x$ through

$$\alpha(\theta) = i\alpha_{x1} = -(i\alpha_{x2})^{-1}, \quad |\alpha(\theta)| \leq 1. \quad (3.4.36)$$

One sees that in the limit where vacuum effects vanish $\delta \to 0$, $\rho \to 1$, the previous pure plasma results of equations (3.2.31)–(3.2.33) are regained.

The combined plasma plus vacuum system differs in a qualitative, as well as quantitative, sense from either of the pure cases. This is due to the introduction of the new dimensionless quantity $\delta$ defined in equation (3.4.6), in addition to the previous $u$ and $v$, to characterize the dielectric properties of the plasma. While in the limit $B \ll B_Q$ it is true that $\delta \ll 1$, in the limit $\omega \ll \omega_{pe}$ we also have $v \ll 1$, and the vacuum effects are comparable to the plasma ones. This has a profound effect on the nature of the normal modes, the refractive indices, and the opacities. The most striking changes are observed near two critical frequencies $\omega_{v1}$, $\omega_{v2}$, where the balance between vacuum and plasma influences switches back and forth. At these critical points, in fact, the "normal" modes cease to be such; they are not orthogonal, and in fact one of the modes vanishes. A discussion of nonorthogonality of the normal modes in a cold plasma plus vacuum requires a more careful treatment, including both the real and imaginary parts of the ellipticity parameter $b$ (Pavlov and Shibanov, 1979; Pavlov and Gnedin, 1984) in the neighborhood of the critical points. This is best done after inclusion of thermal effects, also an important factor in the mode structure, and this will be done in the next two sections. Here we restrict ourselves to the approximate (real) expressions for $\alpha(\theta)$ and $b$, valid everywhere except in narrow frequency bands of order $\gamma$ (the damping constants) around $\omega_{v1}, \omega_{v2}$. Figure 3.4.1a shows a plot of $\alpha(\theta)$ in the absence of vacuum for various values of $\omega/\omega_c = u^{-1/2}$. The typical behavior for such a plasma is different at $\omega$ much less or much greater than $\omega_c$. For $\omega \ll \omega_c$, the modes are nearly linearly polarized at most angles ($b \to 0$, $\alpha \to 0_-$) except close to $\theta = 0°$, $\sin\theta \to 0$, for which they are circular ($b \to 0, \alpha \to -1$) (see eqs. 3.4.26, 3.4.30). For $\omega \gg \omega_c$, the modes are circularly polarized at most angles ($b \to 0$, $\alpha \to -1$) except near $\theta = 90°$, $\cos\theta \to 0$ ($b \to \infty, \alpha \to 0_-$), where they become linear. For intermediate values, including $\omega \sim \omega_c$, the modes are elliptical. In Figure 3.4.1b we have plotted $\alpha(\theta)$ including vacuum polarization for one particular frequency $\omega = 0.5\omega_c$, letting the plasma density $n_e \propto (\omega_p/\omega_c)^2$ vary. For large $n_e$, the vacuum correction factor $V$ in equation (3.4.28) is small and $b$ is close to $b_0$, the plasma value plotted in Figure 3.4.1a. For decreasing $n_e$, the modulus of the vacuum correction $V$ increases, and since $u = (\omega_c/\omega)^2 = 2$, it has negative sign. Thus, $b$ decreases and eventually tends to zero as $V \to -1$, where $\alpha \to -1$. Thus, as the contribution of the plasma decreases relative to that of the vacuum ($\omega_c$ is fixed), the modes change from being elliptical to becoming close to circular at most angles. As we further decrease the plasma density, $V$ becomes less than $-1$, i.e. $b$ suddenly becomes negative and small, $b \sim 0_-$, so that $\alpha \to +1$; the modes are circular but now with the opposite sense of gyration. A further decrease of $n_e$ causes $V \to -\infty$, or $b \to -\infty$, and therefore $\alpha \to 0$ at most angles, i.e. linear polarization, except for $\theta \to 0°$, where $b \to 0$ and $\alpha \to +1$ leads to circular polarization. The vacuum, if unmodified by plasma contributions, pushes the modes toward being linearly polarized, except very close to the longitudinal propagation direction. The physical reason for this is simple: the vacuum consists of equal amounts of positively and negatively charged electrons, which gyrate in the magnetic field in opposite senses. The induced polarization is therefore free of handedness, i.e. linear. It is the presence of a charge imbalance, in the form of (real) plasma electrons, that causes an ellipticity of the normal modes.

It is also interesting to consider fixed $n_e$, $B$, and let the frequency vary. Looking at equations (3.4.26) and (3.4.27) for $\omega \ll \omega_c$ or $u \gg 1$, the vacuum correction is of order $V \sim -3\delta v^{-1}$, and for $\delta \ll v$, the modes are plasma dominated. As $\omega$ increases ($u$ decreases but is still

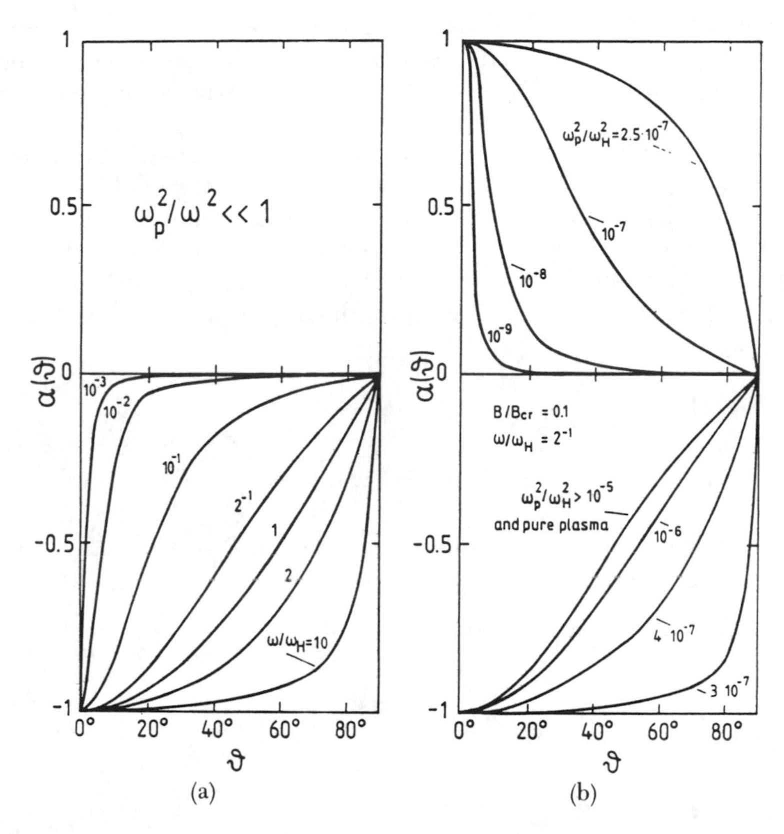


FIGURE 3.4.1 Ellipticity parameter $\alpha(\theta)$ as a function of $\theta$. (a) The case of negligible vacuum polarization for various frequencies. (b) The ellipticity parameter for the plasma plus vacuum case at a particular frequency for a variable plasma density. From P. Mészáros and J. Ventura, 1979, *Phys. Rev.*, D19, 3565.

large) one reaches a first critical frequency

$$\omega_{v1} \sim 3\ \text{keV}\left(n_e/10^{22}\ \text{cm}^{-3}\right)^{1/2}\left(B/0.1B_Q\right)^{-1} \tag{3.4.37}$$

at which $b(\omega, \theta)$ goes through zero and changes sign. Above $\omega_{v1}$, the quantity $V$ dominates the unity in the expression for $b$, and the modes are vacuum dominated for $\omega_{v1} \lesssim \omega \lesssim \omega_{v2}$. As $\omega$ is further increased, a second critical vacuum frequency is reached,

$$\omega_{v2} \sim \omega_c, \tag{3.4.38}$$

where the vacuum and plasma effects cancel. Above $\omega_{v2}$, plasma dominance partially reasserts itself, while at higher frequencies, one needs to consider the higher harmonics, which will be discussed in Chapter 5.

The effects of the vacuum on the real part of the refractive index are equally important. The expression for the refractive index $N_{1,2}$ is given by equation (3.4.21), and the vacuum changes are caused by the changes of the eigenvectors $e_{1,2}$ through $\alpha(\omega,\theta)$ or $b(\omega,\theta)$. An explicit expression for $N$ as a function of $\vec{e}_{1,2}$ may be obtained from the wave equation (3.4.9). Projecting this expression onto the transverse vector $\hat{e}_t = (\overleftrightarrow{1} - \hat{k}\hat{k}) \cdot \hat{e}$, we find

$$N^2\left[1 - 4\delta a^{-1}\left(\hat{e}_t, \hat{\xi}\hat{\xi} \cdot \hat{e}\right)\right]$$
$$= a^{-1}\left(\hat{e}_t, \left[\overleftrightarrow{\epsilon}^{(p)} - \overleftrightarrow{1} + a\overleftrightarrow{1} + 7\delta\hat{b}\hat{b}\right] \cdot \hat{e}\right), \tag{3.4.39}$$

where $\hat{\xi} \equiv \hat{k} \times \hat{b}$ and $\hat{e}$ is a normal mode. To first order in $\delta$, this gives

$$N^2 = 1 - v\left(\hat{e}_t, \overleftrightarrow{\Pi}^{(p)} \cdot \hat{e}\right) + 4\delta\left(\hat{e}_t, \hat{\xi}\hat{\xi} \cdot \hat{e}\right) + 7\delta\left(\hat{e}_t, \hat{b}\hat{b} \cdot \hat{e}\right), \tag{3.4.40}$$

where

$$v\overleftrightarrow{\Pi}^{(p)} = \overleftrightarrow{1} - \overleftrightarrow{\epsilon}^{(p)} \equiv -\overleftrightarrow{\chi}^{(p)} \tag{3.4.41}$$

is the plasma polarization tensor. The last two terms of equation (3.4.40) are real so that the only imaginary part of $N$ is due to $(\hat{e}_t, \overleftrightarrow{\Pi}^{(p)} \cdot \hat{e})$, the plasma polarizability. This is because, for $\hbar\omega < mc^2$, the vacuum polarization diagram (Fig. 2.6.1b) does not lead to the destruction or creation of a photon, only to a modification of its phase velocity and polarization, i.e. to a modification of the real part of the refractive index. On the other hand, in the plasma part of the dielectric tensor, $\overleftrightarrow{\epsilon}^{(p)}$, an imaginary part of it was included formally to represent the effect of "absorption," due to either scattering out of the beam, free-free absorption, or other processes. Using the definitions

$$K_j(\omega,\theta) = E_x^j/E_y^j = b\left[1 + (-1)^j\left(1 + b^{-2}\right)^{1/2}\right],$$
$$\alpha(\omega,\theta) = K_1 = -K_2^{-1}, \quad |\alpha(\omega,\theta)| \le 1, \tag{3.4.42}$$

we may define the normal modes in the system of coordinates I (Fig. 3.2.3) with $z$ along $\vec{B}$. It is useful furthermore to take rotating coordinates

$$e_{\pm} = e_x \pm ie_y, \tag{3.4.43}$$

in which representation we have

$$\vec{e}_j = \left(e_+, e_-, e_z\right) \tag{3.4.44}$$

with

$$\begin{aligned} e^j_{\pm} &= 2^{-1/2} C_j e^{\mp i\varphi}\left[K_j \cos\theta \pm 1\right], \\ e^j_z &= C_j K_j \sin\theta, \end{aligned} \tag{3.4.45}$$

where $C_j = (1 + K_j^2)^{-1/2}$. With these eigenvectors, we can easily evaluate the expression (3.4.40) to yield

$$\begin{aligned} N_j^2 - 1 = -v\left(1 + K_j^2\right)^{-1}\Bigg[&\left(K_j \sin\theta\right)^2\left(1 + i\gamma\right)^{-1} \\ &+ \frac{1}{2}\left(1 - K_j\cos\theta\right)^2\left(1 - u^{1/2} + i\gamma\right)^{-1} \\ &+ \frac{1}{2}\left(1 + K_j\cos\theta\right)^2\left(1 + u^{1/2} + i\gamma\right)^{-1} \\ &- \delta v^{-1}\left(4 + 7K_j^2\right)\sin^2\theta\Bigg]. \end{aligned} \tag{3.4.46}$$

The first three terms in the square brackets are the same as in the vacuumless case if one sets the vacuum correction in the $K_j$ equal to zero (the $V \propto \delta$ correction comes in via the parameter $b$, e.g. eq. 3.4.22). The last term in equation (3.4.46) occurs only in the presence of the vacuum, as evidenced by its explicit proportionality to $\delta$. The pure-plasma limit follows from setting $\delta \equiv 0$. The pure-vacuum limit (see eqs. 2.6.13) also follows from equation (3.4.46) when $v \to 0$. In this case, from equation (3.4.26) we have $b \to 0$ or $K_1 \to 0$, $K_2 \to \infty$, $\alpha \to 0$, i.e. linearly polarized modes with

$$\begin{aligned} N_1 &\to \left(1 + \delta 4 \sin^2\theta\right)^{1/2} \sim 1 + \delta 2 \sin^2\theta, \\ N_2 &\to \left(1 + \delta 7 \sin^2\theta\right)^{1/2} \sim 1 + \delta\left(7/2\right)\sin^2\theta. \end{aligned} \tag{3.4.47}$$

A plot of the real part of the refractive index (3.4.46) is shown in Figure 3.4.2. The dashed lines labeled P1, P2 represent the usual magnetoionic modes (X, O) in the absence of vacuum (see Fig. 3.2.1).

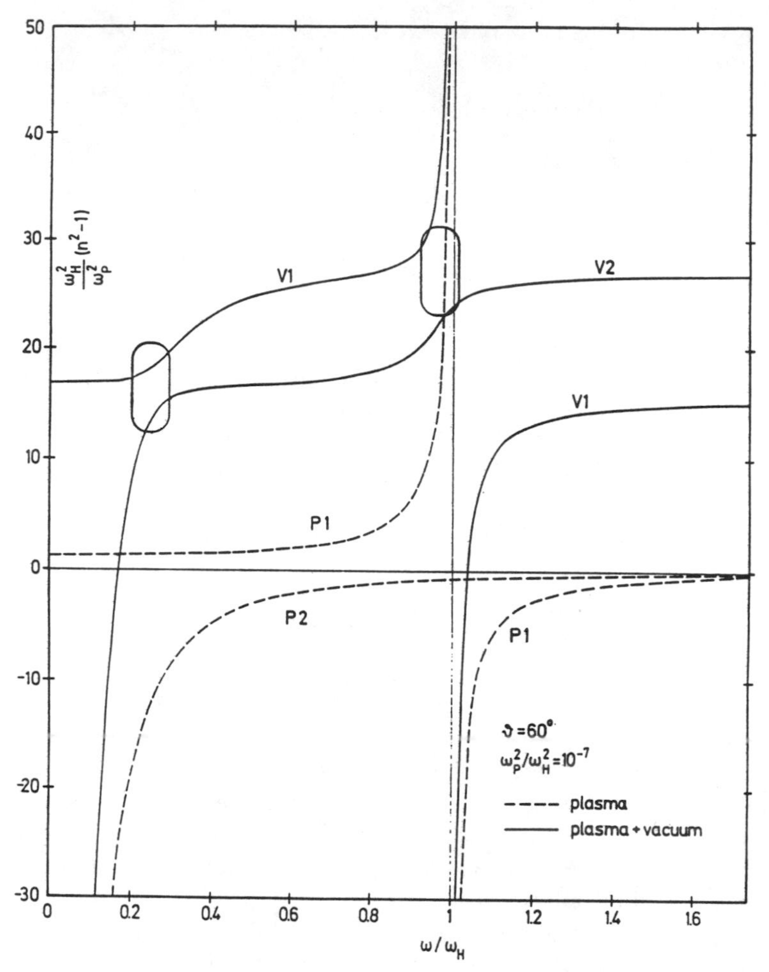


FIGURE 3.4.2 Square of the real part of the refractive index for a particular choice of $(\omega_p/\omega_c)^2 = 10^{-7}$ and a propagation angle of $\theta = 60°$ in the cold-plasma case. The curves P1, P2 represent the cold plasma without vacuum polarization effects, while V1, V2 represent the cold-plasma modes including vacuum polarization. The elliptical closed curves indicate the regions where the mode definition is ambiguous, around the vacuum frequencies $\omega_{v1}$ and $\omega_{v2} \simeq \omega_c$. From P. Mészáros and J. Ventura, 1979, *Phys. Rev.*, D19, 3565.

The full lines marked V1, V2 are the corresponding modes, including vacuum corrections, for a particular value of $(\omega_p/\omega_c)^2$ as might be expected in neutron stars. The change in behavior is large not only quantitatively but also qualitatively, especially in the regions near the critical frequencies $\omega_{v1}$, $\omega_{v2}$, where the two modes approach each other or where, for some values of $\omega_p, \theta$, they even cross each other. In these regions, indicated by an oval in Figure 3.4.2, there is an indeterminacy in the definition of the modes and an increase in the probability of waves transforming into each other by nonlinear processes.

### 3.5 Thermal and Quantum Effects in the Nonrelativistic Limit

So far, we have considered the medium response $\vec{\epsilon}^{(p)}$ to be given by the cold-plasma approximation. As discussed in § 3.3, this is obtained from the quantum mechanical $\vec{\epsilon}^{(p)}$ by letting $\hbar k, p_z \to 0$, i.e. by neglecting thermal motions (represented by $p_z$) and quantum effects (proportional to $\hbar$) in the "real" plasma. We did, however, include quantum effects in the virtual plasma $\vec{\epsilon}^{(v)}$. This approximation is only consistent over part of the frequency range that interests us. Clearly, what we want to do now is to include quantum effects in the plasma response $\vec{\epsilon}^{(p)}$. However, $p_z$ and $\hbar k$ appear in $\vec{\epsilon}^{(p)}$ (eq. 3.3.39) on an equal footing; it will be necessary to include both and to average $p_z$ over $f_\alpha$, the particle distribution function, where $\alpha$ is the set of quantum numbers $(n, p_z)$ characterizing a particle state. The sums over the guiding center quantum number $s$ were already performed in obtaining equation (3.3.39), and for the present purposes spin contributions are neglected, although they can easily be added. In a fully relativistic treatment, which is given in Chapter 5, the spin variable enters naturally, and will be discussed there. The treatment below (following Kirk, 1980) is nonrelativistic, with first-order thermal and quantum terms (see also Pavlov, Shibanov, and Yakovlev, 1980; Kirk and Mészáros, 1980).

For the renormalized susceptibility tensor $\chi_{ij} = \epsilon_{ij} - \delta_{ij}$, we have from equation (3.3.39)

$$\chi_{ij} = -v\left[\delta_{ij} + \sum_{\alpha\alpha'}\left(\frac{F_i^*F_j}{D_1} - \frac{\tilde{F}_i^*\tilde{F}_j}{D_2}\right)f_\alpha\right]. \tag{3.5.1}$$

The matrix elements $F_i$ are given in equation (3.3.34) and depend on the functions

$$I_{nn'}(x^2) = (-1)^{n'}(n!n'!)^{-1/2}\exp(-x^2/2)\,x^{n-n'}Q_n^{n-n'}(x^2), \tag{3.5.2}$$

where

$$x = k\sin\theta\left(Bm^2c^2/B_Q\hbar^2\right)^{-1/2} = (\omega/\omega_c)^{1/2}(\hbar\omega/mc^2)^{1/2}\sin\theta \tag{3.5.3}$$

and the $Q_n^{n-n'}$ are Laguerre polynomials, defined by

$$Q_n^{n-n'}(t) = \sum_{k=0}^{\min(n,\,n')}(-1)^{n'-k}\frac{n'!n!}{k!(n'-k)!(n-k)!}t^{n'-k}. \tag{3.5.4}$$

Since $x \ll 1$, we may expand the $I_{nn'}$, in powers of $x$,

$$\begin{aligned} I_{nn'}(x) = {} & \delta_{nn'}\left[(-1)^{n'}(1 - x^2/2\ldots)\right. \\ & \left.+(-1)^{n'+1}(1 - x^2/2\ldots)n'x^2\right] \\ & + \delta_{n,\,n'+1}\left[(1 - x^2/2\ldots)nx(-1)^{n'}\right] \\ & + \delta_{n,\,n'-1}\left[(1 - x^2/2)n'x(-1)\right] + \cdots \end{aligned} \tag{3.5.5}$$

We can then compute the elements of $\chi_{ij}$ in equation (3.5.1), using $F^*$ for the conjugate of $F$, $\tilde{F}$ for the transpose of $F$, which implies $n \leftrightarrow n'$ for $i, j = 1, 2$, and $\tilde{F}_3 = (\hbar m)^{-1/2}(p_z - \hbar k_\parallel/2)I_{n'n}$ for the $i, j = 3$ component. We have then

$$\vec{\chi}^{(p)} = \begin{pmatrix} S & -iD & iA \\ iD & S & B \\ -iA & B & P \end{pmatrix} + O(x^2). \tag{3.5.6}$$

The terms $A$ and $B$ are the first order in $x$, and may be neglected for fields $B \ll B_Q$. The remaining terms are

$$S = -v\left\{1 + \frac{\omega_c}{2}\sum_\alpha f_\alpha\left[\left(\frac{1}{F^-} - \frac{1}{F^+}\right)\right.\right. \\ \left.\left. + \frac{n\hbar k^2\cos^2\theta}{m}\left(\frac{1}{F^+G^+} + \frac{1}{F^-G^-}\right)\right]\right\},$$

$$D = -v\frac{\omega_c}{2}\sum_\alpha f_\alpha\left[\left(\frac{1}{F^-} + \frac{1}{F^+}\right) + \frac{n\hbar k^2\cos^2\theta}{m}\left(\frac{1}{F^-G^-} - \frac{1}{F^+G^+}\right)\right],$$

$$P = -v\left\{1 + (\hbar m)^{-1/2} \times \sum_\alpha f_\alpha\left[\frac{(p_z + \hbar k\cos\theta/2)^2}{E^-} - \frac{(p_z - \hbar k\cos\theta/2)^2}{E^+}\right]\right\}, \tag{3.5.7}$$

where

$$F^\mp = \omega \mp \omega_c \mp \hbar k^2\cos^2\theta/2m - p_z k\cos\theta/m,$$

$$G^\mp = \omega \pm \omega_c \pm \hbar k^2\cos^2\theta/2m - p_z k\cos\theta/m,$$

$$E^\mp = \omega \mp \hbar k^2\cos\theta/2m - p_z k\cos\theta/m. \tag{3.5.8}$$

The tensor (3.5.6) contains first-order quantum and relativistic terms, unlike in the cold-plasma expansion of equations (3.3.41) to (3.3.47). The Doppler and recoil effects are represented by terms in $p_z k$ and $k^2$ in the susceptibility (3.5.6). The sums over $f_\alpha$ in equations (3.5.7) indicate that the plasma susceptibility is the superposition of the susceptibilities for the individual Landau levels, i.e. electrons in different levels do not interact with each other. This is natural, since the dielectric tensor represents the coherent forward scattering of a wave, i.e. scatterings with $\omega' = \omega$, $\theta' = \theta$, and therefore no energy or momentum is given or taken from the electrons, at least if the damping is set equal to zero as in equation (3.5.6).

*a) Very Strong Field Limit*

We now look at the limiting case where electrons are in the lowest Landau level $n = 0$ (i.e. $kT \ll \hbar\omega_c$, as expected for accreting X-ray pulsars). Let us assume the electrons to be Maxwellian in the longitudinal direction,

$$f_\alpha \equiv \delta_{n,0}(2\pi mkT)^{-1/2}\exp(-p_z^2/2mkT)\,dp_z, \tag{3.5.9}$$

so that the susceptibility tensor will be averaged over this one-dimensional thermal distribution. One then obtains

$$S = -v\left[1 - \frac{1}{2}\omega_c(2kT/m)^{-1/2}|k\cos\theta|^{-1} \times \{W(y_-) - W(y_+)\}\right],$$

$$D = v\frac{1}{2}\omega_c(2kT/m)^{-1/2}|k\cos\theta|^{-1}\{W(y_+) + W(y_-)\}],$$

$$P = -v(z^2/2a)[W(z+a) - W(z-a)], \tag{3.5.10}$$

in which

$$a = \left|\frac{\hbar k\cos\theta}{2m\sqrt{2}\,s}\right|, \quad y_\pm = \frac{\omega \pm \omega_c}{\sqrt{2}\,s|k\cos\theta|} \pm a,$$

$$z = \frac{\omega}{\sqrt{2}\,s|k\cos\theta|}, \tag{3.5.11}$$

where $s = (kT/m)^{1/2}$ is the electron thermal velocity and $W$ is the plasma dispersion function (Fried and Conte, 1961). For $\mathrm{Im}(z) > 0$, the integral representation for the latter is

$$W(z) = \pi^{-1/2}\int_{-\infty}^{\infty} dx\,(x-z)^{-1}\exp(-x^2), \tag{3.5.12}$$

while for $\mathrm{Im}(z) \leq 0$ it is the analytic continuation of the above. One can once more verify that, in the limit $(k\cos\theta) \to 0$, equations (3.5.10) reproduce the cold-plasma limit, while, in the form (3.5.10), the susceptibility contains first-order quantum and thermal effects, in the nonrelativistic limit.

With the thermal and quantum effects included in the plasma susceptibility tensor $\overleftrightarrow{\chi}^{(p)}$, the wave equation can be written

$$\left(\overleftrightarrow{\Lambda} + \overleftrightarrow{\chi}^{(p)} + \overleftrightarrow{\Psi}\right)\cdot\vec{E} = 0, \tag{3.5.13}$$

where $\overleftrightarrow{\Lambda}$ is the Maxwell propagation tensor (3.4.11), $N^2 = c^2k^2/\omega^2$ is the refractive index, and $\overleftrightarrow{\Psi}$ accounts for the magnetic vacuum polarizability (see eq. 3.4.13). In the cold-plasma case, the lack of damping terms in $\overleftrightarrow{\chi}^{(p)}$ would lead to a purely real refractive index (no absorption). Absorption of the waves is mimicked, in the linear one wave–one particle approximation implicit in equation (3.5.13), by a phenomenological damping term which is introduced by replacing

$$v \to v(1 + i\gamma)^{-1}, \quad \omega_c \to \omega_c(1 + i\gamma)^{-1}, \tag{3.5.14}$$

where for radiative damping we put

$$\gamma_r = \frac{2}{3}\frac{e^2\omega}{mc^3}. \tag{3.5.15}$$

This wave damping represents radiation scattered out of the beam so that the imaginary part of the refractive index gives the Thomson scattering cross section. While in the cold-plasma case the imaginary part is given by a delta function in frequency, in the hot-plasma case one has, even in the absence of a damping such as that of equation (3.5.15), an imaginary part to the susceptibility and to the refractive index near the resonances due to purely thermal effects (via the plasma dispersion function).

It is useful to look at the range of validity of the various approximate descriptions of the medium response used so far. The cold-plasma approximation of equations (3.1.27) or (3.3.47) is valid for

$$\begin{aligned} &v \ll 1, \quad v|1 - u|^{-1} \ll 1, \\ &\omega \gg \omega_{pe}, \quad |\omega^2 - \omega_c^2| \gg \omega_p^2, \end{aligned} \tag{3.5.16}$$

provided we neglect all quantum effects in the medium. However, if we take the quantum response of equations (3.5.10) and take the cold-plasma limit $s = \sqrt{kT/m} \to 0$, we get

$$\begin{aligned} &S \to -v(1 - u')\left[1 - u'\{1 - (u/u')^{1/2}\}\right], \\ &D \to -vu^{1/2}(1 - u')^{-1} \\ &P \to -v\omega^2(k\cos\theta)^{-2}\left[\omega^2(k\cos\theta)^{-2} - \hbar^2k^2\cos^2\theta(4m)^{-1}\right], \end{aligned} \tag{3.5.17}$$

where $u' = [\omega_c^2 + \hbar k^2\cos^2\theta(2m)^{-1}]$. In order to obtain the classical cold-plasma response of equations (3.1.27) or (3.3.47), we need to take

$$\omega \gg \hbar k^2\cos^2\theta(2m^{-1}). \tag{3.5.18}$$

This shows that the region of invalidity near $\omega_c$ of the classical cold-plasma approximation is larger than indicated by the last equation of (3.5.16). For the classical cold-plasma response to be valid, we need to satisfy

$$|\omega - \omega_c| \gg \omega_c(\hbar\omega/mc^2), \tag{3.5.19}$$

as well as the nonrelativistic condition

$$\omega_p \ll \omega \ll \hbar^{-1}mc^2. \tag{3.5.20}$$

This condition means that our description of the refractive index and of the polarization eigenvectors in § 3.4 is valid only at frequencies away from the cyclotron resonances, excluding a band of width given by equation (3.5.19). Within this band, the angular dependence of the polarization vectors is no longer of the simple form given by equations (3.2.31) or (3.4.34), since the frequency of the resonance is a function of angle, as can be seen from a look at the resonant denominators of $\vec{\chi}^{(p)}$ in equations (3.5.8). Thermal effects are large whenever $y_{\pm} \ll 1$ or $z \ll 1$ in equations (3.5.10), (3.5.11), i.e. whenever the (angle-dependent) resonant frequency $\omega_R(\sim \omega_c)$ and the wave frequency satisfy

$$|\omega_R - \omega| \lesssim (kT/m)^{1/2}(\omega/c)|\cos\theta|. \tag{3.5.21}$$

For instance, if $\omega_c = 55$ keV and $kT = 10$ keV, this is a band extending from about 45 to 65 keV, while for $kT = 50$ keV, it extends over 35 to 75 keV. The thermal effects on the refractive index $N$ and polarization vectors $\vec{e}$ can be incorporated into the formalism described in the previous section. It will be remembered that, for a vacuum-corrected cold plasma, $N$ and $\vec{e}$ could be expressed to lowest order in $v$ and $\delta$ in terms of the parameter

$$b = \sin^2\theta(2\cos\theta)^{-1}u^{1/2}[1 + 3\delta(1-u)v^{-1}u^{-1}]. \tag{3.5.22}$$

In a similar way, to lowest order in $v/s$ and $\delta$, the thermal and vacuum effects can be incorporated according to

$$b = -\sin^2\theta(2\cos\theta)^{-1}D^{-1}[P - S + 3\delta], \tag{3.5.23}$$

where $S$, $P$, $D$ are given in equations (3.5.10). The refractive index becomes

$$N^2 - 1 = S + 4\delta\sin^2\theta - D\cos\theta\left[b \pm (1+b^2)^{1/2}\right]. \tag{3.5.24}$$

The normal modes for the two high-frequency waves are again

$$e_{1,2} = [iK(b), 1, 0(v)]_{xyz} \tag{3.5.25}$$

in the frame with $z$ along $\vec{k}$, but now $b$ is given by equation (3.5.23), and

$$K_{1,2}(b) = b \pm (1 + b^2)^{1/2}. \tag{3.5.26}$$

Unlike in equations (3.4.25), we adopt here a unique definition of $\alpha$, without the dependence on $\zeta = \text{sign}(\rho\eta_{xx} - \eta_{yy})$. This is an alternative definition of the normal modes, which, unlike in equations (3.4.25), gives a smooth change of the refractive index as $\omega$ moves across the critical frequencies $\omega_{v1}, \omega_{v2}$. Qualitatively, the biggest difference is that in $b$, for the thermal case, we must take into account both the real and imaginary parts, the latter being significant near the resonances and critical frequencies, e.g. near $\omega_{v1}$ and $\omega_{v2} \simeq \omega_c$ or its multiples. In fact, even in the cold-plasma approximation there are narrow bands of relative width proportional to the radiative width $\gamma_r \ll 1$, where $\text{Im}(b) \neq 0$ while $\text{Re}(b) \simeq 0$ (e.g. Pavlov and Shibanov, 1979). In the hot-plasma case, the bands where $\text{Im}(b)$ is significant are, however, large enough that in most circumstances this effect cannot be ignored.

*b) Weak-Field Limit and Classical Correspondence*

In the weak-field limit the number of Landau levels which are populated is large, and it is possible to approximate various sums as integrals (Canuto and Ventura, 1972). It is useful to go back to an earlier expression (3.3.26) for the matrix $\tau_{ij}$, related to $\epsilon_{ij}$ and $\chi_{ij}$. The summations over $f_\alpha$ and over the quantum numbers $\alpha$, $\beta$ can also be extended to include, if desired, the quantum spin number $\zeta = \pm 1$. The inclusion of this modifies the energy spectrum from $E_n(p_z) = \hbar\omega_c(n + \frac{1}{2}) + p_z^2/2m$ to

$$E_{n,p_z} = p^2/2m + n\hbar\omega_c. \tag{3.5.27}$$

The level degeneracy due to the spin is

$$\begin{aligned} a_n &= 1 \qquad \text{for } n = 0, \\ a_n &= 2 \qquad \text{for } n > 0. \end{aligned} \tag{3.5.28}$$

The density of the electrons is given by

$$N_e = V^{-1} \sum_{n,\,p_z,\,s,\,\zeta} f_{np_zs\zeta} = \frac{m\omega_c}{(2\pi\hbar)^2} \sum_{n=0}^{\infty} a_n \int_{-\infty}^{\infty} dp_z f(E_{np_z}), \tag{3.5.29}$$

where $s$ is the quantum number associated with the guiding center variable and $V$ is a macroscopic normalization volume. Using equations (3.3.24) to (3.3.29), one can rewrite equation (3.3.26) as

$$\tau_{ij} = N_e^{-1} \sum_{mnp_z s\zeta} \frac{(t_{ij})_{mn}}{\omega - \omega_{mn}(p_z, p_z + \hbar k_z)} \left[ f(E_{np_z}) - f(E_{mp'_z}) \right], \tag{3.5.30}$$

where

$$(t_{ij})_{mn} = (m\hbar)^{-1} \langle n | \pi_i e^{-ik_\perp r} | m \rangle \langle m | \pi_i e^{ik_\perp r} | n \rangle,$$

$$\omega_{mn}(p_z, p_z \pm \hbar k) = (m - n)\omega_c \pm \frac{p_z}{m} k_z + \frac{\hbar k_z^2}{2m} \tag{3.5.31}$$

and $k_\perp \equiv k_y$, where, as in § 3.3, we took $k$ to lie in the $y$, $z$ plane. Using equations (3.3.27) to (3.3.29), we have for the matrix elements

$$t_{xx} = \frac{1}{2}\omega_c | I_{mn}^{(-)} |^2, \quad t_{yy} = \frac{1}{2}\omega_c \left( I_{mn}^{(+)} + ak_\perp 2^{-1/2} I_{mn} \right)^2,$$

$$t_{xy} = i\frac{1}{2}\omega_c \left( I_{mn}^{(+)} + ak_\perp 2^{-1/2} I_{mn} \right) I_{mn}^{(-)}, \quad t_{xy} = -t_{yx},$$

$$t_{zz} = (m\hbar)^{-1} \left( p_z + \frac{1}{2}\hbar k_z \right)^2 I_{mn}^2,$$

$$t_{zx} = -t_{xz} = -i(\omega_c / 2m\hbar)^{1/2} \left( p_z + \frac{1}{2}\hbar k_z \right) I_{mn} I_{mn}^{(-)},$$

$$t_{zy} = t_{yz} = (\omega_c / 2m\hbar)^{1/2} \left( p_z + \frac{1}{2}\hbar k_z \right) I_{mn} \left( I_{mn}^{(+)} + ak_\perp 2^{-1/2} I_{mn} \right), \tag{3.5.32}$$

where

$$I_{mn} \equiv I_{mn}\left( \frac{1}{2} a^2 k_\perp^2 \right), \quad I_{mn}^{(\pm)} = n^{1/2} I_{m,n-1} \pm (n+1)^{1/2} I_{m,n+1}, \tag{3.5.33}$$

and

$$a^2 = (1/2\gamma) = \lambda_c^2 (B/B_Q)^{-1} = (\hbar / m\omega_c). \tag{3.5.34}$$

The integral over $p_z$ in equation (3.5.30) can be performed using the convention

$$\lim_{\eta\to 0} \frac{1}{\omega \pm \omega_{mn} + i\eta} = P\frac{1}{\omega \pm \omega_{mn}} - i\pi\delta(\omega \pm \omega_{mn}). \tag{3.5.35}$$

In this manner, the principal value $P$ of the integral gives the real (Hermitian) part of the dielectric tensor giving rise to the refractive properties, while the integral over the delta function in equation (3.5.35) gives rise to the absorption.

To proceed to the classical limit of large quantum numbers, one uses the fact that the discrete spectrum approaches a continuum. One can transform the quantum expression into classical ones by using the correspondence

$$\begin{aligned}\sum_{np_z s\zeta} &= m\omega_c(2\pi\hbar)^{-2}\sum_n a_n \int_{-\infty}^{\infty} dp_z \to \\ &2m^2\omega_c(2\pi\hbar)^{-2}\int_0^{\infty} dn \int_{-\infty}^{\infty} dv_z.\end{aligned} \tag{3.5.36}$$

One can then use

$$2m^{-1}\hbar\omega_c n \simeq v_\perp^2 = v_x^2 + v_y^2$$

and

$$\int dv_\perp^2 = \pi^{-1}\iint dv_x\, dv_y,$$

so that

$$m\omega_c(2\pi\hbar)^{-2}\sum_n a_n\int_{-\infty}^{\infty} dp_z \to 2m^3(2\pi\hbar)^{-3}\int d^3v. \tag{3.5.37}$$

Equation (3.5.29) becomes then $N_e = 2m^3(2\pi\hbar)^{-3}\int d^3v f(\vec{v})$. In equation (3.5.30) one can take in this limit

$$\begin{aligned}E_{mp_z'} - E_{np_z} \equiv \Delta E &= \hbar[l\omega_c + k_z v_z(1 + \hbar k_z/2mv_z)] \\ &\simeq \hbar[l\omega_c + k_z v_z],\end{aligned} \tag{3.5.38}$$

where

$$l \equiv m - n, \quad \vec{v} = \vec{\pi}/m, \quad p_z' = p_z + \hbar k_z. \tag{3.5.39}$$

Assuming that $l \ll m$, $\hbar k_z \ll p_z$, we have also

$$f(E_{mp'_z}) - f(E_{np_z}) \simeq \Delta f = \hbar m^{-1}\left(l\omega_c v_\perp^{-1}\,\partial f/\partial v_\perp + k_z\,\partial f/\partial v_z\right), \tag{3.5.40}$$

so that equation (3.5.30) becomes

$$\tau_{ij} = 2m^3 N_e^{-1}(2\pi\hbar)^{-3}\sum_{l=-\infty}^{\infty}\int d^3v$$

$$\times\left(l\omega_c v_\perp^{-1}\,\partial f/\partial v_\perp + k_z\,\partial f/\partial v_z\right)\frac{\hbar m^{-1}t_{ij}}{\omega - l\omega_c - k_z v_z} \tag{3.5.41}$$

For the matrix elements $t_{ij}$ we may use an asymptotic expansion for $n \gg 1$

$$I_{mn}\left(\frac{1}{2}a^2k_\perp^2\right) \simeq J_l(ak_\perp 2^{1/2}n) = J_l(k_\perp v_\perp/\omega_c), \tag{3.5.42}$$

where the $J_l(\rho)$ are Bessel functions. Making use of the identities $J_{l-1} + J_{l+1} = 2(lJ_l/\rho)$ and $J_{l-1} - J_{l+1} = 2(\partial J_l/\partial\rho)$, it is seen that for $n \gg 1$

$$(\hbar\omega_c/2m)^{1/2} I_{mn}^{(-)} \simeq -v_\perp J_l'$$

$$(\hbar\omega_c/2m)^{1/2}\left(I_{mn}^{(+)} + ak_\perp 2^{-1/2}I_{mn}\right) \simeq v_\perp(lJ_l/\rho)\left[1 + a^2k_\perp^2/2l\right], \tag{3.5.43}$$

where the second term in the brackets of the second equation has been dropped, a good approximation at all but the shortest wavelengths. One gets finally

$$\hbar m^{-1}t_{ij} = \begin{pmatrix} v_\perp^2 J_l'^2 & -iv_\perp^2(lJ_l/\rho)J_l' & -iv_\perp v_z J_l J_l' \\ iv_\perp^2(lJ_l/\rho)J_l' & v_\perp^2(lJ_l/\rho)^2 & v_\perp v_z(lJ_l/\rho)J_l \\ iv_\perp v_z J_l J_l' & v_\perp v_z(lJ_l/\rho)J_l & v_z^2 J_l^2 \end{pmatrix}, \tag{3.5.44}$$

where $\rho \equiv k_\perp v_\perp/\omega_c$, $J_l \equiv J_l(\rho)$, $J_l' \equiv dJ_l/d\rho$. One can verify that equations (3.5.41), (3.5.44) are the same as those derived by classical kinetic theory (e.g. Sitenko, 1964). Notice that in equation (3.5.44) we have assumed $k$ in the $y$, $z$ plane. For $k$ in the $x$, $z$ plane, one just interchanges the $x$- and $y$-components.

## 3.6 Validity of the Normal Mode Description

*a) Normal Mode Ambiguities*

The advantage of a treatment of the radiation field based on the normal modes, as will be seen in the next chapter, is that the radiative cross sections are strongly dependent on the polarization of the wave. On the other hand, the direction of the polarization vector of a wave rotates as it propagates (Faraday effect). In calculating mean free paths, escape probabilities, or radiative transfer in general, one would be faced with the problem of dealing with cross sections that vary with the polarization along the photon path. This can be dealt with by calculating the transport of four quantities, the Stokes parameters of the wave, which makes the treatment considerably more cumbersome. On the other hand, for a sufficiently low density plasma, a photon suffers many rotations between scatterings, a condition called Faraday depolarization. This occurs when the refractive indices for the normal eigenmodes satisfy

$$|\mathrm{Re}(N_1 - N_2)| \gg \mathrm{Im}(N_1 + N_2) \tag{3.6.1}$$

(e.g. Gnedin and Pavlov, 1974). In this case, and if the two normal modes are orthogonal to each other, the whole transfer problem may be considered to be one of the transport of the two (transverse) normal mode intensities. The corresponding cross sections are the normal mode cross sections, which remain constant along the path of the photon for a given frequency and direction (provided also the plasma is homogeneous). This greatly simplifies the analysis, since only two quantities need to be calculated, $I_1$ and $I_2$, rather than four. The conditions of validity of the approach using the two normal modes are usually fulfilled, provided that the dissipative effects are small. They are not fulfilled, clearly, when the frequencies or angles approach values such that $\mathrm{Re}(N_1 - N_2) = 0$, as near the critical points

$$P_1^c = (x_1^c, \theta_1^c), \quad P_2^c = (x_2^c, \theta_2^c), \tag{3.6.2}$$

where

$$x = \omega / \omega_c. \tag{3.6.3}$$

In frequency space, these points are called the "vacuum" frequencies $\omega_{v1}$ and $\omega_{v2}$ (e.g. Ventura, Nagel, and Mészáros, 1979; Pavlov, Shibanov, and Yakovlev, 1980). A careful analysis of the mode structure near these critical points has been carried out by Soffel *et al.*

(1983), whose treatment we follow in this section. To determine the values of $P_1^c$, $P_2^c$, it is necessary to include the damping, which is the sum of the radiative and collisional dampings, $\gamma = \gamma_r + \gamma_c$, with

$$\gamma_r = 2/3\left(e^2\omega/mc^3\right) = 4.76 \times 10^{-4} B_{50}\, x,$$
$$\gamma_c = 0.26 \times 10^{-8} Z\mu_e^{-1}\rho_1 B_{50}^{-5/2} x, \qquad (3.6.4)$$

where $\rho_1 = (\rho/1 \text{ g cm}^{-3})$, $\mu_e$ = molecular weight per electron, and $B_{50}$ is the field that gives $\hbar\omega_c = 50$ keV, i.e. $B \cong 0.1 B_Q = 4.414 \times 10^{12}$ G. With this damping we have, to first order in $v$ and $\delta$ (see eqs. 3.5.10),

$$S = -v\lambda^{-1}\left[1 + \frac{\omega_c\lambda^{-1}}{2\sqrt{2}\,\beta\,|\,\omega\cos\theta\,|}\left(W_+ - W_-\right)\right],$$
$$D = v\lambda^{-2}\omega_c\left[\left(2\sqrt{2}\,\beta\,|\,\omega\cos\theta\,|\right)^{-1}\left(W_+ + W_-\right)\right],$$
$$P = -v\lambda^{-1}, \qquad (3.6.5)$$

where $\lambda = (1 + i\gamma)$, $\beta = (kT/mc^2)$, and the functions $W_\pm$ are given by the plasma dispersion function $W(z)$:

$$W_\pm = W(y_\pm), \qquad (3.6.6)$$

which can be evaluated by means of the error function

$$W(z) = i\pi^{1/2}\exp(-z^2)\operatorname{erfc}(-iz). \qquad (3.6.7)$$

The arguments are

$$y_\pm = \frac{\omega \pm \omega_c\lambda^{-1}}{\sqrt{2}\,\beta\,|\,\omega\cos\theta\,|} \equiv B_0\left(x \pm \lambda^{-1}\right), \qquad (3.6.8)$$

with $B_0 \equiv \omega_c(\sqrt{2}\,\beta\,|\,\omega\cos\theta\,|)^{-1} = 2.26\,x^{-1}t^{-1/2}B_{50}^{-1/2}\,|\cos\theta\,|^{-1/2}$ and $t = (kT/\hbar\omega_c)$. The normal modes, defined by equations (3.5.25), (3.5.26), will be in general slightly nonorthogonal, due to the damping and Doppler broadening, since

$$\Delta = |\,\vec{e}_+^{\,*}\cdot\vec{e}_-\,| \neq 0. \qquad (3.6.9)$$

Total orthogonality occurs for $K_\pm$ and $b$ purely real, while total nonorthogonality occurs for

$$K_+(b) = K_-(b) = b = \pm i. \qquad (3.6.10)$$

In this situation, both modes coincide, a situation known as "mode collapse" (Soffel *et al.*, 1983). A plot of the degree of nonorthogonality of modes 1 and 2 near the cyclotron resonance, as a function of propagation angle, is given in Figure 3.6.1 for $\rho_1 = 10^{-2}$, $B_{50} = 1$, $t = 0.2$. As one approaches the second critical frequency $x_2^c = \omega_2^c/\omega_c = 0.9984$ and critical angle $\theta_2^c = 8.1°$ (for these parameters), the overlap builds up toward unity.

The cause of the mode collapse is that $\mathrm{Re}(N)$ and $\mathrm{Im}(N)$ are double-valued functions of $x$ and $\theta$. The surface $\mathrm{Re}(N^2)$ is made up of two Riemann sheets, intersecting along curves starting at each critical point. For instance, for the plasma parameters above, these critical points are $P_1^c = (x_1^c, \theta_1^c) \simeq (\omega_v/\omega_c, \pi/2)$ and $P_2^c = (x_2^c, \theta_2^c) \simeq (1, 8.1°)$, the Riemann sheets being plotted in Soffel *et al.* (1983). It is clear that a straightforward mode designation as "1" or "2" is ambiguous on such a double sheet. If one starts out at a point $A$ near $P_2$, say, and labels that mode 1, a continuous change of $x$ and $\theta$ can lead to either

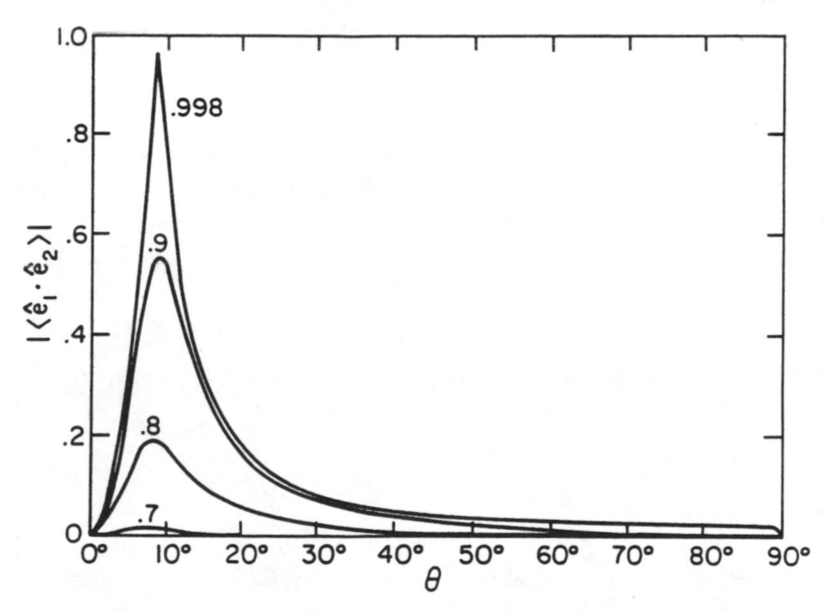


FIGURE 3.6.1 The degree of nonorthogonality of the two normal modes as a function of $\theta$ shown for various values of the frequency $\omega$. The number identifying the curves gives the value $\omega/\omega_c$. The plasma parameters are $\rho = 10^{-2}$ g cm$^{-3}$, $\hbar\omega_c = 50$ keV, and $kT/\hbar\omega_c = 0.2$. From Soffel *et al.* (1983), with permission of *Astronomy and Astrophysics*.

a point $B$ in the upper sheet or a point $B'$ in the lower sheet, depending on how one goes around the critical point $P_2^c$. Thus, one might call either $B$ or $B'$ mode 1, but not both. This ambiguity, however, is purely one of nomenclature. It does not lead to intrinsic difficulties, since $N^2(x, \theta)$ is clearly a continuous and differentiable function everywhere (except at the critical points themselves). This is the requisite for the concepts of geometrical optics and group velocity to be meaningful, when $B$, $\rho$, $T$, etc., vary over scales larger than a photon wavelength. While the designation ambiguity is not a serious problem, an approximate transfer treatment based on propagation of two normal intensities is limited to regions excluding the neighborhood of the critical points, because of the mode collapse and nonorthogonality there.

### *b) Mode Collapse Points*

In order for mode collapse to occur, the condition $\mathrm{Re}(b) = 0$ must be fullfilled (e.g. eq. 3.6.10); that is,

$$\mathrm{Re}(y) = 0$$

$$y \equiv -2b \cos\theta / \sin^2\theta = (P - S + 3\delta)/D. \tag{3.6.11}$$

To find the low-frequency collapse point $P_1^c$, since for $\rho_1 \ll 1$ this lies well below the resonance, we may use the cold-plasma approximation $T \to 0$ but retain damping terms such as $\gamma_r, \gamma_c$. In this case,

$$P = -v\lambda^{-1}; \quad S = -v\lambda(\lambda^2 - u)^{-1}; \quad D = -vu^{1/2}(\lambda^2 - u)^{-1}, \tag{3.6.12}$$

with $u \equiv x^{-2} = (\omega_c/\omega)^2$, $v = (\omega_p/\omega)^2$, and $\lambda = 1 + i\gamma$. To lowest order in $\gamma$, one can write equation (3.6.11) as

$$x^2(1 - x^2) - \xi = 0, \tag{3.6.13}$$

where

$$\xi = \omega_p^2(3\delta\omega_c)^{-1} = (3\delta/v)^{-1}x^2 = 0.22\rho_1 B_{50}^{-4}. \tag{3.6.14}$$

Real solutions for $x$ occur if $\xi < \xi_{cr} = 1/4$. If $\xi \ll 1$, these are

$$x_{1,2}^c \simeq \xi^{1/2}, 1, \tag{3.6.15}$$

or $\omega_{1,2}^c \simeq \omega_v, \omega_c$. We now need to evaluate the critical angles. Along the line Re$(y) = 0$, for $\xi, \gamma \ll 1$ we have

$$\mathrm{Im}(y) = \begin{cases} \gamma\xi^{-1/2}, & \text{for } x = x_1^c, \\ -2\gamma\xi^{-1}, & \text{for } x = x_2^c. \end{cases} \tag{3.6.16}$$

The critical angle $\theta^c$ is formed then from

$$\cos\theta^c/\sin^2\theta^c = -\frac{1}{2}\mathrm{Im}(y)\,|_{x=x_{1,2}^c}. \tag{3.6.17}$$

The cold-plasma approximation yields both a low-frequency critical point $P_1^c(\omega_1^c, \theta_1^c)$ and a high-frequency one $P_2^c(\omega_2^c, \theta_2^c)$ (e.g. Soffel *et al.*, 1983; Pavlov, Shibanov, and Yakovlev, 1980). The latter one, however, is at $\omega_2^c \sim \omega_c$ and clearly should be derived more carefully, since near $\omega_c$ thermal effects are nonnegligible. Near the resonance, the plasma dispersion function can be approximated as

$$W(z) \simeq i\pi^{1/2} - 2z \qquad (\text{for } |z| \ll 1). \tag{3.6.18}$$

Equation (3.6.11) is then dominated near $x \to 1$ by the vacuum term $\delta v^{-1} = 1.5\rho_1^{-1}B_{50}^4 \gg 1$. Thus, in order to satisfy equation (3.6.11) there, we need Re$(W_+ + W_-) = 0$ *or*

$$\begin{aligned} x_2^c &\cong 1 - (\pi/12)v\delta^{-1} - (2B_0)^{-2} \\ &= 1 - 0.17\rho_1 B_{50}^{-4} - 0.049tB_{50}\cos^2\theta. \end{aligned} \tag{3.6.19}$$

At this frequency,

$$\mathrm{Im}(y) \simeq -(\pi^{1/2}\xi B_0)^{-1} = -1.12t^{1/2}\rho_1^{-1}B_{50}^{9/2}\cos\theta, \tag{3.6.20}$$

so that

$$\sin\theta_2^c = 0.89\rho_1 t^{-1/2}B_{50}^{-9/2}. \tag{3.6.21}$$

Using as before $kT = 10$ keV, $\rho_1 = 10^{-2}$, $B_{50} = 1$, we have, as an example,

$$\begin{aligned} &x_1^c \simeq x_v = \omega_v/\omega_c = 0.0473, \\ &|\theta_1^c - \pi/2| \simeq 4.76\times10^{-4}B_{50} \ll 1, \\ &x_2^c \simeq 1, \quad \theta_2^c \simeq 8.1^\circ. \end{aligned} \tag{3.6.22}$$

These points are shown in Figure 3.6.2 as $P_1^c$, $P_2^c$, surrounded by dashed circles.

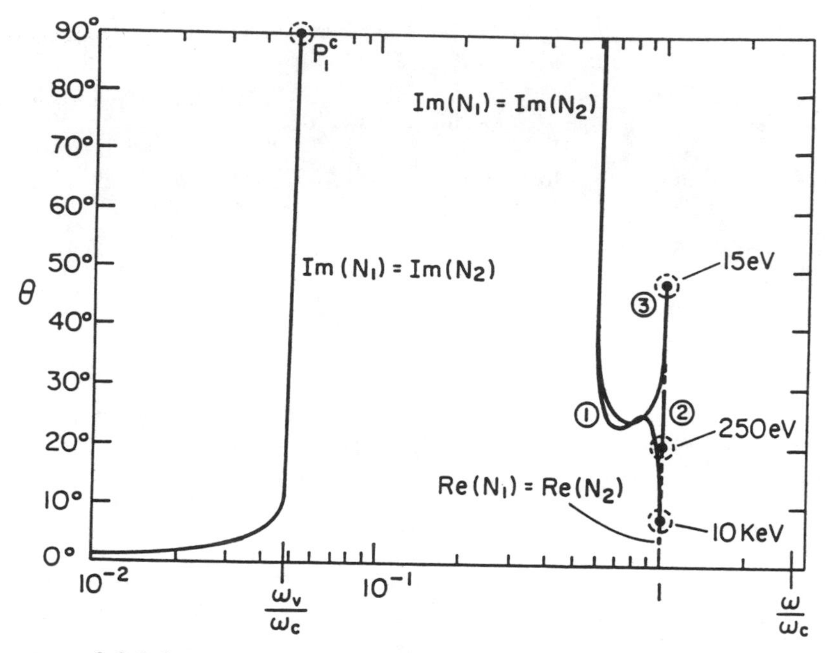


FIGURE 3.6.2 Intersection curves for the Riemannian surfaces $\mathrm{Re}[N^2(\omega,\theta)]$ (*dashed line*) and $\mathrm{Im}(N^2[\omega,\theta])$ (*solid lines*) for three values of the plasma temperature $kT = 15$ (*curve 1*), 250 (*curve 2*), and $10^4$ (*curve 3*) eV. The high-frequency intersection pattern depends on temperature, with the collapse point $P_2^c$ moving toward the limit $\theta = \pi/2$ as the temperature is reduced. The position of the low-frequency collapse point is indicated by $P_1^c$. The high-frequency collapse point $P_2^c$ lies in the resonance $\omega/\omega_c = 1$ and, like $P_1^c$, is indicated by the symbol for the three temperatures considered. The density assumed is $\rho/\mu_e = 10^{-2}$ g cm$^{-3}$. Wave propagation is dominated by the plasma component at low frequencies, $\omega \ll \omega_v$, and at $\theta \simeq 0°$. The vacuum component plays a dominant role at higher frequencies and $\theta \neq 0°$. From Soffel *et al.* (1983), with permission of *Astronomy and Astrophysics*.

The position of the critical points varies with $\rho_1$, $B_{50}$, and $t$ according to equations (3.6.13) to (3.6.21). Thus, for constant $t$ and $B_{50}$ the variation of $P_1^c$, $P_2^c$ as a function of the plasma density $\rho_1$ is found by demanding that $\mathrm{Re}(b) = 0$, $\mathrm{Im}(b) = 1$ in the hot-plasma expression of equation (3.6.11), an example for $t = 0.2$, $B_{50} = 1$ being plotted in Soffel *et al.* (1983). As one increases $\rho$ from $\rho_1 = 10^{-2}$, one has $\theta_1^c \simeq \pi/2 \sim$ constant, while $x_1^c \propto \rho_1^{1/2}$ increases, moving up toward higher frequencies. At the same time, $x_2^c$ remains in the neighborhood (within $\pm\hbar w_c/mc^2$) of $\omega_c$ and $\theta_2^c$ increases until it reaches

$\theta_2^c \sim \pi/2$, after which $x_2^c$ starts moving toward lower values of $x$ with $\theta_2^c \sim \pi/2 \sim$ constant. The two critical points converge to a double solution at $\xi = \xi_{cr} = 0.25$, for $\rho_1 = 1.14$ (at this $t$, $b_{50}$). For $\xi > \xi_{cr}$, there are no mode collapse critical points.

*c) Faraday Depolarization Breakdown*

The regions where the two-normal-mode approach breaks down are given by $\text{Re}(N_1 - N_2) = 0$. These are the curves along which the two Riemann sheets of $N^2$ intersect. One branch of this curve starts at the critical point $P_2^c = (x_2^c, \theta_2^c)$ and extends downward, with $x \sim x_2^c \sim$ constant, $\theta \leq \theta_2^c$ until $\theta = 0°$. The other branch extends from the critical point $P_1^c = (x_1^c, \theta_1^c)$ at $x \sim x_2^c \sim$ constant and $\theta \geq \theta_1^c$ upward. Since, however, $\theta_1^c \sim \pi/2$, this extent is rather small. Figure 3.6.2 shows the curves of intersection $\text{Re}(N_1) = \text{Re}(N_2)$, as well as the curves for the intersection of $\text{Im}(N_1) = \text{Im}(N_2)$. These intersection curves are given by (Soffel *et al.*, 1983)

$$\left\{\begin{matrix}\text{Re}\\ \text{Im}\end{matrix}\right\}\left(N_1^2 - N_2^2\right) = \cos\theta\left\{\begin{matrix}\text{Re}\\ \text{Im}\end{matrix}\right\}\left(D[1 + b^2]^{1/2}\right). \tag{3.6.23}$$

The equation $\text{Re}(b) = 0$ gives two almost vertical lines near $x_1^c \simeq \xi^{1/2} = \omega_v/\omega_c$ and $x_2^c \simeq 1$. At these frequencies, the quantity $(1 + b^2)^{1/2}$ is real for $\theta > \theta^c$ and imaginary for $\theta < \theta^c$. Along $\text{Re}(b) = 0$ we have then polarization modes which are linear with $|\text{Im}(b)| > 1$, elliptical when $|\text{Im}(b)| < 1$, and circular when $|\text{Im}(b)| = 1$, the last occurring at the two points $P_1^c$, $P_2^c$ (see Mészáros and Ventura, 1978, 1979; Gnedin, Pavlov, and Shibanov, 1978a, b). The regions of violation of the Faraday depolarization condition are related to the intersection curves, i.e. are regions of extent

$$|x - x_1^c|/x_1^c \simeq \gamma, \quad |\theta - \theta_1^c|/\theta_1^c \simeq \gamma\xi^{1/2} \tag{3.6.24}$$

near $x \sim x_1^c$, $\theta \sim \pi/2$ and

$$|x - x_2^c|/x_2^c \lesssim \left(kT/mc^2\right)^{1/2}, \quad \theta \lesssim \theta_2^c \tag{3.6.25}$$

near $x \sim x_2^c$.

*d) Wave Propagation Near Mode Collapse Points*

It is possible to rewrite the wave equation (3.6.23) in the semitransverse limit as

$$\left(\vec{\vec{1}} + \vec{\vec{\chi}}\right) \cdot \vec{E}_t = N^2 \vec{E}_t \tag{3.6.26}$$

(Soffel *et al.*, 1983), where $\vec{E}_t$ is the part of $\vec{E}$ transverse to $\vec{k}$, and

$$\chi_{11} = S + (P - S + 7\delta)\sin^2\theta,$$
$$\chi_{21} = -\chi_{12} = iD\cos\theta,$$
$$\chi_{22} = S + 4\delta\sin^2\theta. \tag{3.6.27}$$

The parameter $b$ given by equation (3.5.23) is then

$$b = (\chi_{11} - \chi_{22})/(2i\chi_{21}), \tag{3.6.28}$$

and the eigenvectors $\vec{e}_1, \vec{e}_2$ (see eqs. 3.5.25, 3.5.26) lie in the $\vec{B}, \vec{k}$ plane and perpendicular to it. For a homogeneous medium, the solution of equation (3.6.26) is (e.g. Gnedin and Pavlov, 1974; Melrose and Stoneham, 1976)

$$\vec{E}_t(z) = \exp\left(i\vec{\vec{k}}z\right)\vec{E}_0, \tag{3.6.29}$$

where

$$\vec{\vec{k}} \equiv (\omega/c)\left(\vec{\vec{1}} + \vec{\vec{\chi}}\right)^{1/2} \tag{3.6.30}$$

and $\vec{E}_0$ is a polarization vector. At the parameters corresponding to mode collapse, $\vec{\vec{\chi}}$ becomes degenerate, with $\chi_{12} = \pm(1/2)(\chi_{11} - \chi_{22})$, and can be expressed in terms of the two complex quantities $\chi_\pm$ as

$$\vec{\vec{\chi}} = \chi_+\vec{\vec{1}} + \chi_-\begin{pmatrix} 1 & \pm 1 \\ \pm 1 & -1 \end{pmatrix} \equiv \chi_+\vec{\vec{1}} + \chi_-\vec{\vec{M}}, \tag{3.6.31}$$

where

$$\chi_\pm = \frac{1}{2}(\chi_{11} \pm \chi_{22}),$$
$$\vec{\vec{M}} \equiv \begin{pmatrix} 1 & \pm 1 \\ \mp 1 & -1 \end{pmatrix}. \tag{3.6.32}$$

The matrix $\vec{\vec{M}}$ is nilpotent, i.e. $\vec{\vec{M}}^2 = 0$, so that the solution of equation (3.6.29) at mode collapse can be written as

$$\vec{E}(z) = e^{ik_c z}\left(\vec{\vec{1}} + ik_r z\vec{\vec{M}}\right)\vec{E}_0, \tag{3.6.33}$$

where

$$k_c = (\omega/c)\left(1 + \frac{1}{2}\chi_+\right), \quad k_r = \frac{1}{2}(\omega/c)\chi_-. \tag{3.6.34}$$

At mode collapse, there is only one propagating normal mode, $\vec{e}_c = (\mp 1, 1, 0)_{xyz}$ (since $K_+ = K_- = b = \pm i$), and this one mode is halfway between the $\vec{B}, \vec{k}$ plane and the plane perpendicular to it, at 45°. One can see that the matrix $\overleftrightarrow{M}$ acting on $\vec{e}_c$ gives zero, while $\overleftrightarrow{M}$ acting on a vector perpendicular to $\vec{e}_c$ gives as a result a vector in the direction of $\vec{e}_c$. Thus, for the polarization $e_\perp$ transverse to $e_c$ we have

$$\vec{E}_\perp(z) = e^{ik_c z}\left(\vec{e}_\perp + 2ik_r z\vec{e}_c\right), \tag{3.6.35}$$

where $\mathrm{Im}(k_c)$ gives the attenuation coefficient of the $\vec{E}_\perp$ wave. At the critical points, we can evaluate $k_r$, $k_c$ from the values at $x_1^c, \theta_1^c$,

$$\chi_+ = \left(2/3 + x^2\right)v + iv\gamma/2,$$
$$\chi_- = -iv\gamma/2, \tag{3.6.36}$$

and at $x_2^c, \theta_2^c$,

$$\chi_+ \cong \left(\pi^{1/2}/2\right)vB_0\left(11/3\cos\theta_2^c + i\right),$$
$$\chi_- \cong \left(\pi^{1/2}/2\right)vB_0\cos\theta_2^c. \tag{3.6.37}$$

At both critical points, a wave of polarization $\vec{e}_\perp$ perpendicular to $\vec{e}_c$ is attenuated over a length scale corresponding to the absorption coefficient at this frequency and angle, while a component along $\vec{e}_c$ is created over a length scale comparable to the photon mean free path. On the other hand, a wave initially along $\vec{e}_c$ remains there, since it is a normal mode at the critical points. Thus, a strongly magnetized medium acts as a polarizer for waves of $(\omega, \theta)$ corresponding to the critical values. This phenomenon also occurs in the laboratory at optical wavelengths (e.g. Gnedin and Pavlov, 1974) for the case where the dispersion, instead of being provided by the magnetic vacuum, is due to the presence of atomic resonances.

# 4. Magnetized Radiative Processes: Nonrelativistic Limit

## 4.1 The Radiation Process in an External Field

*a) The Classical Cyclotron Process*

From classical electrodynamics, we know that charged particles subject to an acceleration will radiate. This follows from a consideration of the far-field regime of the retarded (Lienard-Wiechert) potential of a moving charge (e.g. Jackson, 1975; Rybicki and Lightman, 1979). The components of the radiation electric and magnetic fields proportional to $r^{-1}$ which contribute to a nonzero Poynting flux at infinity are, in vacuum,

$$\vec{E}_r(\vec{r}, t) = \frac{q}{c}\left[\frac{\hat{n}}{\left(1 - \hat{n}\cdot\vec{\beta}\right)^3 R} \times \left\{\left(\hat{n} - \vec{\beta}\right) \times \dot{\vec{\beta}}\right\}\right]_{\text{ret}},$$

$$\vec{B}_r(\vec{r}, t) = \hat{n} \times \vec{E}_r. \qquad (4.1.1)$$

Here $\vec{R}(t)$ is a vector between the position of the particle $\vec{r}_0$ and the observation point $\vec{r}$, and $\hat{n} = \vec{R}/R$ is the normalized velocity of the particle of mass $m$ and charge $q$ at time $t$. The suffix "ret" indicates that the bracket is to be evaluated at the retarded time given by $t_{\text{ret}} = t - R(t_{\text{ret}})/c$. From equation (4.1.1), one sees that, if $\dot{\vec{\beta}} = 0$, there is no radiative field in a vacuum. The time-averaged Poynting flux for a harmonic plane wave is

$$\langle S\rangle = \frac{c}{4\pi}\langle \vec{E} \times \vec{B}\rangle = \frac{c}{8\pi}\text{Re}\left(E_0 B_0^*\right) = \frac{c}{8\pi}\,|\,\vec{E}_0\,|\,. \qquad (4.1.2)$$

 The energy per unit frequency per unit solid angle radiated by a single particle is

$$\frac{dW}{d\omega\, d\Omega} = \frac{c}{4\pi}\left|\left[\int \mathrm{Re}\vec{E}(t)\right]_{\mathrm{ret}} e^{i\omega t}\, dt\right|^2$$

$$= \frac{q^2\omega^2}{4\pi c}\left|\int \hat{n}\times\left(\hat{n}\times\vec{\beta}\right)\exp\left[i\omega\left(t' - \hat{n}\cdot\vec{r}_0(t')/c\right] dt'\right|^2, \tag{4.1.3}$$

where $r_0$ is the position of the particle. In the nonrelativistic limit $|\beta| = u/c \ll 1$, the radiation fields become

$$\vec{E}_r = \left[\left(q/Rc^2\right)\hat{n}\times\left(\hat{n}\times\dot{\vec{u}}\right)\right]_{\mathrm{ret}},$$

$$\vec{B}_r = \left[\hat{n}\times\vec{E}_r\right]_{\mathrm{ret}}. \tag{4.1.4}$$

The magnitudes are $|\vec{E}_r| = |\vec{B}_r| = (q\dot{u}/Rc^2)\sin^2\theta$, where $\theta$ is the angle between $\vec{u}$ and $\hat{n}$. The Poynting vector along $\hat{n}$ has magnitude

$$S = \frac{c}{4\pi}E_r^2 = \frac{c}{4\pi}\frac{q^2\dot{u}^2}{R^2c^4}\sin^2\theta, \tag{4.1.5}$$

while the energy per unit solid angle per unit time is (since $dA = R\, d\Omega$)

$$\frac{dW}{dt\, d\Omega} = \frac{q^2\dot{u}^2}{4\pi c^3}\sin^2\theta. \tag{4.1.6}$$

Integrated over solid angles, we get the power as

$$P = \frac{2q^2\dot{u}^2}{3c^3}, \tag{4.1.7}$$

which is Larmor's formula.

For an electron in a magnetic field, the energy radiated per unit time per unit solid angle and per unit frequency is obtained from equation (4.1.3) in the nonrelativistic limit, after dividing by the "time of radiation" $T = 2\pi\delta(\omega_c - \omega[1 - \beta_\parallel\cos\theta])$ (Bekefi, 1966). This yields the cyclotron emissivity per electron at the ground harmonic for $|\beta| \ll 1$,

$$w_\omega(\theta) = \frac{dW}{dt\, d\omega\, d\Omega}$$

$$= \frac{q^2\omega^2}{8\pi c}\beta_\perp^2\left(1 + \cos^2\theta\right)\delta\left(\omega_c - \omega\left[1 - \beta_\parallel\cos\theta\right]\right). \tag{4.1.8}$$

Thus, the frequency at which the radiation will be observed will depend on the angle of observation $\theta$. The angular dependence of the total (unpolarized) emissivity has the characteristic dipole pattern $(1 + \cos^2\theta)$. Looking at equation (4.1.4), one sees that for $\theta \to 0°$ the wave is circularly polarized, while for $\theta \to 90°$ it is linearly polarized with $\vec{E}^w \perp \vec{B}^w$ (i.e. extraordinary polarization). At intermediate angles, the polarization is elliptical. Integrating equation (4.1.8) over solid angle and frequency, one obtains the power emitted

$$P = \frac{dW}{dt} = \frac{2}{3} r_0^2 c \beta_\perp^2 B_0^2. \tag{4.1.9}$$

For an isotropic distribution of electron pitch angles $\alpha = \sin^{-1}(\beta_\perp/\beta)$, an average over $\alpha$ in equation (4.1.9) gives

$$P = \frac{4}{3} \sigma_T c \beta^2 \frac{B_0^2}{8\pi}, \tag{4.1.10}$$

where $\sigma_T = (8\pi/3) r_e^2$ is the Thomson cross section and $B_0^2/8\pi$ is the (static) magnetic field energy density.

The "line width" inherent in equation (4.1.8) is infinitely narrow, as evidenced by the delta function in the $\omega$ variable. A finite width will appear if one considers the radiative width of the line $\Gamma_r$ or the width caused by random collisions of particles $\Gamma_c$, where $\Gamma_c$ is the mean frequency between collisions. As shown in Bekefi (1966), this can be treated by assuming that the particle radiates unhindered for a finite time $\Delta t$, beginning with a random phase, and then stops and restarts again. The probability of occurrence of the interval $\Delta t$ has a Gaussian distribution

$$\Gamma e^{-\Gamma \Delta t} d(\Delta t). \tag{4.1.11}$$

The Fourier analysis of all quantities and fields is then performed only over the interval $\Delta t$,

$$V(\omega) = \int_0^{\Delta t} V(t) e^{i\omega t}\, dt, \tag{4.1.12}$$

and averaged over the distribution of equation (4.1.11). For low values of $\Gamma \ll \omega$ (e.g. Oster, 1960), the result is that the delta function in equation (4.1.8) is replaced by $\pi^{-1}\Gamma([\omega_c - \omega(1 - \beta_{\parallel}\cos\theta])^2 + \Gamma^2)^{-1}$, giving

$$w_\omega(\theta) = \frac{q^2\omega^2\beta_\perp^2(1 + \cos^2\theta)}{8\pi^2 c} \frac{\Gamma}{\left[\omega_c - \omega(1 - \beta_{\parallel}\cos\theta)\right]^2 + \Gamma^2}, \tag{4.1.13}$$

which has the distinctive Lorentz shape, associated with the presence of a damping mechanism, $\Gamma = \Gamma_r + \Gamma_c$.

An additional line broadening is due to a velocity, or momentum, distribution of the electrons. For a Maxwell-Boltzmann distribution isotropic in three dimensions,

$$f(v) = n_e(m/2k\pi kT)^{3/2}\exp(-mv^2/2kT), \tag{4.1.14}$$

and averaging equation (4.1.8) over this distribution one obtains the thermal cyclotron emissivity at the first harmonic in the nonrelativistic limit as

$$j_\omega(\omega) = \frac{\omega_p^2\omega^2}{4\pi c^3}(mv_T^2)(1 + \cos^2\theta)\frac{1}{\Delta}\exp\left[-(\omega - \omega_c)^2/\Delta^2\right], \tag{4.1.15}$$

which has the characteristic Doppler line profile. Here $v_T^2 = kT/m$, $\omega_p = 4\pi e^2 n_e/m_e$, and $\Delta = 2^{1/2}(v_T/c)\omega\cos\theta$ is the Doppler half-width for an isotropic distribution. Note that at $\theta = \pi/2$ the line again becomes infinitely narrow, except for the radiative damping. When damping is important, the emissivity is given by averaging equation (4.1.13) over equation (4.1.14), which for $\beta_{\parallel}\cos\theta \to 0$ and $\Gamma =$ constant gives

$$j_\omega(\theta) = \frac{\omega_p^2\omega^2}{4\pi^2 c^3}(mv_T^2)(1 + \cos^2\theta)\frac{\Gamma}{(\omega - \omega_c)^2 + \Gamma^2}, \tag{4.1.16}$$

which is again a Lorentz profile, of half-width $\Gamma$. When both damping and Doppler effects are important and $\beta_{\parallel}\cos\theta \neq 0$, one obtains the Voigt profile, usually calculated numerically (e.g. Mihalas, 1978; Wasserman and Salpeter, 1980).

The emissivity (4.1.13) or (4.1.16) is for an electron moving in a circular orbit given by the Larmor radius, i.e. it represents the classical cyclotron first harmonic emission. The classical procedure for treating the higher harmonics in general requires abandoning the nonrelativistic limit, since the higher harmonics contribute to the emission proportionally to increasing powers of $(v/c)^2$ (e.g. Bekefi, 1966; Zheleznyakov, 1977; Brainerd and Lamb, 1987). We shall postpone a discussion of the relativistic radiation processes until the next chapter. For the moment, as far as a classical description is concerned, the case of emission at the lowest harmonic is close to the situation encountered in strong-magnetic-field accreting neutron stars for particle energies in the X-ray range or below, since these electrons are found typically in the ground Landau level. Although the classical description is simple and gives answers to some questions of the right order of magnitude, it does have a number of shortcomings. The isotropic distribution used assumes that the typical particle energy $kT$ exceeds significantly the energy difference between two successive Landau levels, $kT \gg \hbar\omega_c$. Also, it assumes that some process (e.g. collisions) keeps the distribution thermal and isotropic, i.e. radiation losses occur on a slower time scale than those processes which reestablish the thermal character of the distribution. Under these assumptions, the cyclotron emissivity $w$ would be connected to the cyclotron absorption coefficient $\kappa_a$ by the usual Kirchhoff relation,

$$\kappa_a(\omega) = w(\omega)/B(\omega), \tag{4.1.17}$$

where $B(\omega)$ is Planck's function

$$B(\omega) = \frac{\hbar\omega^3}{4\pi^3 c^2}\left[\exp(\hbar\omega/kT) - 1\right]^{-1}. \tag{4.1.18}$$

In a very strong magnetic field, however, several of the assumptions used in deriving equations (4.1.10) to (4.1.18) break down. The main reason for this is that the typical time scale for cyclotron radiative energy losses

$$t_r = \frac{(3/2)kT}{P} \sim 2.5 \times 10^{-16} B_{12}^{-2}\ \mathrm{s} \tag{4.1.19}$$

becomes extremely small compared to the collisional (Coulomb) time scale

$$t_c = \Gamma_c^{-1} \sim 10^{-8}(kT/10\ \mathrm{keV})^{1/2}(\rho/10^{-3}\ \mathrm{g\ cm^{-3}})^{-1}, \tag{4.1.20}$$

a fact which leads to the identification of resonant cyclotron scattering ($\propto t_r^{-1}$) as the main radiative opacity, while instead of cyclotron emission one has resonant bremsstrahlung emission ($\propto t_c^{-1}$, so its opacity is much lower than resonant scattering) as a source of photons (Ventura, Nagel, and Mészáros, 1979). The relevance of these same considerations for gamma-ray burst sources was emphasized by

Bussard and Lamb (1982). In equation (4.1.19) we used (4.1.10) with $(v/c)^2 \sim (kT/m_e c^2)$, and the collisional rate in (4.1.20) is based on equation (4.4.17). From this comparison of time scales it is seen that the transverse electron energy is lost much more rapidly than it can be replenished collisionally, and the electrons will rapidly acquire a one-dimensional velocity distribution, with $v_\parallel \neq 0$ and $v_\perp \rightarrow 0$). The radiation losses then become much smaller than those given by equation (4.1.9). In a strong magnetic field, therefore, the usual radiation formulae, obtained under the LTE assumption of an isotropic electron distribution, must be reexamined.

### *b) Quantum Treatment of the Cyclotron Process*

Quantum mechanically, the cyclotron process entails the transition of an electron from a higher Landau level to a lower one. For nonrelativistic electrons, the transitions with $\Delta n > 1$ are strongly suppressed so that the electron preferentially decays in steps of $\Delta n = 1$, emitting each time a photon of frequency $\omega_c$, the cyclotron frequency. In the absorption of a photon of frequency $\omega_c$, the electron in level $n$ makes a transition to higher Landau level $n + 1$. For an electron in the Landau ground level $n = 0$, the only transitions possible are to a higher Landau level $n \geq 1$. Neglecting spin, we consider an electron characterized by the nonrelativistic wave function $|0, s, p\rangle$, which absorbs a photon of wave vector $\vec{k}$ (see Fig. 4.1.1) to make a transition to the higher state $|1, s', p'\rangle$, where $s$ is the guiding center quantum number and $p \equiv p_z$ is the momentum component of the electron along $\vec{B}$. The perturbation Hamiltonian of the electron in the field $\vec{A}$ of the incident photon (e.g. Nagel, 1982; Canuto and Ventura, 1977) is

$$H' = H_1 + H_2 = \frac{e}{2c}\left(\vec{v} \cdot \vec{A} + \vec{A} \cdot \vec{v}\right) + \frac{e^2}{2mc^2}\vec{A}^2. \qquad (4.1.21)$$

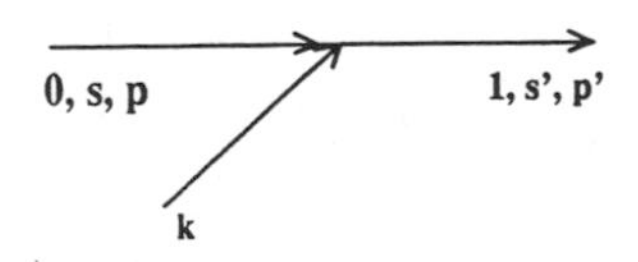


FIGURE 4.1.1 The cyclotron absorption process from the ground state, where the electron is labeled by the nonrelativistic quantum numbers $n$, $s$, $p$ corresponding to the principal quantum number, the guiding center quantum number, and the longitudinal momentum, with $k$ the photon longitudinal momentum.

Calling $a_q^\dagger$ and $a_q$ the creating and annihilation operators for photons in the state $q$, one may rewrite the photon vector potential as

$$\vec{A} = \sum_q \left( \frac{2\pi\hbar c}{\omega L^3} \right)^{1/2} \left[ a_q \vec{e}_q e^{i\vec{q}\cdot\vec{r}} + a_q^+ \vec{e}_q^* e^{-i\vec{q}\cdot\vec{r}} \right], \tag{4.1.22}$$

where the $\vec{e}_q$ are unit polarization vectors and $L^3$ is a normalization volume. The matrix element describing the absorption of a photon in state $\vec{k}$ resulting in an electron transition from $|i\rangle$ to $|f\rangle$ is

$$\langle f | H' | i \rangle = \left( \frac{2\pi\alpha_F \hbar^2 c}{\omega L^3} \right)^{1/2} \frac{1}{2} \langle 1, s', p' | \vec{v} \cdot \vec{e} e^{i\vec{k}\cdot\vec{r}} + e^{i\vec{k}\cdot\vec{r}} \vec{v} \cdot \vec{e} | 0, s, p \rangle. \tag{4.1.23}$$

Using the dipole (Born) approximation $e^{i\vec{k}\cdot\vec{r}} \simeq 1$ and the identity $\vec{v} \cdot \vec{e} = v_r e_- + v_- e_+ + v_z e_z$, this becomes

$$\langle f | H' | i \rangle = \left( \frac{2\pi\alpha_F \hbar^2 c}{\omega L^3} \right)^{1/2} \langle 1, s', p' | v_+ e_- + v_- e_+ + v_z e_z | 0, s, p \rangle. \tag{4.1.24}$$

However, as mentioned in Chapter 2, only the $v_+$ operator represents an upward transition into a higher $n$ state (eq. 2.3.16). Thus, we have

$$\langle 1, s', p' | v_+ e_- + v_- e_+ + v_z e_z | 0, s, p \rangle = \left( \hbar\omega_c / m \right)^{1/2} e_- \delta_{ss'} \delta_{pp'}. \tag{4.1.25}$$

The absorption rate is then obtained using the "Golden Rule" (e.g. Sakurai, 1967):

$$\begin{aligned} k_c &= \frac{2\pi}{\hbar} |\langle f | H' | i \rangle|^2 \delta\left( E_f - E_i \right) \\ &= \frac{4\pi^2 \alpha_F \hbar c}{\omega L^3} \frac{\hbar\omega_c}{m} | e_- |^2 \delta\left( \hbar\omega_c + p'^2/2m - \hbar\omega \right). \end{aligned} \tag{4.1.26}$$

We have used here $p = p'$, $s = s'$, which obviates the summation over the final electronic states. When we further divide by the incoming flux of photons, we obtain the cyclotron absorption cross section

$$\sigma_c = 4\pi^2 \alpha_F \frac{\hbar}{m} |e_-|^2 \delta(\omega - \omega_C), \tag{4.1.27}$$

where $\alpha = e^2/\hbar c$ is the fine structure constant.

It is characteristic of the cyclotron process that it depends only on the $e_-$ component of the photon polarization, the component that rotates in the same sense as the electron. Classically, this was manifested by the fact that only a wave rotating in this sense can resonate with the gyrating electron (give up energy to it). For the usual cold magnetized plasma, at frequencies $\omega = \omega_c$ the $e_-$ component of the ordinary (2) wave vanishes and only the extraordinary (1) wave contains an $e_-$ component. This is why only the extraordinary wave can suffer cyclotron absorption in the cold plasma. However, if one includes vacuum polarization, or spin-flip transitions, or considers thermal motions, the ordinary wave also acquires a nonzero $e_-$ component at $\omega_c$, albeit one that is not as important as for the extraordinary wave.

The energy delta function in equation (4.1.27) says that a cold electron (at rest) can absorb only waves of frequency $\omega = \omega_c$. If the electron is moving at velocity $v$, then the wave frequency that can be absorbed must satisfy this requirement in the electron rest frame, i.e. the laboratory frequency which is absorbed is the Doppler-shifted value $\omega' = \omega_c(1 - (v/c)\cos\theta)$. In a strong field the electron motion will be one-dimensional along $z$, and for a Maxwell-Boltzmann distribution the cyclotron absorption coefficient acquires a frequency dependence proportional to $\exp[-(\omega - \omega_c)/\Delta\omega_D]^2$, where the Doppler width is given by

$$\Delta\omega_D = \omega_c \left( \frac{2kT}{mc^2} \right)^{1/2} |\cos\theta|. \tag{4.1.28}$$

Outside the range $\omega_c \pm \Delta\omega_D$, the cyclotron absorption coefficient falls off exponentially and other effects, such as (nonresonant) Thomson scattering or bremsstrahlung, become important.

The rate at which an excited electron emits radiation into a certain solid angle and at a particular frequency is related to the absorption cross section by the microreversibility, or detailed balance, principle. Unlike its microscopic equivalent, the Kirchhoff law, this principle remains in effect independently of field strength or particle distribution considerations. The emission rate is then given by

$$w_c(\omega,\theta)\,d\omega\,d\Omega = \frac{\omega^2}{(2\pi)^3 c^2}\sigma_c(\omega,\theta)\,d\omega\,d\Omega \tag{4.1.29}$$

so that, using (4.1.27),

$$w_c(\omega,\theta) = \frac{\alpha_F}{2\pi}\frac{\hbar\omega^2}{mc^2}|e_-|^2\delta(\omega-\omega_c). \tag{4.1.30}$$

By integrating over all directions and frequencies, and summing over both polarizations, one obtains the radiative line width $\Gamma_r$, which is the inverse of the typical radiative decay time of an electron in an excited Landau level. Using the relation $|e_-^{(1)}|^2 + |e_-^{(2)}|^2 = 1 + \cos^2\theta$, the radiative width is

$$\Gamma_r = t_r^{-1} = \frac{4}{3}\alpha_F\frac{\hbar\omega_c^2}{mc^2} \simeq 3.846 \times 10^{15} B_{12}^2. \tag{4.1.31}$$

Thus, the radiative deexcitation time scale $t_r$ is inversely proportional to $B^2$, being of the order of $10^{-15}$ s for typical pulsar fields, extremely short by comparison with ordinary atomic transition times or Coulomb collision time scales.

## 4.2 Electron Scattering in a Cold Plasma

### *a) Classical Treatment*

Consider a transverse, plane monochromatic electromagnetic wave of wave vector $\vec{k}$ and frequency $\omega$,

$$\vec{E}(\vec{r},\omega) = \hat{e}E_0 e^{i\vec{k}\cdot\vec{r}}. \tag{4.2.1}$$

If we assume that the plasma is cold, i.e. we neglect any random, thermal velocities of the electrons in the medium, the individual electrons will acquire a velocity due to the electric force $e\vec{E}$ of the wave. The velocity of this oscillation $v_\alpha$, which is related to the polarization current $j_\alpha$ set up in the medium by the wave, is given by (eq. 3.1.8)

$$v_\alpha = (en_e)^{-1} j_\alpha = -\frac{i\omega}{m\omega_p^2}(\epsilon_{\alpha\beta} - \delta_{\alpha\beta})E_\beta, \tag{4.2.2}$$

where $\epsilon_{\alpha\beta}$ is the dielectric tensor of the medium and $\omega_p$ is the electron plasma frequency. If the plasma is isotropic (nonmagnetic),

$\epsilon_{\alpha\beta}$ contains only terms involving $\omega$ and $\omega_p$, and the electron motion $v_\alpha$ is a superposition of an induced motion at the characteristic frequency $\omega$ of the incident wave and a lower-frequency motion at its own characteristic plasma frequency $\omega_p$. The latter is the natural frequency of electrostatic oscillations with which electrons disturbed from their equilibrium position would oscillate with respect to the proton background. If the plasma has a large-scale external magnetic field $\vec{B}_0$, then $\varepsilon_{\alpha\beta}$ also depends on an additional frequency $\omega_c$, the characteristic cyclotron frequency of the field. The motion of the electron will be a superposition of the motions at $\omega$, $\omega_p$ and at the electron gyrofrequency $\omega_c$. This is to be expected, since any transverse (to $\vec{B}_0$) component of $v_\alpha$ induced by the wave will lead to a Lorentz force $\vec{f}_L = (e/c)\vec{v} \times \vec{B}$ which causes the electron to gyrate at $\omega_c$ around the field.

According to the classical radiation theory (see § 4.1), this motion $v_\alpha$ of the electron will result in a secondary wave field $\vec{E}_{sc}(\vec{R}, \omega)$ at large distances $\vec{R}$ from the electron, given by

$$\vec{E}_{sc}(\vec{R}, \omega) = -\frac{i\omega e}{c^2}\frac{e^{i\vec{k}'\cdot\vec{R}}}{R}\left[\hat{k}' \times (\hat{k}' \times \vec{v})\right], \tag{4.2.3}$$

which is a spherical wave propagating radially outward with wave vector $\vec{k}'$. This equation may be reexpressed as

$$E_{sc,\alpha}(\vec{R}, \omega) = E_0\frac{e^{i\vec{k}\cdot\vec{R}}}{R}f_{\alpha\beta}(\hat{k}')\hat{e}_\beta, \tag{4.2.4}$$

where the scattering amplitude $f_{\alpha\beta}$, following Ventura (1979), is a matrix in polarization space

$$f_{\alpha\beta} = -r_0(\delta_{\alpha\gamma} - \hat{k}'_\alpha\hat{k}'_\gamma)\Pi_{\gamma\beta}, \tag{4.2.5}$$

where the summation over repeated indices is assumed. The polarization matrix $\Pi_{\gamma\beta}$ is

$$\Pi_{\gamma\beta} = v^{-1}(\delta_{\gamma\beta} - \epsilon_{\gamma\beta}), \tag{4.2.6}$$

and for a cold plasma this is

$$\vec{\vec{\Pi}} = (1-u)^{-1}\begin{pmatrix} 1 & -iu^{1/2} & 0 \\ iu^{1/2} & 1 & 0 \\ 0 & 0 & 1-u \end{pmatrix}, \tag{4.2.7}$$

with $u = (\omega_c/\omega)^2$ and $v = (\omega_p/\omega)^2$ as usual. The cross section $d\sigma(\hat{k}' \leftarrow \hat{k})/d\Omega$ for scattering radiation from $\hat{k}$ into $\hat{k}'$ is obtained by calculating the power $dP/d\Omega$ radiated into $d\Omega'$ and dividing by the incident flux $\vec{S}$. If the incident wave has polarization $\hat{e}$ and we are interested in the power scattered into polarization $\hat{e}'$, then, before forming the scattered Poynting vector $\vec{E}_{sc} \times \vec{B}_{sc}$ in $dP/d\Omega$, we project $\vec{E}_{sc}$ onto $\hat{e}'$ (e.g. Jackson, 1975). The resulting Thomson scattering cross section is

$$\frac{d\sigma}{d\Omega'}(\hat{e}', \hat{k}' \leftarrow \hat{e}, \hat{k}) = |\hat{e}', \vec{\vec{f}} \cdot \hat{e}|^2. \tag{4.2.8}$$

If we assume that the polarization vectors are transverse, or nearly so, we have

$$\left(\vec{\vec{1}} - \hat{k}\hat{k}\right)^2 \cdot \hat{e} = \left(\vec{\vec{1}} - \hat{k}\hat{k}\right) \cdot \hat{e} = \hat{e}_t, \tag{4.2.9}$$

where $\hat{e}_t$ is the transverse part of the polarization modes of the plasma. The magnetic equivalent of the Thomson scattering cross section is therefore

$$\frac{d\sigma}{d\Omega'}(\hat{e}', \hat{k}' \leftarrow \hat{e}, \hat{k}) = r_0^2 \frac{k'}{k} \left| \left(\vec{e}', \vec{\vec{\Pi}} \cdot \vec{e}\right) \right|^2 / |\vec{e}_t'|^2 |\vec{e}_t|^2, \tag{4.2.10}$$

where $r_0$ is the classical radius of the electron. In the isotropic plasma case, $\Pi_{\alpha\beta} = \delta_{\alpha\beta}$ and we regain the usual Thomson expression (Jackson, 1975)

$$\frac{d\sigma}{d\Omega'} = r_0^2 |\hat{e}'^* \cdot \hat{e}|^2, \tag{4.2.11}$$

where $\hat{e}$ are the normalized polarization vectors.

In a magnetic field, the polarization matrix is diagonal in the rotating coordinates $e_+, e_-, e_z$ defined in equations (3.4.43), with $z$ along $\vec{B}_0$. The eigenvalues of $\vec{\vec{\Pi}}$ in this representation are

$$\pi_\pm = \left(1 \pm u^{1/2}\right), \quad \pi_z = 1. \tag{4.2.12}$$

For the normalized eigenvectors, the differential cross section is then (for $\omega = \omega'$)

$$\frac{d\sigma}{d\Omega'}(\hat{e}', \hat{k}' \leftarrow \hat{e}, \hat{k}) = r_0^2 \left| \hat{e}_z'^* \hat{e}_z + \frac{\hat{e}_+'^* \hat{e}_+}{1 + u^{1/2}} + \frac{\hat{e}_-'^* \hat{e}_-}{1 - u^{1/2}} \right|^2. \tag{4.2.13}$$

The total cross section for the scattering of waves of polarization $\hat{e}^i$ moving in direction $\hat{k}(\theta, \phi)$ is

$$\sigma_i(\theta) = r_0^2 \int d\Omega'_{k'} \sum_{j=1}^{2} \left( \vec{\vec{\Pi}} \cdot \hat{e}^i, \hat{e}_t^j \right) \left( \hat{e}_t^j, \vec{\vec{\Pi}} \cdot \hat{e}^i \right), \tag{4.2.14}$$

where we integrate over outgoing angles and sum over final polarizations. The angular integration is performed using the identities

$$\sum_{j=1}^{2} \hat{e}_t^j \hat{e}_t^{*j} = \vec{\vec{1}} - \hat{k}'\hat{k}', \tag{4.2.15}$$

$$\int d\Omega'_k \hat{k}'_\alpha \hat{k}'_\beta = (4\pi/3)\delta_{\alpha\beta}, \tag{4.2.16}$$

where equation (4.2.16) is a completeness statement specifying that the polarization modes entirely span the two-dimensional space transverse to $\hat{k}'(\theta', \phi')$. Using equations (4.2.15) and (4.2.16), the total cross section is found to be (Ventura, 1979)

$$\sigma_i(\theta) = \sigma_T \left| \vec{\vec{\Pi}} \cdot \hat{e}^i(\theta, \phi) \right|^2 . \tag{4.2.17}$$

This total scattering cross section can also be obtained directly from the wave equation

$$\left[ N_i^2 \left( \vec{\vec{1}} - \hat{k}\hat{k} \right) - \vec{\vec{\epsilon}} \right] \cdot \hat{e}^i = 0 \tag{4.2.18}$$

via the optical theorem (Canuto, Lodenquai, and Ruderman, 1971; Börner and Mészáros, 1979). For this, we project the wave equation (4.2.17) onto the transverse vector $\hat{e}_t$ to obtain

$$N_i^2 = 1 - v \left( \hat{e}_i^t, \vec{\vec{\Pi}} \cdot \hat{e}^i \right). \tag{4.2.19}$$

The optical theorem then gives

$$\sigma_i = \frac{2\omega}{cn_e} \mathrm{Im}(N_i). \tag{4.2.20}$$

In this expression, however, we must include the radiative damping

$$\gamma = \frac{2}{3} \frac{e^2}{mc^3} \omega, \tag{4.2.21}$$

which modifies the elements of $\Pi_{\alpha\beta}$ to

$$\pi_{\pm} = \left(1 \pm u^{1/2} + i\gamma\right)^{-1}, \quad \pi_z = \left(1 + i\gamma\right)^{-1}. \tag{4.2.22}$$

With this, equation (4.2.20) yields the same result as equation (4.2.17).

*b) Quantum Treatment of Thomson Scattering*

Whereas the cyclotron absorption or emission process is a one-vertex, first-order process involving only one photon, scattering has one photon coming in and one going out, i.e. it is a second-order, two-vertex process with two photons. Thus, cyclotron absorption is proportional to the first power of the fine structure constant $\alpha_F$ but scattering is proportional to the second power of $\alpha_F$ when far from resonance. A quantum treatment must reproduce, in the appropriate limits, the classical magnetized scattering cross section derived above, as shown for $\theta = 0$ and $\theta = \pi/2$ by Canuto, Lodenquai, and Ruderman (1971). We consider the case of arbitrary propagation direction in a strong magnetic field with an electron initially in the ground Landau level $n = 0$, which after the scattering process returns to $n = 0$ (Nagel, 1981a, 1982; Kaminker, Pavlov, and Shibanov, 1982). In contrast to the treatment of cyclotron absorption, we must now extend the quantum mechanical perturbation series to the second order to obtain the matrix element of the perturbation Hamiltonian $T$:

$$\langle f \mid T \mid i \rangle = \sum_r \frac{\langle f \mid H_1 \mid r \rangle \langle r \mid H_1 \mid i \rangle}{E_i - E_r} + \langle f \mid H_2 \mid i \rangle. \tag{4.2.23}$$

The transitions are between $\mid i \rangle = \mid 0, s, 0 \rangle \mid k \rangle$ and $\mid f \rangle = \mid 0, s', p' \rangle \mid k' \rangle$, where $\mid k \rangle$, $\mid k' \rangle$ are the initial and final photon states, and in the first term of equation (4.2.23) we sum over all the possible intermediate states $\mid r \rangle$ (not necessarily restricted to $n = 0$). However, if we again make the dipole approximation, only the $\Delta n = 0$ terms remain of the sum over $r$, of which there are two, corresponding to the top and middle diagrams in Figure 4.2.1. The corresponding amplitudes for these two processes in Figure 4.2.1 are

$$T_a = \frac{2\pi\alpha_F \hbar^2 c}{\omega L^3} \frac{\left(\hbar\omega_c/m\right)^{1/2} e_-'^{*} \left(\hbar\omega_c/m\right)^{1/2} e_-}{\hbar\omega - \hbar\omega_c}, \tag{4.2.24}$$

$$T_b = \frac{2\pi\alpha_F \hbar^2 c}{\omega L^3} \frac{\left(\hbar\omega_c/m\right)^{1/2} e_+ \left(\hbar\omega_c/m\right)^{1/2} e_+'^{*}}{\hbar\omega - \left(\hbar\omega_c + 2\hbar\omega\right)}, \tag{4.2.25}$$

The last term of the transition matrix, corresponding to the interaction

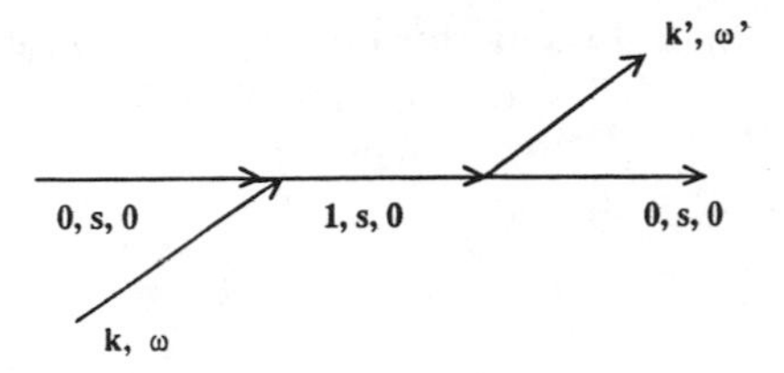


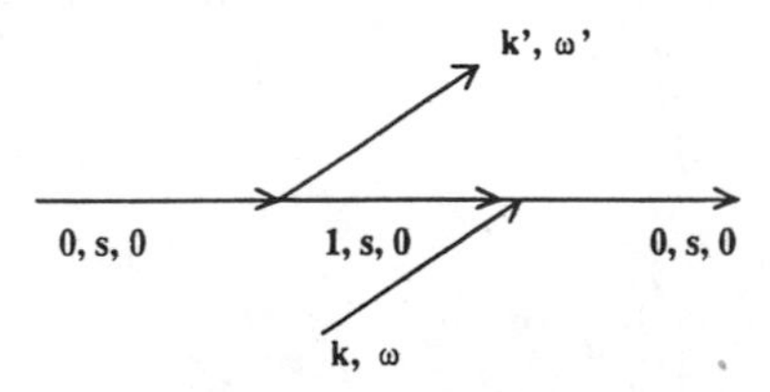


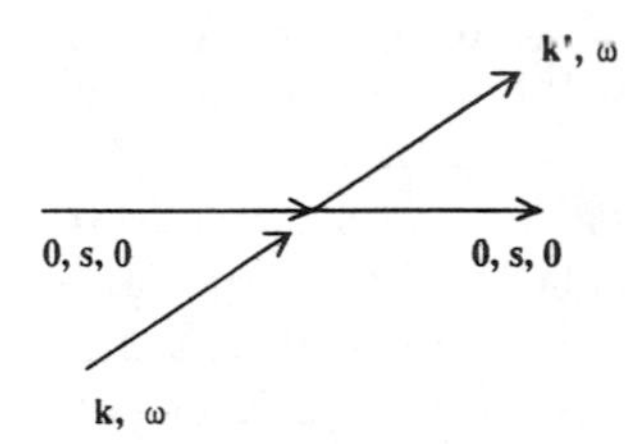


FIGURE 4.2.1 Feynman diagrams for Compton scattering from the ground state.

quadratic in $\vec{A}$, which is described by the seagull diagram at the bottom of Figure 4.2.1, contributes

$$T_c = \frac{\alpha_F \hbar}{2mc} \frac{2\pi \hbar c^2}{\omega L^3} 2\left(e'^*_+ e_+ + e'^*_- e_- + e'^*_z e_z\right). \tag{4.2.26}$$

In the absence of a magnetic field, Thomson scattering is described entirely by this latter amplitude $T_c$, the first-order terms in $\vec{A}$, $T_a$, and $T_b$ being characteristic of the magnetic scattering. The sum of the three amplitudes is then

$$\langle f \mid T \mid i \rangle = \frac{2\pi \alpha_F \hbar^2 c}{m \omega L^3} \left( \frac{\omega}{\omega + \omega_c} e'^*_+ e_+ + \frac{\omega}{\omega - \omega_c} e'^*_- e_- + e'^*_z e_z \right). \tag{4.2.27}$$

Using the Golden Rule and the density of photon states $\rho = L^3\omega^2/(2\pi c)^3\hbar$, we obtain the scattering rate

$$Q = \frac{2\pi}{\hbar}|\langle f|T|i\rangle|^2\rho$$
$$= \frac{\alpha_F^2\hbar^2}{m^2cL^3}\left|\frac{\omega}{\omega+\omega_c}e'^*_+e_+ + \frac{\omega}{\omega-\omega_c}e'^*_-e_- + e'^*_ze_z\right|^2. \qquad (4.2.28)$$

In order to obtain the differential scattering cross section, we divide by the incoming photon flux $c/L^3$ to obtain

$$\frac{d\sigma}{d\Omega}(\hat{e}', \hat{k}' \leftarrow \hat{e}, \hat{k})$$
$$= \alpha_F^2\left(\frac{\hbar}{mc}\right)^2\left|\frac{1}{1-u^{1/2}}e'^*_+e_+ + \frac{1}{1-u^{1/2}}e'^*_-e_- + e'^*_ze_z\right|^2, \qquad (4.2.29)$$

which is identical to the classical expression of equation (4.1.13). This exact agreement between the classical and the quantum scattering cross sections is a special case of a more general result whereby particles in a harmonic oscillator potential yield the same cross section quantum mechanically as classically, as long as $\hbar\omega \ll mc^2$. For relativistic motions, however, the Landau levels are no longer harmonically spaced (equidistant) and the classical calculation ceases to agree with the quantum one, as also seen in the nonmagnetic Klein-Nishina calculation.

The polarization dependence of the cross section is evident from the form of equation (4.2.28) or (4.2.29). Assuming an incoming photon $k(\theta, \varphi)$ and an outgoing photon $k'(\theta', \varphi')$, the $\theta', \varphi'$ dependences can be separated in the form

$$e'^*_\pm(\theta', \varphi') = e'^*_\pm(\theta')\exp(\pm i\varphi'), \quad e'^*_z(\theta', \varphi') = e'^*_z(\theta'). \qquad (4.2.30)$$

Averaging equation (4.2.29) over $\varphi'$, the mixed terms $e^*_+e_-$, etc., disappear, leaving

$$\left\langle\frac{d\sigma}{d\Omega}\right\rangle\varphi' = \alpha_F^2\left(\frac{\hbar}{mc}\right)^2\left[\frac{1}{(1+u^{1/2})^2}|e'_+|^2|e_+|^2 + \frac{1}{(1-u^{1/2})^2}|e'_-|^2|e_-|^2 + |e'_z|^2|e_z|^2\right]. \qquad (4.2.31)$$

From equation (3.2.31) for the normal modes, we see that the following completeness statement is valid:

$$|e_{\pm}^1|^2 + |e_{\pm}^2|^2 = \frac{1}{2}(1 + \cos^2\theta),$$
$$|e_z^1|^2 + |e_z^2|^2 = \sin^2\theta \tag{4.2.32}$$

and similarly for $e'$. Integrating therefore equation (4.2.31) over $\theta'$, we obtain the total cross section

$$\sigma_i(\theta) = \sigma_T\left[\frac{1}{(1 + u^{1/2})^2}|e_+^i(\theta)|^2 + \frac{1}{(1 - u^{1/2})^2}|e_-^i(\theta)|^2 + |e_z^i(\theta)|^2\right], \tag{4.2.33}$$

where the Thomson cross section is

$$\sigma_T = \frac{8\pi}{3}\alpha_F^2\left(\frac{\hbar}{mc}\right)^2 = \frac{8\pi}{3}r_0^2. \tag{4.2.34}$$

The resonance in the $e_-$ term at $u = 1$, characteristic of the cold plasma, does not lead to an infinite result if radiative damping $\gamma_r = (2/3)(e^2/mc^3)\omega_c$ is introduced, i.e.

$$\sigma_i(\theta) = \sigma_T\left[\frac{1}{(1 + u^{1/2})^2}|e_+^i(\theta)|^2 + \frac{1}{(1 - u^{1/2})^2 + \gamma_r^2}|e_-^i(\theta)|^2 + |e_z^i(\theta)|^2\right]. \tag{4.2.35}$$

This equation is identical to that derived classically, equation (4.2.17) or (4.2.20). We can see that, for $\omega \ll \omega_c$, or $u \gg 1$, the cross section is dominated by the $e_z$ terms. This is because a photon of energy $\hbar\omega$ much less than the ground Landau energy $\hbar\omega_c$ is unable to induce transverse motions in the $x$, $y$ plane but has no trouble moving the electron along the magnetic field direction $z$, along which the electron is essentially free. For the same reason, a wave of $\omega \ll \omega_c$ traveling along $\theta = 0°$ will have a very reduced scattering coefficient, since the transversality of the waves causes $|e_z^i|^2 = 0$. Furthermore, at $\omega \ll \omega_c$ the normal modes are almost linearly polarized, such that the $\vec{E}$ vector of the extraordinary wave is perpendicular to $\vec{B}_0$. Thus, at $\omega \ll \omega_c$, the extraordinary wave scattering coefficient is strongly reduced below the

Thomson value, even at large $\theta$. This low-frequency behavior of the cross sections is shown in Figure 4.2.2, where we have used the cold-plasma modes neglecting, for simplicity, vacuum effects.

At frequencies much greater than the cyclotron frequency, $\omega \gg \omega_c$, the effect of the magnetic field becomes less important. One can see this from equation (4.2.35), where in the limit $u \ll 1$ the term in square brackets is just $|e_+|^2 + |e_-|^2 + |e_z|^2 = 1$, so $\sigma_i \to \sigma_T$.

The divergence of the simple expression (4.2.33) for $\omega \to \omega_c$ shows that near the cyclotron frequency the perturbation theory breaks down and one must take into account the finite lifetime of the level, as in (4.2.35). The scattering cross section near the resonance is then approximately given by the Breit-Wigner expression

$$\sigma_{\text{res}} = \sigma_T \frac{\omega^2}{(\omega - \omega_c)^2 + \Gamma_r^2/4} |e_-|^2, \tag{4.2.36}$$

where $\Gamma_r = 2\gamma\omega_c = (4/3)\alpha_F(\hbar/mc^2)\omega_c^2$, which can be rewritten as

$$\sigma_{\text{res}} = 4\pi^2\alpha_F \frac{\hbar}{m} \frac{\Gamma_r/2\pi}{(\omega - \omega_c)^2 + \Gamma_r^2/4} |e_-|^2. \tag{4.2.37}$$

Using the prescription

$$\delta(\omega - \omega_c) = \frac{\Gamma_r/2\pi}{(\omega - \omega_c)^2 + \Gamma_r^2/4}, \tag{4.2.38}$$

one obtains finally

$$\sigma_{\text{res}} = 4\pi^2\alpha_F \frac{\hbar}{m} |e_-|^2 \delta(\omega - \omega_c), \tag{4.2.39}$$

which agrees with expression (4.1.27) for the first-order cyclotron absorption cross section. That is, at resonance, the second-order scattering becomes proportional to the first power of $\alpha_F$. It is not surprising that the cyclotron absorption cross section is the same as that for resonant scattering. This is because an absorbed photon is almost immediately (within $t_r \sim 10^{-15}B_{12}^{-2}$ s) reemitted and the combined absorption plus emission process is equivalent to the resonant scattering process.

The polarization exchange scattering $d\sigma_{ij}/d\Omega$ whereby photons of polarization $i$ are scattered into $j$ is given by equation (4.2.35) with $e' \equiv e_j'$, $e \equiv e_i$. The total (angle-integrated) polarization exchange cross

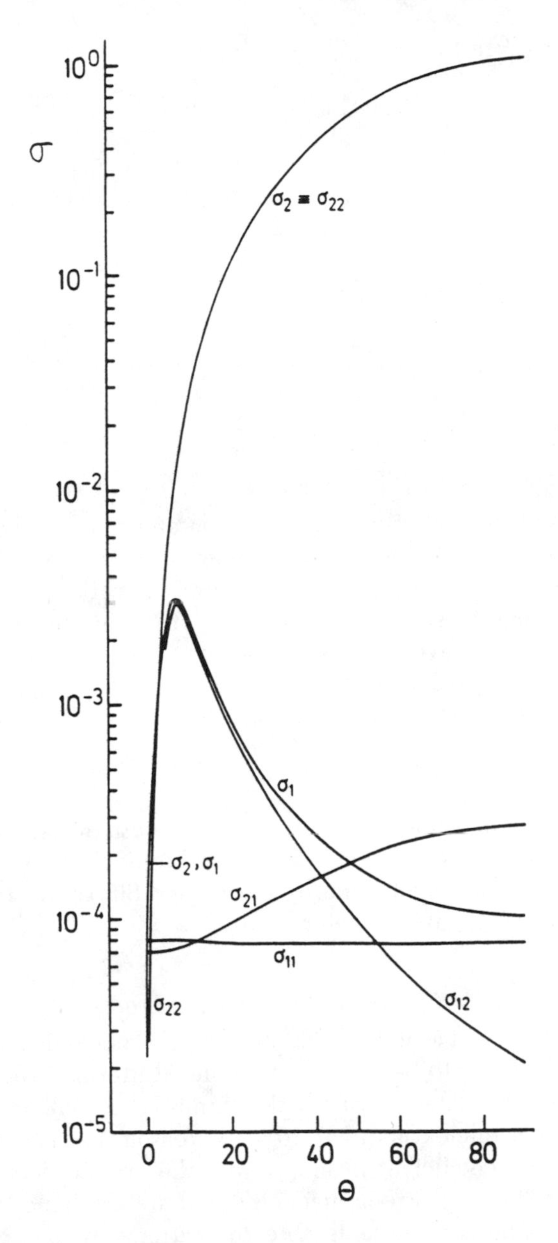


FIGURE 4.2.2 Cold-plasma scattering cross section in a magnetized plasma as a function of angle $\theta$ for $\omega/\omega_c = 10^{-2}$. The cross sections are normalized to the Thomson value. Reprinted with permission from G. Börner and P. Mészáros, 1979, *Plasma Phys.*, 21, 357; © 1979 Pergamon Press, PLC.

section is then given by

$$\frac{\sigma_{ij}}{\sigma_T} = A_z^j \mid e_z^i \mid^2 + \frac{A_+^j \mid e_+^i \mid^2}{\left(1 + u^{1/2}\right)^2} + \frac{A_-^j \mid e_-^i \mid^2}{\left(1 - u^{1/2}\right)^2 + \gamma^2}, \tag{4.2.40}$$

where

$$A_\alpha^j = \frac{3}{4} \int_{-1}^{1} d(\cos\theta') \mid e_\alpha^j(\theta', 0) \mid^2. \tag{4.2.41}$$

From equations (4.2.32), one sees that the $A_\alpha^i$ satisfy

$$\sum_{j=1}^{2} A_\pm^j = 1, \quad \sum_{j=1}^{2} A_z^j = 1, \tag{4.2.42}$$

so that it is sufficient to calculate them for only one of the modes. These integrals have to be evaluated numerically as a function of $\omega/\omega_c$, representative values (without vacuum correction) being listed in Table 4a (from Mészáros and Ventura, 1979) at the end of the present book. The $\varphi$-averaged values of $\sigma_{21}$ and $\sigma_{12}$ are shown in Figure 4.2.2 for a particular choice of $\omega/\omega_c$. By summing (4.2.40) over final polarizations, one again obtains the total cross sections

$$\sigma_i(\theta) = \sigma_{i1}(\theta) + \sigma_{i2}(\theta), \tag{4.2.43}$$

which coincide with equation (4.2.35) and are also shown in Figure 4.2.2. Another discussion of these and related topics with a somewhat different emphasis has been given by Pavlov and Shibanov (1979) and by Wang, Wasserman, and Salpeter (1988).

### *c) Vacuum Polarization Effects in the Scattering Process*

Just as for the refractive index of a strongly magnetized plasma, we expect vacuum effects to be important for the scattering process. The refractive index is related, through the optical theorem, to the forward-scattering coefficient (eq. 4.2.20). The reason the vacuum modifies the scattering coefficient is simply that the virtual electrons and positrons created by the strong magnetic field are also able to scatter waves. While a pure vacuum is able to scatter waves only in the forward direction ($\theta' = \theta$), the scattering by the real plasma electrons, which are able to scatter at $\theta' \neq \theta$, is modified indirectly by virtue of the vacuum corrections on the polarization eigenmodes. Thus, in order to obtain the vacuum-corrected scattering cross sections, in the limit $B/B_Q \ll 1$, one uses directly the expressions derived in subsection (*b*) above and substitutes the vacuum-modified eigenmodes $\hat{e}_i$ discussed in Chapter 3 (see eqs. 3.4.34). This vacuum modification introduces into the numerical values of the polarization exchange coefficients $A^i_\alpha$ of equation (4.2.41) an additional dependence on the plasma density. Some numerical values are given in Table 4b at the end of the present book.

The total Thomson cross section including vacuum effects averaged over $\varphi$ is shown in Figure 4.2.3, as a function of $\theta$, for several values of $\omega/\omega_c$ and for a particular choice of $\omega_p/\omega_c = 10^{-9}$ and $B = 10^{-1}B_Q$. The vacuum-modified cross sections, with the two modes indicated by V1, V2, are significantly different from the usual (unmodified) cross sections, labeled P1, P2, for this choice of parameters. The chosen values of $B/B_Q$ and $\omega_p/\omega_c$ are what might be encountered in accreting magnetized neutron stars and imply a significant vacuum correction at X-ray photon energies (see eq. 3.4.28). The polarization exchange cross sections $\sigma_{ij}(\theta)$ are also significantly affected.

The frequency dependence of the scattering cross section near the resonance, given in equation (4.2.36) or (4.2.39), exhibits qualitatively new features when vacuum modifications are important. Whereas for the usual cold plasma only the extraordinary mode P1 has a nonzero value of $e_-$, one finds that both the extraordinary and the ordinary vacuum-modified modes V1 and V2 contain a nonzero $e_-$ component. It follows that, whereas in the usual cold plasma only the extraordinary mode shows resonant cyclotron absorption (or scattering), in a vacuum-modified plasma both modes are resonant. This is shown in Figure 4.2.4, where the frequency dependence of the cross sections near the resonance is shown for the ordinary and extraordinary modes. In the left panel of Figure 4.2.4, it is seen that the unmodified mode P1 (extraordinary) is resonantly absorbed at $\omega = \omega_c$ for all angles of propagation, having a smooth $(1 + \cos^2\theta)$ angular dependence. On the other hand, the unmodified mode P2 (ordinary) is nonresonant at all angles. By contrast, if vacuum effects are important (right panel of Fig. 4.2.4), the ordinary mode V2 becomes resonant for a range of angles and the extraordinary mode V1 is also resonant, but more so than the ordinary, with a frequency dependence which differs from that in the unmodified case. Thus, in a cold, nonrelativistic plasma, one of the major effects of vacuum polarization is to make both plasma modes resonant at the cyclotron frequency. However, the cold-plasma case is in any case an idealization. Even for those situations where

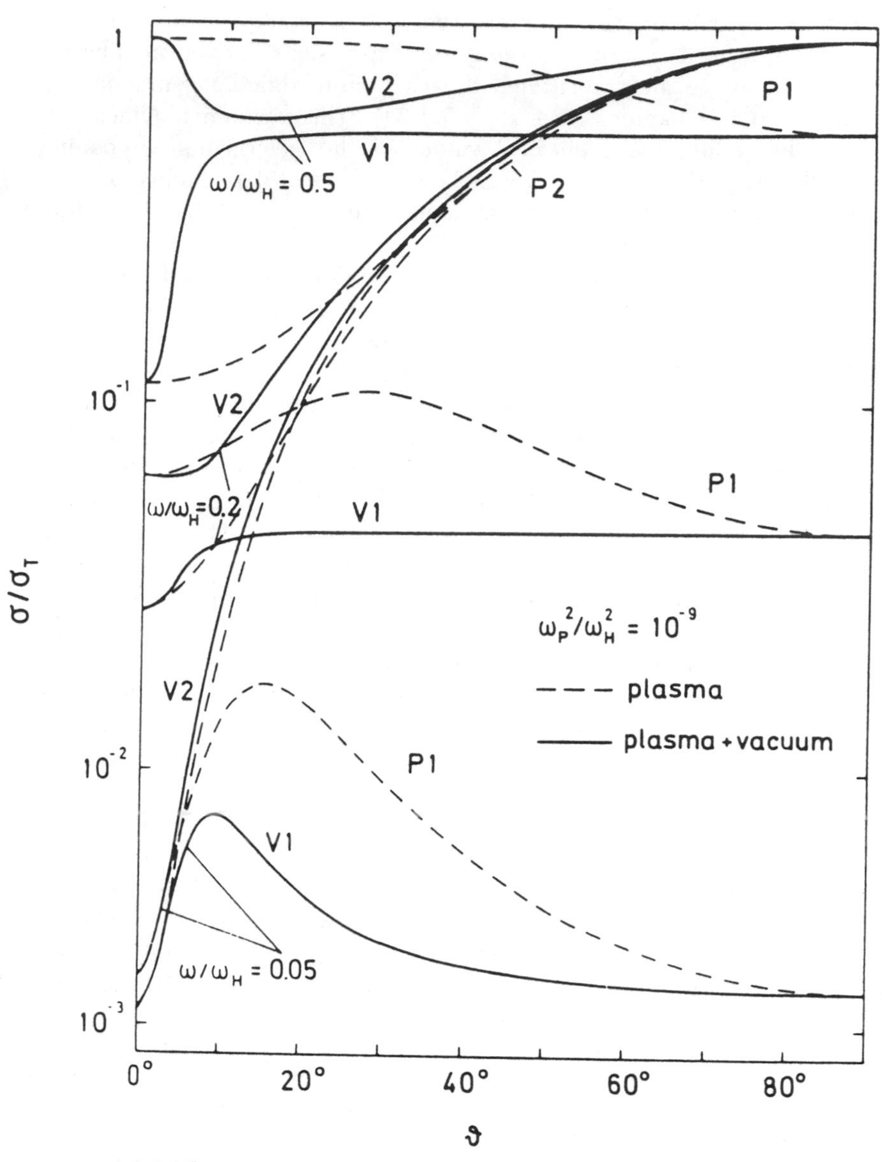


FIGURE 4.2.3 The scattering cross section $\sigma_i$ versus $\theta$ for two values of $\omega/\omega_c$ (where $\omega_{\rm H} \equiv \omega_c$). The modes 1 and 2 are shown for the pure magnetized plasma case without vacuum effects (P1 and P2) and for the magnetized plasma plus vacuum case (V1 and V2) for a particular choice of the plasma frequency $\omega_p$. From P. Mészáros and J. Ventura, 1979, *Phys. Rev.*, D19, 3565.

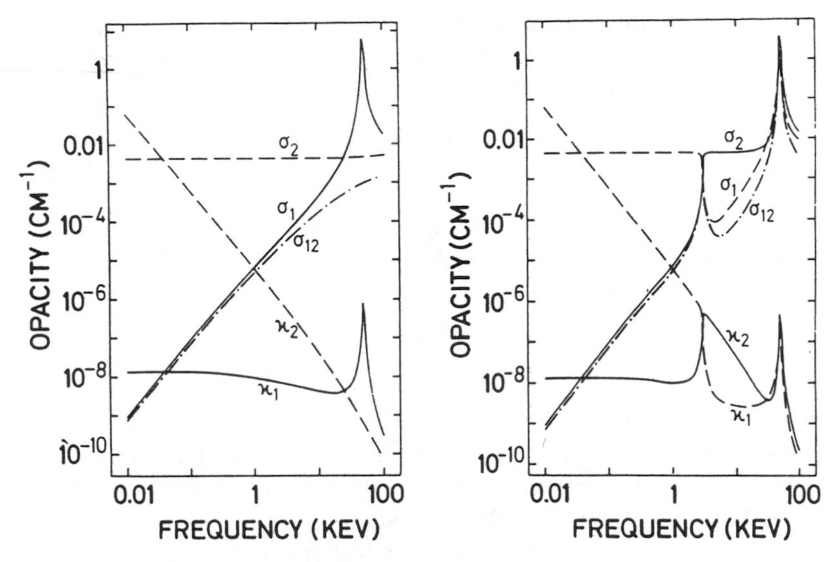


FIGURE 4.2.4 Angle-averaged scattering ($\sigma_i$) and bremsstrahlung ($\kappa_i$) opacities (cross section times density) for the ordinary and extraordinary modes for $n_e = 10^{22}$ cm$^{-3}$ and $\hbar\omega_c = 50$ keV. The left panel is without, and the right panel is with, inclusion of vacuum effects. From J. Ventura, W. Nagel, and P. Mészáros, 1979, *Ap. J.* (*Letters*), 233, L125.

vacuum corrections are unimportant, the inclusion of spin-flip transitions and of thermal motions introduces an additional dissipation which causes the ordinary mode to become resonant, as will be discussed later.

## 4.3 Compton Scattering in a Hot Plasma

### a) *Differential Cross Section*

In a hot plasma, the electrons have thermal motions which may lead to significant photon energy changes because of the difference between the photon frequency in the laboratory and in the electron rest frame. Of course, even in a cold plasma $kT \simeq 0$ there is some photon energy degradation, due to the Compton recoil effect on the electron in a scattering event. Both the Doppler (thermal) and recoil effects are corrections to the classical Thomson cross section. We shall discuss these effects here, using a nonrelativistic quantum approach (Kirk and Mészáros, 1980; Nagel, 1981b, 1982). What is different from the cold-electron case calculated in the previous section is that here (*a*) we no longer assume stationary electrons with $p_z = 0$ but instead will average over an electron longitudinal momentum distribution

$$f(p) = (2\pi mkT_{\parallel})^{-1/2} \exp(-p^2/2mkT_{\parallel}), \tag{4.3.1}$$

where $p \equiv p_z$ is along the magnetic field, and (*b*) we no longer neglect the photon-momentum component $k_z$ along the field, i.e. we no longer use the dipole approximation $e^{i\vec{k}\cdot\vec{r}}$ but rather use

$$e^{i\vec{k}\cdot\vec{r}} \approx e^{ik_z z}. \tag{4.3.2}$$

In addition, we shall continue neglecting spin and shall again assume that both the initial and final states are $n = n' = 0$. That is, we shall keep first-order corrections in the electron and photon momentum in computing the scattering matrix elements.

Including these corrections, the scattering is represented by the five diagrams shown in Figure 4.3.1. Setting $\hbar = c = m = 1$, the respective contributions to the scattering amplitude of the diagrams Figures 4.3.1a–e are

$$T_a = \frac{2\pi\alpha_F}{\sqrt{\omega\omega'}\,L^3} \frac{\sqrt{\omega_c}\,\sqrt{\omega_c}}{\frac{1}{2}p^2 + \omega - \omega_c - \frac{1}{2}(p+k)^2} e_- e_-'^*,$$

$$T_b = \frac{2\pi\alpha_F}{\sqrt{\omega\omega'}\,L^3} \frac{\sqrt{\omega_c}\,\sqrt{\omega_c}}{\frac{1}{2}p^2 + \omega - \omega_c - \frac{1}{2}(p-k')^2 - \omega - \omega'} e_+ e_+'^*,$$

$$T_c = \frac{2\pi\alpha_F}{\sqrt{\omega\omega'}\,L^3} \frac{(p+k)(p+k-k'/2)}{\frac{1}{2}p^2 + \omega - \frac{1}{2}(p+k)^2} e_z e_z'^*,$$

$$T_d = \frac{2\pi\alpha_F}{\sqrt{\omega\omega'}\,L^3} \frac{(p-k')(p-k'+k/2)}{\frac{1}{2}p^2 + \omega - \frac{1}{2}(p-k')^2 - \omega - \omega'} e_z e_z'^*,$$

$$T_e = \frac{2\pi\alpha_F}{\sqrt{\omega\omega'}\,L^3} \left(e_+'^* e_+ + e_-'^* e_- + e_z'^* e_z\right). \tag{4.3.3}$$

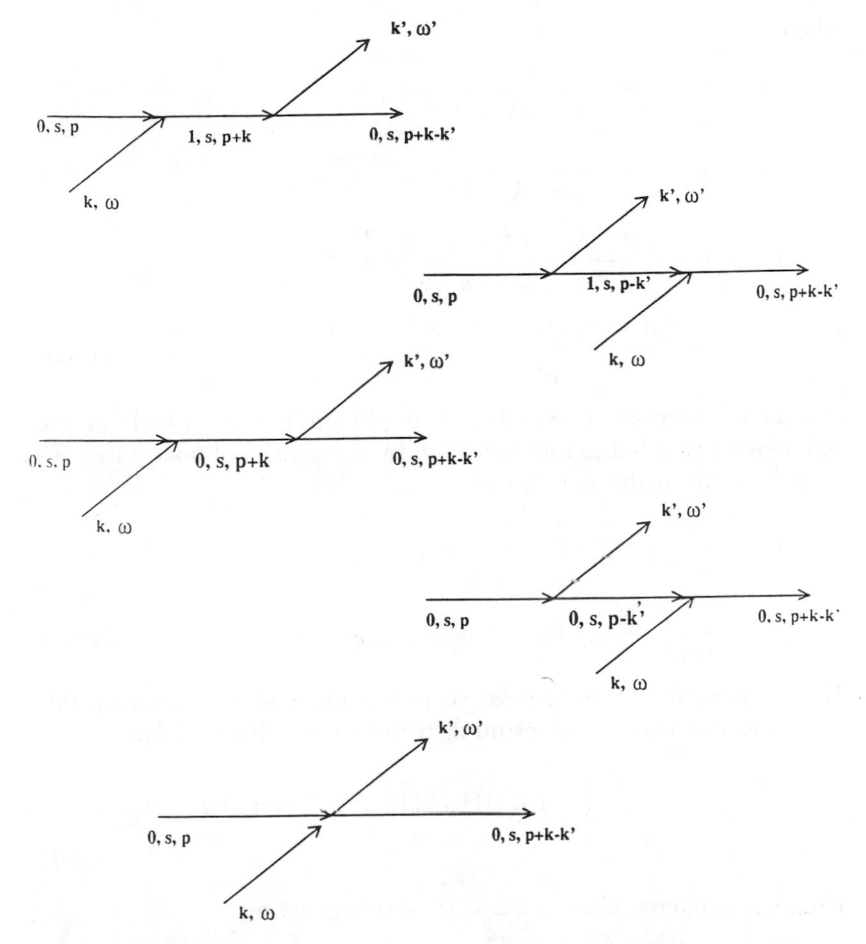


FIGURE 4.3.1 Feynman diagrams for Thomson scattering in a hot nonrelativistic plasma, with the electron starting and ending in the ground state.

Adding the terms in (4.3.3), we obtain the scattering amplitude in the form

$$\langle f \mid T \mid i \rangle = \frac{2\pi\alpha_F}{\sqrt{\omega\omega'}\,L^3}\langle e' \mid \Pi \mid e \rangle$$

$$= \frac{2\pi\alpha_F}{\sqrt{\omega\omega'}\,L^3}\left(\Pi_+ e'^*_+ e_+ + \Pi_- e'^*_- e_- + \Pi_z e'^*_z e_z\right), \qquad (4.3.4)$$

where

$$\Pi_+ = 1 - \frac{\omega_c}{\omega' + \omega_c - pk' + k'^2/2},$$

$$\Pi_- = 1 + \frac{\omega_c}{\omega - \omega_c - pk - k^2/2},$$

$$\Pi_z = 1 + \frac{(p + k/2)(p + k - k'/2)}{\omega - pk - k^2/2} - \frac{(p - k'/2)(p - k' + k/2)}{\omega' - pk' + k'^2/2}. \qquad (4.3.5)$$

Instead of integrating over the final photon energy, which in the cold-plasma case led to multiplication by the density of final states, we write here the scattering rate as

$$k_c = \frac{2\pi}{\hbar}|\langle f|\Pi|i\rangle|^2 \delta(E_i - E_f)$$

$$= \frac{\alpha_F^2 \hbar^2}{m^2 c L^3}\frac{\omega'}{\omega}|\langle f|\Pi|i\rangle|^2 \delta(\omega + \Delta\omega - \omega'). \qquad (4.3.6)$$

The differential scattering cross section is obtained by averaging this scattering rate over the electron distribution function (4.3.1),

$$\frac{d^2\sigma}{d\omega\, d\Omega} = r_0^2 \frac{\omega'}{\omega}\int dp f(p)\,|\langle e'|\Pi|e\rangle|^2 \delta(\omega + \Delta\omega - \omega'). \qquad (4.3.7)$$

The photon energy change $\Delta\omega = \omega' - \omega$ is given by

$$\hbar\,\Delta\omega = (p^2 - p'^2)/2m, \qquad (4.3.8)$$

while the momentum change is

$$p' - p = \hbar\,\Delta k = \frac{\hbar\omega'}{c}\cos\theta' - \frac{\hbar\omega}{c}\cos\theta \qquad (4.3.9)$$

so that

$$\Delta\omega = \Delta k \frac{p}{m} - \hbar\frac{(\Delta k)^2}{2m}. \qquad (4.3.10)$$

The energy and momentum conservation laws (4.3.7) and (4.3.9) determine the final frequency $\omega'$, given an initial $p$ and the incoming and outgoing angles $\theta, \theta'$. Similarly, for a given initial and final state of the photon $(\omega, \theta, \omega', \theta')$, there is only one initial electron momentum $p_0$ which satisfies energy-momentum conservation, namely

$$p_0 = m\frac{\Delta\omega}{\Delta k} + \frac{1}{2}\hbar\,\Delta k. \tag{4.3.11}$$

This facilitates greatly the averaging over $p$ (actually $p_z$), since we have

$$\begin{aligned}
\frac{d^2\sigma}{d\omega\, d\Omega} &= r_0^2\frac{\omega'}{\omega}\int dp f(p)|\langle e'\,|\,\Pi\,|\,e\rangle|^2\delta(\omega + \Delta\omega - \omega') \\
&= r_0^2\frac{\omega'}{\omega}\int dp f(p)|\langle e'\,|\,\Pi\,|\,e\rangle|^2\frac{\delta(p - p_0)}{\left|\dfrac{d(\omega + \Delta\omega - \omega')}{dp}\right|} \\
&= r_0^2\frac{\omega'}{\omega}\frac{m}{|\Delta k|}f(p_0)|\langle e'\,|\,\Pi(p_0)\,|\,e\rangle|^2.
\end{aligned} \tag{4.3.12}$$

The spectrum of the scattered photons is determined mainly by the distribution function $f(p)$ and by the cyclotron resonance appearing in the normal modes $\hat{e}$, as well as in the resonant denominator of the polarization matrix $\Pi$. The scattering probability is largest when the frequency of the incoming wave is such that, after being Doppler-shifted to the electron frame of $p_0(\omega, \theta, \omega', \theta')$, it coincides with the cyclotron frequency. An example shown in the left panel of Figure 4.3.2 uses hot-plasma modes without vacuum effects. In this example, an extraordinary photon (1) is scattered in a plasma of $kT = 10$ keV. The incoming energy is $\hbar\omega = 40$ keV, the incoming angle is $\theta = 45°$, and the cyclotron resonance is at $\hbar\omega_c = 50$ keV. We can see both the broad Doppler hump and the sharp resonance, whose exact position depends on $\theta$ and $\theta'$.

In the case of $\Delta\omega = 0$, which is equivalent to coherent scattering, one can see that equation (4.3.12) reduces to the earlier Thomson expression,

$$\left(\frac{d\sigma}{d\omega\, d\Omega}\right)_{\text{coh}} = r_0^2|\langle e'\,|\,\Pi\,|\,e\rangle|^2\delta(\omega' - \omega) \tag{4.3.13}$$

(see eq. 4.2.29). It is also evident from the expression (4.3.12) that the differential scattering cross section satisfies the principle of

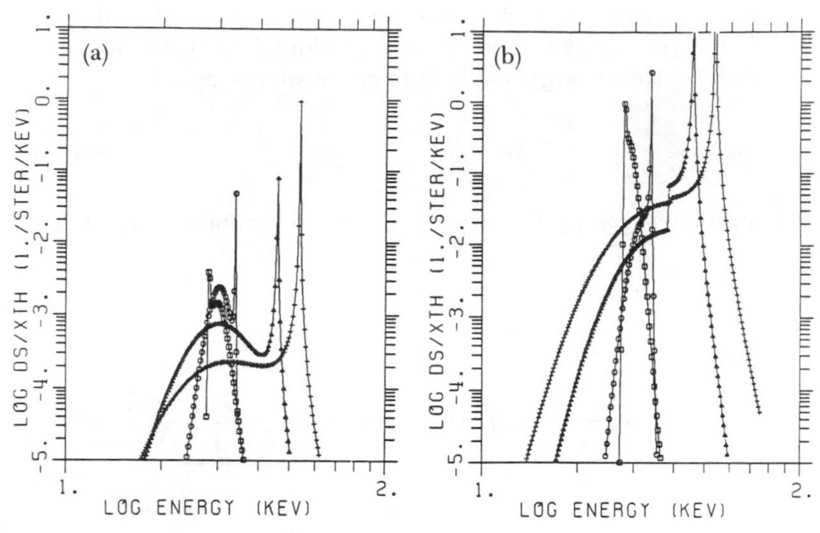


FIGURE 4.3.2 Differential scattering cross section for four different final angles $\theta_f = 20°, 70°, 120°$, and $160°$, from left to right, and incoming angle $\theta_i = 45°$, $kT = 10$ keV, $\hbar\omega_c = 38$ keV. The left panel (a) is without vacuum polarization, while the right panel (b) includes vacuum polarization. Note in the latter the polarization jump for some angles, caused by the mode ambiguity near the second vacuum frequency $\omega_{v2}$. From P. Mészáros and W. Nagel, 1985, *Ap. J.*, 298, 147.

detailed balance, as it should. That is,

$$\frac{d^2\sigma}{d\omega\, d\Omega}(\omega', \theta' \leftarrow \omega, \theta) = \left(\frac{\omega'}{\omega}\right)^2 \exp\left(\frac{\hbar\omega - \hbar\omega'}{kT}\right) \times \frac{d^2\sigma}{d\omega\, d\Omega}(\omega, \theta \leftarrow \omega', \theta') \tag{4.3.14}$$

since the matrix elements $\langle e' | \Pi | e \rangle$ are symmetric with respect to interchange of the initial and final states. The factor $(\omega'/\omega)^2$ represents the ratio of the density of photon states, while $\exp[(\hbar\omega - \hbar\omega')/kT] = f(p')/f(p)$ represents the ratio of the population of electron states. The fact that the scattering cross section satisfies detailed balance ensures that this process leads to the correct equilibrium photon spectrum, which is the Wien spectrum

$$\bar{n}(\omega) \propto \omega^2 \exp(-\hbar\omega/kT). \tag{4.3.15}$$

The equilibrium spectrum is given by the Wien expression as long as stimulated scattering can be neglected, an often encountered situation. Otherwise, if one includes stimulated scattering, the equilibrium spectrum is given by the Planck expression.

*b) The Polarization Modes in a Hot Plasma*

The polarization modes used in equation (4.3.12) for calculating the Compton (as opposed to Thomson) scattering cross section must be consistent with the matrix elements $\Pi_+, \Pi_-, \Pi_z$. Since these depend on $p$, i.e. they are averaged over the electron momentum distribution, so must the polarization eigenmodes include the thermal motions of the electrons. In Chapter 3.3, the quantum mechanical expression for the dielectric tensor, or the related polarization tensor, led to an expression for the hot-plasma eigenmodes which included first-order $p$ and $k$ corrections, averaged over $f(p)$. The dielectric tensor and the refractive index are directly related to the coherent forward-scattering amplitude in the plasma. It is therefore instructive to restate here the normal mode structure in the framework of the scattering amplitudes described in this chapter.

The connection between the refractive index and the forward-scattering cross sections is well known from optics (e.g. Feynman, Leighton, and Sands, 1963). For a given incident wave $(\omega, \theta)$, one will have some probability for scattering out of the beam $(\theta', \varphi') \neq (\theta, \varphi)$ and some probability for forward scattering $(\theta', \varphi') = (\theta, \varphi)$. Whereas the waves scattered out of the beam have random phases relative to that of the incident wave, the forward-scattered waves must combine with the incident wave coherently. This is what leads to a finite change in the phase velocity, that is, to a refractive index $N$ different from unity. The refractive index is related to the forward-scattering amplitude $f(\omega, \theta)$ by the relation (eq. 4.2.20)

$$N(\omega, \theta) = 1 + 2\pi n_e \left(\frac{c}{\omega}\right)^2 f(\omega, \theta). \tag{4.3.16}$$

In general, $N$ and $f$ are complex numbers, the imaginary part describing the damping of the wave. The extinction coefficient, or opacity $\kappa$, is

$$\kappa = 2\frac{\omega}{c}\mathrm{Im}(N) = 4\pi n_e \frac{c}{\omega}\mathrm{Im}(f). \tag{4.3.17}$$

In the system of rotating coordinates $(+, -, z)$, with $\vec{B}_0$ along $z$ and $\vec{k}(\theta, \varphi)$, the coherent scattering amplitude for the cold plasma was computed in equation (4.2.28):

$$\langle j' \mid F \mid j \rangle = -r_0 \left( \frac{\omega}{\omega + \omega_c} e'^*_+ e_+ + \frac{\omega}{\omega - \omega_c} e'^*_- e_- + e'^*_z e_z \right). \tag{4.3.18}$$

In terms of the dimensionless scattering amplitude $\Pi = -F/r_0$, one sees that for a cold plasma in rotating coordinates the components of $\Pi$ are

$$\Pi_\pm = \frac{\omega}{\omega \pm \omega_c}, \quad \Pi_z = 1. \tag{4.3.19}$$

If one includes radiation damping, $\gamma_r = 2e^2\omega^2/3mc^3$ (notice that the $\gamma_r$ defined here is $\omega$ times that defined in eq. 4.2.21), the amplitudes are

$$\Pi_\pm = \frac{\omega}{\omega \pm \omega_c + i\gamma_r}, \quad \Pi_z = \frac{\omega}{\omega + i\gamma_r}. \tag{4.3.20}$$

In a cold plasma, the scattering amplitudes have no angle dependence so that the $\Pi_{+,-,z}$ of equation (4.3.20) are also valid for forward scattering. The plasma eigenmodes are derived in equation (3.2.31) as a function of the parameter

$$b^{-1} = \xi = \frac{2\cos\theta}{\sin^2\theta} \frac{\Pi_+ - \Pi_-}{2\Pi_z - \Pi_+ - \Pi_-}, \tag{4.3.21}$$

which for the cold-plasma values of equation (4.2.7) reduces to

$$b^{-1} = \xi = \frac{2\cos\theta}{\sin^2\theta} \frac{\omega + i\gamma_r}{\omega_c} \simeq \frac{2\cos\theta}{\sin^2\theta} \frac{\omega}{\omega_c}. \tag{4.3.22}$$

For the case of a hot plasma, the eigenmodes are defined in the same way, but in equation (4.3.21) we must use the hot $\Pi_{+,-,z}$ of equations (4.3.5) specialized to the coherent forward-scattering case,

$\omega' = \omega$, $\theta' = \theta$ (Nagel, 1982); that is

$$\Pi'_+ = \int dp f(p) \frac{\omega - kp + k^2/2}{\omega + \omega_c - kp + k^2/2 + i\gamma_r},$$

$$\Pi'_- = \int dp f(p) \frac{\omega - kp - k^2/2}{\omega - \omega_c - kp + k^2/2 + i\gamma_r},$$

$$\Pi'_z = \int dp f(p) \frac{\omega}{k} \left( \frac{p + k/2}{\omega - kp - k^2/2 + i\gamma_r} - \frac{p - k/2}{\omega - kp + k^2/2 + i\gamma_r} \right). \tag{4.3.23}$$

The convergence of these integrals is ensured with the inclusion of the radiation damping $\gamma_r$. With this, the poles of the components $\Pi_{+,-,z}$ of the polarization tensor lie in the lower complex half-plane Im $\omega < 0$ and satisfy the Kramers-Kronig relations as they should. For a one-dimensional Maxwell-Boltzmann distribution $f(p)$, the integrals can be expressed in terms of the plasma dispersion function $W(z)$ (Fried and Conte, 1961)

$$\Pi^f_+ = 1 + i\sqrt{\pi}\, \frac{\omega_c + i\gamma_r}{\Delta\omega} W\left( \frac{\omega + \omega_c + k^2/2 + i\gamma_r}{\Delta\omega} \right),$$

$$\Pi^f_- = 1 - i\sqrt{\pi}\, \frac{\omega_c - i\gamma_r}{\Delta\omega} W\left( \frac{\omega - \omega_c - k^2/2 + i\gamma_r}{\Delta\omega} \right),$$

$$\Pi^f_z = -i\sqrt{\pi}\, \frac{\omega}{\kappa} \frac{\omega + i\gamma_r}{\Delta\omega} \left[ W\left( \frac{\omega - k^2/2 + i\gamma_r}{\Delta\omega} \right) - W\left( \frac{\omega + k^2/2 + i\gamma_r}{\Delta\omega} \right) \right], \tag{4.3.24}$$

where $\Delta\omega = \omega |\cos\theta| (2kT/m)^{1/2}$. These functions are calculated numerically (Hui *et al.*, 1978; Humlicek, 1979). The Doppler broadening is strongest in the $\Pi_-$ component. Because of this, equation (4.3.24) yields a much smaller value of $\Pi_-$ at $\omega \simeq \omega_c$ than the cold-plasma value $\Pi_- = \omega/(\omega - \omega_c + i\gamma_r)$. The hot-plasma polarization eigenmodes are, therefore, different from the cold ones at $\omega \simeq \omega_c$.

Also, in a hot plasma, the imaginary part of

$$b^{-1} = \xi = \frac{2\cos\theta}{\sin^2\theta} \frac{\Pi_+^f - \Pi_-^f}{2\Pi_z^f - \Pi_+^f - \Pi_-^f} \tag{4.3.25}$$

is no longer negligible with respect to the real part. Because of this, even in the absence of vacuum polarization, the eigenmodes $e_1, e_2$ are no longer completely orthogonal. As a result, a nonzero $|e_-|$ component appears in the ordinary wave at $\omega \simeq \omega_c$. This causes the resonance in $\Pi_-$ to appear also in the scattering cross section for the ordinary wave (Kirk and Mészáros, 1980; Nagel, 1981b). The thermal resonance, while present at all angles for the extraordinary wave, appears for the ordinary wave only at angles not too close to 0° or 90°, because in these directions the modes are always orthogonal. For these particular directions, the hot-plasma cross sections are practically identical to the cold-plasma ones. The total scattering cross sections with thermal effects are most easily obtained by means of the optical theorem, which now reads

$$\sigma = -4\pi r_0 \frac{c}{\omega}\left(\text{Im}\,\Pi_+^f |e_+|^2 + \text{Im}\,\Pi_-^f |e_-|^2 + \text{Im}\,\Pi_z^f |e_z|^2\right). \tag{4.3.26}$$

Of course, the same expression can be obtained by integrating equation (4.3.12) over the final angles and frequencies. The net result is to broaden the cyclotron resonances, giving a full width at half-maximum of order $\Delta\omega/\omega \simeq (2kT_\parallel/m_e c^2)^{1/2}|\cos\theta|$, which depends on the angle $\theta$ of observation.

The vacuum correction to the hot normal modes has already been discussed (see eq. 3.4.27). In terms of the polarization tensor $\vec{\vec{\Pi}}$ in rotating coordinates, the correction can be achieved by replacing $\Pi_z^f \to \Pi_z^f - 3\delta/v$, where $\delta$ and $v$ are defined in equations (3.4.6). Thus, the $b$ parameter of equation (4.3.25) becomes

$$b^{-1} = \xi = \frac{2\cos\theta}{\sin^2\theta} \frac{\left(\Pi_+^f - \Pi_-^f\right)}{2\left(\Pi_z^f + 3\delta v^{-1}\right) - \Pi_+^f - \Pi_-^f}\,. \tag{4.3.27}$$

The complete nonrelativistic expressions include both thermal and vacuum effects, both in the resonant denominators and in the polarization eigenmodes. Calculations for particular choices of parameters are given in Mészáros and Nagel (1985a).

## 4.4 The Coulomb and Bremsstrahlung Processes

### *a) The Coulomb Process*

The collision of charged particles in the presence of a strong magnetic field (the Coulomb collision process) has a different character from that in the field-free case. This is caused by the restricting influence of the field in the transverse motions of both electrons and protons, the only remaining truly free motion being along the field (Ventura, 1973). This restriction is, of course, less severe on the protons, which because of their large mass may be considered to have wave functions corresponding to plane waves for most energies and field strengths of interest. However, the electrons must usually be treated allowing for their gyromotion. In general, the interaction between electrons is less important than that between electrons and protons in a magnetic field. This is unlike the field-free situation, where the cross section for electron-electron collision is large by virtue of the small masses of the electrons. However, in a one-dimensional encounter (as is the case for large $B$) of two equal particles, the momenta can only remain the same or be exchanged and since the particles are identical, the final state is unchanged from the initial. Thus, electron-electron scattering is suppressed with respect to electron-proton scattering, at least in the nonrelativistic, spinless quantum treatment. A more careful treatment introduces a finite interaction probability, but it is only at relativistic energies that the electron-electron interaction becomes competitive with the electron-proton collisions (e.g. Bussard, 1980; Langer, 1981).

The Coulomb process, to lowest order, can be treated neglecting the emission of radiation (in the next order, of course, the change in the momentum of the electron leads to the emission of a photon; see e.g. Jackson, 1975). In the radiationless limit, this process relates to at least two important problems in neutron stars: determining the particle distribution relaxation time scale (thermalization time) and studying the energy and momentum deposition by the infalling matter in the quasi-stationary atmosphere of an accreting neutron star. The problem of the equilibrium distribution of particles and the characteristic relaxation and collision time scales has been addressed in general by Ventura (1973), while Langer, McCray, and Baan (1980) have considered the evolution of the particle distribution function by solving the kinetic equation. The results indicated that, in the presence of excitations to first and higher Landau levels, the one-dimensional particle distribution could be expected to depart from a simple Maxwell-Boltzmann distribution. The problem of the stopping length and pitch angle scattering of infalling protons in an accreting pulsar atmosphere was first addressed by Basko and Sunyaev (1975b). The importance of the stopping length is that it gives an upper limit to the depth of the atmosphere which can be heated by the accretion stream. The main process is the interaction of the infalling protons with the atmospheric electrons. The result is that, because of the strong magnetic field, which restricts the phase space available to the electrons, the proton Coulomb stopping length is longer in a magnetic field than in the absence of one by a factor $\sim (m_p/m_e)^{1/2}$. The protons, which initially may be taken to arrive with essentially zero pitch angle, slowly diffuse in angle away from the direction of the magnetic field, and stopping occurs when their distribution no longer differs from that of the atmospheric protons, i.e. when they no longer have a bulk motion component. An upper limit to the proton stopping length is in any case given by the nuclear *pp* collision mean free path, $l_{pp} \sim 55$ g cm$^{-2}$. The Coulomb stopping problem has been addressed in considerably more detail by Kirk and Galloway (1982), Miller and Wasserman (1985), Miller, Salpeter, and Wasserman (1987), and Pakey (1990), using a Fokker-Planck expansion of the kinetic equations and Monte Carlo simulations. The effect of density and temperature gradients in the stopping atmosphere has been considered by Mészáros *et al.* (1983), Harding *et al.* (1984), and Miller, Wasserman, and Salpeter (1989), whose calculations show that the decrease of the stopping length compared to the nonmagnetic case depends on the ratio of the cyclotron energy to the mean ambient temperature and, less sensitively, on parameters such as the initial proton mean velocity and velocity dispersion. These results are of importance for models which attempt to calculate self-consistently the structure of the accreting atmosphere under the effect of heating by the infalling plasma and cooling by radiation. These models are discussed in Chapter 7.4.

*b) Bremsstrahlung Absorption Cross Section*

While the radiationless Coulomb process is of interest in itself and for addressing questions such as those described above, the greatest relevance of these two-particle collisions is to the problem of coupling between matter and radiation, as well as to that of photon emission and absorption. In a true photon absorption process the energy received by the electron from the photon is converted into thermal (kinetic) energy rather than being reemitted as another photon (as occurs, for instance, in the resonant cyclotron scattering process). In a plasma, the simplest radiationless deexcitation of an excited electron comes via Coulomb deexcitation, i.e. the excited electron collides with other particles before radiative deexcitation occurs, and the excitation energy is thus thermalized. The absorption of a photon in the presence of collisions between charged particles is called free-free absorption. Its inverse, whereby collision of charged particles produces a photon, is bremsstrahlung emission. This process requires three particles (a photon and two charges) spatially and temporally close to each other, being a third-order process in an expansion in powers of the fine structure constant. Because of this $t_c \gg t_r$, and this process is not as frequent as radiative excitation or deexcitation (which gives resonant and nonresonant scattering, which are $\propto \alpha_F$ and $\propto \alpha_F^2$). However, bremsstrahlung (either resonant or nonresonant) is the most straightforward process acting as a source or sink of photons (Pavlov and Panov, 1976; Ventura, Nagel, and Mészáros, 1979). In particular, the resonant bremsstrahlung emission (collisional excitation of a Landau level followed by radiative deexcitation) is the usual way to produce cyclotron photons.

We shall here consider a simplified treatment of the collision between an electron and a nucleus of charge $Ze$. The nucleus may be taken to be stationary, at the origin of the coordinate system, with the electron initially in the ground Landau level $n = 0$ moving at a nonrelativistic velocity $v = p/m \ll c$ in the $z$-direction, along $\vec{B}_0$. The collision with the proton is treated in the perturbation (Born) approximation, as is the absorption of the photon (Virtamo and Jauho, 1975). The transition rate, according to the Golden Rule, is

$$w_{fi} = \frac{2\pi}{\hbar} |\langle f | T | i \rangle|^2 \delta\left(E_f - E_i\right) \tag{4.4.1}$$

with matrix element

$$\langle f | T | i \rangle = \sum_r \frac{\langle f | V | r \rangle \langle r | H_1 | i \rangle}{E_i - E_r} + \frac{\langle f | H_1 | r \rangle \langle r | V | i \rangle}{E_i - E_r}, \tag{4.4.2}$$

where

$$H_1 = \frac{e}{2c}\left(\vec{v} \cdot \vec{A} + \vec{A} \cdot \vec{v}\right) \tag{4.4.3}$$

is the perturbation produced by the photon field $\vec{A}$, and the Coulomb potential is $V = Ze^2/r$. The sum over intermediate states $r$ can be significantly reduced. First, since the potential $V$ is rotationally symmetric, it can only connect states with equal angular momentum $l_z = n - s$ so that the transitions must have $\Delta n = \Delta s$. If we again make the dipole approximation, the absorption of a photon can only change the Landau number $n$ by, at most, one unit, $\Delta n = \pm 1, 0$. Second, we restrict ourselves to the same final and initial state, so $n' = n = 0$. An electron left in an excited state would immediately

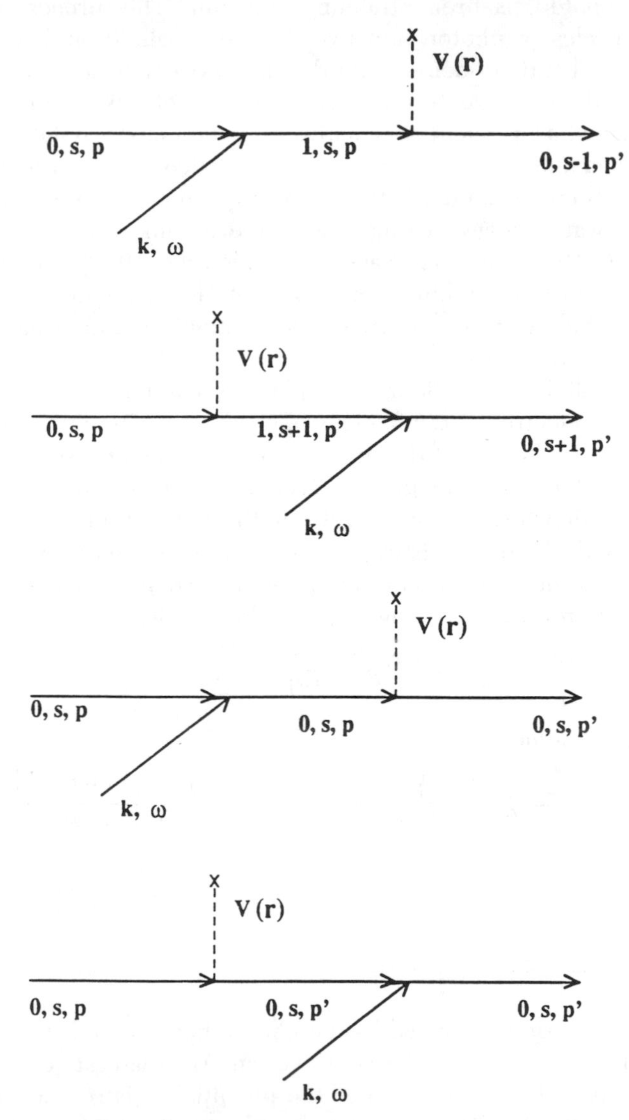


FIGURE 4.4.1 Feynman diagrams for bremsstrahlung in a magnetic field. The dashed line with a cross indicates interaction with a nucleus.

radiatively deexcite itself, which would be a correction to the scattering cross section. After these restrictions, only four contributions to the scattering amplitude are left (Nagel, 1982), whose diagrams are shown in Figure 4.4.1. The corresponding amplitude contributions are

$$T_a = \frac{\langle 0, s-1, p' \,|\, V \,|\, 1, s, p \rangle \left( \dfrac{2\pi\alpha\hbar^2 c}{\omega L^3} \right) \left( \dfrac{\hbar\omega_c}{m} \right)^{1/2} e_-}{\left( \hbar\omega - p^2/2m \right) - \left( \hbar\omega + p^2/2m \right)},$$

$$T_b = \frac{\left( \dfrac{2\pi\alpha\hbar^2 c}{\omega L^3} \right) \left( \dfrac{\hbar\omega_c}{m} \right)^{1/2} e_+ \langle 1, s+1, p' \,|\, V \,|\, 0, s, p \rangle}{\left( \hbar\omega + p^2/2m \right) - \left( \hbar\omega + \hbar\omega_c + p'^2/2m \right)},$$

$$T_c = \frac{\langle 0, s, p' \,|\, V \,|\, 0, s, p \rangle \left( \dfrac{2\pi\alpha\hbar^2 c}{\omega L^3} \right)^{1/2} \dfrac{p}{m} e_z}{\left( \hbar\omega + p^2/m \right) - p^2/2m},$$

$$T_d = \frac{\left( \dfrac{2\pi\alpha\hbar^2 c}{\omega L^3} \right)^{1/2} \dfrac{p'}{m} e_z \langle 0, s, p' \,|\, V \,|\, 0, s, p \rangle}{\left( \hbar\omega + p^2/2m \right) - \left( \hbar\omega + p'^2/2m \right)}. \tag{4.4.4}$$

Only the last two terms interfere with each other, since they lead to the same final state. The energy denominators can be simplified by using energy conservation

$$p'^2/2m - p^2/2m = \hbar\omega \tag{4.4.5}$$

so that the transition rate from $|i\rangle = |0, s, p\rangle |k\rangle$ to $|f\rangle |0, s, p\rangle |0\rangle$ becomes

$$\begin{aligned} w_{fi} = \frac{4\pi^2\alpha c}{\hbar\omega L^3} \Bigg[ & \frac{|e_+|^2}{\left(\omega + \omega_c\right)^2} \frac{\hbar\omega_c}{m} \left| \langle 1, s+1, p' \,|\, V \,|\, 0, s, p \rangle \right|^2 \delta_{s', s+1} \\ & + \frac{|e_-|^2}{\left(\omega - \omega_c\right)^2} \frac{\hbar\omega_c}{m} \left| \langle 0, s-1, p' \,|\, V \,|\, 0, s, p \rangle \right|^2 \delta_{s', s-1} \\ & + \frac{|e_z|^2}{\omega^2} \left( \frac{p - p'}{m} \right)^2 \left| \langle 0, s, p' \,|\, V \,|\, 0, s, p \rangle \right|^2 \delta_{s', s} \Bigg] \delta\left( E_i - E_f \right). \end{aligned} \tag{4.4.6}$$

The summation over the final values $s'$ gives the three terms $s' = s - 1, s, s + 1$. The final impulse may only have the values $p' = \pm(p^2 + 2m\hbar\omega)^{1/2}$, each of which leads, after integration over the delta function, to the density of final states $(L/2\pi)(m/\hbar|p'|)$. The total transition rate $w_i = \Sigma_f w_{fi}$ from the initial state $|i\rangle$ is then

$$w_{fi} = \frac{2\pi\alpha c}{\hbar\omega L^3}\frac{\omega_c}{|p'|}\left[\frac{|e_+|^2}{(\omega+\omega_c)^2}\sum_{\pm p'}|\langle 1, s+1, p'|V|0, s, p\rangle|^2 + \frac{|e_-|^2}{(\omega-\omega_c)^2}\sum_{\pm p'}|\langle 0, s-1, p'|V|0, s, p\rangle|^2 + \frac{|e_z|^2}{\omega^2}\sum_{\pm p'}|\langle 0, s, p'|V|0, s, p\rangle|^2\right]. \quad (4.4.7)$$

This rate must still be averaged over the initial states. The average over $s$ leads (Virtamo and Jauho, 1975) to

$$\frac{1}{\mu}\sum_s|\langle 0, s, p'|V|0, s, p\rangle|^2 = \frac{Z^2e^4}{\mu L^2}C_0(a),$$
$$\frac{1}{\mu}\sum_s|\langle 1, s+1, p'|V|0, s, p\rangle|^2 = \frac{Z^2e^4}{\mu L^2}C_1(a), \quad (4.4.8)$$

where $a = (p - p')^2(2m\hbar\omega_c)^{-1}$ is a measure of the momentum given to the nucleus, $\mu = m\omega_c L^2(2\pi\hbar)^{-1}$ is the multiplicity (degeneracy) of the Landau levels, and the $C_0$, $C_1$ functions are expressible in terms of exponential integral functions (Abramowitz and Stegun, 1964),

$$C_0(a) = a^{-1}\exp(a)E_2(a) = a^{-1} - \exp(a)E_1(a),$$
$$C_1(a) = \exp(a)[E_1(a) - E_2(a)] = (1+a)\exp(a)E_1(a) - 1. \quad (4.4.9)$$

Finally, one must average over the initial electron momentum distribution $f(p)$ (see eq. 4.3.1). The transition rate thus obtained refers to one electron with one ion in a volume $L^3$. In order to obtain the cross section, one must multiply the rate by $n_i L^3$ and divide by the photon flux. The resulting cross section is

$$\sigma_a = \sigma_0\left[\frac{\omega^2}{(\omega+\omega_c)^2}|e_+|^2 g_\perp + \frac{\omega^2}{(\omega-\omega_c)^2}|e_-|^2 g_\perp + |e_z|^2 g_\parallel\right], \quad (4.4.10)$$

where $\sigma_0$ is the absorption cross section in the absence of a field, aside from a factor $4\pi g(3\sqrt{3})^{-1}$ (Karzas and Latter, 1961),

$$\sigma_0 = 4\pi^2 Z^2 \alpha_F^3 \frac{\hbar^2 c^2}{m\omega^3} \frac{n_i}{(\pi mkT/2)^{1/2}}, \tag{4.4.11}$$

and the dimensionless magnetic Gaunt factors $g_\perp$, $g_\parallel$ are given by

$$g_\perp(\omega, T) = \frac{1}{2}\int_{-\infty}^{\infty} dp \exp(-p^2/2mkT)\frac{C_1(a_+) + C_1(a_-)}{(p^2 + 2m\hbar\omega)^{1/2}},$$

$$g_\parallel(\omega, T) = \int_{-\infty}^{\infty} dp \exp(-p^2/2mkT)\frac{a_+C_0(a_+) + a_-C_0(a_-)}{(p^2 + 2m\hbar\omega)^{1/2}},$$

$$a_\pm = \left(p \pm [p^2 + 2m\hbar\omega]^{1/2}\right)^2 (2m\hbar\omega)^{-1}. \tag{4.4.12}$$

The frequency dependence of the Gaunt factors is shown in Figure 4.4.2 for various values of $kT/\hbar\omega_c$.

The similarity of form of the free-free absorption cross section (4.4.10) and the scattering cross section (4.2.33) is evident. In this derivation of $\sigma_a$, as in that of the Thomson (cold) cross section, the photon momentum was neglected in the matrix elements, although in $\sigma_a$ the electron momentum had to be included to get a finite electron-proton collision rate. As in the scattering process, the free-free absorption shows a strong angle, frequency, and polarization dependence, including a strong reduction at $\omega \ll \omega_c$ of the absorption coefficient for the extraordinary wave as well as for propagation along the field. The typical nonmagnetic behavior $\propto \omega^{-3}$ at low frequencies is modified by the different frequency dependences of the ordinary and extraordinary waves, as well as by the presence of the cyclotron resonance.

Near the resonance $\omega_c$, the perturbation treatment again breaks down. To avoid the singularity in the resonant bremsstrahlung (Nagel, 1982), one can introduce a total damping width

$$\Gamma = \Gamma_r + \Gamma_c \tag{4.4.13}$$

made up of radiative and collision width contributions of the first Landau level. At the resonance, then, we have

$$\sigma_a \simeq \sigma_0 \frac{\omega^2}{(\omega - \omega_c)^2 + \Gamma^2/4} |e_-|^2 g_\perp . \tag{4.4.14}$$

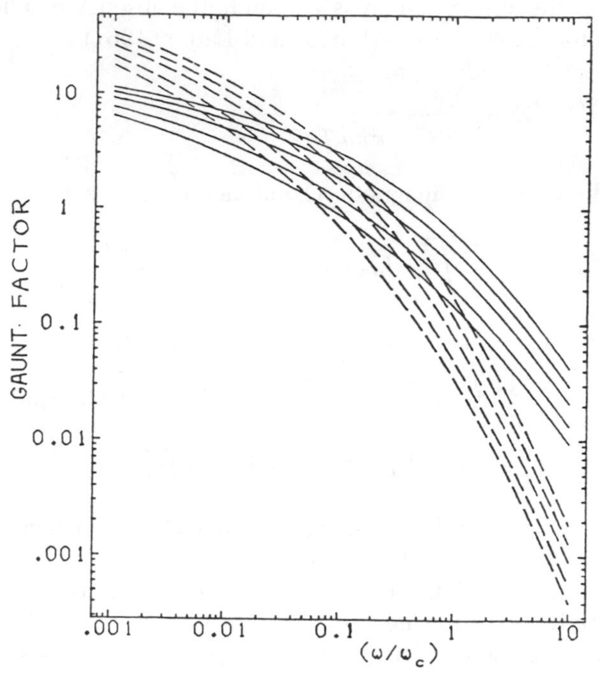


FIGURE 4.4.2 Nonrelativistic magnetic Gaunt factors $g_{\parallel}$ (*dashed*) and $g_{\perp}$ (*full*) as a function of frequency $\omega/\omega_c$ for temperatures $kT/\hbar\omega_c = 0.01$, 0.02, 0.04, 0.1, 0.2 going upward. From Nagel (1982).

It can be seen that the scattering and free-free absorption coefficients near $\omega_c$ can be written in a similar form,

$$\sigma_s \simeq 2\pi\alpha_F \frac{\hbar}{m} |e_-|^2 \frac{\Gamma_r}{(\omega-\omega_c)^2 + \Gamma_r^2/4}, \tag{4.4.15}$$

$$\sigma_a \simeq 2\pi\alpha_F \frac{\hbar}{m} |e_-|^2 \frac{\Gamma_c}{(\omega-\omega_c)^2 + \Gamma^2/4}. \tag{4.4.16}$$

By comparing (4.4.16) with (4.4.10), one sees that the collisional width is

$$\Gamma_c = 2\pi Z^2 \alpha_F^2 \hbar c^2 n_i (\pi mkT/2)^{-1/2} \omega_c^{-1} g_{\perp}(\omega_c), \tag{4.4.17}$$

which agrees with the directly calculated value (Ventura, 1973) for the collisional deexcitation of the first Landau level. In a plasma with density $\rho = 10^{-3}$ g cm$^{-3}$ and temperature $kT = 10$ keV, one finds $\Gamma_c = 8.8 \times 10^7$ s$^{-1}$ so that collisional deexcitations occur far slower than radiative deexcitations $\sim 10^{15} B_{12}^2$ s$^{-1}$. This manifests itself, as seen from equations (4.4.15), (4.4.16), in a free-free absorption cross section many orders of magnitude smaller than the corresponding scattering cross section. This is illustrated in the bottom portion of Figure 4.2.4, for which the cold-plasma eigenmodes were used in equation (4.4.10), consistent with having neglected the photon momentum in the matrix elements.

The actual free-free absorption coefficient in a plasma is somewhat smaller than the value (4.4.10) because one must subtract the photons reemitted by stimulated bremsstrahlung emission. The relative importance of pure absorption and stimulated emission depends on the electron distribution. The assumed electron distribution has all electrons in the ground Landau level $n = 0$, distributed along $z$ with $f(p) = (2\pi mkT_{\|})^{-1/2} \exp(-p^2/2mkT_{\|})$. In the case of absorption, the final electron states $|0, s, p\rangle$ are less occupied than the initial ones $|0, s, p\rangle$ by the factor $\exp[-(p' - p)^2/2mkT_{\|}] = \exp(-\hbar\omega/kT_{\|})$. Therefore, stimulated emission is also weaker than absorption, by the same factor. The net absorption coefficient $\sigma_a'$, corrected for stimulated emission, is therefore

$$\sigma_a' = \sigma_a\left(1 - e^{-\hbar\omega/kT_{\|}}\right), \tag{4.4.18}$$

the quantity plotted in Figure 4.2.4

### *c) Bremsstrahlung Emission*

Spontaneous bremsstrahlung emission is related to the absorption rate via the usual Einstein relations, which follow from detailed balance considerations. The transition rate for an electron in state $|f\rangle = |0, s, p'\rangle$ to spontaneously emit a photon and make a transition to state $|i\rangle = |0, s, p\rangle$ is

$$w_{f\to i}(\omega, \theta) = \frac{\omega^2}{(2\pi)^3 c^2} \sigma_{i\to f}(\omega, \theta). \tag{4.4.19}$$

What is needed in a radiative transfer calculation are the rates or cross sections averaged over electronic states, $\sigma = \sum_{i,f} b_i \sigma_{i\to f}$ and $w_{i\to f} = \sum_{i,f} b_f w_{f\to i}$, where $b_i$ is the probability of finding an electron in state

$i$. Since, for all transitions here considered ($n = n' = 0$) the probability of state $f$ is smaller than that of state $i$ by the factor $\exp[-(p' - p)^2/2mkT_\parallel] = \exp(-\hbar\omega/kT_\parallel)$, we have

$$w(\omega, \theta) = \frac{\omega^2}{(2\pi)^3 c^2} e^{-\hbar\omega/kT_\parallel} \sigma(\omega, \theta). \tag{4.4.20}$$

The absorption cross section corrected for stimulated emission, $\sigma_a'$, satisfies the relation (Nagel, 1982)

$$w(\omega, \theta) = \frac{\omega^2}{(2\pi)^3 c^2} \frac{e^{-\hbar\omega/kT_\parallel}}{1 - e^{-\hbar\omega/kT_\parallel}} \sigma_a'(\omega, \theta). \tag{4.4.21}$$

This is formally equivalent to Kirchhoff's law $w = B_\omega \sigma'$ (e.g. Bekefi, 1966), where $B_\omega$ is the Planck function

$$B_\omega(T) = \frac{\omega^2}{(2\pi)^3 c^2} \frac{e^{-\hbar\omega/kT_\parallel}}{1 - e^{-\hbar\omega/kT_\parallel}} \tag{4.4.22}$$

in units of $\text{cm}^{-2}\ \text{s}^{-1}\ \text{ster}^{-1}\ (\text{rad/s})^{-1}$. However, it must be realized that, to obtain this, we have not assumed thermodynamic equilibrium. Nonetheless, this relation remains valid whenever, as assumed here, the initial and final levels are the same, $n = n' = 0$, and the $z$-distribution is given by a Maxwell-Boltzmann function. Thus, we have, for the population of Landau levels assumed, $\rho_n = \delta_{n0}$ instead of having assumed a population of levels $n$ according to a thermal distribution characterized by some $T_\perp$, such as $b_n = [1 - \exp(-\hbar\omega_c/kT_\perp)]\exp(-\hbar\omega_c/kT_\perp)$. Implicitly, we took here $T_\perp = 0$. However, it should be realized that the Kirchhoff-like relation (4.4.22) would not be valid if we had allowed for final electron states $n' > 0$.

#### *d) Absorption, Scattering, and Level Population*

As discussed previously, in a very strong magnetic field the cyclotron process is essentially equivalent to resonant scattering, while free-free absorption remains a "real" absorption effect. A somewhat better justification of this is offered by a simplified kinetic equation for the electron population among Landau levels in the presence of the previously mentioned processes. Other important true absorption processes, e.g. inverse two-photon scattering, will be discussed later. We restrict the discussion of the bremsstrahlung process to the case of $kT_\parallel \ll \hbar\omega_c$ so that only the levels $n = 0$ and $n = 1$ need be considered

(Nagel and Ventura, 1983; Nagel, 1982). Their populations will be denoted by $N_0(\vec{r}, t)$ and $N_1(\vec{r}, t)$. Evidently, they must satisfy the transport equations

$$\frac{\partial N_0}{\partial t} = AN_1 + BnN_1 - BnN_0 + C_0 N_1 - C_1 N_0,$$

$$\frac{\partial N_1}{\partial t} = -AN_1 - BnN_1 + BnN_0 - C_0 N_1 + C_1 N_0, \qquad (4.4.23)$$

where $A$ is the rate of spontaneous emission, $B$ is the rate for stimulated emission or absorption, $C_0$ and $C_1$ are the rates for collisional excitation and deexcitation, and $n$ is the density of photons in the cyclotron line. The cyclotron photon density will, for its part, satisfy a diffusion equation (since the optical depth is high at $\omega \sim \omega_c$),

$$\frac{\partial n}{\partial t} = cD \frac{\partial^2 n}{\partial r^2} + q, \qquad (4.4.24)$$

where $D$ is the diffusion coefficient and $q$ is a source term given by

$$q = N_1 A + \left(N_1 B \quad N_0 B\right) n. \qquad (4.4.25)$$

We seek a stationary solution $\partial N_0 / \partial t = \partial N_1 / \partial t = \partial n / \partial t = 0$, which from (4.4.23), (4.4.24) implies the relation

$$\frac{N_1}{N_0} = \frac{Bn + C_1}{A + Bn + C_0}. \qquad (4.4.26)$$

The collisions tend to try to establish a transverse distribution with $T_\perp = T_\parallel$, that is

$$\frac{N_1}{N_0} \to \frac{C_1}{C_0} = e^{-\hbar\omega/kT_\perp} = e^{-\hbar\omega/kT_\parallel}. \qquad (4.4.27)$$

However, the collision rates are far smaller under the strong-field conditions than the radiative rates. Thus, the population of the Landau levels will be dictated by the local density of the cyclotron line photons,

$$\frac{N_1}{N_0} \to \frac{Bn}{A + Bn}. \qquad (4.4.28)$$

Only in the deep interior of an atmosphere will the cyclotron photon density be equal to its thermodynamic equilibrium value,

$$\bar{n} = \frac{A}{B} \frac{1}{A + Bn}, \qquad (4.4.29)$$

so that only in these deep regions will the population of Landau levels be characterized by the longitudinal temperature $T_{\parallel}$,

$$\frac{N_1}{N_0} \to \frac{B\bar{n}}{A + Bn} = e^{-\hbar\omega_c/kT_{\parallel}}. \tag{4.4.30}$$

Toward the boundaries of the atmosphere, however, the line photon density must decrease and the excited level occupation number must correspondingly become smaller.

The source term $q$ of equation (4.4.25) can be reexpressed, using the total electron density $N = N_0 + N_1$, as

$$q = N\frac{AC_1 + BnC_1 - BnC_0}{A + 2Bn + C_0 + C_1}. \tag{4.4.31}$$

Since we took $kT_{\parallel} \ll \hbar\omega_c$, we will have $C_1 \ll C_0$, $Bn < B\bar{n} \ll A$, and therefore

$$q \simeq N\frac{A}{A + C_0}C_1 - N\frac{C_0}{A + C_0}Bn. \tag{4.4.32}$$

The first term is the rate of collisional excitation multiplied by the radiative deexcitation probability. The second is the number of radiative absorptions per second multiplied by the (very small) probability that the electron is collisionally deexcited. Thus, both the production and the annihilation of photons are directly proportional to the collision rate. The relation between absorption and emission rates is

$$\frac{BnC_0}{AC_1} = n\frac{B}{A}e^{\hbar\omega_c/kT_{\parallel}} \simeq \frac{n}{\bar{n}}. \tag{4.4.33}$$

Therefore, the time-independent cyclotron line photon diffusion equation can be written as

$$D\frac{\partial^2 n}{\partial r^2} + \alpha\bar{n} - \alpha n = 0, \tag{4.4.34a}$$

where

$$\alpha = \frac{N}{C}BC_0/(A + C_0) \tag{4.4.34b}$$

is the coefficient for free-free absorption at the resonance, which is much smaller than the coefficient for cyclotron absorption, $NB/c$. Since it was assumed that $kT_{\parallel} \ll \hbar\omega_c$, the occupation numbers are practically constant, $N_1 \simeq 0$, $N_0 \simeq N$, and the coefficient $\alpha$ is independent of the photon density. In the opposite situation, however, $\alpha$ would depend on $n$ and one would need to take into account stimulated emission.

*e) Thermal Broadening of Resonant Free-Free Absorption*

As mentioned before, equation (4.4.1) is derived using matrix elements which neglect the photon momentum contributions and is calculated using the corresponding cold-plasma eigenmodes. When the photon momentum contributions are nonnegligible, as in a hot (but still nonrelativistic) plasma, these effects need only be included in the resonant contribution to the rate (Kirk and Mészáros, 1980). Thus, equation (4.4.7) can be rewritten as

$$w_{fi} = \frac{2\pi\alpha_F c}{\hbar\omega L^3}\frac{\omega_c}{|p'|}\left[\frac{|e_+|^2}{(\omega+\omega_c)^2}\sum_{\pm p'}|\langle 1, s+1, p'\,|\,V\,|\,0, s, p\rangle|^2 + \frac{|e_-|^2}{(D+i\gamma_r)^2}\sum_{\pm p'}|\langle 1, s-1, p'\,|\,V\,|\,0, s, p\rangle|^2 + \frac{|e_z|^2}{\omega^2}\sum_{\pm p'}|\langle 0, s, p'\,|\,V\,|\,0, s, p\rangle|^2\right], \tag{4.4.35}$$

where in the term proportional to $|e_-|$ we have replaced $\omega - \omega_c$ with $D + i\gamma_r$, where

$$D = \omega - \omega_c - pk_z/m - \hbar k_z^2/2m \tag{4.4.36}$$

and $k_z = k\cos\theta$. The free-free absorption coefficient therefore becomes

$$\sigma_a = \sigma_0\left[\frac{\omega^2}{(\omega+\omega_c)^2}|e_+|^2 g_\perp + |e_-|^2 g_R + |e_z|^2 g_\parallel\right], \tag{4.4.37}$$

where $g_\perp$ and $g_\parallel$ are, as before, given by equations (4.4.12) and

$$g_R = \frac{1}{2}\int_{-\infty}^{\infty} dp\,\exp(-p^2/2mkT)\frac{[C_1(a_+) + C_1(a_-)]\,\omega^2}{(p^2 + 2m\hbar\omega)^{1/2}\,2i\gamma_r} \times\left[\frac{1}{D - i\gamma_r} - \frac{1}{D + i\gamma_r}\right]. \tag{4.4.38}$$

The absorption cross section (4.4.37) is to be used with the hot-plasma polarization eigenmodes. The additional Gaunt factor (4.4.38) is computed numerically. In the limit $kT_{\parallel} \ll \hbar\omega_c$, the approximation

$$(2m\hbar\omega)^{1/2} \sim |p| \gg |p'| \tag{4.4.39}$$

enables one to perform the integral (4.4.38) analytically, with the result

$$g_R = \pi^{1/2}\left(\frac{kT}{\hbar\omega}\right)^{1/2} C_1(\omega/\omega_c) \times \left[\frac{\omega^2}{(2kT/m)^{1/2}\,|k\cos\theta|}\,\frac{1}{\gamma_r}\,\mathrm{Im}\,W(y)\right] \tag{4.4.40}$$

where

$$y = \frac{\omega - \omega_c + i\gamma_r}{(2kT/m)^{1/2} k\cos\theta} - \frac{\hbar k\cos\theta}{2m(2kT/m)^{1/2}}. \tag{4.4.41}$$

The introduction of hot-plasma effects introduces a resonance in the ordinary mode as well, and an even stronger resonance is introduced when one includes the vacuum effects (e.g. Mészáros and Nagel, 1985a).

# 5. Relativistic Radiation Processes

## 5.1 Relativistic Cross Sections and Rates

### *a) Introduction*

The most complete description of many of the radiative processes in a very strong magnetic field is obtained within the framework of relativistic quantum electrodynamics (QED). In general, for electron or photon energies exceeding about $5 \times 10^{-2} m_e c^2$, relativistic effects start becoming important. In addition, even at lower energies, effects such as the Doppler broadening of lines and the description of higher cyclotron harmonics require relativistic corrections. The asymmetric shape and exact radiative width of the cyclotron scattering lines follow in a straightforward manner from the QED treatment, which also lends itself to the description of multiple-photon scattering and two-photon decay production cross sections. This treatment is also unavoidable for describing processes such as pair formation and annihilation. These processes play a central role in the theory of magnetized neutron stars, especially for radiation in the MeV range or above. Some of the first treatments of such problems using QED in a pulsar type of magnetic field are those of Erber (1964), Daugherty and Ventura (1978), Herold (1979), Bussard (1980), and Daugherty and Bussard (1980). In this chapter, we discuss the relativistic quantum mechanical treatment of these effects and the new features that this introduces.

### *b) Compton Scattering*

For many high-energy astrophysics sources, the most important process is that of Compton scattering. This is the largest source of opacity in the outer parts of the atmosphere of accreting neutron stars, where

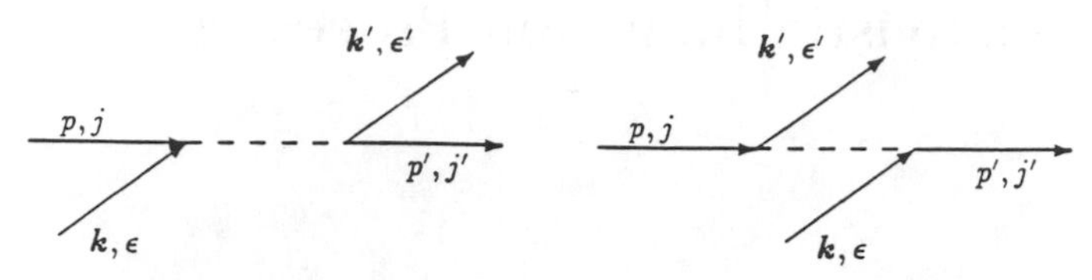


FIGURE 5.1.1 Feynman diagrams for Compton scattering, where $p, j$ are the longitudinal electron momentum and total magnetic quantum number and $\mathbf{k}, \boldsymbol{\epsilon}$ are the photon longitudinal momentum and polarization vectors.

due to the relatively low plasma density the collisional deexcitation rate is far below the radiative deexcitation rate (e.g. eqs. 4.1.19, 4.1.20). It is also the process that determines the refractive properties of the medium in both high-plasma-density accreting objects and low-density rotation-powered pulsars. The scattering process is described by the diagrams of Figure 5.1.1. The S-matrix element for this process (Bjorken and Drell, 1964) can be written in four-dimensional notation as

$$S_{fi} = -e^2 \int d^4x \int d^4x' \overline{\Psi_f(x)} [ \not{A}_f(x) S(x, x') \not{A}_i(x') + \not{A}_i(x) S(x, x') \not{A}_f(x') ] \Psi_i(x'), \tag{5.1.1}$$

where the slash indicates multiplication by the Dirac $\gamma$ matrices, e.g. $\not{A} = \vec{\gamma} \cdot \vec{A}$, and the vector potentials of the final and initial photons are

$$\vec{A}_f(x) = \sqrt{\frac{2\pi}{\omega' V}}\, \hat{\epsilon}'^* e^{i\vec{k}' \cdot \vec{x}}, \tag{5.1.2}$$

$$\vec{A}_i(x) = \sqrt{\frac{2\pi}{\omega V}}\, \hat{\epsilon} e^{-i\vec{k} \cdot \vec{x}}, \tag{5.1.3}$$

where $V$ is the normalization volume. The wave functions are given in terms of a spatial part and a time part,

$$\Psi_\beta(x) = \Psi_{nspq\lambda}(x) e^{-i\lambda E_\beta t}, \tag{5.1.4}$$

where $\lambda = \pm 1$ is used for an electron of positive or negative energy and $\beta$ represents the set of quantum numbers $j, n, s, p, q, \lambda$ with $j = n + s + 1/2$ the total magnetic quantum number, $n$ the Landau principal quantum number, $s$ the spin projection along $\vec{B}$, $q$ the annulus of the guiding center of the electron orbit, and $p$ the electron momentum parallel to $\vec{B}$. Unless otherwise stated, in this chapter we use $c = \hbar = 1$; also, notice the difference in convention compared to Chapter 4, where $s$ stood for guiding center label, unlike here. The treatment of this and the next section follows that of Bussard, Alexander, and Mészáros (1986) and Alexander and Mészáros (1991a). The energy of the electron is given by

$$E = \left[ p^2 + 2m\omega_c j + m^2 \right]^{1/2}, \tag{5.1.5}$$

where $\omega_c$ is the classical cyclotron frequency. The relativistic propagator is constructed from the spatial part of the wave functions and an energy denominator

$$S(x, x') = \int \frac{dp''}{2\pi} e^{ip''(t-t')} \sum_{\beta''} \frac{\psi_{\beta''}(\vec{x})\overline{\psi_{\beta''}(\vec{x}')}}{p'' - \lambda''\left( E'' - \frac{i}{2}\Gamma_{\beta''} \right)}, \tag{5.1.6}$$

where $\Gamma_{\beta''}$ is the inverse of the half-life of the electron state $\beta''$. After integration over $t, t'$ (see Bjorken and Drell, 1964), this becomes

$$\begin{aligned} S_{fi} = &-e^2 \frac{(2\pi)^3}{V\sqrt{\omega\omega'}} \int \frac{dp''}{2\pi} \int d^3x \int d^3x' \overline{\psi_{\beta'}(\vec{x})} \\ &\times \sum_{\beta''} \vec{\epsilon}'^{*} \cdot \vec{\gamma} e^{-i\vec{k}' \cdot \vec{x}} \frac{\psi_{\beta''}(\vec{x})\overline{\psi_{\beta''}(\vec{x}')}}{p'' - \lambda''\left( E'' - \frac{i}{2}\Gamma_{\beta''} \right)} \vec{\epsilon} \cdot \vec{\gamma} e^{i\vec{k} \cdot \vec{x}'} \\ &\times \delta(E' + \omega' + p'')\delta(p'' - \omega - E) \\ &+ \delta(E' - \omega' - p'')\delta(p'' + \omega - E) \\ &\times \vec{\epsilon} \cdot \vec{\gamma} e^{i\vec{k} \cdot \vec{x}} \frac{\psi_{\beta''}(\vec{x})\overline{\psi_{\beta''}(\vec{x}')}}{p'' - \lambda''\left( E'' - \frac{i}{2}\Gamma_{\beta''} \right)} \vec{\epsilon}'^{*} \cdot \vec{\gamma} e^{-i\vec{k}' \cdot \vec{x}'} \psi_\beta(\vec{x}'). \end{aligned} \tag{5.1.7}$$

Defining the Fourier transforms of the transition currents

$$\vec{J}_{\beta\beta'}(\vec{k}) \equiv \int d^3x \overline{\psi_\beta(\vec{x})} \vec{\gamma} \psi_{\beta'}(\vec{x}) e^{i\vec{k} \cdot \vec{x}}, \tag{5.1.8}$$

one can write the matrix element as

$$S_{fi} = -e^2 \frac{(2\pi)^3}{V\sqrt{\omega\omega'}} \int \frac{dp''}{2\pi}$$

$$\times \sum_{\beta''} \left\{ \frac{\left[\vec{\epsilon}'^* \cdot \vec{J}_{\beta'\beta''}(-\vec{k}')\right]\left[\vec{\epsilon} \cdot \vec{J}_{\beta''\beta}(\vec{k})\right]}{p'' - \lambda''\left(E'' - \frac{i}{2}\Gamma_{\beta''}\right)} \right.$$

$$\times \delta(E' + \omega' + p'')\delta(p'' - \omega - E)$$

$$+ \frac{\left[\vec{\epsilon} \cdot \vec{J}_{\beta'\beta''}(\vec{k})\right]\left[\vec{\epsilon}'^* \cdot \vec{J}_{\beta''\beta}(-\vec{k}')\right]}{p'' - \lambda''\left(E'' - \frac{i}{2}\Gamma_{\beta''}\right)}$$

$$\left. \times \delta(E' - \omega' - p'')\delta(p'' + \omega' - E) \right\}. \tag{5.1.9}$$

The integration over intermediate momenta is done using the delta functions. Noting that $[\vec{J}_{\beta\beta'}(\vec{k})]^* = \vec{J}_{\beta'\beta}(-\vec{k})$, we can write the matrix element as

$$S_{fi} = -e^2 \frac{(2\pi)^2}{V\sqrt{\omega\omega'}} \delta(E' + \omega' - E - \omega)$$

$$\times \sum_{\beta''} \left\{ \frac{\left[\vec{\epsilon}'^* \cdot \vec{J}_{\beta'\beta''}(-\vec{k}')\right]\left[\vec{\epsilon} \cdot \left(\vec{J}_{\beta\beta''}(-\vec{k})\right)^*\right]}{E + \omega - \lambda''\left(E'' - \frac{i}{2}\Gamma_{\beta''}\right)} \right.$$

$$\left. + \frac{\left[\vec{\epsilon} \cdot \left(\vec{J}_{\beta''\beta'}(-\vec{k})\right)^*\right]\left[\vec{\epsilon}'^* \cdot \vec{J}_{\beta''\beta}(-\vec{k}')\right]}{E - \omega' - \lambda''\left(E'' - \frac{i}{2}\Gamma_{\beta''}\right)} \right\}. \tag{5.1.10}$$

Using the form of the transition currents as given in Appendix A, the matrix element may be written as

$$S_{fi} = \frac{(2\pi)^3}{LV} e^2 \left[ \frac{E+m}{2\omega E} \frac{E'+m}{2\omega' E'} \right]^{1/2}$$
$$\times \delta(E' + \omega' - E - \omega)\delta(p + k - p' - k') i^{j'-j}(-1)^{q'-q}$$
$$\times \sum_{j'', q''} \{ e^{i(j''-q'')(\phi'-\phi)} e^{-i(j'-q')\phi'} e^{i(j-q)\phi} F_{q'q''}(\xi') F^*_{qq''}(\xi) a_{j''}$$
$$+ e^{-i(j''-q'')(\phi'-\phi)} e^{i(j-q)\phi'} e^{-i(j'-q')\phi} F^*_{q''q'}(\xi) F_{q''q}(\xi') b_{j''} \},$$
$$(5.1.11)$$

where $\phi$ is the azimuthal angle; the $F$ functions are defined in Appendix A, equation (A.10); $L$ is a normalization length; $k$ is the component of the wave vector parallel to the external field; $\xi = k_\perp^2/2m\omega_c$; and

$$a_{j''} = \frac{E''+m}{2E''} \sum_{s'', \lambda''} \left. \frac{\left[\vec{\epsilon}'^* \cdot \vec{G}_{\beta'\beta''}(-\vec{k}'_\perp)\right]\left[\vec{\epsilon} \cdot \vec{G}^*_{\beta\beta''}(-\vec{k}_\perp)\right]}{E + \omega - \lambda''\left(E'' - \frac{i}{2}\Gamma_{\beta''}\right)} \right|_{p''=p+k}, \qquad (5.1.12)$$

$$b_{j''} = \frac{E''+m}{2E''} \sum_{s'', \lambda''} \left. \frac{\left[\vec{\epsilon} \cdot \vec{G}^*_{\beta''\beta'}(-\vec{k}_\perp)\right]\left[\vec{\epsilon}'^* \cdot \vec{G}_{\beta''\beta}(-\vec{k}'_\perp)\right]}{E - \omega' - \lambda''\left(E'' - \frac{i}{2}\Gamma_{\beta''}\right)} \right|_{p''=p-k'}, \qquad (5.1.13)$$

The components of the vector $\vec{G}$ depend on spin and are defined by equations (A.12) and (A.13) in Appendix A. To proceed, the initial guiding center number can be taken to be zero without losing generality, and the sum over intermediate-state guiding center numbers in equation (5.1.13) can be done using equation (A.11) of Appendix A to give

$$S_{fi} = \frac{(2\pi)^3}{LV} e^2 \left[ \frac{E+m}{2\omega E} \frac{E'+m}{2\omega' E'} \right]^{1/2}$$
$$\times \delta(E' + \omega' - E - \omega)\delta(p + k - p' - k')$$
$$\times (i)^{j'-j}(-1)^{q'} \exp\left[ \frac{-\xi + \xi'}{2} + \sqrt{\xi\xi'} \cos(\phi' - \phi) \right]$$

$$\times \frac{(-i)^{q'}}{\sqrt{q'!}} e^{i(j\phi - j'\phi')} \left(\sqrt{\xi'} e^{i\phi'} - \sqrt{\xi} e^{i\phi}\right)^{q'}$$
$$\times \sum_{j''} \{a_{j''} e^{ij''(\phi'-\phi)} \exp[-i\sqrt{\xi\xi'} \sin(\phi'-\phi)]$$
$$+ b_{j''} e^{-ij''(\phi'-\phi)} e^{i(j'+j)(\phi'-\phi)} \exp[i\sqrt{\xi\xi'} \sin(\phi'-\phi)]\}. \tag{5.1.14}$$

The scattering rate, or transition probability per unit time, is obtained by squaring the magnitude of the S-matrix element and multiplying by the appropriate phase space factors. The result must be independent of the electron guiding center, so it is summed over the final guiding center number with the use of equation (A.11) in Appendix A. Thus, the transition rate for a photon $(\vec{k}, \vec{\epsilon})$ to scatter from an electron $(j, p)$ into photon $(\vec{k}', \vec{\epsilon}')$ leaving the electron in $(j', p')$ is

$$w_{jj'}(\vec{k}, \vec{\epsilon}, p \to \vec{k}', \vec{\epsilon}', p') = \sum_{q'} \frac{V}{T} |S_{fi}|^2 \frac{V}{(2\pi)^3} d^3k' \frac{L}{2\pi} dp', \tag{5.1.15}$$

where $T$ is a normalization time. The squares of the delta functions are handled in the usual way (e.g. Bjorken and Drell, 1964):

$$|\delta(E' + \omega' - E - \omega)|^2 = \frac{T}{2\pi} \delta(E' + \omega' - E - \omega), \tag{5.1.16}$$

$$|\delta(p' + k' - p - k)|^2 = \frac{L}{2\pi} \delta(p' + k' - p - k). \tag{5.1.17}$$

Using the definition

$$\Lambda = \frac{j + j'}{2}(\phi' - \phi) + \sqrt{\xi\xi'} \sin(\phi' - \phi), \tag{5.1.18}$$

one obtains then

$$w_{jj'}(\vec{k}, \vec{\epsilon}, p \to \vec{k}', \vec{\epsilon}', p')$$
$$= \frac{e^4}{\omega\omega'} \left[\frac{E + m}{2E} \frac{E' + m}{2E'}\right]$$
$$\times \delta(E' + \omega' - E - \omega) \delta(p' + k' - p - k)$$
$$\times \left| \sum_{j''} \{a_{j''} e^{ij''(\phi'-\phi)} + b_{j''} e^{-ij''(\phi'-\phi)} e^{2i\Lambda}\} \right|^2 d^3k' \, dp', \tag{5.1.19}$$

which is the final expression for the normal magnetic Compton scattering rate per unit time.

*c) Two-Photon Decay*

A phenomenon related to the magnetic scattering described above is that of two-photon emission, which can occur when an electron in an excited state $j > 0$ makes a spontaneous transition to a lower state with the emission of two photons. This is described by the Feynman diagrams in Figure 5.1.2, where the second diagram merely interchanges the order in which the photons are emitted. This process has been discussed by Melrose and Parle (1983) and Melrose and Kirk (1986) for frequencies near the first cyclotron resonance. Here we shall follow the treatment of Alexander and Mészáros (1991a), which deals with both the resonance and continuum region. Because the two photons share the energy of the transition, the spectrum forms a continuum which can go down to arbitrarily soft energies. Having the same number of photons and vertices as in Figure 5.1.1, the $S$-matrix element of two photon decay is similar to that for Compton scattering and is given by equation (5.1.1). The only difference is that here both photons are emitted, and therefore the vector potentials for both photons are given by expressions like equation (5.1.2) for the final (outgoing) photon. This causes some differences in the delta functions and the denominators in the electron propagators. The equivalent of equation (5.1.10) for the two-photon decay process is then

$$S_{fi} = -e^2 \frac{(2\pi)^2}{V\sqrt{\omega\omega'}} \delta(E' + \omega' + \omega - E)$$

$$\times \sum_{\beta''} \left\{ \frac{\left[\vec{\epsilon}'^* \cdot \vec{J}_{\beta'\beta''}(-\vec{k}')\right]\left[\vec{\epsilon}^* \cdot \vec{J}_{\beta''\beta}(-\vec{k})\right]}{E - \omega - \lambda''\left(E'' - \frac{i}{2}\Gamma_{\beta''}\right)} \right.$$

$$\left. + \frac{\left[\vec{\epsilon}^* \cdot \vec{J}_{\beta'\beta''}(-\vec{k})\right]\left[\vec{\epsilon}'^* \cdot \vec{J}_{\beta''\beta}(-\vec{k}')\right]}{E - \omega' - \lambda''\left(E'' - \frac{i}{2}\Gamma_{\beta''}\right)} \right\}. \qquad (5.1.20)$$

The currents $\vec{J}$ involving the unprimed photon in equation (5.1.20) are the same as that used for the Compton scattering case except that

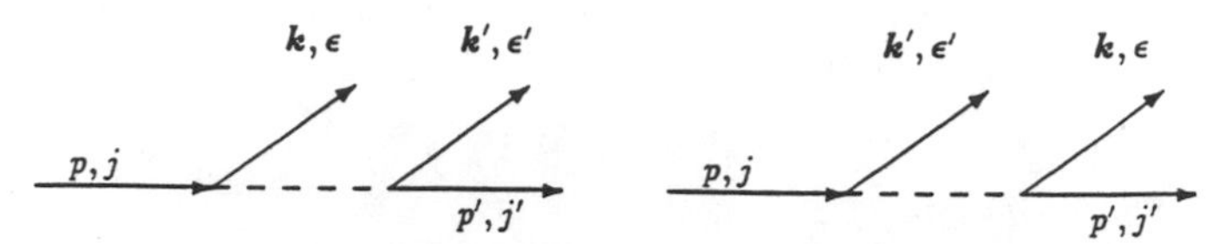


FIGURE 5.1.2 Diagrams for two-photon emission.

now the photon is emitted instead of absorbed. This change is represented by placing $\vec{k} \to -\vec{k}$ in equation (A.9), which causes the azimuthal angle to change to $\phi \to \phi + \pi$. Going through the same steps as between equations (5.1.10) and (5.1.14), the S-matrix element for two-photon decay is found to be

$$
\begin{aligned}
S_{fi} = {} & \frac{(2\pi)^4}{LV} e^2 \left[ \frac{E+m}{2\omega E} \frac{E'+m}{2\omega' E'} \right]^{1/2} \\
& \times \delta(E' + \omega' + \omega - E)\delta(p + k' + k - p')(i)^{j'-j}(-1)^{q'} \\
& \times \exp\left[ \frac{-\xi + \xi'}{2} + \sqrt{\xi\xi'}\cos(\phi' - \phi - \pi) \right] \frac{(-i)^{q'}}{\sqrt{q'!}} \\
& \times \sum_{j''} \Big\{ c_{j''} e^{ij''(\phi'-\phi)} e^{-i(j'-q')\phi'} e^{ij(\phi+\pi)} \\
& \times \left( \sqrt{\xi}\, e^{i(\phi'-\phi-\pi)} - \sqrt{\xi'} \right)^{q'} \\
& \qquad \times \exp\left[ -i\sqrt{\xi\xi'}\sin(\phi' - \phi - \pi) \right] \\
& \qquad + d_{j''} e^{-ij''(\phi'-\phi)} e^{-i(j'-q')(\phi+\pi)} e^{ij\phi'} \\
& \qquad \times \left( \sqrt{\xi'}\, e^{i(\phi'-\phi-\pi)} - \sqrt{\xi} \right)^{q'} \\
& \qquad \times \exp\left[ i\sqrt{\xi\xi'}\sin(\phi' - \phi - \pi) \right] \Big\}.
\end{aligned}
\tag{5.1.21}
$$

where the quantities $c_{j''}$, $d_{j''}$ are defined as

$$
\begin{aligned}
c_{j''} = {} & (-1)^{j''} \frac{E'' + m}{2E''} \\
& \times \sum_{s'', \lambda''} \left. \frac{\left[ \vec{\epsilon}'^{*} \cdot \vec{G}_{\beta'\beta''}(-\vec{k}'_{\perp}) \right] \left[ \vec{\epsilon}^{*} \cdot \vec{G}_{\beta''\beta}(-\vec{k}_{\perp}) \right]}{E - \omega - \lambda'' \left( E'' - \dfrac{i}{2}\Gamma_{\beta''} \right)} \right|_{p'' = p - k},
\end{aligned}
\tag{5.1.22}
$$

$$d_{j''} = (-1)^{j''} \frac{E'' + m}{2E''}$$

$$\times \sum_{s'', \lambda''} \frac{\left[\vec{\epsilon}^{*} \cdot \vec{G}_{\beta'\beta''}(-\vec{k}_{\perp})\right]\left[\vec{\epsilon}'^{*} \cdot \vec{G}_{\beta''\beta}(-\vec{k}'_{\perp})\right]}{E - \omega' - \lambda''\left(E'' - \frac{i}{2}\Gamma_{\beta''}\right)} \Bigg|_{p'' = p - k'} . \tag{5.1.23}$$

The corresponding transition rate for an electron in $(j, p)$ to decay with the simultaneous emission of two photons $(\vec{k}, \vec{\epsilon})$ and $(\vec{k}', \vec{\epsilon}')$ leaving the electron in $(j', p')$ is obtained by squaring the magnitude of the $S$-matrix element, multiplying by the appropriate phase space factors, and summing over the guiding center numbers of the final electron,

$$w_{jj'}(p \to \vec{k}, \vec{\epsilon}; \vec{k}', \vec{\epsilon}'; p') = \sum_{q'} \frac{1}{T} | S_{fi} |^2$$
$$\times \frac{V}{(2\pi)^3} d^3k \frac{V}{(2\pi)^3} d^3k' \frac{L}{2\pi} dp'. \tag{5.1.24}$$

Writing this out explicitly and using the definition

$$\Lambda_2 = \frac{j + j'}{2}(\phi' - \phi - \pi) + \sqrt{\xi\xi'} \sin(\phi' - \phi), \tag{5.1.25}$$

one obtains

$$w_{jj'}(p \to \vec{k}, \vec{\epsilon}; \vec{k}', \vec{\epsilon}'; p')$$
$$= \frac{e^4}{(2\pi)^3 \omega\omega'} \left[\frac{E + m}{2E} \frac{E' + m}{2E'}\right]$$
$$\times \delta(E' + \omega' + \omega - E)\delta(p' + k' + k - p)$$
$$\times \left| \sum_{j''} \{c_{j''} e^{ij''(\phi' - \phi)} + d_{j''} e^{-ij''(\phi' - \phi)} e^{2i\Lambda_2}\} \right|^2 d^3k\, d^3k'\, dp', \tag{5.1.26}$$

for the two-photon decay transition rate.

FIGURE 5.1.3 Diagrams for cyclotron absorption and emission.

*d) Cyclotron Absorption and Emission*

In addition to the second-order processes involving two photons and two vertices, one also needs the simpler one-photon, one-vertex first-order processes which represent the cyclotron (or synchrotron) absorption and emission. These rates have been derived by Daugherty and Ventura (1978) and Bussard (1980) (see also Pavlov, 1986; Latal, 1986; Pavlov and Bezchastnov, 1988; Bezchastnov and Pavlov, 1988, 1989). The Feynman diagrams for cyclotron absorption and emission are shown in Figure 5.1.3. The single-vertex S-matrix element is the same for both processes, namely

$$S_{fi} = -ie\int d^4x\overline{\Psi_f(x)}\gamma^\mu A_\mu \Psi_i(x). \tag{5.1.27}$$

The difference between absorption and emission comes in through the vector potential of the photon. For absorption, the potential is of the form of equation (5.1.3), i.e. an initial photon, while for emission, equation (5.1.2) is used, i.e. a final photon. Replacing these quantities and performing the time integral, the corresponding S-matrix elements are

$$S_{fi}^{(\mathrm{abs})} = -ie\sqrt{\frac{2\pi}{V\omega}}\,2\pi\delta(E'-\omega-E)\vec{\epsilon}\cdot\left(\vec{J}_{\beta\beta'}(-\vec{k})\right)^*, \tag{5.1.28}$$

$$S_{fi}^{(\mathrm{emis})} = -ie\sqrt{\frac{2\pi}{V\omega}}\,2\pi\delta(E'+\omega-E)\vec{\epsilon}^{\,*}\cdot\vec{J}_{\beta'\beta}(-\vec{k}), \tag{5.1.29}$$

where the current vectors $\vec{J}$ are those given in Appendix A. The transition rates are

$$w^{(\mathrm{abs})} = \sum_{q'}\frac{V}{T}\left|S_{fi}^{(\mathrm{abs})}\right|^2\frac{L}{2\pi}\,dp', \tag{5.1.30}$$

$$w^{(\mathrm{emis})} = \sum_{q'}\frac{1}{T}\left|S_{fi}^{(\mathrm{emis})}\right|^2\frac{L}{2\pi}\,dp'\frac{V}{(2\pi)^3}\,d^3k. \tag{5.1.31}$$

One can express therefore the rates for absorption, $w_{jj'}(p; \vec{k}, \vec{\epsilon} \rightarrow p')$, and emission, $w_{jj'}(p \rightarrow \vec{k}, \vec{\epsilon}; p')$, as

$$w_{jj'}\left(p; \vec{k}, \vec{\epsilon} \rightarrow p'\right) = \frac{(2\pi)^2 e^2}{\omega} \delta(E' - \omega - E) \delta(p' - k - p) \times \left[\frac{(E+m)(E'+m)}{4EE'}\right] \times \left|\vec{\epsilon} \cdot \vec{G}^*_{\beta\beta'}\left(-\vec{k}_\perp\right)\right|^2 dp', \tag{5.1.32}$$

$$w_{jj'}\left(p \rightarrow \vec{k}, \vec{\epsilon}; p'\right) = \frac{e^2}{2\pi\omega} \delta(E' + \omega - E) \delta(p' + k - p) \times \left[\frac{(E+m)(E'+m)}{4EE'}\right] \times \left|\vec{\epsilon}^* \cdot \vec{G}_{\beta'\beta}\left(-\vec{k}_\perp\right)\right|^2 dp' \, d^3k. \tag{5.1.33}$$

These rates are used for describing the cyclotron absorption or emission process, and they are also needed for the full description of the total rates of scattering processes which start with or leave the electron in an excited state.

## 5.2 Relativistic Redistribution Functions

### *a) Basic Scattering Redistribution in a Strong Field*

The scattering and emission (or absorption) rates $w$ calculated previously are the probabilities for a specific process to occur in a single electron. In the radiative transport problem, however, one is interested in the statistical behavior of the plasma. One can define in general the Compton scattering redistribution function for initial electrons in state $j$ as

$$\mathcal{R}_j(\alpha \rightarrow \alpha') \equiv \int dp n_j(p) \sum_{j'} \int dp' w_{jj'}(\alpha, p \rightarrow \alpha', p'). \tag{5.2.1}$$

Here $\alpha$ denotes the set of photon parameters $[\vec{k}, \vec{\epsilon}]$, and $n_j(p)$ is the normalized distribution of electrons or positrons in state $j$ with parallel momentum $p$. The latter may be any general distribution, which in the particular case of thermal equilibrium becomes a one-dimensional relativistic Maxwell-Boltzmann distribution,

$$n_j(p) = \frac{1}{2mK_1\left(\frac{m}{T}\right)} e^{-E_j(p)/T}, \tag{5.2.2}$$

where $T$ is the electron temperature in energy units and $K_1$ denotes a modified Bessel function of the second kind. For an arbitrary distribution function, the single-photon Compton scattering redistribution function (e.g. Bussard, Alexander, and Mészáros, 1986) can be calculated by substituting equation (5.1.19) for $w_{jj'}(\alpha, p \to \alpha', p')$ in equation (5.2.1). The momentum delta function allows the integration over final momentum, $p'$, to be done immediately. A further simplification is obtained if azimuthal symmetry is assumed, since then equation (5.1.19) can be integrated over the final photon azimuth, $\phi'$, and averaged over the initial, $\phi$. In this case $\mathcal{R}_j(\alpha \to \alpha')$ becomes

$$\begin{aligned}\mathcal{R}_j(\alpha \to \alpha') = \frac{\pi}{2} m^2 r_0^2 \frac{\omega'}{\omega} \int dp n_j(p) \\ \times \sum_{j'} \left[ \frac{(E+m)(E'+m)}{EE'} \right] \bigg|_{p'=p+k-k'} \\ \times \delta(E' + \omega' - E - \omega) \\ \times \sum_{j''} \Big\{ |a_{j''}|^2 + |b_{j''}|^2 + 2 \\ \times \sum_{j_s} J_{j''+j_s-j'-j}\left(2\sqrt{\xi\xi'}\right) \mathrm{Re}\left(a_{j''} b_{j_s}^*\right) \Big\}. \end{aligned} \tag{5.2.3}$$

Here the term involving the Bessel function $J_j$ comes from the cross product of $a_{j''}$ and $b_{j''}$ in taking the square magnitude of the summation, "Re" designates the real part, and $r_0 = e^2/m$ is the classical electron radius, with $c = \hbar = 1$. Equation (5.2.3) is differential in the frequency $\omega = |\vec{k}|$ and in the direction cosine $\mu$ relative to the field direction, having $d^3k' = \omega'^2 \, d\mu' \, d\omega'$.

In the scattering between a well-defined initial state $\alpha$ and final photon state $\alpha'$, there are only two initial electron momenta that are compatible with conservation of energy and momentum. These are denoted by $p_\pm$ and are given by

$$p_\pm = \frac{k' - k}{2\Delta}\left[\Delta + 2m\omega_c(j' - j)\right] \pm \frac{\omega' - \omega}{2\Delta}\left\{\left[\Delta + 2m\omega_c(j' - j)\right]^2 + 4\Delta\left(m^2 + 2m\omega_c j\right)\right\}^{1/2}, \tag{5.2.4}$$

where $k$ is the parallel component of the photon wave vector and $\Delta = (k' - k)^2 - (\omega' - \omega)^2$. By virtue of the energy delta function, the integration over momentum reduces to a summation:

$$\int dp n_j(p)\delta(E' + \omega' - E - \omega)\cdots \to \sum_{p_\pm} n_j(p_\pm)\frac{\delta(p - p_\pm)}{\left|\frac{\partial}{\partial p}(E' - E)\right|}\cdots. \tag{5.2.5}$$

As a result, the Compton redistribution function may be written as

$$\begin{aligned}
\mathscr{R}_j(\alpha \to \alpha') &= \pi r_0^2 \frac{\omega'}{\omega}\sum_{j'}\sum_{p_\pm} n_j(p_\pm) \\
&\times \left.\frac{m^2(E + m)(E' + m)}{\left[\left(\Delta + 2m\omega_c(j' - j)\right)^2 + 4\Delta\left(m^2 + 2m\omega_c j\right)\right]^{1/2}}\right|_{p' = p + k - k'} \\
&\times \sum_{j''}\Big\{|a_{j''}|^2 + |b_{j''}|^2 \\
&+ 2\sum_{j_s} J_{j'' + j_s - j' - j}\left(2\sqrt{\xi\xi'}\right)\mathrm{Re}\left(\alpha_{j''} b_{j_s}^*\right)\Big\}.
\end{aligned} \tag{5.2.6}$$

*b) Single-Photon Scattering (0 → 0) Redistribution*

We consider now the Compton scattering redistribution case where the initial and final electrons are in the ground state, $j = j' = 0$. Since there is one photon coming in and at the end another one going out, this process will be called single-photon, or one-photon, Compton scattering (as opposed to the two-photon or multiple-scattering discussed later). For some cases of interest, where the excitation energy is small compared to the ground cyclotron energy (e.g. high field accreting pulsars), one expects significant excitations only to the first intermediate excited level so that the initial and final state are the ground state and one-photon Compton scattering is the dominant process. The case where higher levels are excited and the scattering leaves the electron in an excited state will be addressed in the next section.

It will be necessary to consider both the processes that scatter photons out of state $\alpha$ (representing an opacity) and those that do the inverse, i.e. processes that scatter photons from all other states $\alpha'$ into $\alpha$ (representing a source of photons). In terms of the redistribution function, the opacity $\kappa_{c,\alpha}$ and the source function $Q_{c,\alpha}$ for one-photon Compton scattering can be written as

$$\kappa_{c,\alpha} = \int d\alpha' \, \mathcal{R}(\alpha \to \alpha')(1 + N_{\alpha'}), \tag{5.2.7}$$

$$Q_{c,\alpha} = \int d\alpha' N_{\alpha'} \, \mathcal{R}(\alpha' \to \alpha)(1 + N_{\alpha}), \tag{5.2.8}$$

where $\mathcal{R}(\alpha \to \alpha')$ is the redistribution function for $j = j' = 0$, $N_\alpha$ is the occupation number of photons in $\alpha$, and the integration over $\alpha'$ stands for a summation over polarizations and integration over $\mu'$ and $\omega'$. The factors of $(1 + N_\alpha)$ account for stimulated scattering. The source function may be rewritten in terms of $\mathcal{R}(\alpha \to \alpha')$ by using the condition of detailed balance, as was done for the nonrelativistic redistribution function in Chapter 4. This states that, in thermal equilibrium, the number of photons that scatter from $\alpha \to \alpha'$ equals the number going from $\alpha' \to \alpha$, that is

$$\begin{aligned} n_0(p)\overline{N_\alpha} w_{00}(\alpha, p \to \alpha', p')\left(1 + \overline{N_{\alpha'}}\right) \\ = n_0(p')\overline{N_{\alpha'}} w_{00}(\alpha', p' \to \alpha, p)\left(1 + \overline{N_\alpha}\right), \end{aligned} \tag{5.2.9}$$

where $\overline{N_\alpha}$ is the equilibrium photon occupation number [i.e. Planck, $\overline{N_\alpha} = (e^{\omega/T} - 1)^{-1}$]. Therefore, one has in general that

$$\mathcal{R}(\alpha' \to \alpha) = e^{(\omega' - \omega)/T} \mathcal{R}(\alpha \to \alpha'), \tag{5.2.10}$$

and the source function may be rewritten as

$$Q_{c,\alpha} = (1 + N_\alpha) \int d\alpha' N_{\alpha'} e^{(\omega' - \omega)/T} \mathcal{R}(\alpha \to \alpha'). \tag{5.2.11}$$

Equations (5.2.7) and (5.2.8) for the opacity and source include the factors for stimulated scattering. We can decompose $Q_{c,\alpha}$ into two terms: one is independent of $N_\alpha$ and is taken to be the source, while the other, which is proportional to $N_\alpha$, can be treated as a negative opacity. Thus the one-photon Compton opacity and source may be written as

$$\kappa_{c,\alpha} = \int d\alpha'\left[1 + k_{st} N_{\alpha'}\left(1 - e^{(\omega' - \omega)/T}\right)\right] \mathscr{R}\left(\alpha \to \alpha'\right), \quad (5.2.12)$$

$$Q_{c,\alpha} = \int d\alpha' N_{\alpha'} e^{(\omega' - \omega)/T} \mathscr{R}\left(\alpha \to \alpha'\right). \quad (5.2.13)$$

Here $k_{st} = [1, 0]$ is a computational switch such that when it equals 1(0) the calculation includes (neglects) stimulated effects. These opacities and sources are evaluated for a given photon $\alpha$ and integrated over the photon $\alpha'$. The sum over intermediate states $j''$ in (5.2.6) can be carried out to the desired accuracy, and the polarization vectors $\vec{\epsilon}$ used are those derived later in this chapter, corresponding to a relativistic magnetized plasma (see eqs. 5.3.43). The $0 \to 0$ scattering opacity (5.2.12), normalized to the field-free Thomson value, is shown in Figure 5.2.1 for an incoming photon angle with respect to the field of $\theta = 60°$, $\omega_c = 38$ keV, $kT = 8$ keV, and $\rho = 0.5$ g cm$^{-3}$, neglecting stimulated scattering effects (i.e. $k_{st} = 0$). The relativistic Landau level energy structure entails an anharmonicity, given by the relativistic energy dependence of the resonance $\omega_n = (-m + \sqrt{m^2 + 2m\omega_c n \sin^2\theta})/\sin^2\theta$. Notice also that the relativistic cross section gives a faster drop in the high-energy wing of the lines than the corresponding nonrelativistic calculation. The source function (5.2.13) can be similarly obtained, given a particular photon occupation number. It is proportional to the values in Figure 5.2.1 multiplied by the chosen $N_{\alpha'}$.

### *c) Two-Photon and Multiple-Photon Redistribution*

When the scattering process (consisting of one photon in, one photon out) leaves the final electron in a Landau level higher than the initial one, this final excited electron will in a very short time radiatively decay producing one (or more) resonant photons, in addition to the first normal outcoming photon from the scattering event. We call this compound process multiple-photon scattering; it is made up of two (or more) Feynman diagrams in succession, one for the scattering with

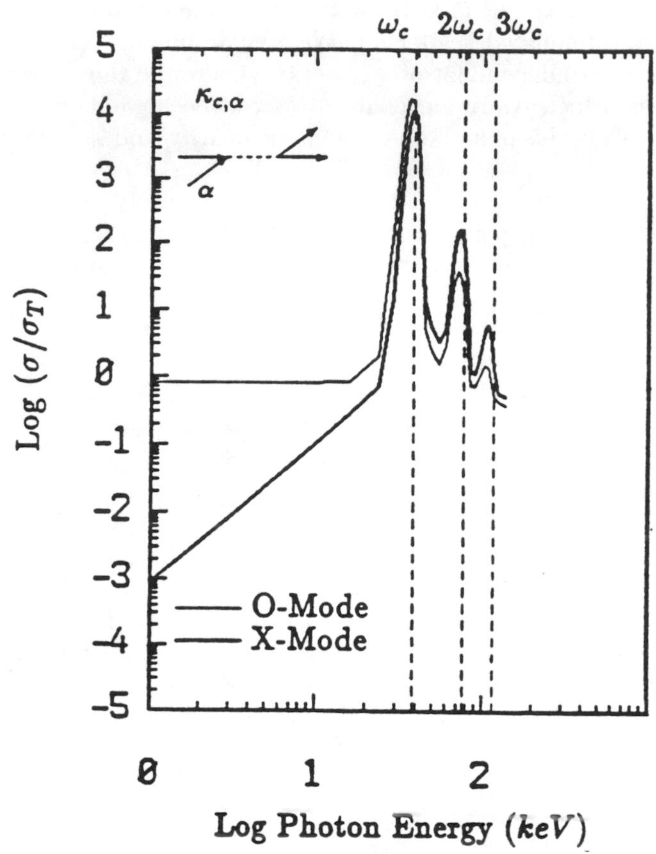


FIGURE 5.2.1 Single-photon Compton scattering opacity normalized to the Thomson value for $j = j' = 0$ and inclusion of the intermediate states $j'' = 0, 1, 2, 3$. From S. Alexander and P. Mészáros, 1991, *Ap. J.*, 372, 545.

excited final state, followed by one or more for the subsequent decays. This is not in general the same as the double (or triple, etc.) Compton scattering, where the initial real electron is excited to a virtual intermediate state (propagator) which decays directly via two or more photons again leaving a ground state final electron; in this case, because there are three (or more) propagator-photon vertices, double (or triple, etc.) Compton scattering is proportional to $\alpha_f^3$ (or $\alpha_f^4$, etc.), a higher-order process. By contrast, in the compound two-photon (or three-photon, etc.) scattering, the fact that the initial single-photon scattering leads to a real excited final electron effectively decouples this from the subsequent cyclotron emission decay. A different but related way of looking at multiple-photon scattering is to concentrate on the resonant regime (e.g. the semirelativistic treatment of Wang *et al.*, 1989, where this process bears resemblance to the well-known Raman scattering effect of atoms in the absence of a field). While at resonance the intermediate states tend to become like real states and the excitation process becomes effectively of order $\alpha_f$ so that in this regime one expects an overlap between the compound multiple-photon scattering and the double (triple, etc.) Compton scattering, a distinction between these processes (and between real and virtual states) is natural and necessary when nonresonant frequencies are also considered (e.g. Alexander and Mészáros, 1989). This is the approach of our description here, following Alexander and Mészáros (1991a), which gives a unified formalism dealing with both the resonant and the nonresonant regions of frequency space.

Because of the presence of real final states before the subsequent decays, one does not have to apply the QED calculation rules to the compound process as a whole. One can calculate separately the rates for one-photon scattering leaving the final electron excited, and that for the subsequent cyclotron emissions which are bound to occur with very high probability. Because of this, the total rate, and hence the opacity for this compound process, is in general $\propto \alpha_f^2$, as for normal single-photon Compton scattering, rather than $\propto \alpha_f^3$, etc. (At resonance, of course, the rate for single-photon scattering is essentially that of cyclotron absorption, $\propto \alpha_f$, since virtual states become essentially real, so that near resonance the previous simple scaling in terms of number of vertices and powers of $\alpha_f$ must be modified.) The diagrams for the particular case of compound two-photon scattering are shown in Figure 5.2.2a. Using the rates for single-photon scattering and cyclotron emission, the general multiple-photon scattering rate can be expressed as

$$\chi(\alpha \to \alpha'; \alpha'') \equiv \int dp n_0(p) \sum_{j'} \sum_{j'' < j} \int dp' \times \frac{\int dp'' \, w_{0j'}(p, \alpha \to p', \alpha') w_{j'j''}(p' \to p'', \alpha'')}{\sum_{j''' < j'} \int d\alpha''' \int dp''' \, w_{j'j'''}(p' \to p''', \alpha''')}, \tag{5.2.14}$$

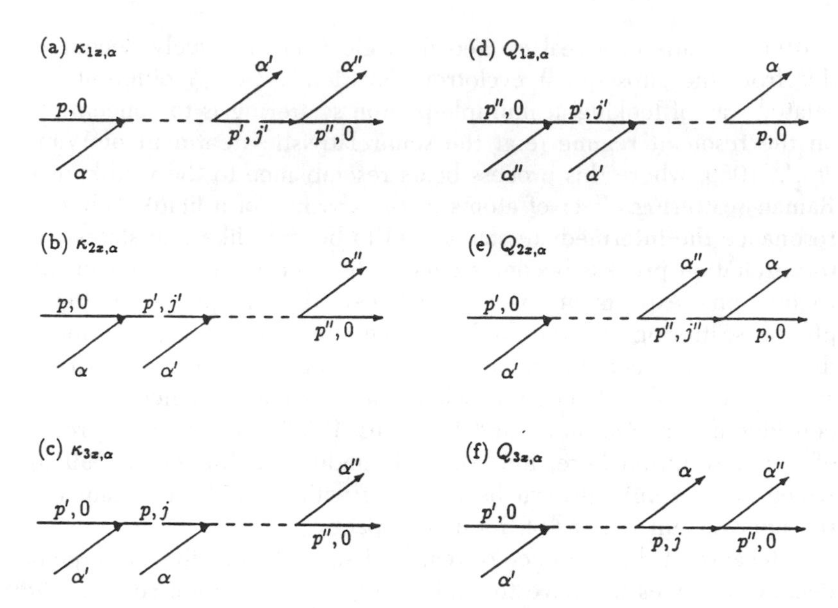


FIGURE 5.2.2 Opacity and source diagrams for two-photon scattering and its inverses. The diagrams with permuted photons (not shown) are also used. This compound process is made up of single-photon scattering involving an excited final state followed by cyclotron emission or absorption. The left-hand diagrams are opacities, while the right-hand diagrams are source terms. For each process, the excited electron line is shown interrupted, to stress that two different Feynman diagrams are involved. From S. Alexander and P. Mészáros, 1991, *Ap. J.*, 372, 545.

which represents the integrated $j = 0 \rightarrow j'$ scattering rate times the probability that the excited electron will radiatively decay with the emission of a particular photon $\alpha''$ (as opposed to any of the other possible photons $\alpha'''$). The denominator should in principle contain all of the possible decay modes, including two-photon decay; for simplicity, the latter has not been written explicitly in equation (5.2.14). Substituting the corresponding rates into equation (5.2.14), one can for most purposes integrate over final azimuth angles and average over initial ones. The integrals over $p'$ and $p''$ can be done with the delta functions, and the integral over $p$ follows the procedure leading to equation (5.2.40).

Addressing now the special case of two-photon scattering (i.e. one photon in, two photons out only), we set $j'' = 0$, and this leads to the two-photon scattering redistribution function

$$\chi(\alpha \to \alpha'; \alpha'')$$
$$= \pi r_0^2 \frac{\omega' \omega''}{\omega} \sum_{j'} \sum_{p_\pm} n_0(p_\pm)$$
$$\times \left. \frac{m^2(E+m)(E'+m)(E''+m)}{\left[(\Delta + 2m\omega_c j')^2 + 4\Delta m^2\right]^{1/2} E''} \right|_{\substack{p'=p+k-k' \\ p''=p'-k''}}$$
$$\times \sum_{j_i} \left\{ |a_{j_i}|^2 + |b_{j_i}|^2 + 2\sum_{j_s} J_{j_i+j_s-j'}(2\sqrt{\xi\xi'})\mathrm{Re}(a_{j_i} b_{j_s}^*) \right\}$$
$$\times \frac{\delta(E'' + \omega'' - E')\left|\vec{\epsilon}''^* \cdot \vec{G}_{\beta''\beta'}(-\vec{k}''_\perp)\right|^2}{\sum_{j''' < j'} \sum_{\vec{\epsilon}'''} \int_{-1}^{1} d\mu''' \frac{\tilde{\omega}(E''' + m)}{E'''} \left|\vec{\epsilon}'''^* \cdot \vec{G}_{\beta'''\beta'}\left(-\tilde{\vec{k}}_\perp\right)\right|^2}, \tag{5.2.15}$$

where the $j_i$ are intermediate virtual electron states. In the denominator, the decay frequency, $\tilde{\omega}$, is given by

$$\tilde{\omega} = \frac{(E' - p'\mu''') - \left[(E' - p'\mu''')^2 - 2m\omega_c j'(1 - \mu'''^2)\right]^{1/2}}{(1 - \mu'''^2)}, \tag{5.2.16}$$

where $E''' = E' - \tilde{\omega}$ by energy conservation. Notice that this two-photon scattering redistribution is accurate for excitations up to the second excited intermediate level $n = 2$ (second harmonic) since in this case the final real state is $n = 1$ and only one more additional resonant photon is expected. The same expression may be used for third harmonic excitations too, but the result will only be approximate, since one of the decays will have to be a less probable $\Delta n = 2$ decay. An accurate treatment of third harmonics requires three-photon redistribution, for which $j'' \neq 0$ must be used.

In a radiative transfer calculation where the optical depth is sufficiently large, one needs to consider not only the direct process of Figure 5.2.2a but also the inverse and exchange processes. This gives rise to three opacity and three source terms. In this case, it is again useful to consider the detailed balance condition for $j = 0 \to j'$ scattering and its inverse, $j = j' \to 0$:

$$\overline{N_\alpha} n_0(p) w_{0j'}(p, \alpha \to p', \alpha')(1 + \overline{N_{\alpha'}})$$
$$= \overline{N_{\alpha'}} n_{j'}(p') w_{j'0}(p', \alpha' \to p, \alpha)(1 + \overline{N_\alpha}). \tag{5.2.17}$$

Now, in thermodynamic equilibrium, the photon occupation numbers are Planckian and the electron distributions are Maxwellian,

$$\frac{n_j(p')}{n_0(p)} = \exp\frac{[E_0(p) - E_j(p')]}{T} = e^{(\omega' - \omega)/T}, \tag{5.2.18}$$

so that the scattering rates are identical,

$$w_{0j'}(p, \alpha \to p', \alpha') = w_{j'0}(p', \alpha' \to p, \alpha). \tag{5.2.19}$$

In a similar manner, one can show that for absorption and emission

$$w_{0j'}(p, \alpha \to p') = w_{j'0}(p' \to p, \alpha). \tag{5.2.20}$$

The three opacity terms for two-photon scattering are shown on the left side of Figure 5.2.2, where the first, (a) $\kappa_{1x,\alpha}$, is the direct process, while (b) $\kappa_{2x,\alpha}$ and (c) $\kappa_{3x,\alpha}$ are the inverse and exchange. Note that the diagrams in (b) and (c) are not identical, since to obtain the opacity for photon $\alpha$, all other photon states are integrated. For the direct diagram in Figure 5.2.2a, $\kappa_{1x,\alpha}$ is given by

$$\kappa_{1x,\alpha} = \int dp n_0(p) \int d\alpha' \sum_{j'} \int dp' w_{0j'}(p, \alpha \to p', \alpha')(1 + N_{\alpha'})$$
$$\times \frac{\int d\alpha'' \int dp'' w_{j'0}(p' \to p'', \alpha'')(1 + N_{\alpha''})}{\sum_{j''' < j'} \int d\alpha''' \int dp''' w_{j'j'''}(p' \to p''', \alpha''')(1 + N_{\alpha'''})}, \tag{5.2.21}$$

where the $(1 + N_\alpha)$ factors account for stimulated scattering. The ratio in equation (5.2.21) is equal to unity, since the numerator is integrated over all photons $\alpha''$; however, it is useful to keep it for consistency in what follows. The stimulated factor in the denominator can be neglected with little loss of accuracy, and comparing this to equation (5.2.14), we can express it as

$$\kappa_{1x,\alpha} = \int d\alpha' \int d\alpha''(1 + k_{st}N_{\alpha''})(1 + k_{st}N_{\alpha'})\chi(\alpha \to \alpha'; \alpha''), \tag{5.2.22}$$

where, as before, $k_{st} = 1(0)$ is a numerical switch allowing us to include (exclude) stimulated scattering.

In a similar manner, Figure 5.2.2b can be expressed as an opacity $\kappa_{2x,\alpha}$:

$$\kappa_{2x,\alpha} = \int dp n_0(p) \int d\alpha' N_{\alpha'} \sum_{j'} \int dp' w_{0j'}(p', \alpha \to p')$$
$$\times \frac{\int d\alpha'' \int dp'' w_{j'0}(p', \alpha' \to p'', \alpha'')(1 + N_{\alpha''})}{\sum_{j''' < j'} \int d\alpha''' \int dp''' w_{j'j'''}(p' \to p''', \alpha''')(1 + N_{\alpha'''})}. \tag{5.2.23}$$

Using the detailed balance conditions and the energy conservation condition $E_0(p) + \omega + \omega' = E_0(p'') + \omega''$ for the entire process,

$$n_0(p) = n_0(p'')\exp\left(\frac{E_0(p'') - E_0(p)}{T}\right) = e^{(\omega + \omega' - \omega'')/T}, \tag{5.2.24}$$

we can write $\kappa_{2x,\alpha}$ as

$$\kappa_{2x,\alpha} = \int d\alpha' \int d\alpha'' N_{\alpha'}(1 + N_{\alpha''})e^{(\omega+\omega'-\omega'')/T} \int dp'' n_0(p'')$$
$$\times \sum_{j'} \int dp' \int dp$$
$$\times \frac{w_{0j'}(p'', \alpha'' \to p', \alpha')w_{j'0}(p' \to p, \alpha)}{\sum_{j''' < j'} \int d\alpha''' \int dp''' w_{j'j'''}(p' \to p''', \alpha''')} \tag{5.2.25}$$

Again, $\kappa_{2x,\alpha}$ can be written as an integral over the two-photon scattering redistribution function,

$$\kappa_{2x,\alpha} = \int d\alpha' \int d\alpha'' N_{\alpha'}(1 + k_{st}N_{\alpha''})e^{(\omega+\omega'-\omega'')/T}\chi(\alpha'' \to \alpha'; \alpha). \tag{5.2.26}$$

Similarly, Figure 5.2.2c gives an opacity $\kappa_{3x,\alpha}$, which is given by

$$\kappa_{3x,\alpha} = \int dp' n_0(p') \int d\alpha' N_{\alpha'} \sum_j \int dp w_{0j}(p', \alpha' \to p)$$
$$\times \frac{\int d\alpha'' \int dp'' w_{j0}(p, \alpha \to p'', \alpha'')(1 + N_{\alpha''})}{\sum_{j''' < j'} \int d\alpha''' \int dp''' w_{jj'''}(p \to p''', \alpha''')(1 + N_{\alpha'''})}, \tag{5.2.27}$$

and following the same steps as above, this can be expressed as

$$\kappa_{3x,\alpha} = \int d\alpha' \int d\alpha'' N_{\alpha'}(1 + k_{st} N_{\alpha''}) e^{(\omega' + \omega - \omega'')/T} \chi(\alpha'' \to \alpha; \alpha'). \tag{5.2.28}$$

For a radiative transport calculation, one also needs expressions for the scattering source terms, i.e. the processes that scatter photons into state $\alpha$ rather than scattering them out from it. For the case of two-photon scattering, the source terms are shown on the right-hand side of Figure 5.2.2, and they can be expressed in terms of the two-photon scattering redistribution function in a manner similar to that for the opacities. Again, these are compound processes, made up of two different processes occurring one after another, symbolized by the interrupted excited electron line. For each of these the cross sections are calculated separately and then combined into a single rate. The first diagram of Figure 5.2.2d represents a source $Q_{1x,\alpha}$, which is

$$Q_{1x,\alpha} = \int dp'' n_0(p'') \int d\alpha'' N_{\alpha''}$$
$$\times \sum_{j'} \int dp' \int d\alpha' N_{\alpha'} w_{0j'}(p'' \to p', \alpha')$$
$$\times \frac{\int dp w_{j'0}(p', \alpha' \to p, \alpha)(1 + N_\alpha)}{\sum_{j''' < j'} \int d\alpha''' \int dp''' w_{j'j'''}(p' \to p''', \alpha''')(1 + N_{\alpha'''})}, \tag{5.2.29}$$

which can be recast as

$$Q_{1x,\alpha} = (1 + k_{\text{st}} N_\alpha) \int d\alpha' \int d\alpha'' N_{\alpha'} N_{\alpha''} \times e^{(\omega'' + \omega' - \omega)/T} \chi(\alpha \to \alpha'; \alpha''). \tag{5.2.30}$$

Similarly, Figure 5.2.2e gives

$$Q_{2x,\alpha} = \int dp' n_0(p') \int d\alpha' N_{\alpha'} \int d\alpha'' \times \sum_{j''} \int dp'' w_{0j''}(p', \alpha' \to p'', \alpha'')(1 + N_{\alpha''}) \times \frac{\int dp w_{j''0}(p'' \to p, \alpha)(1 + N_\alpha)}{\sum_{j''' < j''} \int d\alpha''' \int dp''' \, w_{j''j'''}(p'' \to p''', \alpha''')(1 + N_{\alpha'''})}, \tag{5.2.31}$$

which is

$$Q_{2x,\alpha} = (1 + k_{\text{st}} N_\alpha) \int d\alpha' \int d\alpha'' \times N_{\alpha'}(1 + k_{\text{st}} N_{\alpha''}) \chi(\alpha' \to \alpha''; \alpha). \tag{5.2.32}$$

The last diagram of Figure 5.2.2f gives a term

$$Q_{3x,\alpha} = \int dp' n_0(p') \int d\alpha' N_{\alpha'} \times \sum_{j} \int dp w_{0j}(p', \alpha' \to p, \alpha)(1 + N_\alpha) \times \frac{\int d\alpha'' \int dp'' w_{j0}(p \to p'', \alpha'')(1 + N_{\alpha''})}{\sum_{j''' < j} \int d\alpha''' \int dp''' \, w_{jj'''}(p \to p''', \alpha''')(1 + N_{\alpha'''})}, \tag{5.2.33}$$

which is rewritten as

$$Q_{3x,\alpha} = (1 + k_{\text{st}} N_\alpha) \int d\alpha' \int d\alpha'' \times N_{\alpha'}(1 + k_{\text{st}} N_{\alpha''}) \chi(\alpha' \to \alpha; \alpha''). \tag{5.2.34}$$

The opacities and source functions for two-photon scattering are evaluated in a manner similar to that for single-proton Compton scattering. The magnetized relativistic polarization vectors of § 5.3 are used, and the opacities and sources are calculated for a set of initial or final photon frequencies, directions, and polarization mode (extraordinary or ordinary), assuming a particular set of plasma conditions, e.g. density, temperature, magnetic field strength. The radiation field must be known explicitly in order to calculate the terms that depend on $N_\alpha$, which generally requires an iterative solution of the radiative transport. A frequently used initial guess for the radiation field is that it is an equilibrium spectrum, either Planck or Wien.

Examples of these opacities and sources have been calculated by Alexander and Mészáros (1991a), who give plots of the frequency dependence for various conditions. Both the opacities and the sources show resonances at $\omega_c$, $\sim 2\omega_c$, $\sim 3\omega_c$, etc. The cross section for $\kappa_{1x,\alpha}$ is zero until the photon has sufficient energy to scatter the electron into the first excited state, $j' = 1$, which leads to the production of a scattered photon $\omega'$ having low energy (leading to a soft-photon production source: see Bussard, Mészáros, and Alexander, 1985). The electron then decays to the ground state with the emission of a resonant photon $\omega''$. This opacity has a weak first resonance that peaks near $\omega_c$. When the initial photon has sufficient energy to scatter the electron into either the $j' = 2$ or $j' = 3$ level, strong resonances occur in the cross section at $\omega \sim 2\omega_c$ and $3\omega_c$. This is shown in the left panel of Figure 5.2.3 for $\hbar\omega_c = 38$ keV, $kT = 8$ keV, $\rho = 0.5$ g cm$^{-3}$. Comparing this opacity with that for the one-photon scattering cross section $\kappa_{c,\alpha}$ (Fig. 5.2.1), one sees that, in the vicinity of the second and third cyclotron harmonic, the two-photon scattering cross section exceeds the one-photon scattering by a significant factor near $\omega \sim 2\omega_c$, and $\omega \sim 3\omega_c$ for the physical conditions used (see also Harding and Preece, 1989). Evidently, if the higher-level harmonics are to be studied, two-photon scattering must be included.

The opacity cross sections for the inverse and exchange two-photon scattering, $\kappa_{2x,\alpha}$ and $\kappa_{3x,\alpha}$, also exhibit resonances near $\omega_c$ and $2\omega_c$ that are quite small compared to those for either $\kappa_{1x,\alpha}$ or $\kappa_{c,\alpha}$. A plot of $\kappa_{2x,\alpha}$ given in the middle panel of Figure 5.2.3 shows that this behaves as a resonant absorption cross section, in that it is zero for energies insufficient to cause the electron to be lifted to the first excited state. This behavior is consistent with the diagram for $\kappa_{2x,\alpha}$ in Figure 5.2.2b. On the other hand, $\kappa_{3x,\alpha}$ is nonzero at low photon energies, as seen in the right panel of Figure 5.2.3. This behavior is expected from the corresponding diagram in Figure 5.2.2c,

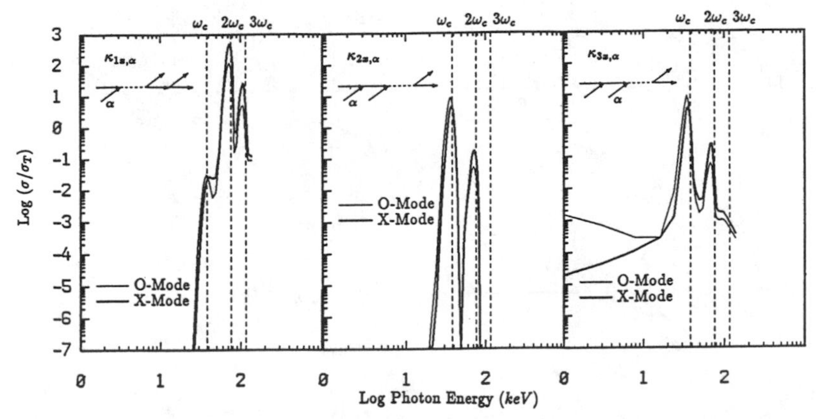


FIGURE 5.2.3 Opacities for two-photon scattering $\kappa_{1,x}$ (*left*), $\kappa_{2,x}$ (*middle*), and $\kappa_{3,x}$ (*right*) as a function of energy for extraordinary and ordinary polarization photons. From S. Alexander and P. Mészáros, 1991, *Ap. J.*, 372, 545.

showing that, for $\kappa_{3x,\alpha}$, the photon $\alpha$ scatters from an excited state electron, which is energetically possible also for a low photon energy. This is what gives the low-energy tail to this cross section.

The redistribution function $\chi(\alpha' \to \alpha; \alpha'')$ is evaluated in a similar manner to calculate the source functions for two-photon scattering. These cross sections for $Q_{1x,\alpha}$, $Q_{2x,\alpha}$, and $Q_{3x,\alpha}$, corresponding to the diagrams on the right-hand side of Figure 5.2.2, have been evaluated by Alexander and Mészáros (1991a) for the case of a Wien occupation number.

*d) Two-Photon Emission Redistribution Function*

The two-photon decay process can be analyzed in a manner similar to the two-photon scattering process. One can envisage the situation of an incoming photon exciting an electron to a real excited state via cyclotron absorption (unlike in two-photon scattering, where this is a virtual excited state), and then this real state decays via two-photon decay (as in Fig. 5.1.2). These two processes, being separated by a real state, are effectively decoupled from each other and the series of events may be considered again as a compound process, for the same reason as for two-photon scattering. This compound two-photon emission process is described by the diagrams in Figure 5.2.4a, which does not show the corresponding photon order permutation diagrams, although these are included in the calculations. As can be seen, these

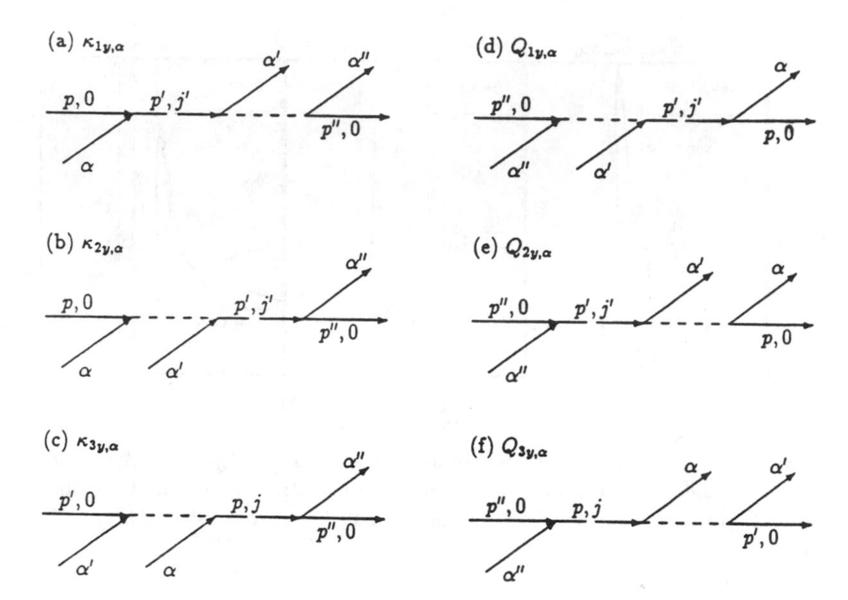


FIGURE 5.2.4 Opacity and source diagrams for two-photon emission and its inverse and exchange processes. These are compound processes made up of cyclotron absorption or emission and a two-photon emission or absorption process. For each opacity or source, the excited electron line is shown interrupted, to stress that two different processes, and hence two Feynman diagrams, are involved in each. From S. Alexander and P. Mészáros, 1991, *Ap. J.*, 372, 545.

diagrams bear a similarity to those of the two-photon scattering shown in Figure 5.2.2; the number of vertices, real and excited states, and photons is the same. The difference lies in an exchange of the time order of the real and virtual intermediate electron states.

In general, the corresponding redistribution function can be constructed from the cyclotron absorption and emission rates and the two-photon decay rate (Alexander and Mészáros, 1991a), and is given by

$$\mathscr{Y}(\alpha \to \alpha'; \alpha'') \equiv \int dp n_0(p) \sum_{j'} \sum_{j''<j'} \int dp' \times \frac{\int dp'' \, w_{0j'}(p, \alpha \to p') w_{j'j''}(p' \to p'', \alpha', \alpha'')}{\sum_{j'''<j'} \int d\alpha''' \int dp''' \, w_{j'j'''}(p' \to p''', \alpha''')}. \tag{5.2.35}$$

This is the integrated $j = 0 \rightarrow j'$ absorption rate times the probability that the excited electron will spontaneously decay by emitting two photons $\alpha'$ and $\alpha''$ instead of a single photon $\alpha'''$. The necessary rates to construct $\mathscr{Y}(\alpha \rightarrow \alpha'; \alpha'')$ have been calculated previously, and the integrations and angle averages proceed in a similar manner to those used for $\mathscr{X}(\alpha \rightarrow \alpha'; \alpha'')$. Specializing again to the case of $j'' = 0$—where there is one photon in, two photons out only—one has

$$\begin{aligned}
&\mathscr{Y}(\alpha \rightarrow \alpha'; \alpha'') \\
&= \pi r_0^2 \frac{\omega' \omega''}{\omega} \sum_{j'} \sum_{p_\pm} n_0(p_\pm) \\
&\quad \times \frac{m^2 (E+m)(E'+m)(E''+m)}{\left[(\Delta + 2m\omega_c j')^2 + 4\Delta m^2\right]^{1/2} E''} \Bigg|_{\substack{p' = p+k \\ p'' = p' - k' - k''}} \\
&\quad \times \sum_{j_i} \Big\{ |c_{j_i}|^2 + |d_{j_i}|^2 + 2(-1)^{j_i} \\
&\quad \times \sum_{j_s} (-1)^{j_s} J_{j_i + j_s - j'}\left(2\sqrt{\xi' \xi''}\right) \mathrm{Re}\left(c_{j_i} d_{j_s}^*\right) \Big\} \\
&\quad \times \frac{\delta(E'' + \omega'' + \omega' - E') \left| \vec{\epsilon} \cdot \vec{G}_{\beta\beta'}^*(-\vec{k}_\perp) \right|^2}{\sum_{j''' < j'} \sum_{\vec{\epsilon}'''} \int_{-1}^{1} d\mu''' \frac{\tilde{\omega}(E''' + m)}{E'''} \left| \vec{\epsilon}'''^* \cdot \vec{G}_{\beta''' \beta'}\left(-\tilde{\vec{k}}_\perp\right) \right|^2},
\end{aligned} \tag{5.2.36}$$

where $p_\pm$ is given by equation (5.2.4) with $\omega' = k' = 0$ and $\Delta = k^2 - \omega^2$, and the decay frequency, $\tilde{\omega}$, is given by equation (5.2.16).

With the use of the detailed balance conditions for the scattering rates, all of the opacities and sources may be expressed as integrals of the redistribution function. The necessary detailed balance conditions are

$$w_{0j'}(p \rightarrow p', \alpha, \alpha') = w_{j'0}(p', \alpha, \alpha' \rightarrow p) \tag{5.2.37}$$

and

$$w_{0j'}(p, \alpha \rightarrow p') = w_{j'0}(p' \rightarrow p, \alpha). \tag{5.2.38}$$

The steps in obtaining the opacities and sources as integrals over $\mathscr{Y}(\alpha \rightarrow \alpha'; \alpha'')$ are identical to those used in the previous section. The opacity terms corresponding to the two-photon decay process diagrams of Figure 5.2.4 are, for Figure 5.2.4a,

$$\kappa_{1y,\alpha}=\int dpn_0(p)\sum_{j'}\int dp'\,w_{0j'}(p,\alpha\rightarrow p')$$
$$\times\frac{\int d\alpha'\int d\alpha''\int dp''\,w_{0j'}(p'\rightarrow p'',\alpha',\alpha'')(1+N_{\alpha'})(1+N_{\alpha''})}{\sum_{j'''<j'}\int d\alpha'''\int dp'''\,w_{j'j'''}(p'\rightarrow p''',\alpha''')(1+N_{\alpha'''})},$$

(5.2.39)

which in its final form becomes

$$\kappa_{1y,\alpha}=\int d\alpha'\int d\alpha''(1+k_{\mathrm{st}}N_{\alpha'})(1+k_{\mathrm{st}}N_{\alpha''})\,\mathcal{Y}(\alpha\rightarrow\alpha';\alpha'').$$

(5.2.40)

Figure 5.2.4b gives the opacity

$$\kappa_{2y,\alpha}=\int dpn_0(p)\sum_{j'}\int dp'\int d\alpha'\,N_{\alpha'}w_{0j'}(p,\alpha,\alpha'\rightarrow p')$$
$$\times\frac{\int d\alpha''\int dp''\,w_{j'0}(p'\rightarrow p'',\alpha'')(1+N_{\alpha''})}{\sum_{j'''<j'}\int d\alpha'''\int dp'''\,w_{j'j'''}(p'\rightarrow p''',\alpha''')(1+N_{\alpha'''})},$$

(5.2.41)

which becomes, after using the detailed balance conditions and energy conservation,

$$\kappa_{2y,\alpha}=\int d\alpha'\int d\alpha''\,N_{\alpha'}(1+N_{\alpha''})e^{(\omega+\omega'-\omega'')/T}\,\mathcal{Y}(\alpha''\rightarrow\alpha;\alpha').$$

(5.2.42)

The last diagram of this type, Figure 5.2.4c, is given by the opacity

$$\kappa_{3y,\alpha}=\int dp'n_0(p')\int d\alpha'\,N_{\alpha'}\sum_{j}\int dpw_{0j}(p,\alpha',\alpha\rightarrow p)$$
$$\times\frac{\int d\alpha''\int dp''\,w_{j0}(p\rightarrow p'',\alpha'')(1+N_{\alpha''})}{\sum_{j'''<j}\int d\alpha'''\int dp'''\,w_{jj'''}(p\rightarrow p''',\alpha''')(1+N_{\alpha'''})},$$

(5.2.43)

which becomes

$$\kappa_{3y,\alpha} = \int d\alpha' \int d\alpha'' N_{\alpha'}(1 + k_{st} N_{\alpha''}) \times e^{(\omega' + \omega - \omega'')/T} \mathscr{Y}(\alpha'' \to \alpha'; \alpha). \tag{5.2.44}$$

The source functions corresponding to two-photon decay are given on the right-hand side of Figure 5.2.4, showing a compound rate made up of two processes. The first of these diagrams, Figure 5.2.4d, is a source given by

$$Q_{1y,\alpha} = \int dp' n_0(p') \int d\alpha' N_{\alpha'} \int d\alpha'' N_{\alpha''} \times \sum_{j''} \int dp'' w_{0j''}(p', \alpha', \alpha'' \to p'') \times \frac{\int dp w_{j''0}(p'' \to p, \alpha)(1 + N_{\alpha})}{\sum_{j''' < j''} \int d\alpha''' \int dp''' w_{j''j'''}(p'' \to p''', \alpha''')(1 + N_{\alpha'''})}, \tag{5.2.45}$$

which, in terms of the redistribution function, is

$$Q_{1y,\alpha} = (1 + k_{st} N_{\alpha}) \int d\alpha' \int d\alpha'' N_{\alpha'} N_{\alpha''} \times e^{(\omega' + \omega'' - \omega)/T} \mathscr{Y}(\alpha \to \alpha'; \alpha''). \tag{5.2.46}$$

The second source term of Figure 5.2.4e is

$$Q_{2y,\alpha} = \int dp'' n_0(p'') \int d\alpha'' \sum_{j'} \int dp' w_{0,j'}(p'', \alpha'' \to p') \times \frac{\int d\alpha' \int dp w_{j'0}(p' \to p, \alpha', \alpha)(1 + N_{\alpha'})(1 + N_{\alpha})}{\sum_{j''' < j'} \int d\alpha''' \int dp''' w_{j'j'''}(p' \to p''', \alpha''')(1 + N_{\alpha'''})}, \tag{5.2.47}$$

which becomes

$$Q_{2y,\alpha} = (1 + k_{st} N_{\alpha}) \int d\alpha' \int d\alpha'' \times N_{\alpha''}(1 + k_{st} N_{\alpha'}) \mathscr{Y}(\alpha'' \to \alpha'; \alpha). \tag{5.2.48}$$

The last source of Figure 5.2.4f is

$$Q_{3y,\alpha} = \int dp'' n_0(p'') \int d\alpha'' N_{\alpha''} \sum_j \int dp w_{0j}(p'', \alpha'' \to p)$$
$$\times \frac{\int d\alpha' \int dp' w_{j0}(p \to p', \alpha, \alpha')(1 + N_{\alpha'})(1 + N_\alpha)}{\sum_{j''' < j} \int d\alpha''' \int dp''' w_{jj'''}(p \to p''', \alpha''')(1 + N_{\alpha'''})}, \tag{5.2.49}$$

which is written as

$$Q_{3y,\alpha} = (1 + k_{st} N_\alpha) \int d\alpha' \int d\alpha''$$
$$\times N_{\alpha''}(1 + k_{st} N_{\alpha'}) \mathcal{Y}(\alpha'' \to \alpha; \alpha'). \tag{5.2.50}$$

The two-photon emission redistribution function can be calculated in the same way as for two-photon scattering, to obtain explicit expressions for the opacities and sources (Alexander and Mészáros, 1991a). As expected from the diagrammatic construction, $\kappa_{1y,\alpha}$ is found to behave as a resonant absorption with a steep threshold, i.e. it is zero for energies insufficient to excite the electron to the first excited level and has a small peak near the first harmonic which is lower than the corresponding peak for two-photon scattering. The process is more efficient for photon energies above the fundamental, the cross section for $\kappa_{1y,\alpha}$ near the second and third harmonics being significantly larger than for the first harmonic, but overall the process is less important than two-photon scattering. A basic difference compared to two-photon scattering is that, unlike that process (e.g. $\kappa_{1x,\alpha}$), $\kappa_{1y,\alpha}$ has no continuum contribution between the harmonics, since the first half of the compound process (absorption) involves two real electrons which must satisfy strict energy momentum conservation. The diagrams for $\kappa_{2y,\alpha}$ and $\kappa_{3y,\alpha}$ are actually identical to each other, except for an interchange of photons $\alpha$ and $\alpha'$ (see diagrams (e) and (f) in Fig. 5.2.4). This identity is also verified in the numerical evaluations. Both $\kappa_{2y,\alpha}$ and $\kappa_{3y,\alpha}$ are nonzero for low photon energies ($\omega < \omega_c$) since the other photon, $\alpha'$, may have sufficient energy to cause the excitation. In a similar manner, the source cross sections for two-photon decay can be calculated for a Wien, or any other, photon distribution.

## 5.3 Relativistic Wave Propagation

*a) Relativistic Polarization Tensor*

The polarization tensor can be calculated from the scattering amplitude or S-matrix in a manner similar to how it was calculated in the nonrelativistic limit (e.g. eqs. 4.3.16 to 4.3.23). That is, we specialize the scattering amplitude to the forward-scattering case and divide by $r_0$ to get the dimensionless forward-scattering amplitude tensor $\vec{\vec{\Pi}} = -\vec{\vec{S}}_{\text{forw}}/r_0$, which is related to the susceptibility tensor $\vec{\vec{\alpha}}$ through

$$\vec{\vec{\alpha}} = \frac{1}{4\pi}\left(\frac{\omega_p}{\omega}\right)^2 \vec{\vec{\Pi}} = -\frac{1}{4\pi}\left(\frac{\omega_p}{\omega}\right)^2 \frac{\vec{\vec{S}}_{\text{forw}}}{r_0}. \tag{5.3.1}$$

The dielectric tensor is obtained as usual from

$$\vec{\vec{\epsilon}} = \vec{\vec{1}} - 4\pi\vec{\vec{\alpha}}. \tag{5.3.2}$$

In this calculation, $\vec{\vec{S}}_{\text{forw}}$ has to be appropriately averaged over the lepton distribution functions $f_{\pm}(p)$. This has been done by Melrose (1974), Svetozarova and Tsytovich (1962), Pavlov, Shibanov, and Yakovlev (1980), Kirk and Cramer (1985), and Bezchastnov and Pavlov (1990). The treatment in this subsection is based on that of Pavlov, Shibanov, and Yakovlev (1980, henceforth PSY), and follows closely the formulation of Bussard, Lamb, and Pakey (1986). The susceptibility tensor can be expressed as

$$\begin{aligned}\vec{\vec{\alpha}}(\omega,\vec{k}) = -\frac{e^2 m\omega_c}{(2\pi)^2\omega^2}\int_{-\infty}^{\infty} dp_{\parallel} \sum_{j,j',s,s'} \Bigg[ & \vec{\vec{Q}}^{(p)} \frac{f_- - f'_-}{E - E' + \omega + i\Gamma/2} \\ & + \vec{\vec{Q}}^{(p)} \frac{f'_+ - f_+}{E' - E + \omega + i\Gamma/2} \\ & + \vec{\vec{Q}}^{(m)} \frac{f_- + f'_+ - 1}{E + E' + \omega + i\Gamma/2} \\ & + \vec{\vec{Q}}^{(m)} \frac{1 - f_+ - f'_-}{\omega - E - E' + i\Gamma/2} \Bigg],\end{aligned} \tag{5.3.3}$$

where $f' = f(p')$ and $p' = p + k_{\parallel}$ ($p$ being the parallel electron momentum), the $\pm$ denotes the sign of the lepton electric charge, and

$$f_{\pm} = \frac{(2\pi)^2 n_e^{\pm}}{m\omega_c} f_{js}^{(\pm)}(p_{\parallel}). \tag{5.3.4}$$

One recognizes in the $f_\pm$ terms the plasma contributions and in the unit terms the vacuum contribution. The $\vec{\vec{Q}}$ tensor is related to the S-matrix elements (see eq. 5.1.11) through

$$\vec{\vec{Q}} = \frac{E+m}{2E}\frac{E'+m}{2E'}\vec{G}_{qq'}(\vec{k})\vec{G}^*_{qq'}(\vec{k}), \tag{5.3.5}$$

where $qq'$ stand for the set of quantum numbers $n$, $s$, $p$ and the $\vec{G}$ functions are given in equations (A.12) and (A.13). In cyclic coordinates, we have

$$\vec{G}\vec{G}^* = \begin{pmatrix} G_-G_-^* & G_-G_+^* & G_-G_z^* \\ G_+G_-^* & G_+G_+^* & G_+G_z^* \\ G_zG_-^* & G_zG_-^* & G_zG_z^* \end{pmatrix} \tag{5.3.6}$$

so that there are only six independent components of $\vec{\vec{Q}}$. These are

$$\begin{aligned} Q(1) &= Q_{11} = \epsilon\left|\left[\vec{G}^{(\lambda\lambda')}_{qq'}(\vec{k})\right]_-\right|^2, \\ Q(2) &= Q_{22} = \epsilon\,|G_+|^2, \\ Q(3) &= Q_{33} = \epsilon\,|G_z|^2, \\ Q(4) &= Q_{12} = Q_{21}^* = \epsilon G_-G_+^*, \\ Q(5) &= Q_{23} = Q_{32}^* = \epsilon G_+G_z^*, \\ Q(6) &= Q_{31} = Q_{13}^* = \epsilon G_zG_-^*, \end{aligned} \tag{5.3.7}$$

where

$$\epsilon \equiv \left[\frac{(E+m)}{2E}\right]\left[\frac{(E'+m)}{2E'}\right]. \tag{5.3.8}$$

One can write out separately the pure parts $(p)$ with $\lambda = \lambda'$, i.e. electrons of equal energy sign (eq. A.21), and the mixed parts $(m)$ with $\lambda = -\lambda'$, i.e. electron and positron or vice versa (eq. A.22), as a function of $\epsilon$ and the parameters

$$u_{\parallel,\perp} = \left(p_\parallel, \sqrt{2m\omega_c j}\right)/(E+m). \tag{5.3.9}$$

The expressions for equations (5.3.7), written out explicitly in Appendix A (eqs. A.21–A.22), are quite general and give the susceptibility for a plasma with (real) electrons and positrons, as well as virtual ones.

A more restricted form of the susceptibility is useful for the case where there are no positrons present, e.g. the case of accreting pulsars, where thus far there is no experimental evidence for electron-positron annihilation lines. If we set $f_+ = 0$ in equation (5.3.3) and consider weak fields ($B/B_Q \ll 1$), to lowest order in the latter parameter the vacuum contribution to the polarization tensor may be written as

$$\vec{\vec{\alpha}}(\text{vac}) = \frac{\delta_B}{4\pi}\begin{pmatrix} -2\sin^2\theta & 0 & -2\sin\theta\cos\theta \\ 0 & 4\sin^2\theta & 0 \\ -2\sin\theta\cos\theta & 0 & 5+2\sin^2\theta \end{pmatrix}, \tag{5.3.10}$$

where $\delta_B = (\alpha_F/45\pi)(\hbar\omega_c/mc^2)^2$. The $f_-$ (plasma) part becomes

$$\vec{\vec{\alpha}}(\text{plas}) = -\frac{\omega_p^2}{4\pi\omega^2}mc^2 \sum_{n,n'=0}^{\infty}\int_{-\infty}^{\infty} dp\left\{\vec{\vec{Q}}^{+}\frac{f_n(p)-f_{n'}(p')}{E+\hbar\omega-E'+i0^+} + \vec{\vec{Q}}^{-}\left[\frac{f_n(p)}{\hbar\omega+E+E'+i0^+} - \frac{f_{n'}(p')}{\hbar\omega-E-E'+i0^+}\right]\right\}, \tag{5.3.11}$$

where $p' = p - k_\parallel$, $E = (p^2c^2 + 2mc^2\hbar\omega_c n + m^2c^4)^{1/2}$, and $\vec{\vec{Q}}^{\pm}$ is given by equation (15) of PSY. For accreting pulsars we may take in the PSY equations for the temperature $T_\perp = 0$, and $T_\parallel$ any arbitrary (relativistic) value. This is valid provided collisional excitations are much less important than radiative deexcitations. We thus have

$$\begin{aligned}
Q^{\pm}_{xx} &= Q^{\pm}_{yy} = \frac{1}{2}\left(1 \pm \frac{m^2+pp'}{EE'}\right)\left(F^2_{n'-1,n} + F^2_{n',n-1}\right),\\
Q^{\pm}_{zz} &= \frac{1}{2}\left(1 \pm \frac{m^2+pp'}{EE'}\right)F^2_{n',n},\\
Q^{\pm}_{xy} &= -Q^{\pm}_{yx} = \frac{-i}{2}\left(1 \pm \frac{m^2+pp'}{EE'}\right)\left(F^2_{n'-1,n} - F^2_{n',n-1}\right),\\
Q^{\pm}_{yz} &= -Q^{\pm}_{zy} = \mp\frac{i}{EE'}\left(\frac{m\hbar\omega_c}{2}\right)^{1/2}\\
&\quad\times\left(p\sqrt{n'}\,F_{n',n}F_{n'-1,n} - p'\sqrt{n}\,F_{n',n}F_{n',n-1}\right),\\
Q^{\pm}_{zx} &= Q^{\pm}_{xz} = \mp\frac{1}{EE'}\left(\frac{m\hbar\omega_c}{2}\right)^{1/2}\\
&\quad\times\left(p\sqrt{n'}\,F_{n',n}F_{n'-1,n} + p'\sqrt{n}\,F_{n',n}F_{n',n-1}\right),
\end{aligned} \tag{5.3.12}$$

where units with $\hbar = c = 1$ are used and the $F_{n,n'}$ are given by equations (A.10), the argument being $u = k_\perp^2 \hbar/(2m\omega_c)$. The $F_{n,n'}$ satisfy the relation

$$F_{n',n} = (-1)^{n'-n} F_{n,n'}, \tag{5.3.13}$$

useful for the case $j' > j$. In the present approximation, either $j$ or $j'$ is zero, and the only values needed are

$$F_{n',0} = (-1)^{n'} \left( e^{-u} \frac{u^{n'}}{n'!} \right)^{1/2},$$

$$F_{0,n} = \left( e^{-u} \frac{u^{n}}{n!} \right)^{1/2}. \tag{5.3.14}$$

Equations (5.3.10) to (5.3.14) are fully relativistic. If the longitudinal electron distribution is thermal (an approximation not always justified), we may use the relativistic Maxwellian distribution,

$$f_n(p) = \delta_{n,0} \exp(-E/T)[2mK_1(m/T)]^{-1}, \tag{5.3.15}$$

where $\delta$ is a Kronecker delta and $K_1$ is the modified Bessel function of the second kind, of order one. For the purposes of calculating the normal modes from the polarization tensor, it is not necessary to include the full damping constant $\Gamma$ in the denominators of $\vec{\vec{\alpha}}$. Thus, in equation (5.3.11) the imaginary parts included describe only resonant cyclotron interaction without any wing opacities, which would follow from the total damping constant $\Gamma$. The latter is harder to calculate relativistically, and its use is only necessary when calculating the scattering cross section, whereas for the normal modes the prescription of equation (5.3.11) is adequate. The numerical evaluation of the resonant denominators in this expression requires special attention (e.g. Bussard, Lamb, and Pakey, 1986).

### *b) Nonrelativistic Limit*

In the nonrelativistic limit, and neglecting any positron component, we have

$$f_{j,s}(p) = \delta_{j,0}\delta_{s,-1/2} N(p/[\sqrt{2}\, m v_T]), \tag{5.3.16}$$

where $N(x)$ is the usual nonrelativistic Maxwellian distribution,

$$N(x) = (\sqrt{2\pi}\, m v_T)^{-1} \exp(-x^2), \quad v_T = \sqrt{k_B T/m}. \tag{5.3.17}$$

We have then for the elements of $\vec{\alpha}$, from equation (5.3.11),

$$\alpha_{ab}(\omega, \vec{k}) = -\frac{e^2 n_e}{\omega^2} \int_{-\infty}^{\infty} dp \left\{ \sum_{j', s'} \frac{Q_{ab}^{(p)}(j', s') N(p/\sqrt{2}\, m v_T)}{\omega - (E' - E_0) + i\Gamma_1/2} \right.$$

$$- \sum_{j,s} \frac{Q_{ab}^{(p)}(j, s) N([p + k_\parallel]/\sqrt{2}\, m v_T)}{\omega - (E' - E_0) + i\Gamma_1/2}$$

$$+ \sum_{j', s'} \frac{Q_{ab}^{(m)}(j', s') N(p/\sqrt{2}\, m v_T)}{\omega + E + E' + i\Gamma_1/2}$$

$$\left. - \sum_{j, s} \frac{Q_{ab}^{(m)}(j, s) N([p + k_\parallel]/\sqrt{2}\, m v_T)}{\omega - E - E' + i\Gamma_1/2} \right\}. \quad (5.3.18)$$

Taking the lowest-order terms only in the $Q_{ab}$, we get

$$Q^{(p)}(1) = Q_{11} \cong 2\delta_{j,0}\delta_{j',1} \frac{2\omega\omega_c}{4m^2} \theta_{-s, -s'} = \frac{\omega_c}{m} \delta_{j,0}\delta_{j',1}\theta_{-s, -s'},$$

$$Q^{(p)}(2) = Q_{22} \cong \frac{\omega_c}{m} \delta_{j,1}\delta_{j',0}\theta_{-s, -s'},$$

$$Q^{(p)}(3) = Q_{33} \cong \frac{(2p + k_\parallel)^2}{4m^2} \delta_{j,0}\delta_{j',0}\theta_{-s, -s'},$$

$$Q^{(p)}(4) = Q^{(p)}(5) = Q^{(p)}(6) \cong 0,$$

$$Q^{(m)}(1) \cong 2\delta_{j,0}\delta_{j',1}\theta_{-s, s'},$$

$$Q^{(m)}(2) \cong 2\delta_{j,1}\delta_{j',0}\theta_{s, -s'},$$

$$Q^{(m)}(3) \cong \delta_{j,0}\delta_{j',0}\theta_{-s, -s'},$$

$$Q^{(m)}(4) = Q^{(m)}(5) = Q^{(m)}(6) \cong 0, \quad (5.3.19)$$

where $\theta_{s, s'} = [1, 0]$ if the sign of $s$ is [equal, opposite] to the sign of $s'$. Thus, in the cyclic representation, the nonrelativistic polarization tensor is diagonal. Expanding the denominators in equation (5.3.14) as

$$E_j(p) \simeq m + \frac{p^2}{2m} + j\omega_c, \quad (5.3.20)$$

we get $\vec{\vec{\alpha}}$ in the cyclic representation as (Bussard, 1986b)

$$\alpha_{11}(\omega, \vec{k}) \cong -\frac{e^2 n_e}{\omega^2}\left[\frac{\omega_c}{m}\int_{-\infty}^{\infty} dp \frac{N(p/\sqrt{2}\,mv_T)}{\omega - \omega_c - \frac{p}{m}k_{\parallel} + \frac{i\Gamma_{1-}}{2}}\right.$$

$$\left. + 2\int_{-\infty}^{\infty} dp \frac{N(p/\sqrt{2}\,mv_T)}{\omega + 2m + \omega_c + \frac{p(p + k_{\parallel})}{m} + \frac{i\Gamma_{1+}}{2}}\right]$$

$$\simeq \frac{\omega_p^2}{4\pi\omega^2}\left[\frac{\omega_c}{\sqrt{2}\,k_{\parallel}v_T} W\left(\frac{\omega - \omega_c + \frac{i}{2}\Gamma_{1-}}{\sqrt{2}\,k_{\parallel}v_T}\right)\right.$$

$$\left. - \frac{1}{1 + \frac{\omega + \omega_c}{2m} + \frac{i}{4}\frac{\Gamma_{1+}}{m}} i\left(1 - \frac{v_T^2}{2}\right)\right],$$

$$\alpha_{22}(\omega, \vec{k}) \cong -\frac{e^2 n_e}{\omega^2}\left[-\frac{\omega_c}{m}\int_{-\infty}^{\infty} dp \frac{N\left(\frac{p + k_{\parallel}}{\sqrt{2}\,mv_T}\right)}{\omega + \omega_c - \frac{pk_{\parallel}}{m} + \frac{i}{2}\Gamma_{1-}}\right.$$

$$\left. - 2\int_{-\infty}^{\infty} dp \frac{N\left(\frac{p + k_{\parallel}}{\sqrt{2}\,mv_T}\right)}{\omega - 2m - \omega_c + \frac{p(p + k_{\parallel})}{m} + \frac{i}{2}\Gamma_{1+}}\right]$$

$$\simeq -\frac{\omega_p^2}{4\pi\omega^2}\left[\frac{\omega_c}{\sqrt{2}\,k_{\parallel}v_T} W\left(\frac{\omega + \omega_c + \frac{i}{2}\Gamma_{1-}}{\sqrt{2}\,mv_T}\right)\right.$$

$$\left. + \frac{1}{\left(1 - \frac{\omega - \omega_c + i/2\Gamma_{1+}}{2m}\right)}\left(1 - \frac{v_T^2}{2}\right)\right],$$

$$\alpha_{33}(\omega, \vec{k}) \cong -\frac{e^2 n_e}{\omega^2}\left(\frac{1}{\omega + 2m + \frac{i}{2}\Gamma_0} + \frac{1}{2m - \omega - \frac{i}{2}\Gamma_0}\right)$$

$$\simeq -\frac{\omega_p^2}{4\pi\omega^2}\frac{4m^2}{(4m^2 - i\Gamma_0\omega)} = -\frac{\omega_p^2}{4\pi\omega^2}\frac{1}{\left(1 - i\frac{\Gamma_0\omega}{4m^2}\right)},$$

(5.3.21)

where $W(z)$ is the plasma dispersion function (Fried and Conte, 1961).

*c) Eigenvectors and Eigenvalues of the Wave Equation*

We can write the wave equation in a manner similar to that in Chapter 3. There, we found that, for $\delta_B \ll 1$, the dielectric tensor and the inverse magnetic susceptibility differed from unity by terms of order $\delta_B$. The polarization tensor $\vec{\vec{\alpha}}$ is itself of order $(\omega_p/\omega)^2$ for the plasma contribution, and of order $\delta_B$ for the vacuum part. To first order in $\delta_B$, we can write the Fourier-transformed wave equation for the wave potential $\vec{A}(\omega, \vec{k})$ as

$$\left(\frac{c^2}{\omega^2}\vec{k}\vec{k} + 4\pi\vec{\vec{\alpha}}\right)\cdot\vec{A}(\omega, \vec{k}) = \left(\frac{c^2k^2}{\omega^2} - 1\right)\vec{A}(\omega, \vec{k}), \qquad (5.3.22)$$

where $\vec{\vec{\alpha}} = \vec{\vec{\alpha}}_{\text{plasma}} + \vec{\vec{\alpha}}_{\text{vac}}$. In what follows we shall assume $\vec{k}$ in the $x$, $z$ plane and $\vec{B}_0$ along $z$. As before, we define the complex refractive index squared,

$$N^2 = \frac{c^2k^2}{\omega^2}, \qquad (5.3.23)$$

which appears in equation (5.3.22), and since $(\omega_p/\omega)^2 \ll 1$, $\delta_B \ll 1$, we expect $N^2 \simeq 1$. In this approximation, the solution of equation (5.3.22) can be written (PSY) as

$$N^2 - 1 \cong 2(n_I - 1) \pm 2\left(n_L^2 + n_C^2\right)^{1/2}, \qquad (5.3.24)$$

where

$$n_I = 1 + \pi\left(\alpha_{yy} + \alpha_{xx}\cos^2\theta + \alpha_{zz}\sin^2\theta - 2\sin\theta\cos\theta\,\alpha_{xz}\right),$$

$$n_L = \pi\left(\alpha_{yy} - \alpha_{xx}\cos^2\theta - \alpha_{zz}\sin^2\theta + 2\sin\theta\cos\theta\,\alpha_{xz}\right),$$

$$n_C = 2\pi i\left(\alpha_{xy}\cos\theta + \alpha_{yz}\sin\theta\right). \qquad (5.3.25)$$

We have therefore, for the two solutions of equation (5.3.24), the refractive indices

$$N_{\pm} \simeq n_I \pm \left(n_L^2 + n_C^2\right)^{1/2}. \tag{5.3.26}$$

The eigenvectors $\vec{e}$ must satisfy

$$\begin{pmatrix} \sin^2\theta + \beta_{xx} - \Lambda\cos^2\theta & \beta_{xy} & \sin\theta\cos\theta(1+\Lambda) + \beta_{xz} \\ -\beta_{xy} & \beta_{yy} - \Lambda & \beta_{yz} \\ \sin\theta\cos\theta(1+\Lambda) + \beta_{xz} & -\beta_{yz} & \cos^2\theta + \beta_{zz} - \Lambda\sin^2\theta \end{pmatrix} \times \begin{pmatrix} e_x \\ e_y \\ e_z \end{pmatrix}, \tag{5.3.27}$$

where $\Lambda = N^2 - 1$ and $\beta_{ij} = 4\pi\alpha_{ij}$. The second line of equation (5.3.27) gives

$$e_y = \frac{\beta_{xy}e_x - \beta_{yz}e_z}{\beta_{yy} - \Lambda}, \tag{5.3.28}$$

and the third line gives

$$e_z = \frac{\beta_{yz}\left(\beta_{xy}e_x - \beta_{yz}e_z\right) - \left(\beta_{yy} - \Lambda\right)\left[\sin\theta\cos\theta(1+\Lambda) + \beta_{xz}\right]e_x}{\left(\beta_{yy} - \Lambda\right)\left(\cos^2\theta + \beta_{zz} - \Lambda\sin^2\theta\right)} \tag{5.3.29a}$$

or

$$\begin{aligned} e_z &= \frac{\beta_{yz}\beta_{xy} - \left(\beta_{yy} - \Lambda\right)\left[\sin\theta\cos\theta\,(1+\Lambda) + \beta_{xz}\right]}{\left(\beta_{yy} - \Lambda\right)\left(\cos^2\theta + \beta_{zz} - \Lambda\sin^2\theta\right) + \beta_{yz}^2} e_x \\ &\simeq -\tan\theta\, e_x \end{aligned} \tag{5.3.29b}$$

to lowest order in $\beta$, $\Lambda$. This is the transversality condition, $\cos\theta\, e_z + \sin\theta\, e_x = \hat{k}\cdot\hat{e} \simeq 0$. If we substitute equation (5.3.29b) in (5.3.28), we get then

$$e_y = \frac{\beta_{xy} + \tan\theta\,\beta_{yz}}{\beta_{yy} - \Lambda} e_x, \tag{5.3.30}$$

and since $\Lambda = N^2 - 1$, using (5.3.24) this becomes

$$e_y^{\pm} = \sec\theta \frac{2n_C}{i}\left(2n_L \mp 2\sqrt{n_L^2 + n_C^2}\right)^{-1} e_x$$
$$= i\sec\theta\left[\left(n_L/n_C\right) \pm \sqrt{\left(n_L/n_C\right)^2 + 1}\right] e_x. \tag{5.3.31}$$

The normalization condition is given by

$$|e_x|^2 + |e_y|^2 + |e_z|^2 = 1. \tag{5.3.32}$$

This yields for $e_x$

$$|e_x^{\pm}|^2\left[1 + \sec^2\theta\left(n_L/n_C \pm \sqrt{\left(n_L/n_C\right)^2 + 1}\right)^2 + \tan^2\theta\right] = 1. \tag{5.3.33}$$

In equations (5.3.31) and (5.3.33) the $\pm$ signs correspond to the two solutions $N_{\pm}$ of the refractive index (5.3.26).

It is interesting to consider the two limiting cases $n_L^2 \gg n_C^2$ and $n_L^2 \ll n_C^2$. In the first case, $n_L^2 \gg n_C^2$, we have

$$\hat{e}^{(+)} \cong \left(\frac{1}{2}\frac{n_C}{n_L}\cos\theta, i, -\frac{1}{2}\frac{n_C}{n_L}\sin\theta\right)_{xyz},$$
$$\hat{e}^{(-)} \simeq \left(\cos\theta, -i\frac{1}{2}\frac{n_C}{n_L}, -\sin\theta\right)_{xyz}. \tag{5.3.34}$$

These are approximately linear modes, with $\hat{e}^{(+)}$ perpendicular to the $\vec{k}, \vec{B}_0$ plane and $\hat{e}^{(-)}$ perpendicular to $\vec{k}$ in the $\vec{k}, \vec{B}_0$ plane. For the other case, $n_C^2 \gg n_L^2$, we have

$$\hat{e}^{(+)} \cong 2^{-1/2}\left(\cos\theta, i, -\sin\theta\right),$$
$$\hat{e}^{(-)} \simeq 2^{-1/2}\left(\cos\theta, -i, -\sin\theta\right). \tag{5.3.35}$$

These are elliptical modes, and from the fact that $\vec{A} \propto \mathrm{Re}\, \hat{e}e^{-i\omega t}$ one sees that $\hat{e}^{(+)}$ is right-handed for $\theta < \pi/2$ while $\hat{e}^{(-)}$ is left-handed for $\theta < \pi/2$. For $\theta \to 0$, these modes become circular.

The modes are usually more useful in the rotating coordinates $e_{\pm} = e_x \pm ie_y$. Following Bussard, Lamb, and Pakey (1986) and Bussard (1986b), the transformation of $\vec{A}$ between Cartesian components $A_j$ and cyclic components $A_a$ is

$$A_a = \sum_j C_{aj} A_j \tag{5.3.36}$$

and the wave equation (5.3.22) can be transformed as

$$\vec{\vec{C}} \cdot \left(c^2 \vec{k}\vec{k}^\dagger + 4\pi\omega^2 \vec{\vec{\alpha}}\right) \cdot \left(\vec{\vec{C}}\right)^{-1} \cdot \vec{\vec{C}} \cdot \vec{A} = \Lambda \vec{\vec{C}} \cdot \vec{A}, \tag{5.3.37}$$

where $\vec{\vec{C}}$ is the unitary matrix

$$C_{aj} = \begin{pmatrix} 1/\sqrt{2} & -i/\sqrt{2} & 0 \\ 1/\sqrt{2} & i/\sqrt{2} & 0 \\ 0 & 0 & 1 \end{pmatrix}. \tag{5.3.38}$$

If we denote with a double bar the tensor and with a single bar the vector quantities expressed in cyclic components, the wave equation is

$$\left(c^2 \bar{k}\bar{k}^\dagger + 4\pi\omega^2 \bar{\bar{\alpha}}\right) \cdot \bar{A} = \Lambda \bar{A}, \tag{5.3.39}$$

where $\bar{\bar{\alpha}}$ and $\vec{\vec{\alpha}}$ are related through the transformations

$$\begin{aligned} \bar{\bar{\alpha}} &= \vec{\vec{C}} \cdot \vec{\vec{\alpha}} \cdot \vec{\vec{C}}^{-1}, \\ \vec{\vec{\alpha}} &= \vec{\vec{C}}^{-1} \cdot \bar{\bar{\alpha}} \cdot \vec{\vec{C}}. \end{aligned} \tag{5.3.40}$$

This gives the $\vec{\vec{\alpha}}$ in Cartesian components as a function of the cyclic components,

$$\vec{\vec{\alpha}} = \begin{pmatrix} (\alpha_{11}+\alpha_{22})/2+\alpha_{12} & i(\alpha_{22}-\alpha_{11})/2 & (\alpha_{23}+\alpha_{31})/\sqrt{2} \\ -i(\alpha_{22}-\alpha_{11})/2 & (\alpha_{11}+\alpha_{22})/2-\alpha_{12} & -i(\alpha_{23}-\alpha_{31})/\sqrt{2} \\ (\alpha_{23}+\alpha_{31})\sqrt{2} & i(\alpha_{23}-\alpha_{31})/\sqrt{2} & \alpha_{33} \end{pmatrix}, \tag{5.3.41}$$

where the $\alpha_{11}$, $\alpha_{12}$, etc., are the cyclic components. From equations (5.3.25) and (5.3.41) we can get the values of $n_I$, $n_L$, and $n_C$ in cyclic coordinates,

$$\begin{aligned} n_I &= 1 + \pi\left[\frac{\alpha_{11}+\alpha_{22}}{2}(1+\cos^2\theta) + (\alpha_{33}-\alpha_{12})\sin^2\theta \right. \\ &\qquad \left. - 2\sin\theta\cos\theta\frac{\alpha_{23}+\alpha_{31}}{\sqrt{2}}\right], \\ n_L &= \pi\left[\left(\frac{\alpha_{11}+\alpha_{22}}{2} - \alpha_{33}\right)\sin^2\theta - \alpha_{12}(1+\cos^2\theta) \right. \\ &\qquad \left. + 2\sin\theta\cos\theta\frac{\alpha_{23}+\alpha_{31}}{\sqrt{2}}\right], \\ n_C &= 2\pi\left[-\frac{\alpha_{22}-\alpha_{11}}{2}\cos\theta + \frac{\alpha_{23}-\alpha_{31}}{\sqrt{2}}\sin\theta\right]. \end{aligned} \tag{5.3.42}$$

The polarization eigenvectors in cyclic coordinates are given by (denoting now with the superscript $\sigma = \pm$ the solution corresponding to $N_{\pm}$, and with the subindex $\pm$ the cyclic component $+$ or $-$)

$$e_{\pm}^{(\sigma)} = \frac{1}{\sqrt{2}} e_x^{(\sigma)} \left( 1 \mp \frac{n_L + (-1)^{\sigma} \sqrt{n_L^2 + n_C^2}}{n_C \cos\theta} \right),$$

$$e_z^{(\sigma)} = -\tan\theta \, e_x^{(\sigma)}. \tag{5.3.43}$$

Using the normalization

$$1 + \tan^2\theta + \left( \frac{n_L + (-1)^{\sigma} \sqrt{n_L^2 + n_C^2}}{n_C \cos\theta} \right)^2 = \frac{1}{| e_x^{(\sigma)} |^2}, \tag{5.3.44}$$

we have

$$| e_x^{(\sigma)} |^2 = \frac{\cos^2\theta}{1 + \left( \dfrac{n_L}{n_C} + (-1)^{\sigma} \sqrt{\left(\dfrac{n_L}{n_C}\right)^2 - 1} \right)^2} \tag{5.3.45}$$

so that the cyclic components of the polarization modes become

$$e_{\pm}^{(\sigma)} = \frac{1}{\sqrt{2}} \frac{\cos\theta \mp \left[ \dfrac{n_L}{n_C} + (-1)^{\sigma} \sqrt{\left(\dfrac{n_L}{n_C}\right)^2 + 1} \right]}{\sqrt{1 + \left[ \dfrac{n_L}{n_C} + (-1)^{\sigma} \sqrt{\left(\dfrac{n_L}{n_C}\right)^2 + 1} \right]^2}},$$

$$e_z^{(\sigma)} = \frac{-\sin\theta}{\sqrt{1 + \left[ \dfrac{n_L}{n_C} + (-1)^{\sigma} \sqrt{\left(\dfrac{n_L}{n_C}\right)^2 + 1} \right]^2}}. \tag{5.3.46}$$

Corresponding to the convention

$$\Lambda^{(\sigma)} = N_{\sigma}^2 - 1 = 2(n_I - 1) + 2(-1)^{\sigma} \sqrt{n_L^2 + n_C^2},$$

$$N_{\sigma} \cong n_I + (-1)^{\sigma} \sqrt{n_L^2 + n_C^2}, \tag{5.3.47}$$

the mode $\sigma = 1$ is the extraordinary, $\sigma = 2$ is the ordinary mode, in the limit where vacuum polarization effects are negligible. The cyclic coordinate representation (5.3.46) is the most useful one for use in the cross section calculations, since the scattering matrix has its simplest form in these coordinates.

*d) Physical and Geometrical Properties of the Eigenmodes*

The previously used system of coordinates, while useful for the purposes of calculations, is not the "natural" system of coordinates for a photon. A clearer insight into the mode properties is provided by recasting them in a right-handed system of Cartesian coordinates $x', y', z'$, with $z'$ along $\vec{k}$ and $x'$ in the direction of the projection of $\vec{B}_0$ on the plane perpendicular to $\vec{k}$. In this system, the modes become

$$\hat{e}^{(\sigma)} = \frac{\left(-i\hat{x} + \alpha_\sigma \hat{y}'\right)}{\left(1 + \alpha_\sigma^2\right)^{1/2}},$$

$$\alpha_\sigma = -\frac{n_L}{n_C} - (-1)^\sigma \left(\frac{n_L^2}{n_C^2} + 1\right)^{1/2} = -\frac{N_\sigma}{n_C}. \quad (5.3.48)$$

The complex quantity $\alpha_\sigma$ is related to the quantity $b$ discussed in Chapter 3 by

$$b = q + ip = -\frac{n_L}{n_C},$$

$$\alpha_\sigma = b^{-1} - (-1)^\sigma \sqrt{b^{-2} + 1}\,. \quad (5.3.49)$$

Writing out $\alpha_\sigma$ explicitly as

$$\alpha_\sigma = r_\sigma e^{i2\chi_\sigma}, \quad (5.3.50)$$

the ellipticity of the normal modes is (Gnedin and Pavlov, 1974; Pavlov, Shibanov, and Yakovlev, 1980)

$$P_\sigma = \frac{r_\sigma - 1}{r_\sigma + 1}, \quad (5.3.51a)$$

where $|P_\sigma|$ is the ratio of the minor to the major axis of the ellipsoid described by the electric vector in the $x', y'$ plane. This is the inverse of the $K_\sigma$ defined in Chapter 3, $|P_\sigma| = K_\sigma^{-1}$. The angle $\chi_\sigma = \frac{1}{2}\arctan[\mathrm{Im}(\alpha_\sigma)/\mathrm{Re}(\alpha_\sigma)]$ is measured between the major axis of the ellipse and the projection of $\vec{B}_0$ on the plane perpendicular to $\vec{k}$ (the plane of the sky for an observer), that is, the $x'$ axis. The sign of $P_\sigma$ gives the direction of rotation of the electric vector, positive $P_\sigma$

corresponding to right-handed helicity. It can be verified that $P_1 = -P_2$ and that the sum of the position angles is always $(n + \frac{1}{2})\pi$, where $n$ is an integer. The two normal modes, as discussed in Chapter 3, are not usually orthogonal, though far from the resonances this departure from orthogonality is small. The degree of nonorthogonality can be calculated from

$$\hat{e}^{(1)} \cdot \hat{e}^{(2)} = \frac{1 + P_1 P_2}{1 - P_2 P_2} \frac{1 + e^{2i(\chi_2 - \chi_1)}}{2}. \tag{5.3.51b}$$

The usual Stokes parameters $I, Q, U, V$ for the normal wave $\sigma$ can be written as

$$I_\sigma = I_\sigma, \quad Q_\sigma = p_Q^\sigma I_\sigma, \quad U_\sigma = p_U^\sigma I_\sigma, \quad V_\sigma = p_V^\sigma I_\sigma, \tag{5.3.52}$$

where $I_\sigma$ is the intensity of the wave $\sigma$, i.e. a quantity proportional to the amplitude squared, and $p_Q^\sigma$, $p_U^\sigma$, $p_V^\sigma$ are defined by (Gnedin and Pavlov, 1974)

$$p_Q^{(\sigma)} = \frac{2\,\mathrm{Re}(\alpha_\sigma)}{|\alpha_\sigma|^2 + 1}, \quad p_V^{(\sigma)} = \frac{|\alpha_\sigma|^2 - 1}{|\alpha_\sigma|^2 + 1},$$

$$p_U^{(\sigma)} = \frac{2\,\mathrm{Im}(\alpha_\sigma)}{|\alpha_\sigma|^2 + 1}. \tag{5.3.53}$$

For a radiation field which is a mixture of both waves, with intensities $I_1$ and $I_2$ present, the total Stokes parameters are

$$I = \sum_\sigma I_\sigma, \quad Q = \sum_\sigma p_Q^{(\sigma)} I_\sigma,$$

$$U = \sum_\sigma p_U^{(\sigma)} I_\sigma, \quad V = \sum_\sigma p_V^{(\sigma)} I_\sigma. \tag{5.3.54}$$

We can also define the total degree of polarization $P$, the degree of linear polarization $P_L$, the degree of circular polarization $P_C$, and the normalized $U$ parameter $P_U$ as

$$P = \frac{\Delta I}{I}, \quad P_L = \frac{Q}{I}, \quad P_C = \frac{V}{I}, \quad P_U = \frac{U}{I}, \tag{5.3.55}$$

where $\Delta I = I_1 - I_2$. The degree of linear polarization is $P_L > 0$, and the degree of circular polarization is $P_C > 0$ for the ordinary (2) wave, which has $\vec{E}^w$ in the $\vec{k}$, $\vec{B}_0$ plane for negligible vacuum effects (but 1 and 2 may be inverted for high-vacuum effects; see Chap. 3). In accreting neutron star atmospheres at X-ray or higher energies (Chap. 7), it usually turns out that $U \simeq V \simeq 0$ is a good approximation.

It is also useful to look at the Stokes parameters in the simpler, cold-plasma limit. In this case we have (see eqs. 3.4.27–3.4.28)

$$b = p + iq = \frac{\sin^2\theta}{2\cos\theta} u^{1/2}\left(1 - W\frac{u-1}{u^2}\right),$$
$$W = \left(n_e/3 \times 10^{28}\ \mathrm{cm}^{-3}\right)^{-1}\left(B/B_Q\right)^4,$$
$$u = \left(\omega_c/\omega\right)^2, \tag{5.3.56}$$

and consequently (Pavlov and Shibanov, 1979)

$$p_Q{}^\sigma = \left(-1\right)^\sigma \frac{|b|}{\sqrt{1+b^2}}, \quad p_V^{(\sigma)} = \left(-1\right)^\sigma \frac{\mathrm{sign}(b)}{\sqrt{1+b^2}}. \tag{5.3.57}$$

The Stokes parameters acquire the form

$$P = \frac{\Delta I}{I} = \frac{I_1 - I_2}{I}, \quad P_L = \frac{Q}{I} = p_Q^1 P,$$
$$P_C = \frac{V}{I} = p_V^1 P, \quad P_U \cong 0, \tag{5.3.58}$$

which satisfy the relation

$$P^2 = P_L^2 + P_C^2. \tag{5.3.59}$$

The cold-plasma expressions (5.3.56)–(5.3.59) are quite good for frequencies more than about one Doppler width from the resonances, and rather simpler than the corresponding hot-plasma values of equations (5.3.53)–(5.3.55).

## 5.4 Synchrotron Radiation

The process of synchrotron (or cyclotron) radiation was discussed in § 5.1*d* to the extent that was necessary to include its rate in the two-photon scattering and two-photon emission and absorption processes. We return now to a discussion of the synchrotron process, going into some further details. The synchrotron emission and absorption is a first-order process, as shown in Figure 5.1.3, so that its cross section will depend on the first power of $\alpha_F$ rather than on $\alpha_F^2$ as in the general case of the scattering processes. Calling $j$ and $j'$ the magnetic (total) quantum number of the (mass shell) electron before and after the interaction with the photon $\omega$, the electrons have the corresponding energies

$$E_j = \left[ p^2 + 2m\omega_c j + m^2 \right]^{1/2}, \quad E_{j'} = \left[ p'^2 + 2m\omega_c j' + m^2 \right]^{1/2}, \tag{5.4.1}$$

where $p$ and $p'$ are the electron longitudinal momenta corresponding to $j$ and $j'$, and $j = (n + s + 1/2)$, $s = \pm \frac{1}{2}$, $n = 0, 1, 2, \ldots,$ etc. The energy and momentum conservation conditions for emission $(n > n')$ are

$$\begin{aligned} E_j - E_{j'} &= \omega, \\ p - p' &= \omega \cos\theta, \end{aligned} \tag{5.4.2}$$

where $\mu = \cos\theta$ is the cosine of the photon angle with respect to $\vec{B}$. The energy of the photon emitted by an electron of initial energy $E = E_j$ is, from (5.4.2),

$$\omega = \frac{(E - p\cos\theta)}{\sin^2\theta}\left[1 - \left\{1 - \frac{2(j - j')m\omega_c \sin^2\theta}{(E - p\cos\theta)^2}\right\}^{1/2}\right], \tag{5.4.3}$$

where $j - j'$ is the harmonic number of the radiation emitted. The electron will experience a recoil $p' = p - \omega\cos\theta$, which is needed to determine its final energy. The $S$-matrix element corresponding to Figure 5.1.3 is given in equation (5.1.27). In order to simplify the discussion, we can constrain ourselves to polarization eigenmodes which are dominated predominantly by the vacuum so that we may take $\hat{\varepsilon}$ to be given by the linearly polarized modes. This is also simpler for applications where we are only interested in polarization averaged rates, i.e.

$$\begin{aligned} \hat{\varepsilon}_\perp &= \hat{k} \times \hat{b}, \\ \hat{\varepsilon}_\parallel &= \hat{\varepsilon}_\perp \times \hat{k}, \end{aligned} \tag{5.4.4}$$

where $\vec{k} = \omega(\sin\theta\cos\varphi\hat{x} + \sin\varphi\hat{y} + \cos\theta\hat{z})$, $\hat{b} = \vec{B}_0/B$, and $\hat{\varepsilon}_\perp$ is perpendicular to the $(\vec{k}, \hat{b})$ plane while $\hat{\varepsilon}_\parallel$ is parallel to that plane. The lepton wave functions $\Psi(x)$ are given by equations (5.1.4), and for synchrotron emission or absorption both leptons have positive energy ($\lambda = 1$, electrons). The differential transition rate from $j$ to $j'$, summed over photon polarizations, is given in terms of $|S_{fi}|^2$,

$$R_{n,n'}^{s,s'}(E,\theta) = \frac{\alpha_f}{2\pi}\int \left(|M_\parallel|^2 + |M_\perp|^2\right)\delta(E - E' - \omega)\,\omega\, d\omega$$

$$= \frac{\alpha_f}{2\pi}\frac{\omega(E-\omega)\left(|M_\parallel|^2 + |M_\perp|^2\right)}{\left(E - p\cos\theta - \omega\sin^2\theta\right)} \qquad (5.4.5)$$

(e.g. Harding and Preece, 1987; see also Sokolov and Ternov, 1986; Herold, Ruder, and Wunner, 1982; Bussard, 1984; Latal, 1986). The $|M_{\parallel,\perp}|^2$ are the squared matrix elements for polarization parallel and perpendicular to $\vec{B}_0$ and are given by

$$|M_\parallel|^2 = G_1^* G_1, \quad |M_\perp|^2 = |G_2\cos\theta - G_3\sin\theta|^2, \qquad (5.4.6)$$

where

$$G_1 = i(1/4)(A_3' A_4 + A_4' A_3)\left[B_3' B_4 I_{j,j'-1}(x) - B_3 B_4' I_{j-1,j'}(x)\right],$$
$$G_2 = (1/4)(A_3' A_4 + A_4' A_3)\left[B_3' B_4 I_{j,j'-1}(x) + B_3 B_4' I_{j-1,j'}(x)\right],$$
$$G_3 = (1/4)(A_3' A_3 - A_4 A_4')\left[B_3' B_3 I_{j-1,j'-1}(x) + B_4 B_4' I_{j,j'}(x)\right], \qquad (5.4.7)$$

and

$$A_3 = (1 + p/E)^{1/2}, \quad A_4 = 2s(1 - p/E)^{1/2},$$
$$B_3 = (1 + 2sm/p_0)^{1/2}, \quad B_4 = 2s(1 - 2sm/p_0)^{1/2},$$
$$I_{m,l}(x) = \frac{\sqrt{l!}}{\sqrt{m!}} e^{-x/2} x^{(m-l)/2} L_l^{m-l}(x),$$
$$x = (\omega^2\sin^2\theta/2m\omega_c), \quad p_0 = (E^2 - p^2)^{1/2}. \qquad (5.4.8)$$

Here the $L_l^{m-l}(x)$ are Laguerre polynomials defined according to Abramowitz and Stegun (1964), and $s = -1/2$ (down), $s = +1/2$ (up) refers to the electron spin. Some results for the angular distribution of synchrotron photons are shown in Figure 5.4.1 for various $(j, j')$ combinations, in the case where $p = 0$ (which is easier to compute; the results can be Lorentz-boosted to arbitrary $p$). For transitions with small $j$ and small $\Delta j$, the photons are emitted almost isotropically. For $\Delta j$ small (e.g. $\Delta j = 1$) and large $j$, an emission peak develops at angles increasingly distant from 90°, which in the classical limit of $B \to B_Q$ and $j \to \infty$ tends to the usual $(1 + \cos^2\theta)$ behavior with maximum at $\theta = 0$. For transitions with large harmonic number, $\Delta j \to \infty$, the emission becomes concentrated toward $\theta = \pi/2$.

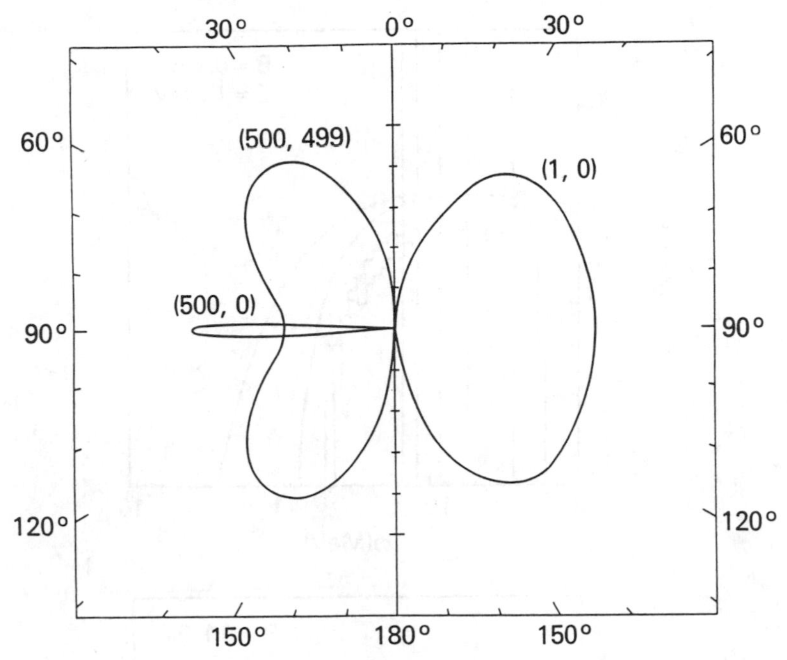


FIGURE 5.4.1 Angular distribution of photons radiated in synchrotron transitions between Landau states $(j, j')$ for $B = 0.1 B_Q$ in the frame where the electron longitudinal momentum is zero. From A. K. Harding and R. Preece, 1987, *Ap. J.*, 319, 939.

The emission spectrum from monoenergetic electrons can be calculated numerically from equation (5.4.5) with a Monte Carlo procedure, as done by Harding and Preece (1987). The rates are calculated in the laboratory frame, after summing over final spins, integrating over photon angle $\theta$, and averaging over the electron pitch angle $\alpha$, defined by

$$\sin \alpha = \left[2 jm\omega_c / \left(E^2 - m^2\right)\right]^{1/2}, \tag{5.4.9}$$

which is assumed to be isotropically distributed. An example is shown in Figure 5.4.2. Again, one can see in this figure the anharmonic character of the relativistic cyclotron "harmonics," which appear at energies

$$\omega_{j,j'} = \left(m_e + 2j\omega_c\right) - \left(m_e + 2j'\omega_c\right). \tag{5.4.10}$$

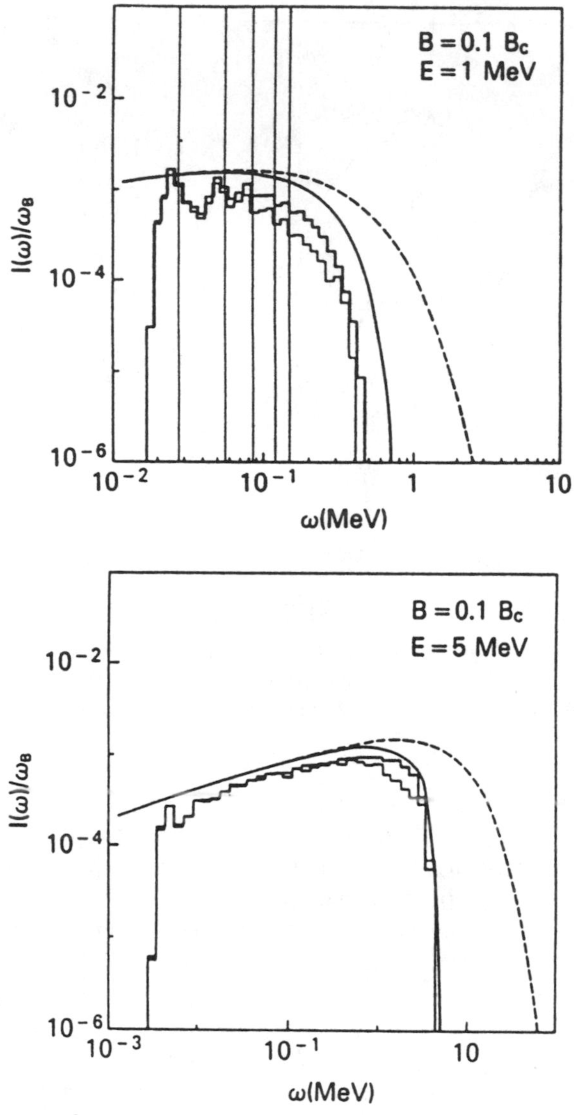


FIGURE 5.4.2 Single-particle synchrotron emissivity (in units of the cyclotron energy) for electrons with initial spin-up (*light histogram*) and spin-down (*dark histogram*). Light vertical lines are the energies of the first five harmonics. The solid curve is the quantum asymptotic emissivity (eq. 5.4.11), and the dashed curve is the classical emissivity. From A. K. Harding and R. Preece, 1987, *Ap. J.*, 319, 939.

These Monte Carlo calculations can be compared to the asymptotic formulae for the synchrotron emissivity calculated by Sokolov and Ternov (1986) in the ultrarelativistic regime $E_j \gg m_e$. In the frame where $p = 0$, and averaging over initial spins, this (quantum) asymptotic expression is

$$\frac{I_Q(\omega)}{\omega_c} \cong \frac{\sqrt{3}}{2\pi}\alpha_F\left(1 - \frac{\omega}{E}\right)\left[\int_y^\infty K_{5/2}(x)\,dx + y^3\left(\frac{3}{2}\Gamma\right)^2 K_{2/3}(y)\right],$$

$$y \equiv \frac{2\omega}{3E\Gamma(1-\omega/E)}, \quad \Gamma \equiv \frac{E\omega_c}{m_e^2}, \tag{5.4.11}$$

For $B \ll B_Q$, $\omega \ll E$, one obtains the classical ultrarelativistic limit,

$$\frac{I_c(\omega)}{\omega_c} = \frac{\sqrt{3}}{2\pi}\alpha_F\int_y^\infty K_{5/3}(x)\,dx. \tag{5.4.12}$$

For an isotropic electron distribution, these can be transformed to a frame where the electron has longitudinal energy $\gamma_\parallel$ by making the substitutions

$$E \rightarrow E/\gamma_\parallel, \quad I(\omega) \rightarrow I(\omega)\gamma_\parallel, \quad \omega \rightarrow \omega/\gamma_\parallel,$$

$$\gamma_\parallel = \left[1 - \frac{(E^2 - m^2)}{E^2}\cos^2\alpha\right]^{-1/2}, \tag{5.4.13}$$

and the resulting expressions are averaged over $\alpha$. The expressions (5.4.11) and (5.4.12) thus calculated are also shown in Figure 5.4.2. In the semirelativistic limit, useful expressions have been calculated by Baring (1988), while the emissivity as a function of photon angle has been given in the ultrarelativistic limit by Pavlov and Golenetskii (1986), Brainerd and Petrosian (1987), and Baring (1988), which show the quantum cutoff effect at $\omega \lesssim E$ imposed by the kinematics of equation (5.4.1). This cutoff is absent in the classical asymptotic limit (5.4.12), but arises naturally in the quantum expressions (5.4.11) and (5.4.5).

The asymptotic expressions (5.4.11) and (5.4.12) depend on the pitch angle via (5.4.13), being integrated over the final photon angle. However, the asymptotic expressions integrated over pitch angle will depend on the final photon angle through equations similar to (5.4.11)–(5.4.13), but with $\theta$ instead of $\alpha$. The angle-dependent photon emissivity in conventional units (ergs $s^{-1}$ $ster^{-1}$) in the ultra-relativistic quantum limit (Pavlov and Golenetskii, 1986; Baring, 1988) is therefore

$$j(\gamma, \omega, \theta) = \frac{1}{\sqrt{3}\,\pi} \alpha_F \frac{(mc^2)^2}{\hbar}$$

$$\times \frac{\omega}{\gamma^2} \left\{ \int_y^\infty K_{5/3}(s)\, dx + \frac{\omega^2}{\gamma(\gamma - \omega)} K_{2/3}(y) \right\},$$

$$y \equiv \frac{2}{3} \frac{m}{\omega_c} \frac{1}{\sin\theta} \frac{\omega}{\gamma(\gamma - \omega)}, \tag{5.4.14}$$

where $\gamma = E/m_e$. This reduces to the classical single-electron emissivity (e.g. Bekefi, 1966) for $\omega \ll \gamma$, $\omega_c \ll m_e$. For thermal and power law distributions, closed expressions based on (5.4.14) have been given by Brainerd and Petrosian (1987) and Pavlov and Golenetskii (1986), and Monte Carlo simulations for arbitrary energies (i.e. not necessarily $\gamma \gg 1$) have been performed by Bussard (1984) and Harding and Preece (1987). Calculations using semirelativistic closed expressions have been given by Baring (1988) and Brainerd (1987).

The cyclotron emission rates were calculated in § 5.1*d*, and the corresponding absorption rate can be obtained from detailed balance. Detailed expressions for cyclotron absorption between the ground state and arbitrary excited levels for arbitrary electron parallel momenta have been calculated by Daugherty and Ventura (1978). A simple expression is obtained by Harding and Daugherty (1991) for the absorption cross section from $j = 0$ to $j$ in the rest frame of the electron, which is useful for many of the applications encountered in practice,

$$\sigma_{\rm abs}^j(\theta) = \alpha_F \pi^2 \hbar^2 c^2 \frac{e^{-Z} Z^{j-1}}{E_j (j-1)!} \left[ (1 + \cos^2\theta) + \frac{Z}{j} \sin^2\theta \right]$$

$$\times\, \delta(E_j - m - \omega), \tag{5.4.15}$$

where $Z = (\omega^2 \sin^2\theta / 2\omega_c)$. In this case, the energy-momentum conservation requirement expressed by the delta function gives a resonance energy

$$\omega_j = m\left[ \left(1 + 2j \frac{\omega_c}{m} \sin^2\theta\right)^{1/2} - 1 \right] \Big/ \sin^2\theta. \tag{5.4.16}$$

In the limit $j\omega_c \ll m$, one has $\omega \simeq j\omega_c$, $Z \simeq j^2 \omega_c \sin^2\theta / 2m$, $E_j \simeq m$, and one regains the nonrelativistic expression (Canuto and Ventura, 1977)

$$\sigma_{\rm abs}^{j}(\theta) \simeq \frac{\alpha_F \pi^2 \hbar^2 c^2}{m} \left[ \frac{j^2 \omega_c}{2m} \sin^2\theta \right]^{j-1} \frac{(1 + \cos^2\theta)}{(j-1)!}. \qquad (5.4.17)$$

The frequency behavior of the cross section for a plasma in the laboratory frame is obtained by averaging equation (5.4.15) over the electron momentum distribution. However, even in the electron rest frame, there is a broadening due to the natural line width caused by the finite lifetime of the excited levels, giving rise to a Lorentz profile given by

$$L_{j,s}(\omega) = \frac{\Gamma_{j,s}/2\pi}{(\omega - \omega_j)^2 + \Gamma_{j,s}^2/4}, \qquad (5.4.18)$$

where the radiative width $\Gamma_{j,s}$ depends also on the spin variable $s$. The cyclotron absorption cross section in the rest frame is then

$$\sigma_{\rm abs}^{j}(\omega, \theta) = \sum_s L_{j,s}(\omega) \sigma_{\rm abs}^{j,s}(\theta), \qquad (5.4.19)$$

where the spin-dependent absorption cross section is used, expressions for which are given in Harding and Daugherty (1991). The spin-dependent radiative widths have been computed by Herold, Ruder, and Wunner (1982) and Latal (1986). In the limit $j\omega_c \ll m$, or fields such that $jB \ll B_Q$, Herold *et al.* (1982) give the nonrelativistic limit for the $j \to 0$ inverse lifetime $t_j^{-1} = \Gamma_j/\hbar$,

$$t_j^{-1} = \hbar^{-1}\Gamma_j = \hbar^{-1} \frac{4}{3} \alpha_F j m c^2 \left( \frac{B}{B_Q} \right)^2 \simeq 3.846 \times 10^{15} j B_{12}^2 \ {\rm s}^{-1}. \qquad (5.4.20)$$

This expression is fairly accurate up to $jB/B_Q \lesssim 0.1$, while it overestimates the width at higher fields, where the exact spin-dependent expressions must be used. As mentioned in Chapter 4, this radiative deexcitation rate (see eq. 4.1.31) is much shorter than the Coulomb collision rate (4.1.20), (4.4.17).

## 5.5 Magnetic Pair Production and Annihilation

### *a) One-Photon Pair Processes*

The pair production and annihilation process in a strong magnetic field can proceed via either two-photon or one-photon absorption or creation. Whereas in the absence of an external field energy-

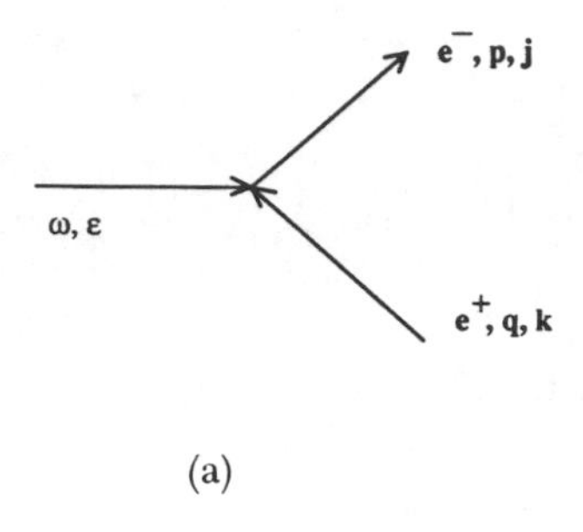


(a)

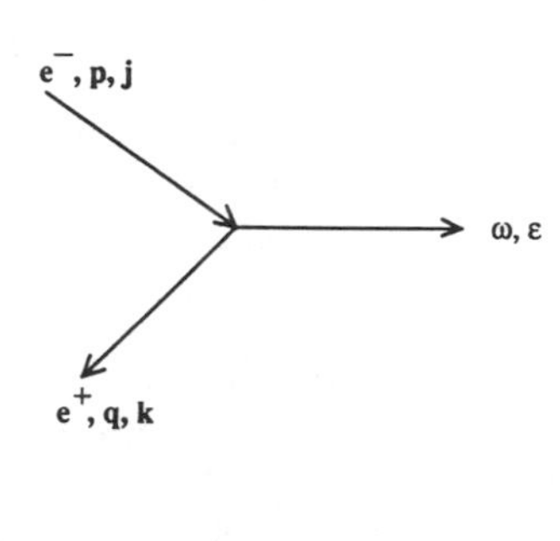


(b)

FIGURE 5.5.1 The one-photon pair production (a) and annihilation (b) process in the presence of a strong magnetic field.

momentum conservation requires two photons to be involved, a magnetic field can also exchange momentum with a photon, and the collision kinematics can be satisfied also when only one photon is involved (the magnetic field can be thought of as consisting of virtual photons, which can also take up momentum and energy). This latter process was first considered in pulsars by Sturrock (1971). The one-photon pair production and annihilation diagrams are shown in Figure 5.5.1. This process satisfies the conservation equations

$$E_p + E_q - \omega = 0,$$
$$p - q - \omega \cos \theta = 0, \tag{5.5.1}$$

where $p$ and $q$ are the $z$-momenta of the two leptons. These conditions and the Feynman diagrams are similar (but not identical) to those for the synchrotron process (eqs. 5.4.2), with Figure 5.5.1 the same as Figure 5.1.3 rotated, and one of the leptons here being a positron. As before, the S-matrix is given by equation (5.1.27) and the rates and cross sections differ from those of the synchrotron process only by a different choice of initial and final states over which one averages and integrates, and by a different energy delta function.

*b) 1γ Pair Annihilation*

Pair annihilation with emission of a single photon has been discussed by Wunner (1979), Daugherty and Bussard (1980), Harding (1986), Wunner *et al.* (1986), and Baring (1988). Calling $j, k$ the magnetic quantum numbers corresponding to the momenta $p, q$, the annihilation rate for positrons and electrons with occupation number distributions $n_- f_j^-(p)$, $n_+ f_k^+(q)$, and the $f_j(p)$ normalized to unity, is given by

$$Q_{1\gamma}(\omega, \theta)\, d\omega\, d\Omega = \alpha_F \lambda_c^2 \frac{c\omega}{4\pi B'} n_+ n_- \times \sum_j \sum_k \int dp \Gamma\{|T_\perp|^2 + |T_\parallel|^2\} \times f_j^-(p) f_k^+(q) \delta(\omega - E_j - E_k), \quad (5.5.2)$$

where $B' = B/B_Q$, $\omega \equiv \hbar\omega/mc^2$, $\Gamma = (1+\delta_{j0})(1+\delta_{k0})$, $\lambda_c$ is the Compton wavelength, and for each $(\omega, \theta)$ the energies and momenta obey equation (5.5.1). The corresponding cross section $d\sigma_{1\gamma}$ is related to the rate $Q_{1\gamma}$ by

$$Q_{1\gamma} = n_+ n_- \int dp f_-(p) \int dq f_+(q) \left| \frac{p}{E_p} + \frac{q}{E_q} \right| d\sigma_{i\gamma}. \quad (5.5.3)$$

The squares of the matrix elements corresponding to photons of $\parallel$ and $\perp$ polarization are given by

$$\begin{aligned} 4E_j E_k |T_\perp|^2 &= (E_j E_k - pq + m^2)[I_{j,k-1}^2 + I_{j-1,k}^2] \\ &\quad + 2\sqrt{2jB'}\sqrt{2kB'}[I_{j,k-1} I_{j-1,k}], \\ 4E_j E_k |T_\parallel|^2 &= \cos^2\theta \{(E_j E_k - pq + m^2)[I_{j,k-1}^2 + I_{j-1,k}^2] \\ &\quad - 2\sqrt{2jB'}\sqrt{2kB'}[I_{j,k-1} I_{j-1,k}]\} \\ &\quad - 2\cos\theta \sin\theta \{-p\sqrt{2kB'} \\ &\quad \times [I_{j-1,k} I_{j-1,k-1} + I_{j,k-1} I_{j,k}] \\ &\quad + q\sqrt{2jB'}[I_{j,k} I_{j-1,k} + I_{j-1,k-1} I_{j,k-1}]\} \\ &\quad + \sin^2\theta \{(E_j E_k + pq + m^2)[I_{j-1,k}^2 + I_{j,k}^2] \\ &\quad + 2\sqrt{2jB'}\sqrt{2kB'}[I_{j-1,k-1} I_{j,k}]\}, \end{aligned} \quad (5.5.4)$$

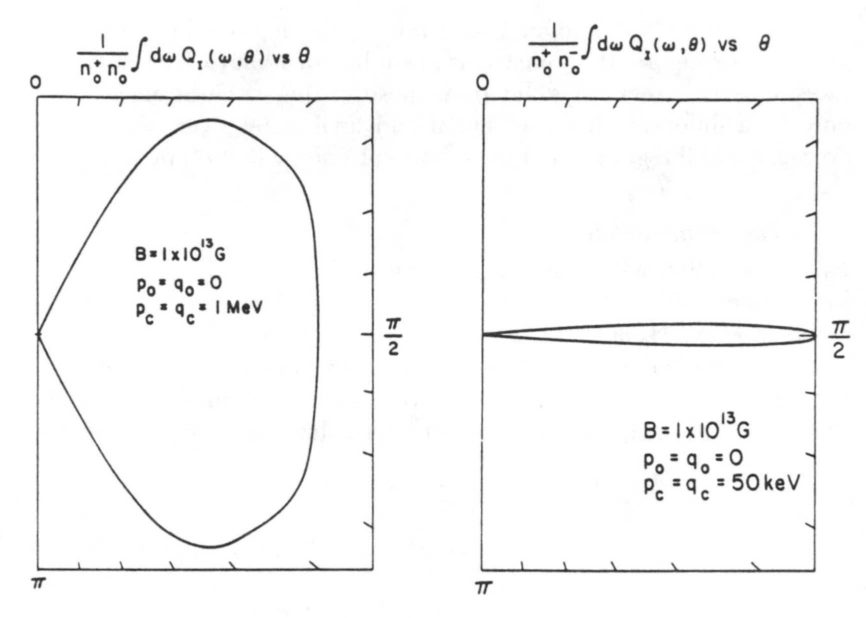


FIGURE 5.5.2 The frequency-integrated 1$\gamma$ annihilation rate from the ground state as a function of $\theta$ for $\Delta p = 1$ MeV (*left*) and $\Delta P = 50$ keV (*right*) at $B = 10^{13}$ G. From J. K. Daugherty and R. W. Bussard, 1980, *Ap. J.*, 238, 296.

where the $I_{j,k}(x)$ are defined in equation (5.4.8), with $x = \omega^2 \sin^2\theta / 2m\omega_c \equiv \omega^2 \sin^2 / 2B'$. The energy-momentum conservation conditions (5.5.1) are satisfied for two solutions $p_\pm$ of the electron momentum,

$$p_\pm = \frac{1}{2}\left[\omega \cos\theta \pm \left(\omega^2 - 4m^2/\sin^2\theta\right)^{1/2}\right]. \tag{5.5.5}$$

For either of these solutions to exist, one needs $\omega^2 - 4m^2/\sin\theta^2 \geq 0$, and since here $\omega, \sin\theta$ are nonnegative, the threshold condition for one-photon pair annihilation is

$$\omega > \omega_{\text{thr}} = 2m/\sin\theta. \tag{5.5.6}$$

As an example, for a ground state Gaussian distribution $f_\pm(p) = (\pi^{1/2}\,\Delta p)^{-1}\exp[-p^2/(\Delta p)^2]$ and $B = 10^{13}$ G, the frequency-integrated rate as a function of angle is shown in Figure 5.5.2. In the limit of electrons and positrons at rest, the annihilation pattern tends to a fan beam at $\theta \to \pi/2$, which broadens at higher lepton energies.

In the center of mass frame, where $p - q = 0$, the total ground state annihilation cross section and rate have a simple form (Wunner, 1979),

$$\sigma_{1\gamma}^{(0,0)} = \frac{\alpha_F}{2} \frac{m^2}{pE_p} \lambda_c^2 \frac{1}{B'} \exp\left[ -\frac{2(E_p/m)^2}{B'} \right], \qquad (5.5.7)$$

while the angle- and frequency-integrated total ground state annihilation rate is

$$Q_{1\gamma}^{(0,0)}(p) = n_+ n_- c \alpha_F \left( \frac{m}{E_p} \right)^2 \lambda_c^2 \frac{1}{B'} \exp\left[ -\frac{2}{B'} \left( \frac{E_p}{m} \right)^2 \right]. \qquad (5.5.8)$$

When one or both of the annihilating leptons are not in the ground Landau level, the expression is more complicated (5.5.2). Harding (1986) has presented calculations for this case and finds that, for fields $\lesssim 10^{13}$ G, the rate of annihilation from excited states can exceed by an order of magnitude the ground state annihilation rate. Considering isotropic thermal distributions, where the transverse levels are also populated according to a Boltzmann factor $T_{\parallel} = T_{\perp}$, the annihilation spectrum shows resonances, or peaks, corresponding to discrete $(j, k)$ pairs of levels, the width and spacing of the peaks depending on the angle of emission with respect to $\vec{B}$. An example is shown in Figure 5.5.3. An asymptotic approximation for $(j, k) \gg 1$ for thermal pairs and for arbitrary isotropic distributions is given by Harding (1986) and Baring (1988). This latter asymptotic rate can be written as

$$Q_{1\gamma}(\omega, \theta) = \frac{n_+ n_-}{8\pi\sqrt{3}} \alpha_F \lambda_c^2 c \times \int_0^{\omega} d\gamma \frac{\bar{R}(\omega, \gamma, \theta) f(\gamma) f(\omega - \gamma)}{\gamma(\omega - \gamma)\sqrt{\gamma^2 - 1}\sqrt{(\omega - \gamma)^2 - 1}}, \qquad (5.5.9)$$

where

$$\bar{R}(\omega, \gamma, \theta) = -\int_y^{\infty} K_{5/3}(x)\, dx + \frac{\omega^2}{\gamma(\omega - \gamma)} K_{2/3}(y) \qquad (5.5.10)$$

and $y = 2\omega[3B' \sin\theta\, \gamma(\omega - \gamma)]^{-1}$, $\omega \equiv \hbar\omega/mc^2$, and $\gamma \equiv E/mc^2$.

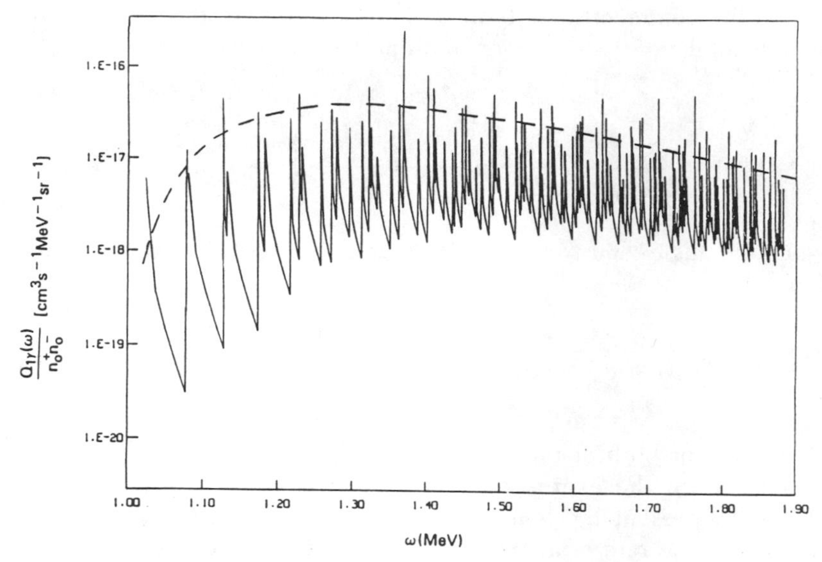


FIGURE 5.5.3 Exact $1\gamma$ annihilation spectrum including excited states for an isotropic Maxwellian of $kT = 0.2mc^2$, $B = 5 \times 10^{12}$ G at $\theta = 90°$. The dashed line is the asymptotic approximation. From A. K. Harding, 1986, *Ap. J.*, 300, 167 and 462 (corrigendum).

### *c) 1γ Pair Production*

The $1\gamma$ pair production process is sometimes called magnetic photon absorption, and it occurs in a strong field when photons exceed the kinematical threshold condition (5.5.6). In a frame where $\vec{k} \cdot \vec{B} = 0$ or $p = -q$, the energy-momentum conservation equations (5.5.1) give for the longitudinal momentum $p$ the relation

$$p = p(j, k) = \pm m\left[\left(\frac{\omega}{2}\right)^2 - 1 - (j + k)B' + (j - k)^2\frac{B'^2}{\omega^2}\right]^{1/2}, \tag{5.5.11}$$

where $\omega \equiv \omega/m$. Since $p^2 \geq 0$, the allowed transitions to $(j, k)$ states must lie below the curve $p^2 = 0$, which requires

$$j_{\max} = k_{\max} \leq \frac{\omega}{B'}\left(\frac{\omega}{2} - 1\right). \tag{5.5.12}$$

The number of integer $(j, k)$ pairs enclosed within the boundary of the $p^2 = 0$ curve is $N_{\text{states}} \sim \omega(\omega + 4)(\omega - 2)^2/24B'^2$, which for $\omega \gg 1$ is

$N_{\text{states}} \sim \omega^4/24B'$. At fixed $B'$ one has an increment $dN_{\text{states}} \sim (\omega^3/6B'^2)\,d\omega'$, which corresponds to twice the number of singularities appearing in the attenuation coefficient between energies $\omega$ and $\omega'$. Still in the same frame, the attenuation coefficient $\kappa$ ($\text{cm}^{-1}$) for photons polarized parallel or perpendicular to the filled is (Daugherty and Harding, 1983)

$$\kappa_{\parallel,\perp} = \frac{1}{T}\sum_{j\geq 0}\sum_{k\geq 0}\int L\frac{dp}{2\pi}\int L\frac{dq}{2\pi}\int L\frac{da}{2\pi\lambda^2} \times \int L\frac{db}{2\pi\lambda^2}\left\{\sum_{s_+}\sum_{s_-}|S_{fi}|^2_{\parallel,\perp}\right\}, \tag{5.5.13}$$

where $L^3T$ is the space-time volume element, the sums over $s_+$, $s_-$ range over the spin states of the electron and positron, and the $j$, $k$ sums range over the kinematically allowed values for which $p^2(j,k) \geq 0$. This can be written as

$$\kappa_{\parallel}(\omega, B') = \frac{\alpha_F}{2\xi}\sum_j\sum_k\frac{1}{|p_{jk}|} \times\left\{\left(E_jE_k + m^2 - p^2\right)S^2_{1,\parallel} + 2\sqrt{jk}\,B'mS^2_{2,\parallel}\right\},$$

$$\kappa_{\perp}(\omega, B') = \frac{\alpha_F}{2\xi}\sum_j\sum_k\frac{1}{|p_{jk}|} \times\left\{\left(E_jE_k + m^2 - p^2\right)S^2_{1,\perp} + 2\sqrt{jk}\,B'mS_{2,\perp}\right\}, \tag{5.5.14}$$

where $\xi = \omega^2/2B'$, $\omega \equiv \omega/m$, and $S_{i,\parallel,\perp}$ are combinations of matrix elements related to the $T_{\parallel,\perp}$ of equation (5.5.4) (e.g. Daugherty and Harding, 1983; Baring, 1988). The presence of $|p_{jk}|$ in the denominator of (5.5.14) indicates that the $\kappa_{\parallel,\perp}$ have singularities or absorption edges, occurring at

$$\omega_{jk} = \left(\sqrt{1 + 2jB'} + \sqrt{1 + 2kB'}\right). \tag{5.5.15}$$

Only the $\parallel$ polarization is able to interact with the lowest Landau resonance, while the $\perp$ interacts with the second and higher states. A plot of the polarization-averaged $1\gamma$ absorption rate is shown in Figure 5.5.4. Well above the threshold $\omega_{\text{thr}}$ of equation (5.5.6), in the limit $\gamma \gg 1$, the rate of production (5.5.14) integrated over electron energies gives the asymptotic attenuation coefficient $\bar{\kappa}$ for parallel and perpendicular polarization as (Tsai and Erber, 1974; Daugherty and Harding, 1983)

$$\kappa_{\parallel} \simeq 0.35 \frac{\alpha_F}{\lambda_c} B'_{\perp} \exp\left[ -\frac{8}{3\omega B'_{\perp}} \right], \quad \kappa_{\perp} = \frac{1}{2}\kappa_{\parallel} \qquad \text{for } \omega B'_{\perp} \ll 1,$$

$$\kappa_{\parallel} \simeq 0.5 \frac{\alpha_F}{\lambda_c} \sin\theta \left(\omega B'_{\perp}\right)^{-1/3}, \quad \kappa_{\perp} = \frac{2}{3}\kappa_{\parallel} \qquad \text{for } \omega B'_{\perp} \gg 1.$$

(5.5.16)

These asymptotic coefficients are also shown in Figure 5.5.4 for comparison, which shows how the asymptotic expressions smooth over the sawtooth behavior. They are quite good at $\omega \gg \omega_{\text{thr}}$, $\omega B' \ll 1$, but in general they overestimate the cross section near threshold. Near the threshold for $B = 10^{13}$ G, the asymptotic expressions give an overestimate of several orders of magnitudes, but for $B \gtrsim 5 \times 10^{12}$ G, $\omega \gtrsim 3$ MeV the fit starts becoming quite adequate.

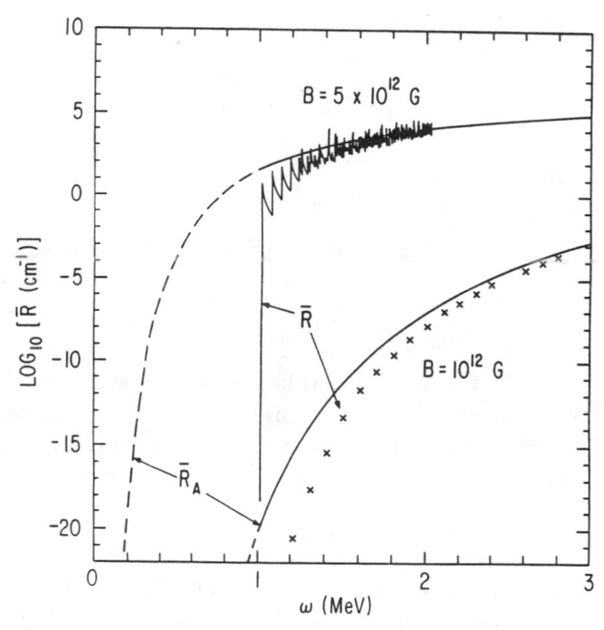


FIGURE 5.5.4 Exact $1\gamma$ attenuation coefficient $\bar{\kappa} = \frac{1}{2}(\kappa_{\parallel} + \kappa_{\perp})$ for $\sin\theta = 1$ and $B = 5 \times 10^{12}$ G and $B = 10^{12}$ G. Also shown are the asymptotic attenuation coefficients. From J. K. Daugherty and A. K. Harding, 1983, *Ap. J.*, 273, 761.

*d) 2γ Pair Annihilation*

The process of $2\gamma$ pair annihilation has the same *S*-matrix form as the Compton scattering process, being given by the same pair of diagrams as Figure 5.1.1, but with time flowing from bottom to top, and one electron and one positron in and two photons out. Expressions for the annihilation rate have been calculated only for the case when the annihilating leptons are in the ground Landau state (Wunner, 1979; Daugherty and Bussard, 1980; Kaminker, Pavlov, and Mamradze, 1986, 1990). The energy and momentum conservation conditions are

$$E_p + E_q - \omega_1 - \omega_2 = 0,$$
$$p - q - \omega_1 \cos\theta_1 - \omega_2 \cos\theta_2 = 0. \tag{5.5.17}$$

These can be solved for the momentum $p$, which has two solutions:

$$p_{\pm} = \frac{\zeta_1 \zeta_2 \pm \left[\zeta_2^2 - m^2\left(1 - \zeta_1^2\right)\right]^{1/2}}{1 - \zeta^2}, \tag{5.5.18}$$

where $\zeta_1 = \sum_i k_{zi} \sum_i \omega_i$, $\zeta_2 = \frac{1}{2}[\sum_i \omega_i - (\sum_i k_{zi})^2/(\sum_i \omega_i)]$, and $k_{zi} = \omega \cos\theta_i$. For each $(\omega_1, \theta_1, \omega_2, \theta_2)$ one has two electron momenta $p_{\pm}$, which also determine the positron $q_{\pm}$ and the $E_p^{\pm}$, $E_q^{\pm}$, and the cross section and annihilation rates are given by the sum of the contributions corresponding to each $p_{\pm}$ momentum solution. For a given set of angles $(\theta_1, \theta_2)$ there is an energy threshold given by $(\sum_i \omega_i)^2 - (\sum_i k_{zi})^2 = 4m^2$. At a given angle $\theta$, the maximum photon energy is

$$\omega_{\max}\left(p_+, p_-, \theta\right) = \left(E_+ + E_-\right)\left(1 \pm \beta\right)/\left(1 \pm \cos\theta\right), \tag{5.5.19}$$

where the $\pm$ sign corresponds to $\cos\theta$ greater or less than $\beta = (p_+ + p_-)/(E_+ + E_-)$, where $\beta$ is the center of momentum velocity. In the center of momentum (CM) frame, the differential cross section for annihilation with both leptons in the ground state has an analytic expression (Wunner, Herold, and Ruder, 1983). Denoting by $\pm p_0$, $E_0$ the momentum and energy of the $e^+e^-$, and by $k'_{z1}$, $k'_{z2}$ the decay photon momenta in the CM frame (with $k'_1 + k'_2 = 2E_0$ and $k'_{z1} + k'_{z2} = 0$), this is

$$\frac{d\sigma_{2\gamma}^{\text{CM}}}{dk'_1\, d\Omega'_1\, d\varphi'_{12}} = \frac{\alpha_F^2 k'_1}{4 p_0 E_0} \frac{\exp\left(-\left[\vec{k}_{\perp 1} + \vec{k}'_{\perp 2}\right]^2 / 2eB\right)}{\pi eB} |M|^2, \tag{5.5.20}$$

where

$$M = m_e\{(2p_0 - k'_{z1})e_{1z}e_{2z}F(1, Q_1 + 1, k'_{1+}k'_{2+}/2eB)/2eBQ_1$$
$$+ (e_{1+}e_{2-}k'_{1z} + e_{1+}e_{2z}k'_{2-} - e_{1z}e_{2-}k'_{1+})$$
$$\times F(1, Q_1 + 2, k'_{1+}k'_{2-}/2eB)/(2eBQ_1 + 1)$$
$$+ [\text{terms with subscripts } (1 \leftrightarrow 2)]\}, \qquad (5.5.21)$$

where $Q_i = [m_e^2 + (p_0 - k'_{zi})^2 - (E_0 - k'_i)^2]/2eB$, $\vec{e}_i$ is the polarization vector of mode $i(i = 1, 2)$, and $(+, -, z)$ refers to circular coordinates. The sum over intermediate states is analytical, giving the hypergeometric functions $F \equiv {}_1F_1$. In the laboratory frame, using

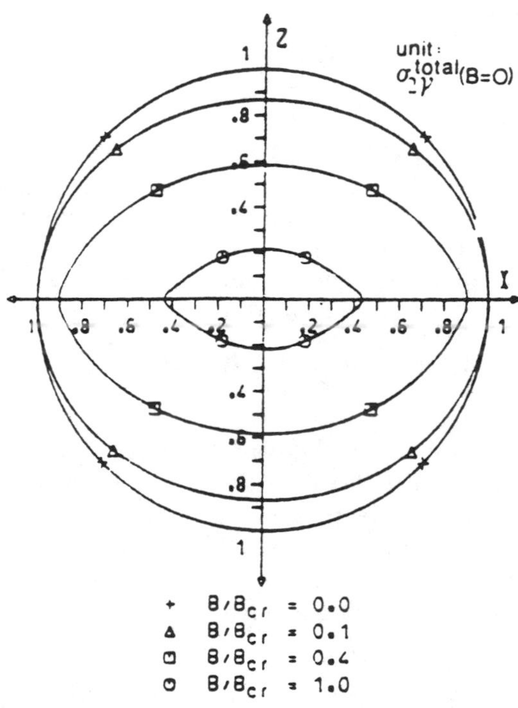


FIGURE 5.5.5 Angular distribution of the $2\gamma$ annihilation cross section $d\sigma_{2\gamma}/d\Omega_1$ for various values of $B/B_Q = 0, 0.1, 0.4$, etc. From G. Wunner, H. Herold, and H. Ruder, 1983, in *Positron-Electron Pairs in Astrophysics*, ed. M. Burns *et al.* (AIP, New York), p. 411.

$\beta = (p_+ + p_-)/(E_+ + E_-)$ for the CM velocity, we have the relation $k' = k\gamma(1 - \beta\cos\theta)$, $\cos\theta' = (\cos\theta - \beta)/(1 - \beta\cos\theta)$ and the cross section is

$$\frac{d\sigma_{2\gamma}}{dk_1\, d\Omega_1\, d\varphi_{12}} = \frac{d\sigma_{2\gamma}^{\mathrm{CM}}}{dk_1'\, d\Omega_1'\, d\varphi_{12}'} \cdot \frac{1}{\gamma\left(1 - \beta\cos\theta\right)}. \tag{5.5.22}$$

The angular distribution of the $2\gamma$ annihilation cross section from the ground level is shown in Figure 5.5.5 for $E_e \simeq m_e$ and various values of the field strength $B$. One sees that, for increasing $B/B_Q$, the two photons are preferentially emitted perpendicular to $\vec{B}$ in the CM frame. Integrating the differential cross section (5.5.22) over $d\Omega_1$ gives the average line profile, which shows an increasing broadening as $B' = B/B_Q \to 1$ which for pairs at rest is $\Delta E \sim mB'/2$ for $\sin\theta < (2B')^{1/2}$, or $\Delta E \sim m(B'/2)^{1/2}\sin\theta$ for $\sin\theta > (2B')^{1/2}$ (see Daugherty and Bussard, 1980; Wunner, Herold, and Ruder, 1983). This intrinsic broadening is due to the field and, if not blurred by other effects, could be used as a diagnostic of the field strength. However, as the field strength is increased beyond about $B \sim 10^{13}$ G, the cross section for $1\gamma$ annihilation becomes much larger than that for $2\gamma$ annihilation. A comparison of the $1\gamma$ annihilation rate from the ground (0, 0) and excited states $(j, j')$ with the $2\gamma$ annihilation from ground (0, 0) states made by Harding (1986) is shown in Figure 5.5.6. While at low fields $B \lesssim$ few $10^{12}$ G it is $R_{2\gamma}$ that dominates, at intermediate fields $B \lesssim 10^{13}$ G the $1\gamma$ annihilation from excited states dominates, and at very high fields $B \gtrsim 10^{13}$ G it is the $1\gamma$ annihilation from ground states that dominates. We are not aware of a magnetic $2\gamma$ annihilation calculation from excited states having been made. However, for low fields, or for $(j, j') \gg 1$, it must tend to the field-free value (Ramaty and Mészáros, 1981) so that probably $\sigma_{2\gamma}(j, j')$ does not increase much for $j, j' \gg 1$ and $B \lesssim B_Q$.

*e) $2\gamma$ Pair Production*

The $2\gamma$ pair production process, $\omega_1 + \omega_2 \to e^+ + e^-$, is again described by the diagrams of Figure 5.1.1, with time now running from top to bottom and the S-matrix the same as for magnetic Compton scattering. Calculations of this process have been presented by Kozlenkov and Mitrofanov (1987). The threshold condition for

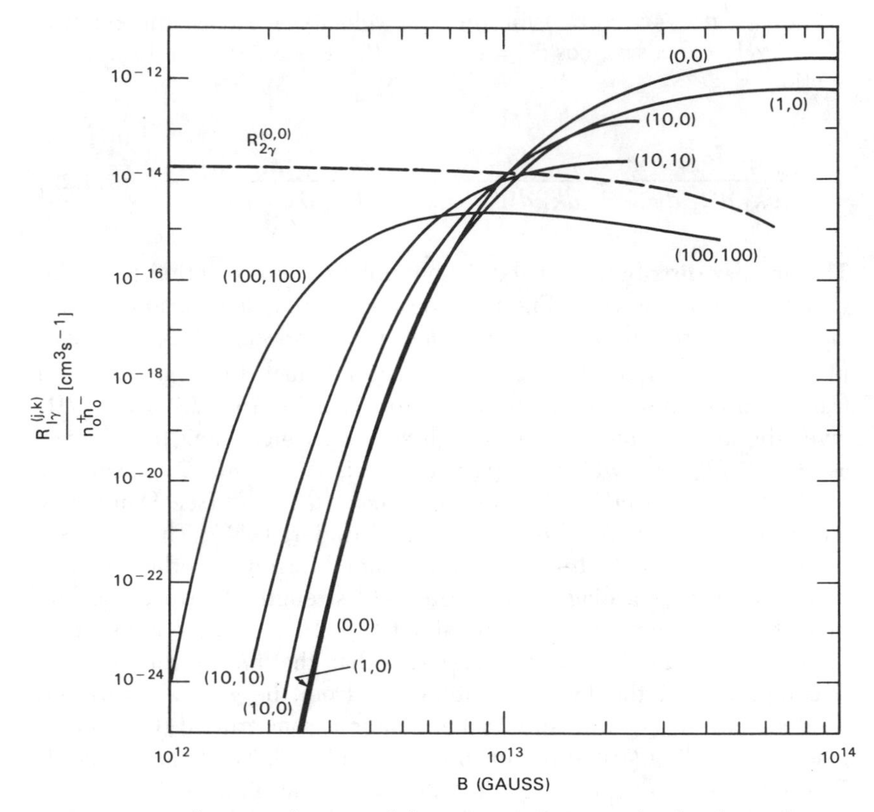


FIGURE 5.5.6 The $1\gamma$ and $2\gamma$ annihilation rates from Landau levels $(j, j')$ as a function of $B$. After A. K. Harding, 1986, *Ap. J.*, 300, 167 and 462 (corrigendum), and A. K. Harding, 1990, private communication.

creating pairs in levels $(j, j')$ which follows from equations (5.5.17) is

$$\left(\omega_1 + \omega_2\right)^2 + \left(k_{1z} + k_{2z}\right)^2 \geq \left[\left(1 + 2jB'\right)^{1/2} + \left(1 + 2j'B'\right)^{1/2}\right]^2, \tag{5.5.23}$$

where $m_e = e = \hbar = 1$ and $B' = B/B_Q$. With decreasing photon energy, the number of outgoing channels decreases and the minimum permissible threshold is for $j = j' = 0$, giving

$$\left(\omega_1 + \omega_2\right)^2 - \left(k_{1z} + k_{2z}\right)^2 \geq 4. \tag{5.5.24}$$

For any $k_{1z}$, $k_{2z}$ the region of permissible values of $\omega_1$, $\omega_2$ is broader than the region of permissible values of $\omega$ in the $B = 0$ case, $\omega_1\omega_2(1 - \cos\theta) \geq 2$, which again leads to a purely magnetic broadening of the absorption. The cross section has the form (Kozlenkov and Mitrofanov, 1987)

$$\begin{aligned} &\sigma_{e^+e^+}\left(\vec{k}_1, \vec{e}^{(1)}, \vec{k}_2, e^{(2)} \rightarrow j', p, j, q\right) \\ &= \frac{3}{16}\sigma_T B' \frac{1}{|\, pE_j + qE_{j'} \,| \left(1 - \cos\theta\right)\omega_1\omega_2} \\ &\quad \times \exp\left[-\frac{\lambda^2}{2}\left(k_{1\perp}^2 + k_{2\perp}^2\right)\right] \\ &\quad \times \sum_{i=1}^{4} \left| N_i^{(1)} \exp\left(-i\lambda^2 k_{1y} k_{2x}\right) + N_i^{(2)} \exp\left(-i\lambda^2 k_{2y} k_{1x}\right)\right|^2, \end{aligned} \tag{5.5.25}$$

where $\lambda = \lambda\!\!\!^{-}_c B'^{-1/2}$, $\theta$ is the angle between $\vec{k}_1$ and $\vec{k}_2$, the $\vec{e}^{(i)}$ are the polarization vectors, and the matrices $N_i^{(1,2)}$ are again expressible in terms of those of equation (5.5.3). The presence of the $|\, pE_j + qE_{j'} \,|$ denominator in (5.5.25) again leads to a sawtooth behavior, as for the $1\gamma$ pair production process. Detailed numerical calculations have been performed only for a few special cases, e.g. below the $1\gamma$ threshold (Kozlenkov and Mitrofanov, 1987), so that the relative importance of the $2\gamma$ and $1\gamma$ pair production rate is only approximately known. The cross sections per photon for $1\gamma$ and $2\gamma$ are equal at $B \sim 10^{13}$ G for pair states $j = k = 0$, with the $1\gamma$ process dominating above (Burns and Harding, 1984), as for the $2\gamma$ and $1\gamma$ pair annihilation processes. However, the total photon absorption rate per photon depends on the density of virtual photons ($1\gamma$) and of real photons ($2\gamma$), so that one can expect the $2\gamma$ process to be important in high-radiation-density environments, especially if $B \lesssim 10^{13}$ G, while the $1\gamma$ pair creation is expected to dominate in low-radiation-density, high-magnetic-field regions.

## 5.6 Other Magnetic Effects

### *a) Coulomb and Bremsstrahlung Processes*

Electrons and protons in a magnetic field can undergo transitions between continuum or discrete energy and momentum states by colliding with each other, i.e. by interacting with the electric field of another charged particle. To lowest order in $\alpha_f$ these transitions are radiationless (Coulomb process), and to next order they occur accompanied by emission or absorption of a photon (bremsstrahlung process). These processes in a strong magnetic field have been discussed in the nonrelativistic limit in Chapter 4.4. In the relativistic regime, the (radiationless) Coulomb process for electron-electron and electron-proton collisions has been studied by Bussard (1980), Langer (1981), and Storey and Melrose (1987). For relative energies below the electron cyclotron ground energy, the collisions result mainly in longitudinal (continuum) momentum changes, while for energies approaching or exceeding it they can result in exciting electrons to higher Landau levels (transverse momentum excitations). For relative kinetic energies below the cyclotron energy, (*ee*) collisions where both electrons are in the ground Landau state are very inefficient in exchanging energy. This is because electrons constrained to move along one dimension can only keep or swap their momenta, which does not lead to any appreciable change. For this reason electrons in the ground state interact mainly with protons. This provides an obvious mechanism leading to the thermalization of the electrons and (*pe*) collisions are the main mechanism for exciting electrons to higher Landau states, with (*ee*) collisions providing at most 10–20% of the total excitation rate. For this reason the (*ee*) bremsstrahlung, which in the nonmagnetic case can be very significant, plays a much smaller role than (*ep*) bremsstrahlung in the magnetic case. As in the nonrelativistic case (Chap. 4.4), the relativistic bremsstrahlung in a magnetic field leads to a continuum spectrum associated with longitudinal momentum changes, and to resonances associated with the excitation of the electron discrete transverse Landau energy levels. Some of the properties of the resonance behavior of (*ep*) bremsstrahlung expected from a QED calculation in the semirelativistic limit (including spin-flip transitions) have been discussed by Lieu (1983). So far, no detailed calculations of the photon spectrum for this process in the fully relativistic strong-magnetic-field regime have been published. However, as in the nonrelativistic limit discussed in Chapter 4.4, the spectrum is expected to be qualitatively similar to that of one-photon Compton scattering, except for an additional factor $\propto \alpha_f n_e \omega^{-3}[1 - \exp(\hbar\omega / kT_{\parallel})]$ which changes the slope by a factor $\omega^{-2}$ in the Rayleigh-Jeans regime.

### *b) Curvature Radiation*

Curvature radiation is important in rotation-powered pulsars, although it can occur anywhere, even in nonmagnetic situations, whenever the charged particle paths are curved. In pulsars, the curvature arises by virtue of the longitudinal motion of the particles along the dipole (or multipole) field lines (e.g. Sturrock, 1971). The particular importance of this effect is that it occurs even when the particles are in their ground Landau level, i.e. when synchrotron or cyclotron radiation does not occur. Of course, if excitation to higher Landau levels occurs, the latter radiation arises, in addition to the curvature radiation. Whereas the synchrontron process implies a change of the transverse energy (with respect to $\vec{B}$) of the particles, the curvature radiation is connected to a change in the longitudinal (along $\vec{B}$) energy.

In general, the radius of curvature of the pulsar field lines is large enough that, for most longitudinal particle energies, the classical radiation formulae are adequate. In this case the particle energy spectrum is (Jackson, 1975)

$$\frac{dI_\omega}{d\omega} = \sqrt{3}\,\frac{e^2}{c}\gamma f\left(2\frac{\omega}{\omega_0}\right), \tag{5.6.1}$$

where

$$\omega_0 = \frac{3}{2}\frac{c\gamma^3}{\rho_c} \tag{5.6.2}$$

is the typical maximum frequency obtained for a curvature radius $\rho_c$ and the function $f(x)$ is

$$f(x) \equiv 2x\int_{2x}^{\infty} K_{5/3}(x')\,dx' = \begin{cases} 2.149\,x^{1/3}, & x \ll 1 \\ 1.253\,x^{1/2}e^{-x}, & x \gg 1 \end{cases}. \tag{5.6.3}$$

This is the same spectrum as that of classical synchrotron radiation of the same radius of curvature. To convert (5.6.1) from energy to power, one must divide by a typical time $c/2\pi\rho_c$. The frequency-integrated power is then given by

$$P = \frac{2}{3}\frac{e^2c}{\rho_c^2}\beta^3\gamma^4 \text{ ergs s}^{-1}. \tag{5.6.4}$$

For typical pulsar values $B \sim 10^{12}$ G, $R \sim 10^6$ cm, and $e^+e^-$ energies $10^{12}$–$10^{13}$ eV, the curvature photons of energy around the value given by (5.6.2) are in the GeV range, which is well above the $1\gamma$ and $2\gamma$ pair creation threshold and can lead to an $e^+e^-$ cascade.

*c) Photon Splitting*

Just as in the absence of a magnetic field there is the possibility of elastic photon-photon scattering, $\omega_1 + \omega_2 \rightarrow \omega_3 + \omega_4$, in a magnetic field a similar process is also possible which uses the virtual photons of a strong magnetic field $B$ as the target photons, $\omega_1 + B \rightarrow \omega_2 + \omega_3$. This process is called photon splitting (e.g. Erber, 1964; Adler, 1971). Its Feynman diagram is shown in Figure 2.1.6c. Unlike the $1\gamma$ photo-pair production, $\omega_1 + e^+ + e^-$, photon splitting does not have a threshold energy, being kinematically unrestricted at all energies $\omega$, whether above or below $2mc^2/\sin\theta$. This makes it particularly interesting at $\omega \leq 2mc^2/\sin\theta$, where pair production is forbidden so that photon splitting becomes the only photon absorption process possible in a magnetized vacuum. Actually the photon splitting can also be viewed as a soft photon source, since it increases the total number of photons. The effective absorption coefficient has been calculated in lowest order by Adler (1971), who gives the following expressions for linearly polarized photons with electric vector parallel and perpendicular to the $(\vec{k}, \vec{B}_0)$ plane:

$$\kappa^{(s)}(\perp \rightarrow \parallel_1 + \parallel_2) = 0.12\left(\frac{B}{B_Q}\right)^6 \left(\frac{\omega}{m}\right)^5 \sin^6\theta \text{ cm}^{-1},$$

$$\kappa^{(s)}(\perp \rightarrow \perp_1 + \perp_2) = 0.39\left(\frac{B}{B_Q}\right)^6 \left(\frac{\omega}{m}\right)^5 \sin^6\theta \text{ cm}^{-1},$$

$$\kappa^{(s)}(\parallel \rightarrow \perp_1 + \parallel_2) + \kappa^{(s)}(\parallel \rightarrow \parallel_1 + \perp_2) = 2\kappa^{(s)}(\perp \rightarrow \parallel_1 + \parallel_2). \tag{5.6.5}$$

These are the only combinations of polarization channels allowed by charge and parity (CP) invariance. When the parameter $\chi = (\omega/2m)(B\sin\theta/B_Q) \ll 1$, the kinematic selection rules allow only the first line of (5.6.5), while the second and third lines become $\kappa^{(s)}(\perp \rightarrow \perp_1 + \perp_2) \simeq \kappa^{(s)}(\parallel \rightarrow [\parallel, \perp]) \simeq 0$. Notice that our definition of $\parallel, \perp$ refers to the electric vector of the wave, as also used in Chapter 2 (eqs. 2.6.10, 2.6.11). This convention is the opposite of that used by Adler (1971) and Mitrofanov *et al.* (1986) for this process. The incident photon $\omega_1$, which is a $\perp$ in the first line of (5.6.5), gives two photons $\omega_2, \omega_3$ which are both $\parallel$, polarized with their electric vector parallel to the $(\vec{k}, \vec{B})$ plane. As emphasized by Adler *et al.* (1970), this would act as a polarizer of the $\gamma$-rays below about 1 MeV in strong-field, rotation-powered pulsars. For instance, for $\omega/m \sim B/B_Q \sim \sin\theta \sim 1$, one has $\kappa_\perp^{(s)} \sim 0.1 \text{ cm}^{-1}$, which implies about $10^5$

absorption lengths over a typical pulsar magnetosphere dimension of $R \sim 10^6$ cm. However, above pair production threshold, the photon-splitting coefficient is far smaller than either the $1\gamma$ or $2\gamma$ pair production coefficient.

### d) Positronium Decay and Photon Conversion

Positronium is the "atom" formed by an electron and a positron. This system, just like the hydrogen atom in a sufficiently strong magnetic field $B \gg \alpha_F^2 B_Q = 2.35 \times 10^9$ G, has a level structure which is entirely different from that of the field-free hydrogenic level structure (Wunner, Ruder, and Herold, 1981). Its structure is dominated entirely by the magnetic effects, which constrain the phase space perpendicular to $\vec{B}$. Because of this, the Coulomb binding becomes stronger along the one remaining free direction, that along $\vec{B}$, and the strength of the binding along this direction increases logarithmically with $B$. As a consequence, in fields of $B \gtrsim 0.1 B_Q$ the binding energy of the ground state can be of order $\gtrsim 70$ eV instead of the field-free value $\frac{1}{2}$Ry $\sim 6.8$ eV. Whereas in the field-free case the positronium ground state can decay into two or, less probably, into three photons, in the magnetic field case there arises the question of the possibility of one-photon decay. As shown by Wunner and Herold (1979), for a vanishing total momentum of the system one-photon decay could only occur with strict momentum conservation, and therefore one-photon decay in this case is forbidden. The two-photon decay rate calculated by these authors in the magnetic case is given by

$$
\begin{aligned}
\frac{dw}{d\Omega_1} &= |f(0)|^2 \frac{\alpha_F^2 c}{\pi}\left[1 - (\vec{e}_1 \cdot \vec{e}_2)^2\right] \\
&\quad \times \frac{B}{B_Q}\left[\frac{F(1, p, q)}{1 + \cos^2\theta_1 + 2B/B_Q}\right]^2 \\
&\simeq w_0 \frac{1}{\alpha_F}\frac{B}{B_Q} 2\sqrt{\frac{2}{\pi}}\,\lambda a_0,
\end{aligned}
\tag{5.6.6}
$$

where $|f(0)|^2 = \lambda(2/\pi)^{1/2}$, $p = 2 + (1 + \cos^2\theta)(B/B_Q)$, $q = -\sin^2\theta(B_Q/2B)$, $\Omega$ is the solid angle into which photon 1 is emitted, and $e_1, e_2$ are the polarization vectors of the two photons. The parameter $\lambda$ is the solution of $\lambda a_0 = \sqrt{2/\pi}\,(\ln[\sqrt{B/B_Q}/\alpha_F \lambda_0] - 1)$, $a_0$ is the Bohr radius, and $w_0 = \alpha_F^5 m_e c^2/2\hbar$ is the field-free decay rate. The two photons are polarized at right angles to each other, and carry off in opposite directions an equal energy equivalent to the electron rest mass minus the binding energy. The total decay rate can be estimated as $w_{2\gamma} \sim 8 \times 10^{12}(B/10^{12}\ \mathrm{G})\ \mathrm{s}^{-1}$.

An interesting possibility, raised by Shabad and Usov (1982, 1984) and modified by Herold, Ruder, and Wunner (1985), Shabad and Usov (1985), and Usov and Shabad (1985), is the adiabatic mixing of photon and positronium states. In this case, the positronium energy levels must be calculated for a finite total perpendicular momentum $\vec{K}_\perp$. For $\vec{K}_\perp = \vec{p}^{\,+} + \vec{p}^{\,-}$ in the $x, z$ plane ($B_z = B$, $B_x = B_y = 0$, $A_x = -B_y$, $A_y = A_z = A_0 = 0$) the distance between the $y$ coordinates of the gyrocenters is $(y_0^- - y_0^+) = (p_x^- + p_x^+)c/eB$, and the discrete positronium energy spectrum is $\varepsilon_{nn'}(n_c, K_\perp)$, labeled by the quantum number $n_c = 0, 1, \ldots$ for each Landau state $n, n'$ of the $e^+e^-$, and $K_\perp^2 = K_x^2 = p_x^+ + p_x^-$. These energies are plotted as the thin, nearly horizontal lines in Figure 5.6.1. For a positronium, the abscissa is the square of the perpendicular total momentum $\hbar^2 K_\perp^2 = p_x^+ + p_x^-$, while the ordinate is $\varepsilon_{nn'}(n_c, K_\perp)$. For a photon related to the positronium states through the energy-momentum conservation laws, the abscissa represents the square of the transverse photon momentum, while the

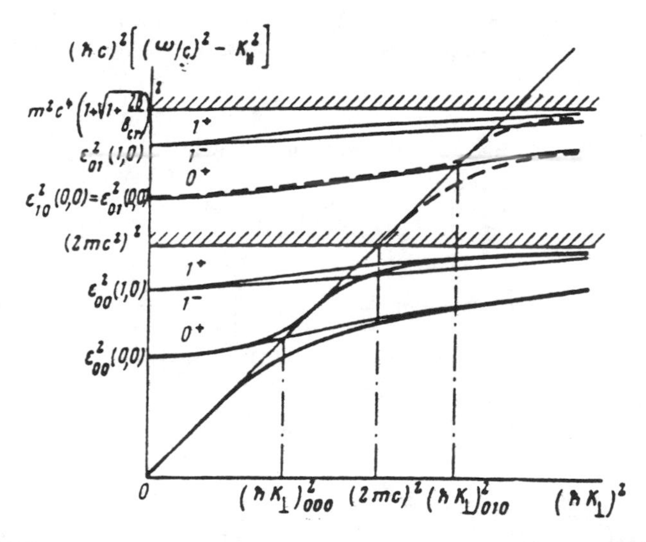


FIGURE 5.6.1 Dispersion curves and energy levels for the photon and positronium and their mixed state. The positronium curves are labeled with $n_c^\pm$, where the suffix $n_c = 0, 1, \ldots$, and the superscript indicates the parity of the state under reflection across the $z$ axis. From V. V. Usov and A. E. Shabad, 1985, *Sov. Phys. JETP Letters*, 42, 19.

ordinate is the square of the photon energy $(\hbar\omega)^2$ minus the square of its longitudinal momentum $(c\hbar K_{\parallel})^2$, which from energy conservation is $c^2\hbar^2(K^2 - K_{\parallel}^2) = \varepsilon_{nn'}^2(n_c, K_{\perp})$. For the lowest spectral state, with $n_c = 0$, one has for each $n, n'$ series the expression (Shabad and Usov, 1985)

$$\begin{aligned} c\hbar\left[(\omega/c)^2 - K_{\parallel}^2\right]^{1/2} &= \epsilon_{nn'}(0, \hbar K_{\perp}) \\ &= (m_n + m_{n'})c^2 \\ &\quad - 2\alpha_F^2 M_{nn'} c^2 \left( \ln \frac{mB'}{2\alpha_F M_{nn'}\sqrt{B' + \eta^2}} \right)^2, \end{aligned} \tag{5.6.7}$$

valid as long as the logarithm is large. Here $B' = B/B_Q$, $\eta = \hbar K_{\perp}/mc$, $M_{nn'} = m_n m_{n'}/(m_n + m_{n'})$, and $m_j = m\sqrt{1 + 2jB'}$ is the mass of the lepton in Landau levels $j = n, n'$. The dashed horizontal lines in Figure 5.6.1 show the continua for $n = n' = 0$ (lower) and $n = n' = 1$ (upper).

The pure photon states and the mixed photon-positronium (photo-positronium) states can also be drawn (Fig. 5.6.1). The dispersion curve for a pure photon (no vacuum effects) is $(\omega/c)^2 - K_{\parallel}^2 = K_{\perp}^2$, drawn as the thin straight diagonal line starting from the origin. As discussed by Shabad and Usov (1985) and Herold, Ruder, and Wunner (1985), the regions where this straight photon meets the even-parity positronium lines and the continua are quasi intersections, near which perturbation theory cannot be used; the photon and positronium states repel each other and reconnect. To investigate this, they use a mixed photon-positronium dispersion relation, $K^2 - K_{\parallel}^2 - K_{\perp}^2 = (\omega/c)^2 - K_{\perp}^2 = \pi_i(\omega, K_{\perp})$. Here $\pi_i(\omega, K_{\perp})$ is an eigenvalue of the polarization operator for the normal mode of polarization $i$, calculated near the quasi intersections as a sum of the pole contributions of the positronium states. For $n = n' = 0$ it is only the $i = 2$ (ordinary, $\vec{e}$ in the $\vec{K}$, $\vec{B}$ plane) polarization photon that interacts with the positronium to form a mixed state. This mixed dispersion relation is given by

$$\begin{aligned} (c\hbar)^2(K^2 - K_{\parallel})_{\pm} = 2m^2c^4 \Bigg\{ E^2 + \frac{\eta^2}{4} \pm \Bigg[ \left( E^2 - \frac{\eta^2}{4} \right)^2 \\ + 4\alpha_F B' E \left( \frac{1 - E}{2} \right)^{1/2} \exp\left( \frac{\eta}{2B'} \right) \Bigg]^{1/2} \Bigg\}, \end{aligned} \tag{5.6.8}$$

where $\eta = \hbar K_\perp/mc$ and $E = \varepsilon_{00}(0, \hbar K_\perp)/2mc^2$. The thick curved lines in Figure 5.6.1 show the two branches of this mode 2 (ordinary) photon dispersion curve. The thick curved dashed lines show the dispersion curves for the mode 1 (extraordinary) photon which interacts only with positronia states of either $n$ or $n' \geq 1$.

The interpretation of the mixed dispersion curves requires some care. Following Herold, Ruder, and Wunner (1985), Usov and Shabad (1985), and Shabad and Usov (1985), in the limit of geometrical optics, a photon emitted tangent to the field ($K_\perp = 0$ initially) would experience (if the field curves, as in a dipole) a gradual change in its $K_\parallel$, $K_\perp$ components. A photon of mode 2, say, would move along the lower mixed branch which starts parallel to the pure (diagonal) photon state, but gradually curve down to the right, undergoing a gradual conversion into a pure positronium state. This is associated with a decrease and eventual vanishing of the longitudinal component of the group velocity vector of the wave packet. The necessary condition for this bending to occur smoothly is tied directly to the validity of the adiabatic passage from the photon to positronium stage, and this is shown to be satisfied provided $B \gtrsim 0.1 B_Q$, in which limit the geometric optics approximation is also valid. Since this transition starts occurring already below the threshold $\hbar K_\perp = 2mc^2$ for one-photon production of free pairs (§ 5.5), because the positronium binding energy must be subtracted, this effect could have an influence upon the effectiveness of free-pair production by $1\gamma$ photo-pair production. On the other hand, free pairs could still be obtained if the positronium thus formed is dissociated by the pulsar electric field or by $2\gamma$ decay or by photodissociation with thermal photons from the star.

# 6. Radiation Transport in Strongly Magnetized Plasmas

## 6.1 The Transport Equation

### *a) The Boltzmann Equation for Photons*

Let us denote by $n(k, r, t)$ the photon occupation number, which is the density of photons in the quantum state $k$, of wave vector $\vec{k}$, at the point $r$ and time $t$. A phenomenological description of the change of $n$ is given by the Boltzmann equation (e.g. Pomraning, 1978),

$$\frac{1}{c}\frac{\partial n}{\partial t}(k, r, t) + \frac{c}{\omega}\vec{k} \cdot \vec{\nabla} n(k, r, t)$$
$$= q(k, r, t)[1 + n(k, r, t)] - \alpha(k, r, t) n(k, r, t)$$
$$+ \int d^3k' \frac{d^3\sigma}{dk^3}(k \leftarrow k', r, t)[1 + n(k, r, t)] n(k, r, t)$$
$$- \int d^3k \frac{d^3\sigma}{dk^3}(k' \leftarrow k, r, t)[1 + n(k', r, t)] n(k, r, t), \tag{6.1.1}$$

which can be derived from Liouville's theorem in the usual way. The factors $[1 + n]$ account for spontaneous plus induced processes, and we have assumed in equation (6.1.1) that the refractive index of the medium $N$ does not depart significantly from unity (e.g. Chaps. 3 and 5), at least for the purposes of solving the transfer equation (6.1.1). That is, we assume that photons continue moving in a straight line with a speed given by $c$ in the geometrical optics limit. Otherwise, $N$ would enter the transport equation explicitly (e.g. Zheleznyakov, 1977; Bekefi, 1966). The $q$, $\alpha$, and $d^3\sigma/dk^3$ are the emission, absorption, and differential scattering coefficient, which depend on the density $\rho_i$ of electrons in the initial state $i$. As a function of the cross sections $\sigma^{\text{abs}}_{f \leftarrow i}$, $\sigma^{\text{sca}}_{f \leftarrow i}$, these are

$$\alpha(k, r, t) = \sum_{i, f} \rho_i(r, t) \sigma^{\text{abs}}_{f \leftarrow i}(k),$$

$$q(k, r, t) = \sum_{i, f} \rho_i(r, t) \sigma^{\text{abs}}_{i \leftarrow f}(k),$$

$$\frac{d^3\sigma}{dk^3}(k, r, t) = \sum_{i, f} \rho_i(r, t) \sigma^{\text{sca}}_{f \leftarrow i}(k \leftarrow k'). \tag{6.1.2}$$

The general equation (6.1.1), including terms for induced radiation processes, is nonlinear in the radiation field and rather complicated to solve. In many circumstances, however, simplifications are possible. For instance, if the distribution of electrons is an equilibrium one, $\rho_\rho(E_i)$, and if $d^3\sigma/dk^3(k \leftarrow k') = d^3\sigma/dk^3(k' \leftarrow k)$, as for instance in coherent scattering, then the terms $d^3\sigma/dk^3(k \leftarrow k')n(k')n(k)$ and $d^3\sigma/dk^3(k' \leftarrow k)n(k)n(k')$ cancel and the transport equation becomes linear. For incoherent scattering ($\omega' \neq \omega$), it is still possible to neglect induced scattering, provided that $n(\omega) \ll 1$. Writing $\alpha' = \alpha - q$ and $\sigma(k) = \int dk'^3\, d^3\sigma/dk^3(k' \leftarrow k)$, in the steady state $\partial n/\partial t = 0$ we obtain then

$$\begin{aligned} \hat{k} \cdot \vec{\nabla} n(k, r, t) = q(k, r) - \alpha'(k, r) n(k, r) \\ + \int d^3k' \frac{d^3\sigma}{dk^3}(k \leftarrow k', r) n(k', r) \\ - \sigma(k, r) n(k, r). \end{aligned} \tag{6.1.3}$$

In equation (6.1.3), the new absorption coefficient $\alpha'$ has been corrected (reduced) by the amount of induced emission, as discussed in Chapter 4. Even if the induced scattering ($\propto n^2$) is neglected, the induced emission ($\propto n$), a larger effect, must be kept in order to ensure that an equilibrium spectrum is obtained, for the case of local thermodynamic equilibrium (LTE). In astrophysics, it is common to use the specific intensity $I$ per unit frequency $\nu$, for propagation direction $\hat{\Omega} = \hat{k}$, instead of the photon occupation number $n(k, r, t)$. They are related by

$$I(\nu, \Omega, r, t) = hc^{-2}\nu^3 n(k, r, t). \tag{6.1.4}$$

Defining then total opacity coefficient $\kappa$ of the material through

$$\kappa = \alpha + \sigma, \tag{6.1.5}$$

the steady-state transport equation becomes

$$\vec{\Omega} \cdot \vec{\nabla} I(\nu, \Omega, r) = \alpha'(\nu, \Omega, r) B_\nu - \kappa(\nu, \Omega, r) I(\nu, \Omega, r) + \int d\nu' \int d\Omega' \frac{d^2\sigma}{d\nu \, d\Omega}(\nu, \Omega \leftarrow \nu', \Omega'; r) \times \frac{\nu}{\nu'} I(\nu', \Omega', r), \tag{6.1.6}$$

where $B_\nu = h\nu^3 c^{-2} q/\alpha'$, which in LTE is equal to the Planck function.

*b) The Transport Equation in Plane and Cylinder Geometrics*

A significant simplification is obtained if we consider media where the plasma properties vary only along a single spatial coordinate. For the magnetized neutron star case, the slab and cylindrical symmetries are the most interesting. For a slab of infinite transversal extent or a cylinder of infinite height, the radiation field can then depend solely on the single spatial dimension $r$. The general transfer problem in these configurations has been studied by Nagel (1982), whose treatment we follow. In the slab, the space variable $r$ goes from $-R$ to $+R$, being 0 at the mid-slab, and in the cylinder it goes from 0 at the axis to $R$ at the surface. The direction of propagation $\vec{\Omega}$ is best measured with respect to the stellar magnetic field direction $\vec{B}_0$. The simplest realistic configuration has $\vec{B}_0$ perpendicular to the slab surface or along the cylinder axis. Calling $\theta = \cos^{-1}(\mu)$ the angle between $\vec{k}$ and $\vec{B}$, and $\phi$ the corresponding azimuthal angle, the transport equation reads

$$\cos\theta \frac{\partial I}{\partial r} = \kappa(S - I) \quad \text{(slab)} \tag{6.1.7a}$$

or

$$\sin\theta \cos\phi \frac{\partial I}{\partial r} - \sin\theta \sin\phi \frac{1}{r} \frac{\partial I}{\partial \phi} = \kappa(S - I) \quad \text{(cylinder)}, \tag{6.1.7b}$$

where the source function $S$ has been defined by

$$\kappa S(\nu, \theta, \phi) = \int d\nu' \int d\Omega' \frac{d^2\sigma}{d\nu \, d\Omega}(\nu, \theta, \phi \leftarrow \nu'\theta'\phi') \times \frac{\nu}{\nu'} I(\nu', \theta', \phi') + \alpha(\nu, \theta) B_\nu. \tag{6.1.8}$$

It is useful to consider the symmetric and antisymmetric averages of $I$

with respect to the exchange $\vec{\Omega} \to -\vec{\Omega}$. We may define the intensities in the outward (+) and inward (−) directions

$$I_+ = I(\nu, \theta, \phi),$$
$$I_- = I(\nu, \pi - \theta, \pi + \phi), \tag{6.1.9}$$

where $0 < \theta < \pi/2$ for slab and $-\pi/2 < \theta < \pi/2$ for cylinder geometries. We have then the symmetric and antisymmetric quantities

$$u = \frac{1}{2}(I_+ + I_-), \tag{6.1.10a}$$

$$v = \frac{1}{2}(I_+ - I_-), \tag{6.1.10b}$$

which are related, respectively, to the radiation density and the net flux. Adding and subtracting equations (6.1.7) for $I_+$ and $I_-$, we obtain the equivalent set of transport equations:

$$\cos\theta \frac{\partial v}{\partial r} = \kappa \frac{1}{2}(S_+ - S_-) - \kappa u,$$
$$\cos\theta \frac{\partial u}{\partial r} = \kappa \frac{1}{2}(S_+ - S_-) - \kappa v \quad \text{(slab)} \tag{6.1.11}$$

and

$$\sin\theta\cos\phi \frac{\partial v}{\partial r} - \sin\theta\sin\phi \frac{1}{r}\frac{\partial v}{\partial r} = \kappa \frac{1}{2}(S_+ + S_-) - \kappa u,$$
$$\sin\theta\cos\phi \frac{\partial u}{\partial r} - \sin\theta\sin\phi \frac{1}{r}\frac{\partial u}{\partial r} = \kappa \frac{1}{2}(S_+ - S_-) - \kappa v$$
$$\text{(cylinder)}. \tag{6.1.12}$$

Further simplification is possible if $S_+(\nu, \theta, \phi) = S_-(\nu, \pi - \theta, \pi + \phi)$. This requires not only that the scattering cross section but also that the emission coefficient be symmetric. The second equations of (6.1.11) and (6.1.12) allow $v$ to be replaced in the first ones, to obtain a second-order equation for $u$,

$$\frac{\cos^2\theta}{\kappa}\frac{\partial^2 u}{\partial r^2} = \kappa(u - S) \quad \text{(slab)}, \tag{6.1.13a}$$

$$\frac{\sin^2\theta}{\kappa}\left[\cos^2\phi \frac{\partial^2 u}{\partial r^2} + \frac{\sin^2\phi}{r}\frac{\partial u}{\partial r} + \frac{\sin 2\phi}{r^2}\frac{\partial u}{\partial \phi} + \frac{\sin^2\phi}{r^2}\frac{\partial^2 u}{\partial \phi^2} - \frac{\sin 2\phi}{r}\frac{\partial^2 u}{\partial r\,\partial \phi}\right] = \kappa(u - S) \quad \text{(cylinder)}. \tag{6.1.13b}$$

These equations can be solved, subject to appropriate boundary conditions specifying $I_+$ and $I_-$. For instance, $I_+ = I_-$, or $v = 0$ at $r = 0$ gives

$$\left.\frac{\partial u}{\partial r}\right|_{r=0} = 0 \quad \text{(slab or cylinder)} \tag{6.1.14}$$

and if there is no radiation incident from outside, $I_- = 0$ at $r = R$, or $u(R) = v(R)$, so that

$$\left.\frac{\cos\theta}{\kappa}\frac{\partial u}{\partial r}\right|_{r=R} + u(R) = 0 \quad \text{(slab)},$$

$$\sin\theta \left.\frac{\cos\phi}{\kappa}\frac{\partial u}{\partial r}\right|_{r=R} - \sin\theta \left.\frac{\sin\phi}{\kappa r}\frac{\partial u}{\partial \phi}\right|_{r=R} + u(R) = 0 \quad \text{(cylinder)}. \tag{6.1.15}$$

The "flux" $v$ at any point $r$ inside the atmosphere follows from $u$, by the second of the equations (6.1.11) and (6.1.12):

$$v = -\frac{\cos\theta}{\kappa}\frac{\partial u}{\partial r} \quad \text{(slab)},$$

$$v = -\sin\theta\frac{\cos\phi}{\kappa}\frac{\partial u}{\partial r} + \sin\theta\frac{\sin\phi}{\kappa r}\frac{\partial u}{\partial \phi} \quad \text{(cylinder)}. \tag{6.1.16}$$

For the cylinder, by symmetry one expects $\partial u/\partial\phi = 0$. It is seen therefore that equations (6.1.16) correspond to Fick's law for diffusion processes (Pomraning, 1978), $v = -D\nabla u$, where $D$ is the diffusion coefficient. The second-order equations (6.1.13) are the diffusion equations for $u$, the photon density quantity. Notice, however, that $u$ and $v$ are functions of $\nu, \theta, \phi$, and so are the opacity $\kappa$ and the scattering coefficients $\sigma$. Therefore, equations (6.1.14) to (6.1.16) contain more information than does what is usually called the "diffusion" or Eddington approximation.

*c) Diffusion Approximation*

The angle and frequency dependence of equations (6.1.7) or (6.1.13) makes it necessary in general to resort to a numerical solution. The gross features of the radiation field, however, can be obtained in a simpler way, by making the problem frequency and angle independent. The first simplification may be achieved by considering some average over frequencies, or in a scattering nonrelativistic atmosphere by assuming that the scattering is approximately coherent, $\omega' \simeq \omega$. In fact, this assumption is compatible with the neglect of stimulated scattering, and strengthens the justification of $S_+ = S_-$ used to derive the second-order equation for $u$. For coherent scattering, these equations can be solved independently for each frequency. A further simplification can be achieved by assuming that, at least in the interior of the medium, the photon density $u(\nu, \theta, \phi)$ is approximately independent of the angle, $u = u(\nu)$. In the case of the slab, we may then average the transport coefficients over $\mu = \cos\theta$. Defining the average $\langle\ \rangle \equiv \int_0^1 d(\cos\theta)$ and taking the $\phi$-averaged cross sections, we get (e.g. Kaminker, Pavlov, and Shibanov, 1982)

$$D_{\parallel} = \left\langle \frac{\cos^2\theta}{\kappa(\theta)} \right\rangle, \quad D_{\perp} \left\langle \frac{\sin^2\theta}{\kappa(\theta)} \right\rangle, \quad A = \langle \alpha(\theta) \rangle,$$

$$S = \langle \sigma(\theta) \rangle, \quad K = \langle \kappa(\theta) \rangle = A + S, \tag{6.1.17}$$

where $D_{\parallel}$ and $D_{\perp}$ are the spatial diffusion coefficients parallel/perpendicular to $\vec{B}_0$ (slab/cylinder case). The angle-averaged transfer equation is

$$D_{\parallel} \frac{d^2 n}{dr^2} = Kn - Sn - A\bar{n} = A(n - \bar{n}) \quad \text{(slab)}, \tag{6.1.18a}$$

$$D_{\perp} \left( \frac{d^2 n}{dr^2} + \frac{1}{r}\frac{dn}{dr} \right) = A(n - \bar{n}) \quad \text{(cylinder)}, \tag{6.1.18b}$$

with boundary condition (after a similar averaging)

$$D_{\parallel, \perp} \left. \frac{dn}{dr} \right|_{r=R} + 2n(R) = 0, \quad \left. \frac{dn}{dr} \right|_{r=0} = 0 \tag{6.1.19}$$

for centrally symmetric, homogeneous boundary conditions (zero incoming intensity). Here $n$ is the photon density at a particular frequency ($\mathrm{cm}^{-3}\ \mathrm{Hz}^{-1}$), which is the angle integral of $u$ divided by the speed of light $c$, and $\bar{n} = (4\pi/h\nu c)B_\nu$ is the thermal equilibrium photon density. The solution of equation (6.1.18a) is (e.g. Nagel, 1982; Rybicki and Lightman, 1979)

$$n(r) = \bar{n}\left[1 - \beta \frac{\cosh(r/\lambda)}{\cosh(R/\lambda)}\right], \tag{6.1.20}$$

where

$$\lambda = (D/A)^{1/2},$$
$$\beta = [1 + (2D/\lambda)\tanh(R/\lambda)]^{-1}. \tag{6.1.21}$$

The photon flux escaping from one side of the slab is

$$F = -cD\frac{dn}{dr}\bigg|_{r=R} = \bar{n}c\beta\frac{D}{\lambda}\tanh\left(\frac{R}{\lambda}\right). \tag{6.1.22}$$

For a thin slab $R \ll \lambda$, $F \to \bar{n}cDR/\lambda^2 = \bar{n}cAR$, which is one-half of the total photons produced per second. For a very thick slab $R \gg \lambda$,

$$F(R \to \infty) \to \bar{n}c(D/\lambda)[1 + (2D/\lambda)]. \tag{6.1.23}$$

Taking the value $D$ as $D \simeq \langle\cos^2\theta\rangle/\langle\kappa\rangle = (3K)^{-1}$, this limiting value of $F$ is below that of a blackbody $F_{bb} = \bar{n}c/4$,

$$\frac{F}{F_{bb}} \simeq \frac{4D}{\lambda} = \frac{4}{\sqrt{3}}\varepsilon^{1/2}, \tag{6.1.24}$$

where $\varepsilon = A/K$. Near the center of the slab, $n \to \bar{n}$, as expected, while at the surface, $n(R) \to (2/\sqrt{3})\sqrt{\varepsilon}\,\bar{n}$. The result that for high slab depths the flux saturates to a maximum value less than the blackbody value is due to the fact that photons originating beyond a certain depth are, after a number of scatterings, reabsorbed without escaping the medium. In fact, the absorption probability of a photon originating at position $r$ is

$$p_a(r) = n(r)/\bar{n} = 1 - \beta\exp[-(R-r)/\lambda] \tag{6.1.25}$$

whereas the corresponding photon escape probability is (Sobolev, 1963)

$$p_e(r) = \beta\exp[-(R-r)/\lambda]. \tag{6.1.26}$$

If photons are created inward and outward with equal probability, at the surface the escape probability must be $p_e(R) = \frac{1}{2}(1 + r^*(R))$, where $r^*$ is the reflectivity of the slab. Defining also the absorptivity $a^* = 1 - r^*$ of the surface, one has therefore

$$r^* = \frac{1 - 2D/\lambda}{1 + 2D/\lambda}, \quad a^* = \frac{4D/\lambda}{1 + 2D/\lambda}. \tag{6.1.27}$$

Since $\varepsilon \sim (D/\lambda)^2 \sim A/K \ll 1$, if scattering dominates, we see that $r^*$ is large and $a^*$ is small. Also, since the limiting luminosity for

$R \to \infty$ is equation (6.1.24), we see that Kirchhoff's relation follows, as it should, $F = a^* F_{\rm bb}$. The thermalization length $\lambda$ acquires meaning by inspecting equation (6.1.26), where one sees that photons can escape only from a depth $(R - r) \lesssim \lambda$. Photons born deeper than that are reabsorbed or thermalized. For $\varepsilon \ll 1$, the photons perform a random walk, suffering on average $N \sim \varepsilon^{-1}$ scatterings between absorptions, each step of average length $\sim D$. The thermalization length is seen to be, then,

$$\lambda \sim D\varepsilon^{-1/2} \sim N^{1/2} D. \tag{6.1.28}$$

This result is fairly accurate for the angle-averaged coherent scattering $\omega' \sim \omega$, but becomes less accurate when the change of $\omega$ or $\mu = \cos\theta$ leads to a change of the step-size $D$ during the photon lifetime.

### *d) The Polarized Transfer Equations*

A description of partially polarized radiation requires the use of a correlation matrix, or density matrix,

$$\rho_{\alpha\beta} \propto \begin{pmatrix} \langle E_x E_x \rangle & \langle E_x E_y \rangle \\ \langle E_y E_x \rangle & \langle E_y E_y \rangle \end{pmatrix}, \tag{6.1.29}$$

where $E_x$, $E_y$ are the components of the electric field along two arbitrary but orthogonal directions, perpendicular to the direction of propagation. A more traditional, but equivalent, description is in terms of the Stokes parameters (Chandrasekhar, 1960) $I, Q, U, V$, given by

$$\rho_{xx} = \frac{1}{2}(I + Q), \quad \rho_{xy} = \frac{1}{2}(U + iV),$$
$$\rho_{yx} = \frac{1}{2}(U - iV), \quad \rho_{yy} = \frac{1}{2}(I - Q). \tag{6.1.30}$$

If the refractive index differs little from unity, the transfer equation for $\rho_{\alpha\beta}(\vec{k})$ is of the form (Gnedin and Pavlov, 1974)

$$(\hat{k} \cdot \vec{\nabla})\rho_{\alpha\beta} = -\frac{1}{2}\sum_{\gamma}\left(T_{\alpha\gamma}\rho_{\alpha\beta} + \rho_{\alpha\gamma}T^{+}_{\gamma\beta}\right) + S_{\alpha\beta}, \tag{6.1.31}$$

where $T_{\alpha\beta}(\vec{k})$ is the transfer matrix. The anti-Hermitian part of $T_{\alpha\beta}$ describes the transition of one polarization to the other, while the Hermitian part describes the extinction and outscattering of radiation. $S_{\alpha\beta}$ is a source term including both inscattering and thermal (or other) emission mechanisms. The inscattering term is an integral depending on all four components of $\rho_{\alpha\beta}(\vec{k})$. The basis vectors used to describe

$\rho_{\alpha\beta}$ and $T_{\alpha\beta}$ are an arbitrary set $e_x, e_y, e_z$. It is, of course, possible to find the eigenvectors of $T_{\alpha\beta}$, $\vec{e}_j$, of components $e_{\alpha j}$, and the eigenvectors $T_j$, with $j = 1, 2$. The index $j = 1$ is the extraordinary normal mode and 2 the ordinary normal mode in the usual description, and the real and imaginary parts of $T_j$ are the coefficients of extinction $\kappa_j$ and refractive indices $\omega N_j / c$ for the $j$th normal wave. The transfer equation can be expressed in the normal wave representation as

$$\left(\hat{k} \cdot \vec{\nabla}\right) R_{jk} = -g_{jk} R_{jk} + S_{jk}, \tag{6.1.32}$$

where

$$\begin{aligned} g_{jk} &= \left(T_j + T_k^*\right)/2 = \left(\kappa_j + \kappa_k\right)/2 + i\omega\left(N_j - N_k\right)/c, \\ S_{jk} &= e_{j\alpha}^{-1} S_{\alpha\beta}\left(u^+\right)_{\beta k}^{-1}, \\ R_{jk} &= u_{j\alpha}^{-1} \rho_{\alpha\beta}\left(u^+\right)_{\beta k}^{-1} \end{aligned} \tag{6.1.33}$$

are the corresponding transfer, scattering, and density matrix in the normal wave representation. The diagonal elements give the normal intensities, $R_{jj} = I_j$. The source matrix $S_{jk}$ still depends, generally, on the quantities $R_{jk}$ so that equations (6.1.31) and (6.1.32) represent a system of four coupled equations, although (6.1.32) is somewhat simpler than (6.1.31), since there is no summation over indices.

The transfer equation (6.1.32) can be rewritten formally as

$$\begin{aligned} R_{jk}(z) &= R_{jk}(z_0)\exp\left[-\int_{z_0}^{z} g_{jk}(z')\, dz'\right] \\ &\quad + \int_{z_0}^{z} S_{jk}(z_1)\exp\left[-\int_{z_1}^{z} g_{jk}(z')\, dz'\right] dz_1, \end{aligned} \tag{6.1.34}$$

where $z$ is along $\vec{k}$ and $z_0$ is some initial value. Looking at the exponential in the second term of (6.1.34), we see that the main contribution comes from those $z$ that satisfy $\mathrm{Re} \int_{z_1}^{z} g_{jk}\, dz' \leq 1$. In the case when

$$\mathrm{Im} \int_{z_1}^{z} g_{jk}\, dz' \gg \mathrm{Re} \int_{z_1}^{z} g_{jk}\, dz' \tag{6.1.35}$$

or for a homogeneous medium when

$$\omega\left(N_j - N_k\right)/c \gg \left(\kappa_j - \kappa_k\right)/2, \tag{6.1.36}$$

the integral over $z_1$ of $g_{jk}$ for $j \neq k$ is much smaller than for $j = k$, because of the rapidly oscillating factor $\exp[i\, \mathrm{Im} \int_{z_1}^{z} g_{jk}\, dz']$. This situation occurs, as seen from equation (6.1.36), when the phase shift between the two normal waves is very large between two scattering events. For circular polarization modes, this is called the phenomenon of large Faraday depolarization.

In the above limit (6.1.36), the transfer equation becomes

$$(\hat{k}\cdot\vec{\nabla})R_{jk} = -g_{jk}R_{jk} + S_{jj}\delta_{jk}. \tag{6.1.37}$$

The off-diagonal elements of $R_{jk}$ in equation (6.1.37) are damped out over a distance $\sim 2/(\kappa_j + \kappa_k)$, whereas the diagonal elements are not, because of the presence of the term $S_{jj}\delta_{jk}$. Consequently, if one is interested in the radiation field from a sufficiently large atmosphere, and the large Faraday depolarization condition (6.1.35) is valid, the transfer equation may be taken to be

$$(\hat{k}\cdot\vec{\nabla})I_j = -\kappa_j I_j + S_{jj}, \tag{6.1.38}$$

where $S_{jj}$ depends only on the diagonal elements of $R_{ij}$, that is, on $I_1$ and $I_2$. There is therefore a coupling between $I_1$ and $I_2$ due to scattering from one mode to the other, but the system of equations is now reduced to two instead of four. It is fortunate that in pulsars, at X-ray frequencies the condition (6.1.36) is usually valid, which makes things rather easier to treat than, say, at radio frequencies (where all four Stokes parameters are coupled). The transport equation for the normal mode intensities, (6.1.38), can be written in full (see eq. 6.1.6) as

$$\begin{aligned}\hat{k}\cdot\vec{\nabla}I_i(\nu,\Omega) &= -\kappa_i(\nu,\Omega,r)I_i(\nu,\Omega) + \alpha_i'(\nu,\Omega,r)B_\nu \\ &\quad + \sum_{j=1,2}\int d\nu'\int d\Omega'\frac{d^2\sigma_{ij}}{d\nu\, d\Omega}(\nu,\Omega\leftarrow\nu',\Omega') \\ &\quad \times\frac{\nu}{\nu'}I_j(\nu',\Omega'),\end{aligned} \tag{6.1.39}$$

where $d\sigma_{ij}(\nu,\Omega\leftarrow\nu',\Omega')/d\nu\, d\Omega$ describes scattering from $(\nu,\vec{\Omega},\vec{e}_j)$ into $(\nu',\vec{\Omega}',\vec{e}_i)$.

## 6.2 Approximate Solutions of the Polarized Transfer Equations

### *a) The Polarized Diffusion Problem*

For an unpolarized radiation field, the diffusion equations for the slab and cylinder obtained in equation (6.1.18) can be written as

$$D\frac{d^2n}{dr^2} + \frac{\gamma}{r}\frac{dn}{dr} = Kn - Sn - A\bar{n}, \tag{6.2.1}$$

where for slabs the constant $\gamma = 0$ and for cylinders $\gamma = 1$, and $D \equiv (1 - \gamma)D_{\parallel} + \gamma D_{\perp}$. These equations are angle-averaged, and assume coherent scattering. There is no difficulty in generalizing equation (6.2.1) to the case of two polarizations (see eq. 6.1.39). In this case, one has two coupled equations

$$D_1\left(\frac{d^2 n_1}{dr^2} + \frac{\gamma}{r}\frac{dn_1}{dr}\right) - K_1 n_1 + S_{11} n_1 + S_{12} n_2 + A_1 \bar{n} = 0,$$

$$D_2\left(\frac{d^2 n_2}{dr^2} + \frac{\gamma}{r}\frac{dn_2}{dr}\right) - K_2 n_2 + S_{21} n_1 + S_{22} n_2 + A_2 \bar{n} = 0, \tag{6.2.2}$$

where for slabs $D_i = D_{i\parallel} = \langle \frac{1}{2}\cos^2\theta / \kappa_i(\theta)\rangle$, and for cylinders $D_i = D_{i\perp} = \langle \frac{1}{2}\sin^2\theta / \kappa_i(\theta)\rangle$. The $S_{ij}$ are coefficients of scattering from polarization $j$ into polarization $i$, averaged over incoming angle and integrated over outgoing angle,

$$S_{ij} \int \frac{d\Omega}{4\pi} \int d\Omega' \frac{d\sigma_{ij}}{d\Omega'}(\Omega' \leftarrow \Omega). \tag{6.2.3}$$

From detailed balance, one has here $S_{12} = S_{21} = S$, and the total opacity satisfies $K_i = A_i + S_{1i} + S_{2i}$, so the system of equations is

$$D_i\left(\frac{d^2 n_i}{dr^2} + \frac{\gamma}{r}\frac{dn_i}{dr}\right) + A_i(\bar{n} - n_i) + S(n_{3-i} - n_i) = 0, \tag{6.2.4}$$

where $i = 1, 2$, with boundary conditions (for the homogeneous, symmetric case) of

$$\left.\frac{dn_i}{dr}\right|_{r=0} = 0, \quad n_i(R) + 2D_i \left.\frac{dn_i}{dr}\right|_{r=R} = 0. \tag{6.2.5}$$

The diffusion equations now contain the coupling term linking $n_1$ and $n_2$. Its effect is to try to equalize the two photon densities, which leads to the obvious equilibrium solution $n_1 = n_2 = \bar{n}$ for high enough optical depths.

The system of equations (6.2.4) can be solved analytically. For slab symmetry in the forward-backward scattering case ($\theta = 0, -\pi$) this was obtained by Mészáros, Nagel, and Ventura (1980), while for the more general two-stream case (arbitrary $\theta, \pi - \theta$ directions) the solutions for slab and cylinder were obtained by Nagel (1980). The homogeneous system of equations obtained from (6.2.4) by dropping the constant term $A_i\bar{n}$ can be solved with the anzats

$$\begin{pmatrix} n_1 \\ n_2 \end{pmatrix} = \begin{pmatrix} f_1 \\ f_2 \end{pmatrix} I(r/\lambda), \tag{6.2.6}$$

where

$$I(x) = \begin{cases} \cosh(x), & \text{(slab);} \\ I_0(x), & \text{(cylinder)} \end{cases} \tag{6.2.7}$$

and $I_0$ is the modified Bessel function of order zero. This solution already satisfies $dn_i/dr\,|_{r=0} = 0$. Replacing (6.2.6) into the inhomogeneous equations (6.2.4), we obtain a set of algebraic equations

$$\begin{aligned} &\frac{D_1}{\lambda^2} f_1 - A_1 f_1 + S(f_2 - f_1) = 0, \\ &\frac{D_2}{\lambda^2} f_2 - A_2 f_2 + S(f_1 - f_2) = 0 \end{aligned} \tag{6.2.8}$$

whose solutions are the eigenvalues $\lambda$ and $\mu$,

$$\begin{pmatrix} \lambda \\ \mu \end{pmatrix} = \left[ \frac{A_1 + S}{2D_1} + \frac{A_2 + S}{2D_2} \mp \left\{ \left( \frac{A_1 + S}{2D_1} - \frac{A_2 + S}{2D_2} \right)^2 + \frac{S^2}{D_1 D_2} \right\}^{1/2} \right]^{-1/2}. \tag{6.2.9}$$

The corresponding eigenvectors are

$$\begin{pmatrix} f_1 \\ f_2 \end{pmatrix} = \begin{pmatrix} S/D_1 \\ \dfrac{A_1+S}{2D_1} - \dfrac{A_2+S}{2D_2} + \left\{ \left( \dfrac{A_1+S}{2D_1} - \dfrac{A_2+S}{2D_2} \right)^2 + \dfrac{S^2}{D_1 D_2} \right\}^{1/2} \end{pmatrix},$$

$$\begin{pmatrix} g_1 \\ g_2 \end{pmatrix} = \begin{pmatrix} \dfrac{A_1+S}{2D_1} - \dfrac{A_2+S}{2D_2} + \left\{ \left( \dfrac{A_1+S}{2D_1} - \dfrac{A_2+S}{2D_2} \right)^2 + \dfrac{S^2}{D_1 D_2} \right\}^{1/2} \\ S/D_2 \end{pmatrix}. \tag{6.2.10}$$

The general solution of the inhomogeneous system (6.2.4) is then (Nagel, 1981a, 1982; Kaminker, Pavlov, and Shibanov, 1982)

$$\begin{pmatrix} n_1 \\ n_2 \end{pmatrix} = \begin{pmatrix} \bar{n} \\ \bar{n} \end{pmatrix} - a \begin{pmatrix} f_1 \\ f_2 \end{pmatrix} \frac{I(r/\lambda)}{I(R/\lambda)} - b \begin{pmatrix} g_1 \\ g_2 \end{pmatrix} \frac{I(r/\mu)}{I(R/\mu)}, \tag{6.2.11}$$

where $a$ and $b$ are constants. These are determined from the boundary conditions to be

$$a = \frac{A_1 - A_2}{A_1 B_2 - A_2 B_1} \bar{n}, \quad b = \frac{B_2 - B_1}{A_1 B_2 - A_2 B_1} \bar{n}, \tag{6.2.12}$$

with

$$\begin{aligned} A_i &= f_i \left[ 1 + \frac{2 D_i}{\lambda} G(R/\lambda) \right], \\ B_i &= g_i \left[ 1 + \frac{2 D_i}{\lambda} G(R/\mu) \right], \\ G(x) &= \begin{cases} \tanh(x), & \text{(slab);} \\ I_1(x)/I_0(x), & \text{(cylinder),} \end{cases} \end{aligned} \tag{6.2.13}$$

where $I_1$ is a modified Bessel function of order 1. An equivalent solution of the magnetic transfer problem for the slab case in the forward-backward scattering limit (Mészáros, Nagel, and Ventura, 1980) leads to the slab case above with $\cos \theta = 1$. An interesting limit of the solution (6.2.11) is at $R \gg (\lambda, \mu)$, where one can use the asymptotic expressions for $I(x)$, which shows that for an optically thick atmosphere the solution tends to

$$\begin{pmatrix} n_1 \\ n_2 \end{pmatrix} \to \begin{pmatrix} \bar{n} \\ \bar{n} \end{pmatrix} - a \begin{pmatrix} f_1 \\ f_2 \end{pmatrix} e^{-(R-r)/\lambda} - b \begin{pmatrix} g_1 \\ g_2 \end{pmatrix} e^{-(R-r)/\lambda}. \tag{6.2.14}$$

That is, deep inside the density tends to the saturation values $\bar{n}$, and it falls off exponentially toward the edges.

The solution (6.2.11) of the polarized diffusion equation exhibits some new features not present in the solution (6.1.20) of the unpolarized problem. This becomes apparent when we consider the two following limits (Mészáros, Nagel, and Ventura, 1980):

i) Weak coupling, $S \ll A_1, A_2$. Clearly, in this case photons have no time to change their polarization between birth and an absorption event. Therefore, the eigenvalues $\lambda$ and $\mu$ are the thermalization lengths defined in (6.1.21) for the individual polarizations,

$$\lambda, \mu \simeq \sqrt{\frac{D_1}{A_1}}, \sqrt{\frac{D_2}{A_2}} \tag{6.2.15}$$

with eigenvectors

$$\begin{pmatrix} f_1 \\ f_2 \end{pmatrix}, \begin{pmatrix} g_1 \\ g_2 \end{pmatrix} \simeq \begin{pmatrix} 1 \\ 0 \end{pmatrix}, \begin{pmatrix} 0 \\ 1 \end{pmatrix}. \tag{6.2.16}$$

ii) Strong coupling, $S \gg A_1, A_2$. Here, a typical photon is scattered many times from one polarization into the other before it is absorbed (or escapes). One expects therefore that the photon has been in each polarization state about an equal number of times, independently of what its original polarization was, and therefore the penetration distance should not depend on its initial state. In this case the eigenvalue $\lambda$ determines the thermalization length for both photons:

$$\lambda \simeq \left[\frac{D_1 + D_2}{A_1 + A_2}\right]^{1/2}, \quad \begin{pmatrix} f_1 \\ f_2 \end{pmatrix} \simeq \begin{pmatrix} 1/2 \\ 1/2 \end{pmatrix}. \tag{6.2.17}$$

This case is illustrated in Figure 6.2.1. The parameter $\lambda$ determines the length scale over which the total photon density varies. The other eigenvalue $\mu$ quantifies the difference in the concentration of photons

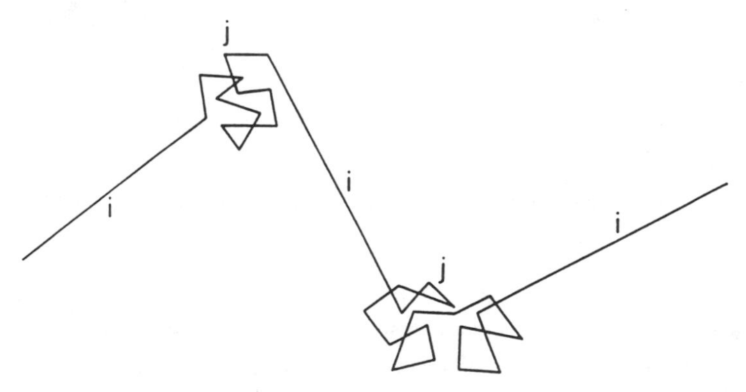


FIGURE 6.2.1 Photon random walk with changes of the polarization between states $i$ and $j$.

according to their polarization, being given by

$$\mu \simeq \left[ \frac{D_1 D_2}{S(D_1 + D_2)} \right]^{1/2}, \quad \begin{pmatrix} g_1 \\ g_2 \end{pmatrix} \simeq \begin{pmatrix} D_2/(D_1 + D_2) \\ -D_1/(D_1 + D_2) \end{pmatrix}. \tag{6.2.18}$$

One can interpret $\mu$ as a mixing length (see Ventura, Nagel, and Mészáros, 1979; Mészáros, Nagel, and Ventura, 1980; and Bussard and Lamb, 1982) within which distance the two photon densities tend to become equal.

The two limits of weak and strong coupling are shown in Figure 6.2.2, adapted from Nagel (1981a), for a set of arbitrary coefficients $K_1 = 10^4\ \text{cm}^{-1}$, $K_2 = 1\ \text{cm}^{-1}$, $\varepsilon = A/K = 10^{-6}$, $D = (3K)^{-1}$, and either $S = 0$ (weak) or $S = 0.5\ \text{cm}^{-1}$ (strong). In the weak-coupling case, each polarization behaves as if it were oblivious to the other (no mixing), and the photon density $n_i$ saturates to $\bar{n}$ at depths larger than

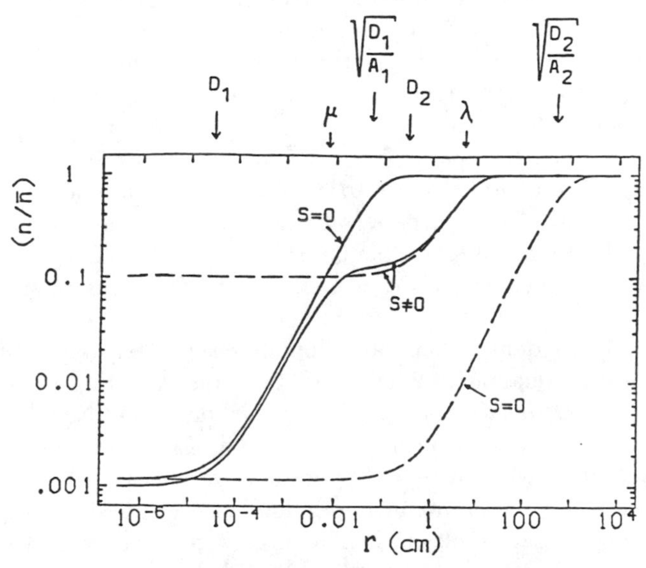


FIGURE 6.2.2 Photon density as a function of depth given by the analytic solution (6.2.11). *Full lines*: polarization 1; *dashed lines*: polarization 2. The curves with $S = 0$ and $S \neq 0$ compare the effect of not including and including the polarization exchange scattering. From W. Nagel, 1982, Ph.D. thesis, University of Munich.

$\lambda_i = \sqrt{D_i / A_i}$ while at lower depths it decays exponentially, until the atmosphere becomes transparent at a depth $\sim D_i$, from then on remaining constant (free flight, for the plane case). The escaping flux is unpolarized in the weak-coupling limit. In the strong-coupling limit, the effective thermalization length $\lambda$ becomes intermediate between the two weak-coupling thermalization lengths, since it is determined simultaneously by the large mean free path of mode 1 and the high absorption coefficient of mode 2, $\lambda \sim [(D_1 + D_2)/(A_1 + A_2)]^{1/2}$. The longer mean free path mode becomes transparent at depth $D_2$, so these photons tend toward a constant, free flight density, but they are still coupled to mode 1 until the depth $\mu \sim [D_1 D_2 / S(D_1 + D_2)]^{1/2}$, where the modes decouple. At this point, the photons 2 escape freely with constant density, while photons 1 continue to diffuse with diminishing density, until eventually they also become transparent at $D_1$ and escape. The result is that the emerging flux is highly polarized, unlike in the weakly coupled case. In the case $S = 0$, the photons 1 could only diffuse out slowly, being retained until the depth $\sqrt{D_1 / A_1}$ due to their shorter mean free path, at which point they had reached the same density as the photons 2. Instead, in the $S \neq 0$ case, the photons 1 have the option of scattering into the long mean free path mode 2, at a deeper level, where their density is higher. It is this that allows these photons to reach a flux level much higher than the $\sqrt{\varepsilon}$ limit present for the weak case. One can also interpret the higher flux of the long mean free path photons 2 as being due to a decreased reflectivity of the atmosphere to this mode. Such photons, when incident upon the surface, penetrate deep into the medium and are converted by polarization-exchange scattering into short mean free path photons, which are absorbed before they can reemerge from the medium.

The photon density can be calculated separately for every frequency, using equation (6.2.11), and the emergent flux is obtained from $F = -cD(dn/dr)_{r=R}$, as has been done by Nagel (1980), Mészáros, Nagel, and Ventura (1980), and Kaminker, Pavlov, and Shibanov (1982), who presented spectra in various interesting limits. These results are important in understanding the physical properties of scattering magnetic plasmas. The use of the coherent scattering approximation restricts the range of usefulness of the spectra thus obtained to the continuum region outside the Doppler width of cyclotron resonances. We shall return to a fuller discussion of the spectrum after we have considered the more realistic incoherent scattering transfer problem. It turns out that this is also intimately linked to a treatment of the angular dependence of the radiation.

*b) Approximate Angular Dependence in the Coherent Scattering Case*

In the diffusion approximation, one obtains an angle-averaged photon density as a function of depth. In a second approximation, it is possible to deduce a rough angular distribution of the escaping photons by a method similar to that originally developed by Eddington, Barbier, and others (e.g. Sobolev, 1963). The essence of the method can be illustrated by means of a simplified example (Nagel, 1982). Consider an unpolarized, pure scattering atmosphere, where the angle-averaged photon density is calculated using the simplified diffusion equation (6.1.18),

$$D\frac{d^2n}{dr^2} = 0. \tag{6.2.19}$$

At the surface, the boundary condition is given by the first of equations (6.1.19), namely $2D(dn/dr) + n = 0$ at $r = R$, showing that the photon density as a function of depth $t$ is

$$n(t) = \text{const.}(2D + t); \qquad t = R - r. \tag{6.2.20}$$

This isotropic photon density at depth $t$ is rescattered (with isotropic scattering coefficient $\sigma$), giving rise to a source function $q$ [$\text{cm}^{-3}\ \text{s}^{-1}\ \text{ster}^{-1}$] of the form

$$q = (c/4\pi)\sigma n. \tag{6.2.21}$$

These photons have a probability $p_e(t)$ of escaping,

$$p_e(t) = e^{-\sigma t/\cos\theta}, \tag{6.2.22}$$

which, when multiplied by the source function (6.2.21), gives the (approximate) flux of photons escaping in direction $\theta$,

$$F(\theta) = \frac{c}{4\pi}\sigma\int_0^\infty dt n(t) e^{-\sigma t/\cos\theta}\ \text{cm}^{-2}\ \text{s}^{-1}\ \text{ster}^{-1}. \tag{6.2.23}$$

For the linear dependence of $n(t)$, this integral is well known, while for more general $n(t)$, a one-point Gauss-Laguerre integration can be used. For the linear case, even this one-point integration gives the exact result

$$\int_0^\infty dt \frac{\sigma}{\cos\theta} n(t) e^{-\sigma t/\cos\theta} = n(t^*), \tag{6.2.24}$$

where $t^*$ is the mean escape depth

$$t^* = \frac{\int dtp_e(t)t}{\int dtp_e(t)} = \frac{\cos\theta}{\sigma}. \tag{6.2.25}$$

The intensity $I(\theta) = F(\theta)/\cos\theta$ is given by the Eddington-Barbier relation,

$$I(\theta) = \frac{c}{4\pi}n(t^*) = \text{const.}(2D + t^*). \tag{6.2.26}$$

This method was first applied to a two-polarization magnetized atmosphere by Kanno (1975), who used a polarization-averaged diffusion coefficient $D = \frac{1}{2}(D_1^{\|} + D_2^{\|})$ and a penetration depth $t^* = \frac{1}{2}\cos\theta[\sigma_1^{-1}(\theta) + \sigma_2^{-1}(\theta)]$, whereby the angular dependence of the penetration depth is no longer given by the simple $\cos\theta$ proportionality.

A discussion of the approximations involved in this type of treatment is useful in understanding the overall characteristics of the problem (e.g. Nagel, 1981a). The approximation relies on the use of two simplifications. First, one extracts a mean intensity $\langle I\rangle$ from the real scattering source term,

$$\int d\Omega_0 \frac{d\sigma}{d\Omega}(\Omega \leftarrow \Omega_0)I(\Omega_0) \rightarrow \int d\Omega_0 \frac{d\sigma}{d\Omega}(\Omega \leftarrow \Omega_0)\langle I\rangle = \sigma(\theta)\langle I\rangle, \tag{6.2.27}$$

where the last step uses the symmetry of the coherent scattering cross section, namely $d\sigma/d\Omega(\Omega \leftarrow \Omega_0) = d\sigma/d\Omega(\Omega_0 \leftarrow \Omega)$. Second, the average intensity $\langle I\rangle$, which is $c/4\pi$ times the photon density, is replaced by its approximate value $n$, as deduced from the diffusion approximation. The first approximation (6.2.27) is exact for the case of isotropic scattering, which is a special case of complete redistribution, whereby the scattering coefficient is of the form

$$\frac{d\sigma}{d\Omega}(\theta \leftarrow \theta_0) = \frac{1}{4\pi\langle\sigma\rangle}\sigma(\theta)\sigma(\theta_0). \tag{6.2.28}$$

When the scattering cross section has the more general form (6.1.8), the photon forgets upon scattering the direction from which it came.

In this case, a variant of equation (6.2.27) is exactly valid,

$$\int d\Omega_0 \frac{d\sigma}{d\Omega}(\theta \leftarrow \theta_0) I(\theta_0) = \sigma(\theta) \int \frac{d\Omega_0}{4\pi} \frac{\sigma(\theta_0)}{\langle\sigma\rangle} I(\theta_0), \quad (6.2.29)$$

so that one can solve the problem by calculating the depth dependence of the average $J = \langle\sigma I\rangle / \langle\sigma\rangle$. In fact, for $\omega \ll \omega_c$ the magnetic scattering cross section is of the form

$$\frac{d\sigma}{d\Omega}(\theta' \leftarrow \theta) \simeq r_0^2 |e_z'(\theta')|^2 |e_z(\theta)|^2, \quad (6.2.30)$$

which is of the type (6.2.28). This method was used by Basko (1976) to calculate an approximate angular dependence in X-ray pulsars. At higher frequencies, the complete redistribution property of the cross sections is less accurate but still useful for qualitative estimates. The second approximation, using the photon density computed from the diffusion equation, instead of the angle average of the exact intensity, is more difficult to justify but leads to a rather direct calculation of the angular dependence by means of a simple integration. In the case of a finite slab, one needs only to use min[$t^*$, $R$], instead of $t^*$, in equation (6.2.26).

The extension of this method to two polarizations is due to Nagel (1981a) and Kaminker, Pavlov, and Shibanov (1982). Including thermal emission, one has the source functions

$$\frac{4\pi}{c} q_1 = \sigma_{11}(\theta) n_1 + \sigma_{12}(\theta) n_2 + \alpha_1(\theta) \bar{n},$$

$$\frac{4\pi}{c} q_2 = \sigma_{21}(\theta) n_1 + \sigma_{22}(\theta) n_2 + \alpha_2(\theta) \bar{n}, \quad (6.2.31)$$

where from symmetry ($\omega' = \omega$) one has

$$\sigma_{ij}(\theta) = \int d\Omega_0 \frac{d\sigma_{i \leftarrow j}}{d\Omega}(\theta \leftarrow \theta_0) = \int d\Omega_0 \frac{d\sigma_{j \leftarrow i}}{d\Omega}(\theta_0 \leftarrow \theta), \quad (6.2.32)$$

$\sigma_{ij}$ being the coefficient of scattering of $j$ into $i$, and $\sigma_i(\theta) = \sum_{j=1,2} \sigma_{ij}(\theta)$ being the total scattering coefficient for polarization $i$. For an isothermal medium, the thermal emission terms $\alpha_i(\theta)\bar{n}$ are independent of depth, whereas the scattering source terms $\sigma_{ij}(\theta) n_j(\theta)$ depend on depth through the $n_j$. The total emission of photons $j$ is given by

$$\int dVq_j(r,\theta)\,p_j(r,\theta) \quad (\text{ph s}^{-1}\ \text{ster}^{-1}), \tag{6.2.33}$$

evaluated over the whole emitting volume. For a slab of surface $A$ we have $dV = A\,dr$, while for a cylinder of length $h$ we have $dV = h2\pi r\,dr$. Dividing this by the observer projection of the emitting surface, $A\cos\theta$ (slab) or $2Rh\sin\theta$ (cylinder), one obtains the intensity $I_j(\theta)$. The escape probability, which in general is different for the two polarizations, is

$$p_j(r,\theta) = \exp\left[-\kappa_j(\theta)\frac{(R-r)}{\cos\theta}\right] \quad (\text{slab}),$$

$$p_j(r,\theta) = \int_0^{2\pi}\frac{d\phi}{2\pi}\exp\left[-\kappa_j(\theta)\frac{\sqrt{R^2 - r^2\sin\phi} - r\cos\phi}{\sin\theta}\right]$$

$$(\text{cylinder}), \tag{6.2.34}$$

where it is assumed for the second that the emission is uniform in $\phi$. With the explicit solution (6.2.11) for the polarized photon densities $n_j$, we obtain for the intensity normalized to the blackbody value, in the slab case,

$$\frac{I_i(\theta)}{B} = \int_{-R}^{R}\frac{dr}{\cos\theta}e^{-\kappa_i([R-r]/\cos\theta)}\left[\alpha_i + \sigma_{i1}n_1(r)/\bar{n} + \sigma_{i2}n_2(r)/\bar{n}\right]$$

$$= 1 - e^{-2\kappa_i R/\cos\theta} - (\sigma_{i1}f_1 + \sigma_{i2}f_2)\left[aE_i(\lambda)/(1 + e^{-2R/\lambda})\right]$$

$$-(\sigma_{i1}g_1 - \sigma_{i2}g_2)\left[bE_i(\mu)/(1 + e^{-2R/\mu})\right], \tag{6.2.35}$$

where

$$E_i(\lambda) = \int_{-R}^{R}\frac{dr}{\cos\theta}e^{-\kappa(R-r)/\cos\theta}\left(e^{-(R-r)/\lambda} + e^{-(R+r)/\lambda}\right)$$

$$= \frac{1 - e^{-2R/\lambda - 2\kappa_i R/\cos\theta}}{\kappa_i + \cos\theta/\lambda} + \frac{e^{-2R/\lambda} - e^{-2\kappa_i R/\cos\theta}}{\kappa_i - \cos\theta/\lambda}. \tag{6.2.36}$$

For the cylinder, the corresponding expression is best integrated numerically,

$$\frac{I_i(\theta)}{B} = \int_0^R \frac{\pi\, dr}{\sin\theta} \frac{r}{R} p_i(r,\theta)\left[\alpha_i + \sigma_{i1} n_1(r)/\bar{n} + \sigma_{i2} n_2(r)/\bar{n}\right]. \tag{6.2.37}$$

The angular dependence thus obtained will be discussed later, when comparing the results with those of more detailed numerical calculations.

An interesting extension of this approximate method of calculating the angular distribution is due to Kaminker, Pavlov, and Shibanov (1982), who relaxed the assumption of having the magnetic field either perpendicular or parallel to the surface. The diffusion coefficient is then a tensor of $zz$ component perpendicular to the surface given by

$$D_j = D_{j\parallel}\cos^2\theta_B + D_{j\perp}\sin^2\theta_B, \tag{6.2.38}$$

where $\theta_B$ is the angle between the normal and the magnetic field, and

$$D_{j\parallel} = \int_0^{\pi/2} \cos^2\theta \sin\theta\left[\kappa_j(\theta)\right]^{-1} d\theta,$$

$$D_{j\perp} = \frac{1}{2}\int_0^{\pi/2} \sin^3\theta\left[\kappa_j(\theta)\right]^{-1} d\theta. \tag{6.2.39}$$

This approach is useful for studying nondipole magnetic field configurations, or off-center dipoles, which give rise to asymmetric emission patterns. Since asymmetric pulse profiles are not uncommon in X-ray pulsars, this type of calculation is of some interest. Another approximate method, of use in large optical depth media, has been discussed by Silant'ev (1981).

## 6.3 Numerical Treatments of the Transport Equation

### *a) Introduction*

Of particular interest among the various methods for treating more accurately the radiation transport problem are those methods which allow for frequency redistribution and angular anisotropies (see e.g. Pomraning, 1978; Kalkofen, 1987). The magnetic radiation transport problem using all four Stokes parameters (Chandrasekhar, 1960; Sobolev, 1963; Kalkofen, 1987) is in general needed for treating the radio emission of rotation-powered pulsars. However, in the case of high-energy radiation in magnetized neutron stars (i.e. $\hbar\omega \gtrsim$ keV), the large Faraday depolarization limit is approximately valid, and the transfer approach with two normal modes described above has been used. A numerical method frequently used is the discrete ordinate method (Nagel, 1981a, b; Mészáros and Nagel, 1985a, b; Burnard, Klein, and Arons, 1988, 1990; Alexander, Mészáros, and Bussard, 1989; Alexander and Mészáros, 1989, 1991b). Another frequently used method is the Monte Carlo simulation (Yahel, 1980; Pravdo and Bussard, 1981; Wang, Wasserman, and Salpeter, 1988, 1989; Wang *et al.*, 1989). A method based on considering the diffusion in frequency (energy) space as well as in physical space has been employed by Bonazzola, Heyvaerts, and Puget (1979), Wang and Frank (1981), Bonazzola (1982), Melrose and Zheleznyakov (1981), Melrose and Padden (1986), and Lyubarsky (1986). The escape probability method, which is semianalytic, has been used by Kanno (1975), Kaminker, Pavlov, and Shibanov (1982), and Kii *et al.* (1986). There are various advantages and disadvantages to these different methods, which has led in some cases to the use of mixed approaches (Burnard, Klein, and Arons, 1988; Brainerd and Mészáros, 1991). In what follows we consider mainly the results obtained by the first of these methods, which provides a compromise between optical depth, angle, and frequency resolution (including resonances) on the one hand, and computation time on the other. The discrete ordinate method builds upon an expansion scheme used by Chandrasekhar (1960) in which the full integro-differential transport equation is transformed into a system of coupled differential equations. This is achieved by representing the full radiation field through the values it adopts for some particular discrete directions, frequencies, and polarizations. The integral for the scattering source term is then replaced by a quadrature sum, which is responsible for the simplification.

### *b) The Discrete Ordinate Method*

The simplest application of this method is to the more restricted problem where the scattering is still assumed to be coherent, but unlike in the diffusion approximation, the radiation field is calculated at each depth for a discrete set of angles (and polarizations). The source term

$$S(\Omega) = \frac{1}{\kappa(\Omega)} \int d\Omega_0 \frac{d\sigma}{d\Omega}(\Omega \leftarrow \Omega_0) I(\Omega) + \frac{\alpha(\Omega)}{\kappa(\Omega)} B \qquad (6.3.1)$$

is replaced by the quadrature sum

$$S(\Omega_i) \approx \frac{1}{\kappa_i} \sum_j w_j \frac{d\sigma}{d\Omega}(\Omega_i \leftarrow \Omega_j) I(\Omega_j) + \frac{\alpha(\Omega_i)}{\kappa(\Omega_i)} B, \tag{6.3.2}$$

where $w_i$ is an appropriate weight factor. Usually, the simplest type of quadrature is the Gaussian (e.g. Abramowitz and Stegun, 1964). The indices $i$ and $j$ denote one of the discrete channels (ordinates), given by one of the discrete angle integration points, for a particular polarization. The accuracy, of course, increases with the number of discrete integration points. Applying this method to the symmetrized transfer equations (6.1.13) for the slab, one obtains the set of equations

$$D_i \frac{d^2 u_i}{dr^2} - K_i u_i + \sum_{j=1}^{N} S_{ij} u_j + A\bar{u} = 0 \tag{6.3.3}$$

for $i = 1, \ldots, N$, where $N/2$ is the number of angle integration points and two polarizations are considered. This is a generalization of the diffusion equations (6.2.1), allowing now for scattering between different directions of propagation. The diffusion, absorption, and scattering coefficients are now

$$D_i = \cos^2\theta_i / \kappa_i(\theta_i),$$
$$A_i = \alpha_i(\theta_i),$$
$$K_i = \kappa_i(\theta_i) = \alpha(\theta_i) + \sigma(\theta_i). \tag{6.3.4}$$

The $N$ equations are coupled by the redistribution matrix $S$, whose elements $S_{ij}$ represent the probability of scattering from channel $j$ into channel $i$,

$$S_{ij} = 4\pi w_j \frac{d\sigma}{d\Omega}^{(i\leftarrow j)} (\theta_i \leftarrow \theta_j), \tag{6.3.5}$$

and the $w_j$ are Gaussian weights for the interval [0, 1] of the variable $\cos\theta$. The reason one can use this interval ($0 < \theta < \pi/2$) rather than $[-1, 1]$ is that the transfer equation (6.1.13) for $n$ was derived assuming the symmetry property $S_+ = S_-$. For the slab, the transfer equation is independent of $\phi$, and one can use the $\phi$-averaged coherent scattering cross sections, since by symmetry the radiation field cannot depend on $\phi$.

For the cylinder case, the transfer equation (6.1.3b) has $\phi$-derivatives, and one expects a limb-darkening effect between $\phi = 0$ and $\phi = \pi/2$. Since, however, the opacities are independent of $\phi$, this effect is not expected to be large compared to the $\theta$-variations. The $\phi$-dependence of the cylinder equation (6.1.3b) can be suppressed by ignoring the $\phi$-derivatives and taking for $\sin \phi$ its mean value $1/\sqrt{2}$, leading to the approximate cylinder equations (Nagel, 1981a)

$$D_i \left( \frac{d^2 u_i}{dr^2} + \frac{1}{r} \frac{du_i}{dr} \right) - K_i u_i + \sum_{j=1}^{N} S_{ij} u_j + A_i \bar{u} = 0 \tag{6.3.6}$$

for $i = 1, \ldots, N$, with the diffusion coefficient

$$D_i = \frac{1}{2} \sin^2\theta_i / \kappa_i(\theta_i). \tag{6.3.7}$$

The boundary conditions for the slab are given by (6.1.14), (6.1.15), being $\phi$-independent. For the cylinder, one can use the same treatment of the $\phi$-variable as for equation (6.3.6). The boundary conditions are therefore

$$\left. \frac{du_i(\theta_i)}{dr} \right|_{r=0} = 0, \tag{6.3.8}$$

$$\left. \frac{\cos \theta_i}{\kappa_i(\theta_i)} \frac{du_i(\theta_i)}{dr} \right|_{r=R} + u_i(\theta_i) \,|_{r=R} = 0 \quad (\text{slab}), \tag{6.3.9a}$$

$$\left. \frac{1}{\sqrt{2}} \frac{\sin \theta_i}{\kappa_i(\theta_i)} \frac{du_i(\theta_i)}{dr} \right|_{r=R} + u_i(\theta_i) \,|_{r=R} = 0 \quad (\text{cylinder}), \tag{6.3.9b}$$

for $i = 1, \ldots, N$.

In principle, the system of $N$ coupled differential equations (6.3.3) or (6.3.6) could be solved by the same method used for the system of two diffusion equations (6.2.2), that is, by calculating the eigenvalues and eigenvectors of the problem and by choosing the constants of the linear combination of eigensolutions so as to satisfy the boundary conditions. For a calculation of, say, 2 polarizations and 16 angles, the eigenvalues and eigenvectors of the resulting $32 \times 32$ matrix are rather unwieldy for writing down explicitly, and a numerical evaluation of the eigenvalue problem becomes preferable. Once a numerical solution is envisaged, however, there are other, alternative methods which are more flexible, e.g. in relaxing the assumption of isothermality or uniform density. For this, except in the simplest cases, one usually has to treat the differential equations themselves numerically instead of just the eigenvalue equations resulting from an analytical solution.

*c) Solution by Feautrier's Method*

This method, of wide application in stellar atmospheres (Feautrier, 1964; Mihalas, 1978), consists of solving by finite differences the set of coupled differential equations (6.3.3) or (6.3.6). These have the form

$$D_i\left(\frac{d^2u_i}{dr^2}+\frac{\gamma}{r}\frac{du_i}{dr}\right)-K_iu_i+\sum_{j=1}^{N}S_{ij}u_j+A_i\bar{u}=0 \qquad (6.3.10)$$

with $\gamma = 0, 1$ for slab or cylinder. The idea is to consider the variable $u$ as an $N$-dimensional vector, $\vec{u} = (u_1, \ldots, u_N)^\dagger$, and to write the space derivatives in terms of difference operators, e.g.

$$\left(\frac{d^2\vec{u}}{dr^2}+\gamma\frac{d\vec{u}}{dr}\right)\approx\frac{2}{r_{d+1}-r_{d-1}}\left(\frac{\vec{u}_{d+1}-\vec{u}_d}{r_{d+1}-r_d}-\frac{\vec{u}_d-\vec{u}_{d-1}}{r_d-r_{d-1}}\right.$$
$$\left.+\gamma\frac{\vec{u}_{d+1}-\vec{u}_{d-1}}{2r_d}\right), \qquad (6.3.11)$$

where $d = 0, 1, \ldots, D$ represents the finite number of depth points in the atmosphere. The transfer equation (6.3.10) takes the general form

$$-\mathscr{A}_d\cdot\vec{u}_{d-1}+\mathscr{B}_d\cdot\vec{u}_d-\mathscr{C}_d\cdot\vec{u}_{d+1}=\mathscr{Q}_d, \qquad (6.3.12)$$

involving the matrices

$$\mathscr{A}_d=\frac{2}{r_{d-1}-r_{d+1}}\left(\frac{1}{r_{d-1}-r_d}+\frac{\gamma}{2r_d}\right)\operatorname{diag}\mathscr{D}_i,$$

$$\mathscr{B}_d=\frac{2}{r_{d-1}-r_{d+1}}\left(\frac{1}{r_d-r_{d+1}}+\frac{1}{r_{d-1}-r_d}\right)\operatorname{diag}\mathscr{D}_i$$
$$+\operatorname{diag}\mathscr{K}_i-\mathscr{S}_{ij},$$

$$\mathscr{C}_d=\frac{2}{r_{d-1}-r_{d+1}}\left(\frac{1}{r_d-r_{d+1}}-\frac{\gamma}{2r_d}\right)\operatorname{diag}\mathscr{D}_i,$$

$$\mathscr{Q}_d=\operatorname{diag}\mathscr{A}_i, \qquad (6.3.13)$$

where $\operatorname{diag}\mathscr{A}_i$ stands for a matrix which has only diagonal components. At the inner boundary $r = 0$, we can use an appropriate boundary condition—for instance, equation (6.3.8)—to link $\vec{u}_{D-1}$ with $\vec{u}_D = \vec{u}(0)$, which leads to

$$-\mathscr{A}_D \cdot \vec{u}_{D-1} + \mathscr{B}_D \cdot u_D = \mathscr{Q}_D \tag{6.3.14}$$

with

$$\mathscr{A}_D = \frac{2(1+\gamma)}{r_{D-1}^2} \operatorname{diag} \mathscr{D}_i,$$
$$\mathscr{B}_D = \frac{2(1+\gamma)}{r_{D-1}^2} \operatorname{diag} \mathscr{D}_i + \operatorname{diag} \mathscr{K}_i - \mathscr{S}_{ij}. \tag{6.3.15}$$

At the surface, the boundary conditions could be (6.3.9), linking $\vec{u}_0 = \vec{u}(R)$ with $\vec{u}_1$. Because the depth points need to be chosen closer near the surface, it is useful to use a combination of (6.3.9) and a series expansion to second order of $u$ (see Mihalas, 1978) for greater accuracy. The surface condition (no outside incident radiation) is then

$$\mathscr{B}_0 \cdot \vec{u}_0 - \mathscr{C}_0 \cdot \vec{u}_1 = \mathscr{Q}_0 \tag{6.3.16}$$

with

$$\mathscr{B}_0 = \left(\frac{2}{r_0 - r_1} + \frac{\gamma}{R}\right) \operatorname{diag} \beta_i + \frac{2}{(r_0 - r_1)^2} \operatorname{diag} \mathscr{D}_i$$
$$+ \operatorname{diag} \mathscr{K}_i - \mathscr{S}_{ij},$$
$$\mathscr{C}_0 = \frac{2}{(r_0 - r_1)^2} \operatorname{diag} \mathscr{D}_i. \tag{6.3.17}$$

In equations (6.3.13)–(6.3.17) we have used the definitions

$$D_i = \beta_i^2 / K_i,$$
$$\beta_i = (1-\gamma)\cos\theta_i + \gamma 2^{-1/2} \sin\theta_i. \tag{6.3.18}$$

The solution of the transport equations consist therefore in the inversion of the following matrix equation:

$$\begin{pmatrix} \mathscr{B}_0 & -\mathscr{C}_0 & \cdots & \cdots & 0 \\ -\mathscr{A}_1 & \mathscr{B}_1 & -\mathscr{C}_1 & \cdots & 0 \\ \vdots & \vdots & \vdots & \vdots & \vdots \\ 0 & \cdots & -\mathscr{A}_{D-1} & \mathscr{B}_{D-1} & -\mathscr{C}_{D-1} \\ 0 & \cdots & \cdots & -\mathscr{A}_D & \mathscr{B}_D \end{pmatrix} \begin{pmatrix} \vec{u}_0 \\ \vec{u}_1 \\ \vdots \\ \vec{u}_{D-1} \\ \vec{u}_D \end{pmatrix} = \begin{pmatrix} \mathscr{Q}_0 \\ \mathscr{Q}_1 \\ \vdots \\ \mathscr{Q}_{D-1} \\ \mathscr{Q}_D \end{pmatrix}. \tag{6.3.19}$$

This is achieved by successive inversions of matrices, starting from $d = 0$ down to $d = D$, and by successive substitutions for the radiation field $u_d$, starting back from $d = D$ to $d = 0$ (Mihalas, 1978). The computation time, for large $N$, is proportional to $N^3$. On the other hand, due to the band structure of the matrix equations (6.3.19), the time increases only linearly with $D$, the number of depth points. These need to be chosen closer together than the largest diffusion coefficients near the surface, but deep inside, where the radiation field varies slowly, they can be chosen much farther apart than the diffusion coefficients. A logarithmic grid in $r$ is usually appropriate, with the smallest step near the surface set equal to $10^{-2} \times \min[D_i]$, $i = 1, \ldots, N$. As a result, atmospheres of arbitrarily high optical depth are only slightly more time consuming than low-depth ones. This is a definite advantage over some other numerical methods, such as Monte Carlo or $\Lambda$-iterations.

*d) Directionality in the Coherent Scattering Limit*

As the simplest example, it is useful to consider the homogeneous, isothermal infinite slab and cylinder models with the previous homogeneous boundary conditions, as was done in the more approximate treatment of § 6.2*b*. One example of such a calculation (Nagel, 1981a) uses 16 angles ($N = 32$) for homogeneous atmospheres of $\hbar\omega_c = 50$ keV, $kT = 10$ keV, $\rho = 0.1$ g cm$^{-3}$, neglecting for simplicity vacuum effects, and the angle dependence of the intensity was calculated for a number of frequencies in the coherent scattering limit. It is interesting to compare results for $I_1(\theta)$, $I_2(\theta)$ obtained by the Feautrier method with those from the approximate method of § 6.2*b* at large and small optical depths for a slab (Figs. 6.3.1b and c). The penetration depths $\cos\theta/\kappa_i(\theta)$ are also shown in Figure 6.3.1a. It is seen that, in the optically thick case (b), for both polarizations the intensity behaves with angle in the same way as the penetration depth. This is indeed the polarized equivalent of the Eddington-Barbier relation (6.3.26). It is caused by the fact that, for an optically thick medium, photons escape preferentially along those directions where the escape mean free path (or penetration depth) is large, $I(\theta) \propto \cos\theta/\kappa_i(\theta)$. In the optically thin case, (c), the behavior is the opposite, because all photons escape, and the intensity is maximum in the directions where the product of the emission cross section $\alpha(\theta)B$ and the path length $2R/\cos\theta$ is maximum. Since the absorption and scattering coefficient have approximately the same angular behavior, $\alpha(\theta) \propto \kappa(\theta)$, in the

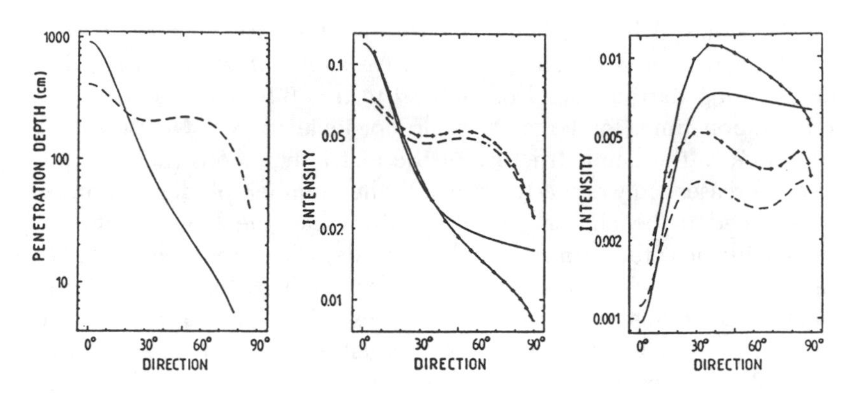


FIGURE 6.3.1 Angle dependence of the penetration depths (*left*), the intensity (*center*) for a thick slab of $2 \times 10^4$ cm, and the intensity (*right*) for a thin slab of $2 \times 10^2$ cm. The approximate diffusion equation (escape probability) method is shown by the smooth curves, and the Feautrier solution is shown by the curves with dots. The density is $\rho = 0.1$ g cm$^{-3}$, $kT = 10$ keV, and $\hbar\omega_c = 50$ keV. The ordinary polarization is the solid line; the extraordinary, the dashed line. From W. Nagel, 1981, *Ap. J.*, 251, 278.

optically thin case one expects $I(\theta) \propto \kappa(\theta)/\cos\theta$; i.e. $I(\theta)$ is inversely proportional to the penetration depth. In Figure 6.3.1c, the slab is thin in the direction $\theta \sim 0$ for the ordinary photons, and is thin in all directions for the extraordinary photons. One also sees that, in the optically thin case, the ordinary photons dominate, since they are produced with higher cross section. In the optically thick case, on the other hand, it is the extraordinary photons that dominate, since they have the longer mean free path. At frequencies near or above the cyclotron frequency, the above conclusions do not apply, since other effects (e.g. vacuum, thermal motions) become important.

The behavior of the beaming pattern in a magnetized atmosphere is determined largely by the effect of $\vec{B}_0$ on the cross sections $\kappa_i(\theta)$. For frequencies $\omega \ll \omega_c$, both polarizations have a much reduced cross section near $\theta \sim 0°$, where $\sigma(\theta) \sim (\omega/\omega_c)^2$ (see Chap. 3). The corresponding large mean free path leads in the optically thick case to a collimation of the radiation along this direction. As the depth of the slab is decreased, it also becomes transparent first along $\theta \sim 0°$, and later along progressively larger $\theta$. This leads, in an optically thin slab, to a minimum in the intensity around $\theta \sim 0°$ so that the beaming pattern resembles a hollow inverted cone. Such a behavior was also deduced by Basko (1976), using the approximate method and a photon source distributed in depth according to $e^{-\tau/\tau_0}$, which resembles the radiation from a finite slab of width $\tau_0$ which first becomes transparent near $\theta \sim 0°$. It is important to note that the same slab may be thick at one frequency and thin at a lower frequency, in a particular direction, due to the $(\omega/\omega_c)^2$ behavior of the opacity. Thus, at some frequencies one may have a full beam, and at a lower frequency a hollow beam. This is a potential method for determining the value of $\vec{B}_0$ if the width is known.

A comparison with the beaming obtained by the approximate method of § 6.2*b* is also shown in Figure 6.3.1. These are marked by the solutions without dots on them. As seen in this figure, the cross sections induce a fairly strong anisotropy, since in this example $(\omega/\omega_c)^2 = 1/25$, which also reflects on the photon density. The discrepancy between the two methods is due mainly to the use of the diffusion treatment to derive an isotropized density. For larger frequencies, $\omega \gtrsim \omega_c$, the cross sections tend to become more isotropic and the agreement between the two methods is consequently better. One of the weaknesses of the diffusion approximation is that near the surface it gives a shallower dependence of the photon density on depth than is actually the case. The Marshak boundary condition (6.2.5) is also responsible for giving a somewhat higher photon density at the surface than it should (Pomraning, 1978). This is confirmed by looking at the ordinary (short mean-free-path) photons in Figure 6.3.1b, where the Feautrier results lie below the diffusion results at angles $\theta$ close to 90°.

In the optically thin case, the diffusion approximation is, as expected, less accurate than at higher optical depths (see Fig. 6.3.1c). While the angular distribution agrees qualitatively with the more accurate numerical results, the intensities are systematically lower by about 40%. The diffusion coefficient for the ordinary photons is dominated by the contribution near $\theta \sim 0°$, where these photons have the longest mean free path (Fig. 6.3.1a). In the optically thin case, however, these photons are practically absent, since they are produced with a low cross section. The diffusion coefficient is therefore too large, leading in the diffusion method to an overestimate of the escape of photons. As a consequence, the diffusion method predicts an ordinary photon density toward the surface which is too low. This is seen in Figure 6.3.2, which compares the Feautrier and diffusion photon densities. Since the extraordinary photons arise mostly from polarization-exchange scattering of ordinary photons, it follows that

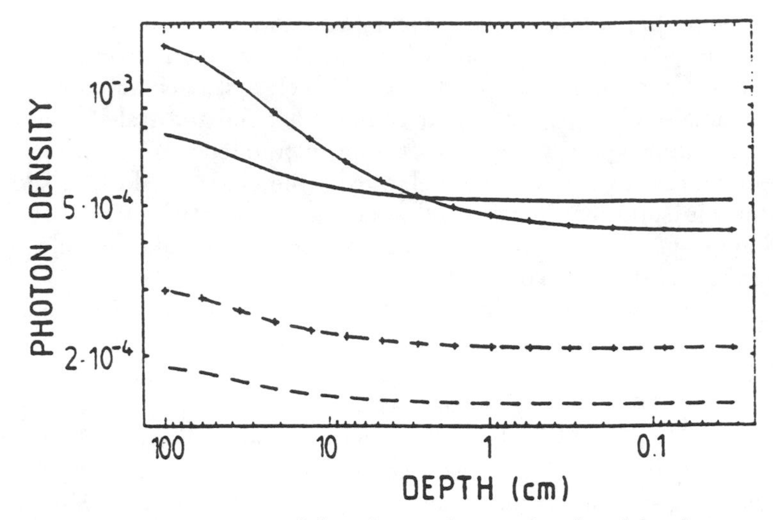


FIGURE 6.3.2 Comparison of the photon density for the slab of Fig. 6.3.1 (*right*) in the diffusion approximation (*smooth curves*) and the Feautrier calculations (*curves with dots*). The full lines are the ordinary, and the dashed are the extraordinary polarization. From W. Nagel, 1981, *Ap. J.*, 251, 278.

the density of extraordinary photons is also too low using the diffusion method.

For the case of the infinite cylinder and using the same assumptions as for the slab, a comparison between the Feautrier and the diffusion results has been done in Nagel (1981a). The penetration depth is now $\sin\theta/\kappa_i(\theta)$ rather than $\cos\theta/\kappa_i(\theta)$, since the emitting surface is parallel to $\vec{B}_0$. The results can again be expressed qualitatively as modified Eddington-Barbier relations: For optically thick cylinders, $I_i(\theta) \propto 2D_i + \sin\theta/\kappa_i(\theta)$, and for optically thin cylinders, $I_i(\theta) \propto \kappa_i(\theta)/\sin\theta$. Because the factor $\cos\theta$ has been replaced by $\sin\theta$, the mean free path along the principal direction of escape is less steeply dependent on $\theta$ (except near 0°). For this reason, the diffusion and Feautrier methods are in better agreement for cylindrical symmetry than for slab symmetry.

In general, for very thick or very thin cylinders, the directionality of the radiation is broader and less concentrated in angle than the corresponding slab cases. The flux $F(\theta) = I(\theta)\sin\theta$ is maximum at $\theta = \pi/2$ and decreases monotonically toward $\theta = 0°$. For columns of intermediate depth, however, the beaming pattern becomes less uniform, with the extraordinary photons transparent at $\theta \leq \pi/2$ but opaque below $\theta \sim (\omega/\omega_c)^{1/2}$ (where the cross section has a maximum). In this case the flux is no longer maximal at $\theta = \pi/2$ but rather near the angle where the column changes from optically thin to thick. The maximum consists of a relatively narrow bundle of extraordinary photons, while toward $\theta \sim \pi/2$ one has a smoother distribution of ordinary photons (Nagel, 1981a). If the column is larger than the mean free path of the extraordinary photons (for $\theta \leq \pi/2$), then these dominate at all angles. The same change of behavior of the beaming pattern is observed, for a column of fixed radius, at different frequencies.

*e) Incoherent Scattering Effects*

A more accurate treatment of the radiative transfer problem takes into account the photon energy changes involved in scattering events. Even in the nonrelativistic regime, these changes for the nonmagnetic case are roughly of the order (Rybicki and Lightman, 1979)

$$\frac{\Delta\omega}{\omega} \sim \frac{4kT_e - \hbar\omega}{mc^2}, \tag{6.3.20}$$

where the first term arises from the Doppler effect and the second from the recoil of the electron. At X-ray energies, photons may change their energy by 5–10% in individual scatterings, so that for a large optical depth the typical escaping photon has had its energy drastically altered before escape. At $\gamma$-ray energies, of course, the effect could be more drastic, since electrons of Lorentz factor $\gamma$ can boost a photon to energies $\omega' \sim \gamma^2\omega$. In the magnetic case, the equivalent of equation (6.3.20) would involve an average over scattering angles, which is obtained by a solution of the magnetic Kompaneets equation (e.g. Gnedin and Sunyaev, 1973; Lyubarsky, 1986). However, a Kompaneets approach is mostly constrained to the frequency region $\omega \ll \omega_c$, since near the resonances the cross section varies too sharply and the small momentum change requirement used in the Fokker-Planck expansion is not valid in that region. A more detailed numerical approach is therefore necessary. While in the nonrelativistic limit the Comptonization does not much alter the angular dependence obtained using the coherent scattering approximation (except of course near the resonances), since the directionality is determined largely by the last scattering; it is nonetheless clear that it has a significant effect on the formation of the frequency spectrum arising from multiple scatterings.

The transfer equations (6.3.3) do not include stimulated scattering effects. In this case, if scattering dominates over emission and absorption, the frequency redistribution will cause the radiation field to achieve a Wien equilibrium distribution rather than a Planck distribution (Pomraning, 1978). In the coherent scattering case, there is no frequency redistribution and the equilibrium spectrum is dictated by the Kirchhoff relation, $q_\omega = \alpha_\omega B_\omega$, so that, if $B_\omega$ is the Planck function, at high optical depths the radiation field will tend toward that equilibrium value. The inclusion of frequency redistribution due to incoherent scattering is incompatible with this, since for dominant scattering it would give a Wien spectrum and for dominant absorption a Planck spectrum. Short of including stimulated scattering, which would make the equations nonlinear, the most consistent procedure is to modify the Kirchhoff relation for emission and absorption, to read $q_\omega = \alpha_\omega W_\omega$. Here $W_\omega$ is the Wien function,

$$W_\omega = \frac{\hbar\omega^3}{(2\pi)^3 c^2} e^{-\hbar\omega/kT}, \tag{6.3.21}$$

and $\alpha_\omega$ is the free-free absorption coefficient without the $(1 - e^{-\hbar\omega/kT})$ factor, i.e. uncorrected for stimulated emission. This modification to the Kirchhoff relation is important only at low frequencies $\hbar\omega \ll kT$, where the correction is significantly different from unity. The result will be that the spectra thus calculated will, at low frequencies, saturate to the Wien slope $I_\omega \propto \omega^3$ rather than the Rayleigh-Jeans slope $\omega^2$. The Feautrier equations with frequency redistribution acquire in this case the form (Nagel, 1981b)

$$\begin{aligned} &\frac{\beta_i^2}{\kappa_i(\omega)}\left[\frac{d^2u_i(\omega)}{dr^2} + \frac{\gamma}{r}\frac{du_i(\omega)}{dr}\right] - \kappa_i(\omega)u_i(\omega) + \alpha_i(\omega)W_\omega \\ &\quad + \sum_{j=1,2}\int d\omega' 2\int_0^1 d\mu' \int_0^{2\pi} d\phi'\left[\frac{d^2\sigma_{ij}}{d\Omega\, d\omega}(\omega,\mu \leftarrow \omega',\mu')\right] \\ &\quad \times \frac{\omega}{\omega'}u_j(\omega') = 0, \end{aligned} \tag{6.3.22}$$

where $\gamma = (0, 1)$ is for (slab/cylinder), $\mu = \cos\theta$, and $\beta_i = (1 - \gamma)\cos\theta_i + \gamma 2^{-1/2}\sin\theta_i$. The scattering coefficient is $n_e$ times the incoherent scattering cross section. In the nonrelativistic limit, this is

$$\frac{d^2\sigma}{d\omega\, d\Omega}(\omega'\mu' \leftarrow \omega\mu) = n_e r_0^2 \frac{\omega'}{\omega}\frac{m_e}{|\Delta k|} f(p_0)|\langle e'|\Pi(p_0)|e\rangle|^2 \tag{6.3.23}$$

with the definitions $\hbar\,\Delta k = \hbar\omega'\mu'/c - \hbar\omega\mu/c$, $p_0 = m\,\Delta\omega/\Delta k + \frac{1}{2}\hbar\,\Delta k$, and $f(p_0) = (2\pi mkT)^{-1/2}\exp(-p_0^2/2mkT)$. In writing equation (6.3.22), we have continued to use the assumption $S_+ = S_-$ made to derive the Feautrier equations in second-order form. That is why we have $2\int_0^1 d\mu$ rather than $\int_{-1}^1 d\mu$. This assumption, however, implies that scattering into $\mu$ and $-\mu$ is equally probable, which is not strictly true. Taking the full integral from $-1$ to 1 would double the number of angle points and require a set of two first-order equations rather than one of second order. To retain the second-order treatment, we must take some appropriate symmetrized version of the scattering probability. The simplest approach is to take for this the average of the forward and backward $(+\mu, -\mu)$ scattering probabilities (Mészáros and Nagel, 1985a). One can rewrite equation (6.3.22), making use of the property of microreversibility of the cross section

$$\frac{d^2\sigma_{ji}}{d\omega'\,d\Omega'}(\omega'\mu' \leftarrow \omega\mu) = \left(\frac{\omega'}{\omega}\right)^2 e^{-(\hbar\omega' - \hbar\omega)/kT}\frac{d^2\sigma_{ij}}{d\omega\,d\Omega}(\omega\mu \leftarrow \omega'\mu'), \tag{6.3.24}$$

and converting from the radiation density $u_\omega$ to the photon occupation number $\eta_\omega$,

$$\eta_\omega = \frac{u_\omega}{\hbar\omega^3/[(2\pi)^3c^2]}. \tag{6.3.25}$$

The transport equation for the occupation number is then

$$\frac{\beta_i^2}{\kappa_i(\omega)}\left[\frac{d^2\eta_i(\omega)}{dr^2} + \frac{\gamma}{r}\frac{d\eta_i(\omega)}{dr}\right] - \kappa_i(\omega)\eta_i(\omega) + \alpha_i(\omega)e^{-\hbar\omega/kT}$$
$$+4\pi\sum_{j=1,2}\int d\omega'\int d\mu' e^{-(\hbar\omega-\hbar\omega')/kT}$$
$$\times\frac{d^2\sigma_{ji}}{d\Omega'\,d\omega'}(\omega'\mu' \leftarrow \omega\mu)\eta_j(\omega) = 0, \tag{6.3.26}$$

where the cross section is taken to be the $\phi$-averaged value and the integral over $\phi$ has been performed on this. It remains to choose a discrete set of angles and frequencies. If one chooses the frequency points logarithmically, it is useful to convert to the logarithmic frequency variable $f$,

$$f = \ln\omega. \tag{6.3.27}$$

Changing also $\omega\, d\sigma/d\omega = d\sigma/df$, we have

$$\frac{\beta_i^2}{\kappa_i(f)}\left[\frac{d^2\eta_i(f)}{dr^2} + \frac{\gamma}{r}\frac{d\eta_i(f)}{dr}\right] - \kappa_i(f)\eta_i(f) + \alpha_i(f)e^{-\hbar\omega/kT}$$

$$+4\pi \sum_{j=1,2} \int df' \int_0^1 d\mu'$$

$$\times \frac{d^2\sigma}{df'\, d\mu'}(f'\mu' \leftarrow f\mu)e^{-\hbar(\omega-\omega')/kT}\eta_j(f') = 0\,. \qquad (6.3.28)$$

The discretization of the variables proceeds as before, giving

$$D_i\left[\frac{d^2\eta_i}{dr^2} + \frac{\gamma}{r}\frac{d\eta_i}{dr}\right] - K_i\eta_i + A_i e^{-\hbar\omega/kT} + \sum_{j=1}^{N} S_{ij}\eta_j = 0, \qquad (6.3.29)$$

where now $i, j = 1, \ldots, N$, and $N = N_P N_A N_F$ is the product of the number of polarizations $N_P$ times the number of angle points $N_A$ times the number of frequency points $N_F$. The definition of the redistribution matrix $S_{ij}/K_i$ requires some care, as far as the frequency variable is concerned. The angular discretization is, as before,

$$S(f_i, \mu_i) = 4\pi \int_{-\infty}^{\infty} df' \sum_j w_j \frac{d\sigma_{i\leftarrow j}}{df'}(f', \mu_j \leftarrow f_i\mu_i)e^{-\hbar(\omega'-\omega_i)/kT}, \qquad (6.3.30)$$

where $w_j$ is the angular weight, here taken to be Gaussian. However, the cross section varies strongly at the resonances over frequency ranges which can be much smaller than any reasonable frequency grid spacing, which is limited not only by economical factors but also by a need not to unduly increase the size of the redistribution matrix, for numerical accuracy. The compromise procedure (Nagel, 1982) is to perform the integral over $df'$ on an auxiliary $f'$ grid, which is much denser than the transfer grid $f$, but concentrated over a smaller neighborhood ($\pm N_r kT/mc^2$) of the initial frequency $f_i$, where $N_r$ is an integer of order several (6 in the examples here). The result of the integral is then reapportioned to the two neighboring points of the coarser transfer grid $f_j$ by some appropriate weighting procedure, represented by $\Psi_j(f')$. We have thus

$$S_{ij} = w_j 4\pi \int_{-\infty}^{\infty} df' \frac{d\sigma_{j\leftarrow i}}{df'}(f', \mu_j \leftarrow f_i, \mu_i)\Psi_j(f')e^{-\hbar(\omega_j-\omega_i)/kT}$$

$$= 4\pi w_j \Delta f_j \frac{\Delta\sigma_{ij}}{\Delta f_j}(f_j\mu_j \leftarrow f_i\mu_i)e^{[-\hbar(\omega_j-\omega_i)/kT]}. \quad (6.3.31)$$

Due to the finiteness of the grid, in general the sum $\sum_{j=1}^{N} = S_i$ is not exactly equal to $\sigma_i(\omega_i, \mu_i)$ as calculated directly from the optical theorem. In order to avoid spurious spectral features and remain self-consistent, one can accept the small inaccuracy inherent in the numerical integration and redefine correspondingly the other transfer coefficients as, e.g., $A_i \equiv (S_i/\sigma_i)\alpha_i$, $K_i = A_i + \sum_j S_{ji}$, $D_i = \beta_i^2/K_i$.

*f) Generalized Boundary Conditions*

In order to deal with a wider variety of astrophysical models, it is of interest to introduce more general boundary conditions (e.g. Mészáros and Nagel, 1985a, b) than the simple homogeneous ones of equations (6.3.8) and (6.3.9), which represent a symmetric (reflection) boundary condition at $r = 0$ and no incident radiation inward at $r = R$ or $r = -R$. Because of the symmetry about $r = 0$, it was sufficient to calculate only from $r = 0$ to $R$, even in the slab case. Alternatively, one may redefine the slab to extend from $r = 0$ (bottom) to $r = R$ (rather than $-R$ to $R$). In this case, one may consider the case of radiation $I_*^+$ incident at the bottom, or $I_*^-$ incident at the top. If the two boundaries are otherwise free, in the sense that radiation escapes freely, then, by considering the first-order equations (6.1.11) and the definitions $u, v = \frac{1}{2}(I_+ \pm I_-)$, one can represent these boundary conditions as

$$\frac{\beta_i}{k_i}\frac{du_i}{dr} = -u_i + I_{i*}^-, \quad r = R, \quad (6.3.32a)$$

$$\frac{\beta_i}{k_i}\frac{du_i}{dr} = u_i - I_{i*}^+, \quad r = 0, \quad (6.3.32b)$$

where $I_{i*}$ is the external incident radiation inward. In a realistic neutron star emitting region, if this is a flat slab on the neutron star polar cap, the $I_*^+$ irradiation from below could be the blackbody radiation from the deeper, inner regions of the polar cap. The $I_*^-$ irradiation from above could be e.g. soft radiation from other parts of the magnetosphere or the companion star. When the emitting region is better represented by a cylinder standing above the polar cap, the external $I_*^-$ could be due to other parts of the neutron star or magnetosphere. For the cylinder, however, at $r = 0$ a reflecting boundary condition $du_i/dr = 0$ is required by symmetry. Alternatively, if one imagines some kind of central source along $r = 0$, since the diffuse intensity must satisfy $I_-(0) = 0$, or $u = v$, the condition is

$$\frac{\beta_i}{k_i}\frac{du_i}{dr} = -\frac{1}{2}I_{i*}^+ \qquad \text{at } r = 0. \tag{6.3.33}$$

For a slab the condition (6.3.33) would represent radiation incident from below, $I_*^+$, but no escape of diffuse radiation back downward, $I_-(0) = 0$, i.e. some kind of one-way membrane. If $I_*^+$ is due to the denser portion of the neutron star polar cap, it is unrealistic to assume that it would not reabsorb some of the radiation sent toward it. By the same token, the second of equation (6.3.32b) is also unrealistic, if $r = 0$ is the neutron star surface, since it implies that the neutron star swallows freely the diffuse radiation $I_-(0)$ from the slab. The bottom conditions (6.3.32a) and (6.3.33) are the two extreme examples for a neutron star surface which is fully absorbent or fully reflective. In that sense, these equations bracket any realistic representation of neutron star surface properties. Let us introduce three constants $C_{\text{ref}}$, $C_{\text{incb}}$, $C_{\text{inct}}$, which if set equal to 1 imply that we have a reflecting condition (at $r = 0$), bottom external incident radiation ($r = 0$), and top external incident radiation. If set equal to zero, there is a free bottom, no bottom incident and no top incident radiation. The Feautrier matrices for the bottom and top are still given by equations (6.3.14) and (6.3.16), but for the top depth $r = R$ we must use, instead of the simple thermal value of $\mathscr{Q}$ given by the last of equations (6.3.13), the value

$$\mathscr{Q}_0 = \text{diag}\,\mathscr{A}_i + C_{\text{inct}}I_{i*}^-\left(\frac{2}{r_0 - r_1} + \frac{\gamma}{R}\right)\text{diag}\,\beta_i. \tag{6.3.34}$$

The Feautrier matrices for the bottom depth ($r = 0$) are (6.3.14), but with $B_D$, $Q_D$ replaced by

$$\mathscr{B}_D = \frac{2(1+\gamma)}{r_{D-1}^2}\text{diag}\,\mathscr{D}_i + \text{diag}\,\mathscr{K}_i - \mathscr{S}_{ij}$$

$$+\left(1 - C_{\text{ref}}\right)\frac{2}{r_{D-1}}\text{diag}\,\beta_i, \tag{6.3.35}$$

$$\mathscr{Q}_D = \text{diag}\,\mathscr{A}_i + C_{\text{inct}}\left(1 - \frac{1}{2}C_{\text{ref}}\right)\frac{2}{r_{D-1}}I_{i*}^+\text{diag}\,\beta_i. \tag{6.3.36}$$

Thus, if $C_{\rm ref} = 1$, we regain the homogeneous, symmetric boundary conditions of equations (6.3.8). If there is radiation incident from top or bottom, setting $C_{\rm inct}$ or $C_{\rm incb}$ equal to 1 corresponds to equation (6.3.32a) or (6.3.23b), provided $C_{\rm ref} = 0$ (free escape). If $C_{\rm incb} = 1$ but we set $C_{\rm ref} = 1$ (which refers only the bottom surface), we have the condition (6.3.33). In the case $C_{\rm inct} = 0$ we have $I_i(R) = 2u_i(R)$, but if $C_{\rm inct} = 1$ then $I_i = 2u_i(R) - I^-_{i*}$.

*g) Inclined Magnetic Field and Azimuthal Dependence*

The analytic methods of § 6.2 and the second-order Feautrier method described above in this section both rely, among other things, on the approximation of averaging over the azimuthal variable $\phi$. This is strictly justified for the slab symmetry case, where the magnetic field is at an angle $\Theta_B = 0$ with respect to the surface normal. It is less justified, although expedient, for the case of cylinder symmetry, where the magnetic field is parallel to the surface, i.e. perpendicular to the surface normal, $\Theta_B = \pi/2$. In this case, there is a partial symmetry given by the fact that an observer at arbitrary $\phi$ will always be seeing the field at the same inclination $\Theta_B$. However, when the magnetic field is at some intermediate inclination angle $\Theta_B$, there can be no avoiding an intrinsic $\phi$-dependence of the intensity.

The problem of an arbitrary inclination magnetic field has been approached using the escape probability method by Kaminker, Pavlov, and Shibanov (1982). Essentially, they obtain the radiation density and the radiation flux in the diffusion approximation (i.e. using effectively an angle-averaging procedure); once this is known, they obtain the flux as a function of direction by the escape probability method, using the coherent scattering cross sections, in a manner similar to that in Kanno (1975). The latter approximation restricts the validity of the solution to the frequency region $\omega \lesssim 0.5\omega_c$ but allows for a relatively simple semianalytical procedure. This can be done for a parallel or perpendicular field, as well as for arbitrary $\Theta_B \neq (0, \pi/2)$. The result is an intensity which is asymmetric with respect to the direction $\Theta = 0$, with a maximum flux along the direction $\theta = \Theta_B$.

A more detailed numerical treatment based on a variant of this approach is that of Burnard, Klein, and Arons (1988), who use a combination of the moments equations including generalized Eddington factors and the coherent scattering Feautrier equations. The former are used to find the angle-averaged photon density, while the latter give the angular dependence for frequencies $\omega \lesssim 0.5\ \omega_c$. The $\phi$-dependence is explicitly included in the latter so that the Feautrier

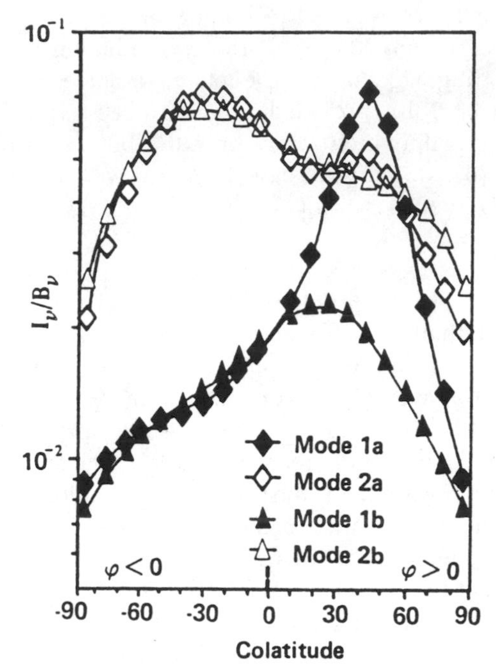


FIGURE 6.3.3 Intensities as a function of angle $\theta$ for an inclined magnetic field that makes an angle $\Theta_B = 45°$ with respect to the surface normal for two different values of the azimuth. (a) *Diamonds*: $\phi = 15°$, (b) *triangles*: $\phi = \pi/3$. From D. J. Burnard, R. I. Klein, and J. Arons, 1988, *Ap. J.*, 324, 1001.

equations cannot be cast into second-order form, as they are in equations (6.3.3) and (6.3.6). Instead, they use the two first-order equations for the symmetric and antisymmetric Feautrier variables $u$ and $v$, and use a numerical grid that has a number of discrete values of the azimuthal variable $\phi$, as well as of $\theta$ and the optical depth. The moment and Feautrier equations are then solved simultaneously by an iteration method. An example is shown in Figure 6.3.3. The possible use of such inclined magnetic field calculations is that, because of the lack of reflection symmetry about the direction $\theta = 0$, they naturally produce asymmetric pulse shapes, a feature which is not uncommon in observed pulse profiles of accreting pulsars. It is also useful if the accretion flow remains azimuthally symmetric but departs from a simple slab or cylinder geometry, as in the accretion mounds discussed in Chapter 7.

## 6.4 Magnetic Comptonization Effects

*a) Frequency Redistribution in the Diffusion Case*

The simplest problem involving frequency redistribution is when the angular dependence is suppressed. This can be done by angle-averaging the Feautrier equations for frequency redistribution, which for $S_{ij}$ would mean a double integral over $\Omega$ and $\Omega'$. Alternatively, since such a numerical average is approximate in any case, one may use the Feautrier equations as in § 6.3*d*, but in the two-stream, or Schuster-Schwarzschild, approximation (e.g. Chandrasekhar, 1960). That is, one can assume that the only angles are $\pm\mu_1$, and this in effect is a sort of diffusion approximation (Nagel, 1981b). Since the scattering $+\mu_1 \to +\mu_1$ cannot lead to a frequency change because of the kinematics (forward scattering is coherent), there is only backscattering in this approximation. In order to understand better the physics of the Comptonization and line formation process, it is also useful to neglect initially the vacuum polarization, $V = 0$. In this nonrelativistic limit and without spin-flip transitions, the ordinary mode is essentially nonresonant (e.g. Fig. 4.2.4), although a very small degree of resonance is introduced by thermal effects. Let us take as an example a medium with $\rho = 10^{-2}$ g cm$^{-3}$, $kT = 10$ keV, $\hbar\omega_c = 50$ keV, and homogeneous symmetric boundary conditions (eqs. 6.3.8–6.3.9a). For an optically thin situation of Thomson optical depth $\tau_T = 0.1$ the spectrum is shown in Figure 6.4.1. The continuum flux is seen to be proportional to the free-free emission, $\alpha_\omega W_\omega$. Near the cyclotron resonance, however, the slab becomes opaque to the extraordinary photons. They can only escape after they have diffused in frequency into the wings, as a consequence of one or more Compton scatterings. There the opacity is lower, and their escape is achieved. The effect of this is to reduce the intensity of the line near the core and increase it at the wings: the extraordinary line shows a central self-reversal. Whereas the thermalization probability $p_\omega = u_\omega / W_\omega$ is symmetrical about the core, the photon density $u_\omega = p_\omega W_\omega$ is larger in the red wing because of the Wien factor. This represents a bias for the core photons to scatter into the red wing, due to the electron recoil: for $kT < \hbar\omega_c$ as here, the core photons on average give up energy to the electrons (Wasserman and Salpeter, 1980). If one had $2kT > \hbar\omega_c$ so that the resonance was in the climbing portion of the Wien curve, the blue wing would be stronger due to the larger phase space there, the density of final states being proportional to $\omega^2$. Another way in which the extraordinary photons trapped in the core can escape is by

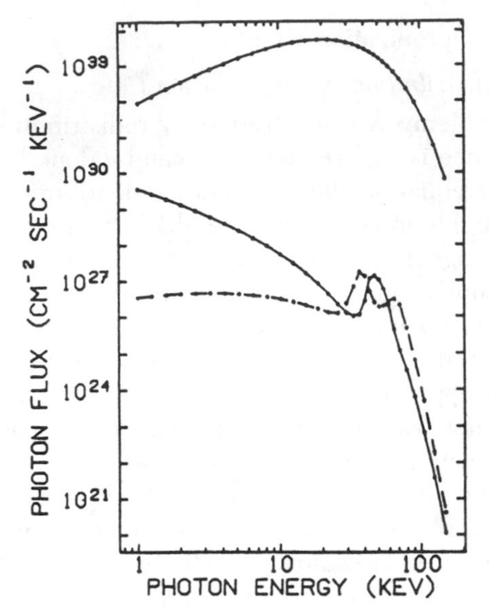


FIGURE 6.4.1 Spectrum of a magnetized slab for $\tau_T = 10^{-1}$, $\hbar\omega_c = 50$ keV, $\rho = 10^{-2}$ g cm$^{-3}$, showing the ordinary (*full line*) and extraordinary (*dashed line*) polarizations. The top curve is the Wien spectrum corresponding to the temperature $kT = 10$ keV. From W. Nagel, 1981, *Ap. J.*, 251, 288.

converting into long mean-free-path ordinary photons, which at $\tau_T = 0.1$ escape freely even in the core. In the absence of spin flip and vacuum polarization the polarization exchange term is small, but since the ordinary photons can escape from deeper layers, where the photon density is higher, they still produce a significant intensity at the core (Fig. 6.4.1), which is strong enough to overwhelm the tendency toward self-reversal.

As the optical depth increases, the emission line in Figure 6.4.1 starts to disappear and is replaced by a photon-deficiency which looks like an absorption line. For instance, for an optical depth $\tau_T = 10$, the slab is opaque to both polarizations, and over the whole range the escaping photons are mostly ordinary due to their longer mean free path. The shape of the absorption feature is given by the weak resonance of the ordinary photons. In the case of a cold plasma, where the ordinary mode is completely nonresonant, even this weak absorption line would be absent. In the more realistic case, where vacuum effects (and spin effects) are included, the ordinary mode and the exchange cross section are more strongly resonant, and a stronger absorption line is obtained, as in the calculations discussed in the next subsection.

The mechanism for the formation of this cyclotron "absorption" line (which in reality is due to the scattering of photons out of the core) is related to that of the classical Fraunhofer mechanism for stellar absorption lines. Although here there is no temperature gradient, there is an outward decreasing photon density. Near the line, we see only a short distance into the atmosphere, where the density is low, leading to a low intensity. Near the wings or in the continuum, we see deeper, where the larger density leads to a higher intensity. This is another example of the fact that, for an optically thick atmosphere, the intensity increases with the mean free path (or penetration depth), $I_\omega \propto W_\omega(\text{const.} + 1/\kappa_\omega)$. The changeover from an emission line to an "absorption" cyclotron line is shown in Figure 6.4.2 for various optical depths.

The shape of the continuum in the presence of repeated Compton

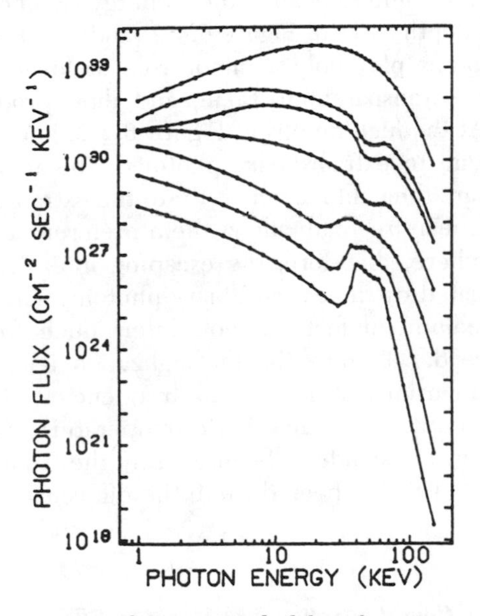


FIGURE 6.4.2 Spectrum of magnetized slabs of increasing Thomson optical depth $\tau_T = 10^{-2}, 10^{-2}, \ldots, 10^3$, from bottom to top, showing the total flux in both polarizations. The plasma parameters are as in Fig. 6.4.1. From W. Nagel, 1981, *Ap. J.*, 251, 288.

scattering resembles the nonmagnetic Comptonized spectra (e.g. Felten and Rees, 1972; Sunyaev and Titarchuk, 1980), with some significant differences. As in the nonmagnetic case, the spectrum is a power law at low frequencies, and at high optical depths develops a hump due to the accumulation of low-energy photons which have been upscattered but cannot go beyond $\hbar\omega \sim 3kT$. This hump has the shape of a diluted Wien spectrum, $aW_\omega$, where $a$ is a small number representing the thermalization probability. As the optical depth increases, the Wien hump increases in strength and fills in more of the spectrum toward lower frequencies, while the power law becomes flatter. Such a behavior is also seen in the magnetized spectrum for $\tau_T = 10^3$ shown in Figure 6.4.2, at least for the ordinary photons, whose cross section is not too different from the nonmagnetic case. There is of course the additional phenomenon of the cyclotron absorption line, which modifies the Wien hump. There are differences in the low-frequency part of the spectrum too due to the effects of the polarization exchange scattering. Ordinary photons scattering into the extraordinary mode also escape, due to the longer mean free path, before they have had time to change their energy significantly. In the two lower optical depth cases of Figure 6.4.2 (and also Fig. 6.4.1), the extraordinary photons play only a minor role at lower frequencies, since the slabs are transparent to them, and their production cross section is small. At the high depths of Figure 6.4.2, however, the slab has become opaque to extraordinary photons. The more numerous ordinary photons are then able to convert into the extraordinary mode, where the longer mean free path allows them preferential escape. In a very thick atmosphere, therefore, the escaping intensity at low frequencies is dominated by the extraordinary photons. The results from a semiinfinite medium, in fact, do not differ much from those of $\tau = 10^3$ in Figure 6.4.2, since the thermalization length is smaller than the slab dimensions at almost all frequencies. The previous calculations assumed a fixed atmospheric temperature; this condition can be relaxed (e.g. to include self-consistently the Compton cooling of the atmosphere). This has been done in the diffusion approximation by Riffert (1987).

### *b) Angle-dependent Comptonization and Vacuum Effects*

The energy-momentum conservation equations for a photon-electron system undergoing scattering (see eqs. 4.3.9) show that the scattered frequency $\omega'$ is directly dependent on the incoming and outgoing angles $\mu$ and $\mu'$. It is therefore expected that a simultaneous treatment of the angle and frequency redistribution is necessary for a realistic calculation of the spectrum and beaming of a magnetized atmosphere. The formalism has been described in § 6.3*d* for stationary slab and cylinder atmospheres. With the increased accuracy of this calculation, it is also worthwhile to include vacuum polarization effects in the lowest-order nonrelativistic limit. The scattering cross section in this limit is given by equation (4.3.12), where the hot-plasma normal modes $\vec{e}, \vec{e}'$ are corrected for vacuum polarization (see eq. 4.3.27). A plot of the differential cross section including vacuum effects was shown in Figure 4.3.2b, and the scattering cross sections integrated over outgoing angles, as well as the absorption cross section at various incoming angles, where shown in Figure 4.2.4b. The inclusion of a finite temperature has the effect of Doppler-broadening the resonance by an amount that depends on the angle of scattering, examples having been calculated in Mészáros and Nagel (1985a) and Bussard, Alexander, and Mészáros (1986). As discussed in Chapter 4.3, the discontinuity in these cross sections near the lower vacuum frequency $\omega_{v1}$ (eq. 3.4.37) indicates a transition from plasma-dominated modes at $\omega \lesssim \omega_{v1}$ to vacuum-dominated modes at $\omega_{v1} \lesssim \omega \lesssim \omega_c$. At $\omega < \omega_{v1}$, the electric vectors of the extraordinary and ordinary modes (1 and 2, respectively) are oriented as if vacuum were absent, whereas for $\omega_{v1} \lesssim \omega \lesssim \omega_c$, the electric vectors of 1 and 2 have switched positions.

For comparison with the previous calculations, we consider again the homogeneous symmetric boundary conditions, corresponding to a self-emitting atmosphere with free boundaries (eqs. 6.38–6.39). These calculations (Mészáros and Nagel, 1985b) have been done with 32 frequencies, eight angles, and two polarizations, for $\rho = 0.5$ g cm$^{-3}$, $kT = 8$ keV, and $\bar{\omega}_c = 38$ keV. For this high density, $\omega_{v1} \simeq 30$ keV and the modes are mostly plasma dominated, except for the stronger, vacuum-induced resonance in the ordinary mode. The angle-integrated flux $F(\omega)$ and the differential flux $F(\omega, \mu) = I(\omega, \mu)\mu$ for the case of a slab of $\tau_T = 20$ are shown in Figure 6.4.3. The angle-integrated flux and the differential flux for a cylinder of the same parameters and depth are shown in Figure 6.4.4. As seen in Figures 6.4.3 and 6.4.4, for a moderate optical depth $\tau_T = 20$ the ordinary photons dominate at low energies due to their higher emission rate. At higher frequencies, however, the mean free path against polarization exchange becomes smaller than the depth, and these photons escape after transforming into the longer mean-free-path extraordinary mode. At the line, extraordinary photons are preferentially produced, but

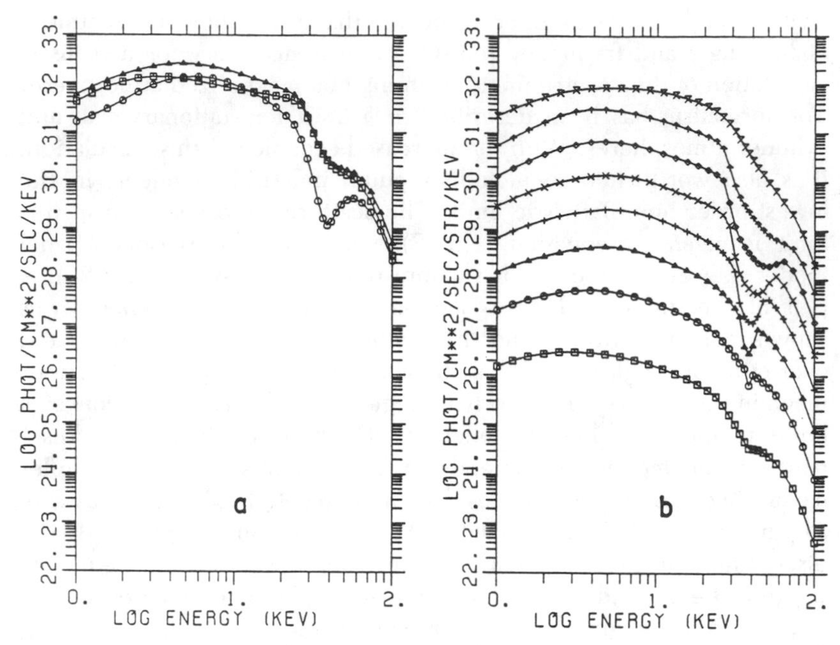


FIGURE 6.4.3 Spectrum of a magnetized slab with $\tau_T = 20$, $\rho = 0.5$ g cm$^{-3}$, $kT = 8$ keV, $\hbar\omega_c = 38$ keV. Panel (a) shows the angle-integrated flux, squares being ordinary polarization, circles being extraordinary, and triangles the total. Panel (b) shows the differential flux (sum of polarizations 1 and 2) for various angles $\theta$ between 11° (*top*) and 89° (*bottom*), offset by increasing powers of $10^{-1/2}$. From P. Mészáros and W. Nagel, 1985, *Ap. J.*, 299, 138.

they have again short mean free paths, so they escape instead after scattering into ordinary photons, both in the wings and in the core. If one performs a one-angle (two-stream) calculation including vacuum, one sees a deep absorption line if the angle $\mu_i$ is some intermediate value, e.g. $\mu_i = 1/\sqrt{3}$, whereas the line is less deep for the corresponding one-angle and no-vacuum calculation of § 6.3*d*. However, in the eight angle calculations the total flux is dominated by the direction components close to the normal, $dF(\mu) = \mu I(\mu)$, as can be seen from the right panels of Figures 6.4.3 and 6.4.4. For the slab (Fig. 6.4.3), the normal direction is along $\vec{B}_0$, which leads to a large Doppler broadening, since

$$\left(\frac{\Delta\omega}{\omega}\right)_D \approx \left(\frac{2kT}{mc^2}\right)^{1/2} |\cos\theta| \tag{6.4.1}$$

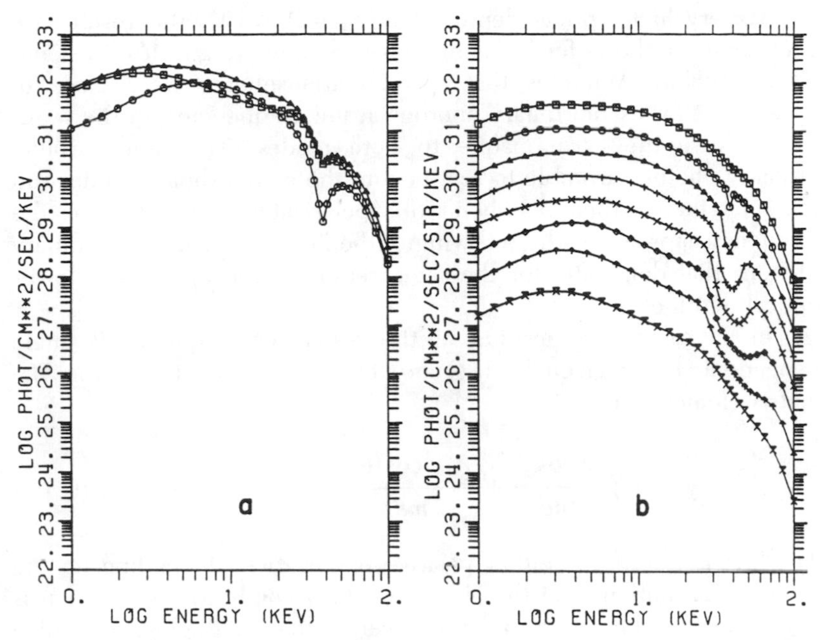


FIGURE 6.4.4 Spectrum for a magnetized cylinder of $\tau_T = 20$, $\rho = 0.5$ g cm$^{-3}$, $kT = 8$ keV, $\hbar\omega_c = 38$ keV. Panel (a) shows the angle-integrated flux, squares being ordinary polarization, circles being extraordinary, and triangles being the total. Panel (b) shows the differential flux (sum of polarizations 1 and 2) for various angles $\theta$ between 11° (*bottom*) and 89° (*top*), offset by increasing powers of $10^{-1/2}$. From P. Mészáros and W. Nagel, 1985, *Ap. J.*, 299, 138.

and the electrons are moving nearly along the line of sight. After integration over eight angles, the ordinary flux and the total flux do not show very pronounced absorption lines, despite the fact that the ordinary mode is now resonant due to the inclusion of vacuum (although still less strongly than the extraordinary). (The use of higher relativistic corrections, however, can lead to a line deeper than in this case, which includes only first-order relativistic corrections, e.g. Alexander, Mészáros, and Bussard, 1989). For the cylinder (Fig. 6.4.4), the normal is perpendicular to $\vec{B}_0$ and the line is very narrow, since only lowest-order relativistic effects are included. In cylinders, therefore, the line tends to be more visible in the total intensity for optically thick self-emitting atmospheres.

At very high optical depths, e.g. $\tau_T = 2 \times 10^4$, the results are equivalent to those for a semiinfinite medium (e.g. Mészáros and Nagel, 1985b). Whereas the $\tau_T = 20$ atmospheres were close to transparent for extraordinary photons at low frequencies, in the semi-infinite case they are opaque to both modes. The more prolific ordinary photons are able to scatter into the extraordinary mode, and the total flux is then mainly in this polarization. However, as the frequency approaches the red wing of the line, it is again the ordinary photons that dominate, for the same reason as for $\tau_T = 20$ or lower optical depth cases.

In all cases, the position of the resonance frequency is angle dependent, being given by the zero of the denominator of the $\Pi_-$ matrix element, or

$$\omega_{\text{res}} \simeq \omega_c + p\frac{\omega \cos\theta}{mc} + \frac{\hbar\omega^2\cos^2\theta}{2mc^2}. \tag{6.4.2}$$

The first correction is either positive or negative, depending on the direction of motion $p$ of the electron. The second-order correction is always positive, however, so that on average the angles close to 0° give a higher resonant frequency than the angles close to 90°. This effect is seen in the position of the minimum of the line, as a function of angle, in Figures 6.4.3 and 6.4.4. However, in slabs the flux is $F = \cos\theta\, I(\theta)$ whereas in columns it is $F = \sin\theta\, I(\theta)$. This leads to a small correlation of the line frequency with the continuum (wing) intensity in the slab case, and to a corresponding anticorrelation in the cylinder case.

The photon density as a function of the optical depth is shown for a deep slab of $\tau_T = 2 \times 10^4$ in Figure 6.4.5. This is essentially a semiinfinite medium, and it shows features similar to those in Figure 6.2.2. There is a thermalization region at high optical depths where the photon densities saturate to the Wien value, a diffusion region where the photon density drops, and a free flight region at low optical depths. At the cyclotron energy (*diamonds*), one can also see in the diffusion region the intermediate change of slope discussed in connection with Figure 6.2.2, where the polarization exchange scattering converts most of the cyclotron photons. These are produced mainly as extraordinary photons, but the atmosphere is opaque to this mode until close to the surface, so the escape occurs in the longer mean-free-path ordinary mode. In a less optically thick slab, such as the $\tau_T = 20$ case discussed in Figure 6.4.3, the thermalization length is longer than the slab depth at most frequencies, and the photon

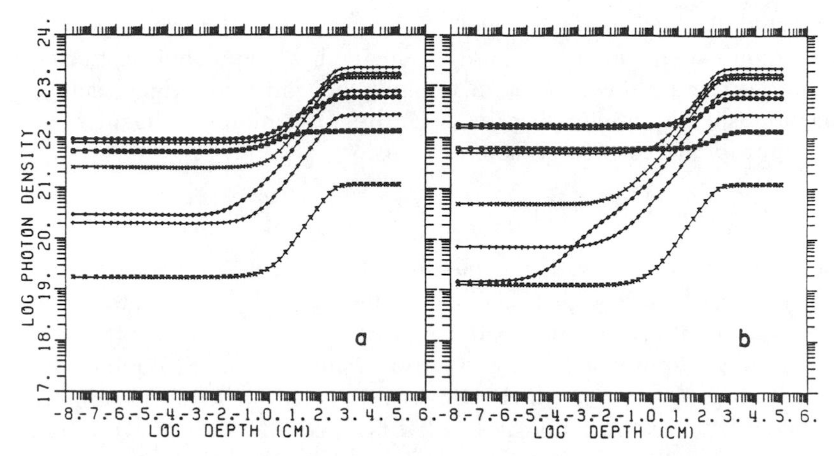


FIGURE 6.4.5 Photon density as a function of depth for a slab of $\tau_T = 2 \times 10^4$. The various photon energies are 1.6, 3.8, 9.0, 18, 29, 38, 52, and 85 keV (*square, circle, triangle, plus, cross, diamond, arrow*, and *closed cross*, respectively). The conditions are the same as in Fig. 6.4.3. The cyclotron frequency is 38 keV (shown by the diamond symbol). Panel (a) is ordinary, (b) is extraordinary. From P. Mészáros and W. Nagel, 1985, *Ap. J.*, 299, 138.

densities do not saturate to the Wien equilibrium density deep in the slab (or, rather, at the center, since a reflecting boundary condition is used at the midpoint). Otherwise, the behavior is similar (e.g. Mészáros and Nagel, 1985b), with a diffusion and free flight region, and a change of slope for the cyclotron energy photons in the extraordinary polarization at the depth where they switch over to the ordinary mode and escape.

## 6.5 Nonlinearities in Radiation Transport

### a) Stimulated Scattering Effects

In previous discussions we neglected stimulated scattering effects, and to be entirely consistent we should in that case also neglect stimulated emission. This of course leads to a Wien equilibrium spectrum in the limit of very large opacities $\tau \rightarrow \infty$. This approximation linearizes the transfer problem and is attractive because of the inherent economy of computation, but it is only useful at energies $\hbar\omega \gtrsim 0.5kT$ in an effectively thick medium (of depth greater than the thermalization depth), since at lower energies the calculation does not reproduce the expected Rayleigh-Jeans behavior, $F \propto \omega^2$. If the medium is absorption dominated, one may consider neglecting stimulated scattering while correcting the absorption for induced emission, which can be accomplished by multiplying (in LTE) the absorption coefficient $\alpha$ by the appropriate correction factor, i.e. using

$$\alpha' = \alpha\left(1 - e^{-\hbar\omega/kT}\right) \tag{6.5.1}$$

(see § 6.1*a*). However, if the medium is scattering dominated, as is often the case, at energies significantly less than the temperature one has to deal with stimulated scattering explicitly.

A simple approximate way to ensure that a blackbody equilibrium is achieved at $\tau \to \infty$, which still avoids solving the full nonlinear problem, is to use an artificial detailed balance (Mészáros, Pavlov, and Shibanov, 1989). The motivation for this method follows from a consideration of the transfer equation in a plane-parallel, magnetized LTE plasma with magnetic field along the surface normal to the slab (e.g. Mészáros and Nagel, 1985a, b). Similar considerations would apply to nonmagnetized situations where an angle, frequency, and polarization dependence is used explicitly. The latter could be eliminated in a nonmagnetic situation by summing over polarization indices:

$$\begin{aligned}\mu\frac{\partial n_m(x,\mu)}{\partial\tau} = &\sum_{m'=1,2}\int d\mu'\,dx' \\ &\times\Big\{\sigma_{m\to m'}(x,\mu\to x',\mu')n_m(x,\mu)\left[1+n_{m'}(x',\mu')\right] \\ &-\sigma_{m'\to m}(x',\mu'\to x,\mu)\left(\frac{x'}{x}\right)^2 \\ &\times n_{m'}(x',\mu')\left[1+n_m(x,\mu)\right]\Big\} \\ &+\kappa_m(x,\mu)\left[n_m(x,\mu)-n_{\rm bb}(x)\right],\end{aligned} \tag{6.5.2}$$

where $\tau$ is the (nonmagnetic) Thomson scattering depth, $\sigma_{m\to m'}$ is the differential scattering cross section averaged over the electron distribution function, and $\kappa_m(x,\mu)$ is the inverse bremsstrahlung coefficient. Both $\sigma$ and $\kappa$ are in Thomson units. The indexes $m, m' = 1, 2$ denote polarization modes; in a nonmagnetic case these would be the

$(l, r)$ modes used e.g. in Chandrasekhar (1960). The unity term in square brackets describes normal scattering, while the additional $n_m$ represents stimulated scattering. The scattering cross sections averaged over the LTE distribution functions obey the usual detailed balance relation

$$\sigma_{m \to m'}(x, \mu \to x', \mu') = \left(\frac{x'}{x}\right)^2 e^{(x-x')} \sigma_{m' \to m}(x', \mu' \to x, \mu), \tag{6.5.3}$$

which ensures that equation (6.5.2) leads to an equilibrium photon spectrum of the Planck form, $n_m(x, \mu) \to n_{bb} = [e^x - 1]^{-1}$. This is seen by substitution, in the limit $\tau \to \infty$, for which the $\tau$ derivative can be set equal to zero. This, however, implies keeping the full $n[1 + n]$ terms, which are time consuming to calculate. However, if we set the $[1 + n]$ terms in the transfer equation equal to unity, as we did in the previous sections when we used the linear transport equation, we may still obtain a blackbody equilibrium function $n_m \to n_{bb}$ in the limit $\tau \to \infty$ if we resort to the trick of using an artificially defined detailed balance relation,

$$\sigma_{m \to m'}(x, \mu \to x', \mu') = \left(\frac{x'}{x}\right)^2 \left[\frac{e^x - 1}{e^{x'} - 1}\right] \sigma_{m' \to m}(x', \mu' \to x, \mu), \tag{6.5.4}$$

instead of the usual detailed balance (6.5.3). With this artificial relation (6.5.4) the spontaneous scattering terms in equation (6.5.2) (i.e. those which are linear in $n$) are sufficient for yielding a Planckian equilibrium distribution. That is, we can use a linearized equation (6.5.2) with the stimulated scattering terms ($\propto n^2$) left out, and this equation will correctly yield $n \to n_{bb}$ in the appropriate limit, provided that equations (6.5.1) and (6.5.4) are explicitly used. This is usually done in numerical Feautrier treatments by using the direct cross section in the outscattering term and the detailed balance relation in the inscattering term. Figure 6.5.1 shows the photon spectra obtained for a magnetized atmosphere with parameters typical of an accreting X-ray pulsar, (A) without allowance for stimulated scattering, and (B) with stimulated scattering according to the prescription (6.5.4). Both solutions include stimulated corrections to the free-free opacity so that both converge to Planck at low frequencies,

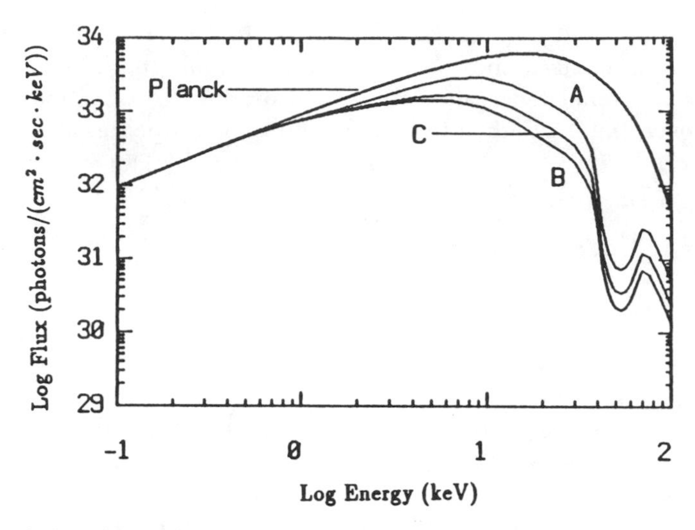


FIGURE 6.5.1 Photon flux for a magnetized slab with $\rho = 0.5$ g cm$^{-3}$, $kT = 8$ keV, $\hbar\omega_c = 38$ keV. Shown are the total flux (A) with no stimulated scattering, (B) with stimulated scattering using the artificial detailed balance prescription of eq. (6.5.4), and (C) with stimulated scattering calculated from an iteration solution of the full nonlinear transfer equation. From S. Alexander, P. Mészáros, and R. W. Bussard, 1989, *Ap. J.*, 342, 928.

the difference between stimulated scattering and its absence being manifest at higher frequencies and near $\omega_c$. The approximate solution (B) of the stimulated scattering problem gives, as seen in Figure 6.5.1, a deeper cyclotron line, which is closer to the full solution (C) than the curve (A) without any stimulated scattering. The solution presents no new numerical difficulties beyond what is involved in solving the linearized equations, which is considerably simpler and faster than the nonlinear problem. In Mészáros, Pavlov, and Shibanov (1989), plots are also given of the photon density as a function of optical depth for a semiinfinite atmosphere that show the approach to the equilibrium blackbody density in the case of the inclusion of stimulated effects according to the above prescription. The effects of stimulated scattering are important both very deep in the atmosphere and near the surface for the escaping flux at low frequencies. It is also important at frequencies where the optical depth is large, such as near cyclotron resonances.

While the artificial detailed balance method is simple, and at low frequencies it reproduces the exact solutions quite well, near the cyclotron energy the agreement is only approximate, when the flux is not close to the blackbody value for that energy. In this case, a full solution of the nonlinear transport equation (6.5.2) is necessary. This can be done, with methods similar to those used in non-LTE numerical radiative transport programs (e.g. as described in Mihalas, 1978), by deriving a second-order version of equation (6.5.2) in the usual Feautrier procedure, including now the finite difference expressions for the second-order nonlinear terms (Alexander, Mészáros, and Bussard, 1989). These can be linearized and considered to be a small perturbation on the equation, which has spontaneous processes only. The solution is found then by successive iterations of the linearized equations, using the linear perturbing terms calculated using the solution from the previous iteration, the zeroth perturbation being calculated from e.g. the unperturbed solution, which has only the spontaneous processes. This involves creating a number of additional matrices for the perturbation terms, and the calculation usually requires at least 10–15 iterations (more if the initial guess is far from the final solution) so that the computation time is increased by about an order of magnitude. Such solutions to the magnetic transfer problem in accreting pulsar conditions have been calculated by Nagel (1983), and more extensively by Alexander, Mészáros, and Bussard (1989). Due to the strong frequency dependence of the scattering cross section near the resonances, it is necessary for good convergence to increase the density of points in the neighborhood of the line, in comparison to the coarser distribution which can be used in the linear solution. An example is shown in Figure 6.5.1, where the exact iterated solution of the nonlinear problem, curve (C), is compared with the approximate stimulated scattering solution, curve (B), obtained by means of the artificial detailed balance method. While the latter method is quite good at low frequencies, it exaggerates the effects of stimulated scattering at higher frequencies, being based on replacing the occupation number in the $[1 + n_\omega]$ term by its equilibrium value, which is the maximum it can have. However, the higher the optical depth or the closer the spectrum to a blackbody, the closer the artificial detailed balance solution to the exact nonlinear solution. It provides a qualitative estimate of the maximum effect of stimulated scattering, which can be done in a linear computation. Since typically 10–15 iterations are the minimum needed to achieve a converged nonlinear spectrum, this is also the order of magnitude of the economy in the computation time.

*b) Two-Photon Scattering Nonlinearities and Higher Harmonics*

Another set of effects which are obviously nonlinear in the photon density are the multiple-photon processes, such as the two-photon scattering and the two-photon emission and their inverses discussed in Chapter 5.1. The calculations described in §§ 6.3 and 6.4 included only the usual magnetic (one-photon) Compton scattering, which conserves photon number. This made the equations linear if one neglected stimulated scattering effects, which made the solution of the equations relatively simpler. However, the two-photon effects (as well as the stimulated scattering), because they do not conserve photon number, require a special Feautrier scheme and need to be treated with care in order not to violate thermodynamical equilibrium. Despite the increased complexity, the inclusion of the multiple-photon effects is necessary if one wants to solve the transfer equation at the higher cyclotron harmonics accurately. This is because the depth and shape of the lower harmonics are controlled by photons that cascade down from the higher harmonics. In the treatments of accreting pulsar atmospheres described in § 6.4, it was assumed that the supply of photons at energies above the first harmonics was limited by e.g. a thermal cutoff, or an intrinsically low energy photon source, so that the problem of higher harmonics did not arise. However, in gamma-ray bursters (see Chap. 9), higher cyclotron harmonics have been observed, and theoretical calculations (Harding and Preece, 1989; Alexander and Mészáros, 1989; Wang *et al.* 1989) show that the relative ratio of the first to the second harmonic is controlled by the effect of two-photon scattering.

The cross sections and the redistribution functions for the various diagrams representing two-photon scattering and emission as well as their inverses were discussed in Chapter 5.1 and 5.2. The same method that was described in subsection *a* above to deal with the stimulated scattering nonlinearities can be applied to this case, i.e. the linearization method of the Feautrier equations. This solves the nonlinear equations by successive iterations, considering the nonlinear terms to be perturbations on the initial linear guess solution. Alexander and Mészáros (1991b) have done this for the case of accreting pulsars using parameters similar to those of Her X-1. An example using one angle is shown in Figure 6.5.2. The final solution is again achieved fairly quickly, in 15 or less iterations, provided a good initial guess is used. However, because the various two-photon processes have slightly different resonant denominators, the choice of

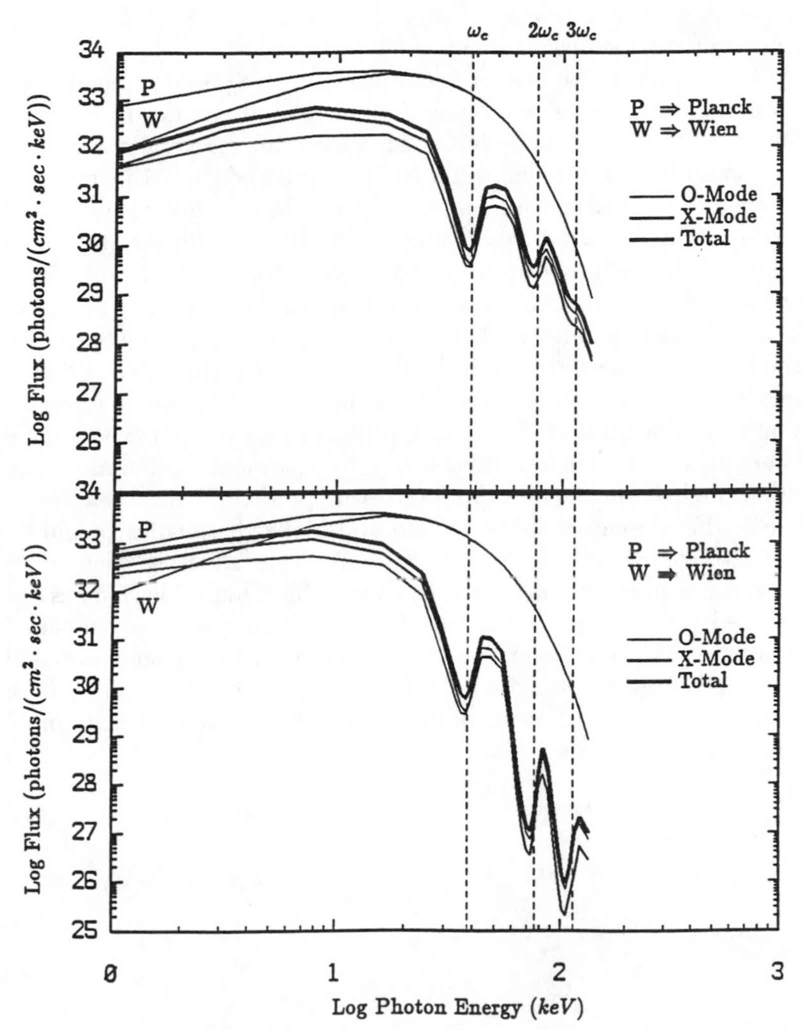


FIGURE 6.5.2 A calculation of the continuum and cyclotron line spectrum including three harmonics for $\rho = 0.5$ g cm$^{-3}$, $kT = 9$ keV, $\tau_T = 2 \times 10^4$, and $\hbar\omega_c = 38$ keV. The top panel includes only the linear one-photon scattering processes, while the bottom panel includes all the one- and two-photon scattering and emission processes discussed in Chap. 5.2 These calculations use fully relativistic cross sections and one angle of scattering. The curves marked P and W are the Planck and Wien spectra. From S. Alexander and P. Mészáros, 1991, *Ap. J.*, 372, 554.

frequency grids is a delicate matter that has to be experimented with before getting good convergence.

A comparison of the one-photon and two-photon calculations shows that the latter give a much deeper second and third harmonic. This is because in the one-photon case, even after several scatterings, the photon has not wandered too far from the line core. However, for two-photon scattering and emission, the photons are redistributed directly from the higher harmonics to the lower, with for example a $\sim 2\omega_c$ photon splitting up into two $\sim \omega_c$ photons. This replenishes the ground harmonic and depletes the higher harmonics. The inclusion of higher harmonics has, among other things, the effect of modifying the blue shoulder of the first harmonic, which falls off more steeply when a second harmonic is included. This finding is in agreement with observations of accreting pulsars (see Chap. 7), which show a steep falloff much better fitted with a two-harmonic assumption than with a single harmonic one. An interesting prediction of these calculations is that the fluxes at the second and higher harmonics should be very strongly polarized (see Fig. 6.5.2). This effect is even more pronounced in gamma-ray burst sources, where higher harmonics are observed (see Chapter 9 and Fig. 9.3.3). These results are obtained for simple static and homogeneous atmospheres. Inhomogeneities and departures from plane symmetry may be expected to reduce these estimates, but one still expects the polarization to remain substantial in more complicated models.

# 7. Accreting X-Ray Pulsars

## 7.1 Observational Overview

### *a) Introduction*

Accreting X-ray pulsars (AXPs) were discovered by the *Uhuru* satellite when regular X-ray pulsations reminiscent of pulsars were detected in the source Cen X-3 (Giacconi *et al.*, 1971; Schreier, 1972) and Her X-1 (Tananbaum *et al.*, 1972). This came shortly after the discovery of X-ray pulsations from the rotation-powered radio pulsar (RPP) in the Crab, but it became evident immediately that these new objects must be powered by accretion rather than rotation, since their X-ray luminosity was much larger, they were associated with a normal star binary companion, and they appeared to show a long-term overall spin-up (shortening) of their period rather than the spin-down (lengthening) characteristic of the rotational energy losses of isolated RPPs. The obvious interpretation was that these sources were accreting material from the binary companion, which liberated gravitational energy upon falling on the X-ray source. The latter, in order to explain the amount of energy liberated and the short pulsation periods, could only be magnetized neutron stars, and the spin-up should be due to the accretion of angular momentum associated with the infalling gas.

Since their discovery, the observational material on AXPs has grown significantly with data from *Ariel V*, *SAS 3*, *HEAO 1*, *Einstein*, *Hakucho*, *Tenma*, *EXOSAT*, *Ginga*, and *ROSAT*, as well as from many balloon observations, extending from the soft to hard X-ray ranges. The X-ray detectors most commonly used have included gas proportional counter (PC) devices, which in some cases are position-sensitive proportional counter (PSPC) devices used for imaging; solid-state detectors such as the *Einstein* solid-state spectrometer (SSS); and microchannel plate detectors. Sometimes these are used together with grazing incidence reflecting mirrors that focus the X-rays on a focal plane (e.g. the HRI and IPC systems on *Einstein* and *ROSAT*). At higher energies ($\gtrsim$ 20 keV) scintillation detectors are also used, e.g. NaI. Using such instruments, a rich variety of pulse shapes and spectra of AXPs have been measured, and spin and orbital periods have been tracked for long periods of time, showing among other things that these objects do not always spin up regularly. While the basic phenomenological interpretation has remained essentially unchanged, a number of interesting details and complications have emerged which bear upon fundamental questions of stellar evolution, accretion flow dynamics, and radiation physics, many aspects of which remain poorly or not understood to this day. Excellent and more extensive reviews of the observational material have been given by, among others, White, Swank, and Holt (1983), Joss and Rappaport (1984), and Nagase (1989a). Here we give only an overview of the main observational facts, referring the reader to the above references for further details.

As of late 1990, the list of known accreting X-ray pulsars numbered over 30 (e.g. Nagase, 1989a, 1990; and Table 1 near the end of the present book). A number of them are transients, having been observed only in brief periods of outburst, which, however, were long enough to allow detection of the characteristic X-ray pulsations. A larger number are quasi-steady sources, visible over very long periods of time, which makes them easier to detect and to study. In most AXPs, the presence of a binary companion star has been inferred from the Doppler shift caused by periodic orbital variations ($P_0 \gtrsim$ days) of the spin period ($P \gtrsim$ s), and in many cases these companions have also been detected optically. The companion stars generally fall into three categories:

a) Massive early-type companions (in Cen X-3, Vela X-1, etc.),
b) Be-star companions (in 4U 0115 + 63, A0535 + 26, etc.), and
c) Low-mass companions (in Her X-1, GX 1 + 4, etc.).

The two former types of binary systems belong to the broader class of high-mass X-ray binaries (HMXBs), while the latter type belongs to the broader class of low-mass X-ray binaries (LMXBs). The masses of the companions are not distributed uniformly, most of the companions being either larger than 15–20 solar masses or smaller than a few solar masses. The masses of the neutron stars themselves can be determined in a number of cases by using the dynamics of binary star systems, essentially Kepler's laws. This requires good determinations

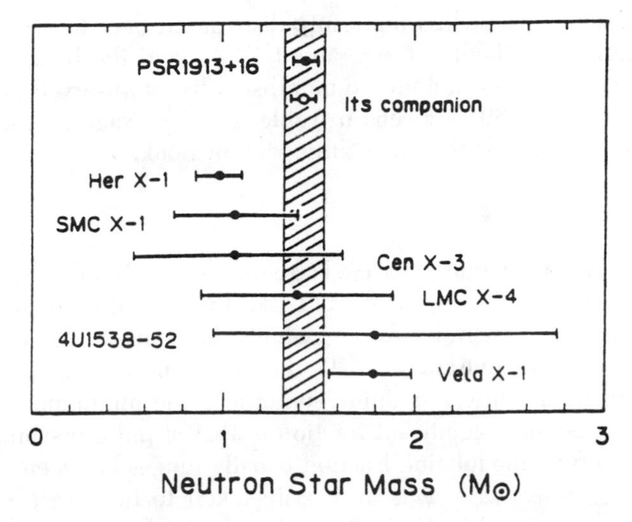


FIGURE 7.1.1 Mass distribution of neutron stars determined empirically from their binary orbital parameters. The six lower ones are AXPs, while the two upper objects are neutron stars in a binary rotation-powered pulsar. The hatched region represents the range $1.42 \pm 0.1\ M_\odot$. From F. Nagase, 1989, *Pub. Astron. Soc. Japan*, 41, 1.

of the orbital Doppler amplitude of the pulse variation to get the velocity, as well as an orbital period. The inclination of the system can be estimated in those systems where an eclipse by the companion is detected, which allows a determination of the mass function of the system. If spectroscopic observations of the companion are feasible, the mass of the neutron star is then determined to reasonable accuracy (see Joss and Rappaport, 1984, or Nagase, 1989a, for details). The neutron star masses for six AXP binary systems thus analyzed have yielded values which are compatible with the canonical Chandrasekhar value of $M_{NS} \simeq 1.4\ M_\odot$, except possibly for Her X-1, whose latest value is in the 1 $\sigma$ range 1.11–1.36 $M_\odot$, and Vela X-1, which is in the range 1.5–1.98 $M_\odot$ (Nagase, 1989a). These neutron star masses and errors are plotted in Figure 7.1.1. The distribution of AXPs in space, especially those with early-type, high-mass companions, is concentrated along the galactic plane. Some, like Her X-1, as well as those associated with the Magellanic Clouds (LMC X-4, etc.), are outside the plane. The typical X-ray luminosities $L_x$ range roughly between $10^{34}$ ergs $s^{-1}$ and $10^{39}$ ergs $s^{-1}$. The pulse periods $P$

detected are more or less uniformly distributed over four decades in the period, from about 0.1 s to about $10^3$ s, and the binary (orbital) periods $P_0$ extend from hours to months. A list of observed properties of AXPs as of 1989 is given in Table 1 (after Nagase, 1989a, and Ögelman, 1988) near the end of the present book.

*b) Pulse Shapes*

Typically, the X-ray pulses have a large duty cycle, $\gtrsim 50\%$, as opposed to radio pulsars, which have $\lesssim 10\%$, this quantity denoting the fraction of the cycle time during which the emission is high. Also, unlike many radio pulsars, in AXPs the emission never quite drops to zero intensity at pulse minimum, sometimes the minimum having an intensity which is a significant fraction of that at pulse maximum. The pulse amplitude modulation fraction usually ranges between 20% and 90%. The pulse phase $\Phi$ is arbitrarily taken to be zero (or one) at pulse maximum, and typically the pulse minimum occurs near phase 0.5. Some typical AXP pulse shapes are shown in Figure 7.1.2. From a study of the entire catalog of pulse morphologies observed so far, the pulse shape types can be roughly grouped into the following classes (Nagase, 1989a):

a) single quasi-sinusoidal pulses with only a weak dependence on the photon energy (GX 304-1, X Per, etc.);
b) double-peaked quasi-sinusoidal pulses, the two peaks usually having different amplitudes and little dependence on energy (SMC X-1, 4U 1538 − 52, etc.);
c) asymmetric single peaks with some features (Cen X-3, GX 1 + 4, etc.);
d) single quasi-sinusoidal pulses at high and low energies, and double (often asymmetric) peaks at intermediate energies (Her X-1, 4U 1626 − 67, etc.), with the phase of maximum at low energies shifted by 0.5 ($\sim 108°$) with respect to that at high energies in some sources (e.g. Her X-1);
e) double quasi-sinusoidal pulses at high energies and complex five-peaked pulses at low energies (Vela X-1, A0535 + 26, etc.).

The regular pulsations are thought to be connected to the presence of a strong magnetic field which, through its effect on the radiative opacities and the particle distributions, causes a beaming of the radiation (e.g. § 7.6). Unlike the situation in radio pulsars, most of the emission in AXPs comes almost certainly from near the neutron

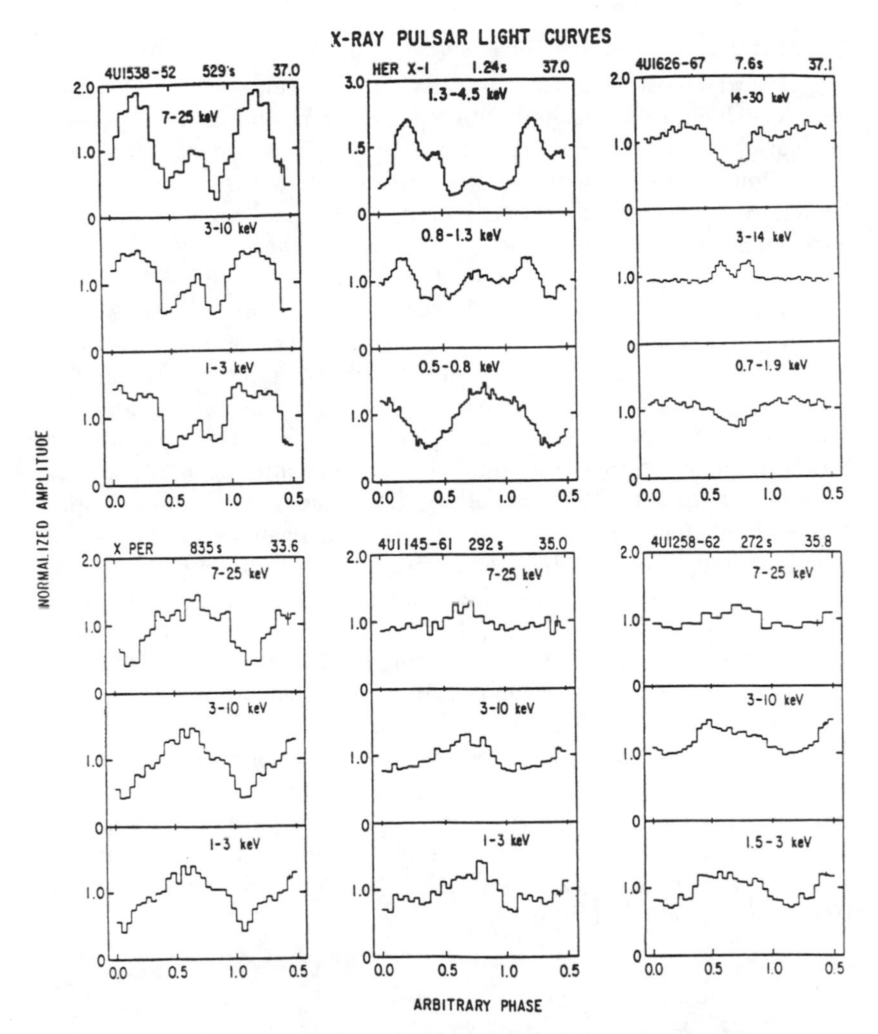


FIGURE 7.1.2 Various AXP pulse shapes in three different energy bands. The pulse period in seconds is the number at the top middle of each figure, while the logarithm of the X-ray luminosity is at the top right. From N. White, J. Swank, and S. S. Holt, 1983, *Ap. J.*, 270, 711.

star surface, and therefore the strong angle and frequency dependence of the opacities promotes the formation of an anisotropic emission spectrum, although there is controversy as to whether this is a radiation pencil beam along the magnetic axis or a fan beam parallel to the magnetic equatorial plane. A discussion of this issue is given in § 7.6.

*c) Medium and Long-Term Periodicities*

Superposed on the regular pulsations are longer period variations, some of which are associated with the orbital period or perhaps with regular changes in the accretion rate, while some may be due to precession of the neutron star or the accretion disk which brings matter from the companion to the neutron star, and some objects also show long time scale aperiodic variations. The most noticeable variations are the quasi-regular ON-OFF periods of objects such as Her X-1, with a time scale of about 35 days, during which the pulsations are ON for about 11 days, then OFF (but near the middle of the OFF period there is usually another small ON period), and this is thought to be related either to a regular modulation of the accretion from the companion or to precession of the accretion disk or the neutron star, which regularly brings the emission region into occultation (e.g Trümper *et al.*, 1986). Additional regular changes in the pulse shapes detected in Her X-1 have been ascribed to changes in the beam-sampling angles caused by stellar precession, and there could be additional contributions from other regions, such as the accretion disk or the Alfvén surface, which could reflect or reprocess some of the neutron star radiation into different angles. In the case of Her X-1, various features such as marching preeclipse dips and preferred turn-ons of the 35-day cycle at orbital phases 0.2 and 0.7 have been explained within the framework of interaction between the X-ray source and the heated surface of the companion star (Boynton, Crossa, and Deeter, 1980). Other AXPs with regular long period variations are LMC X-4 (Skinner *et al.*, 1982; Heemskerk and van Paradijs, 1989), with a period of 30.5 days, and SMC X-1 (Gruber and Rothschild, 1984), with a period of 60 days.

*d) Spectra and Luminosities*

The pulse-averaged spectra of AXPs are typically nonthermal and extend into the medium to hard X-ray energies. A fairly flat power law shape (photon index $\alpha \sim 0.8-1.5$) is often encountered above a few keV, which becomes much steeper above a break or turnover energy $E_b \sim 15-25$ keV, this drop-off sometimes appearing exponential (e.g. Joss and Rappaport, 1984). The turnover occurs at energies generally somewhat (but not much) higher than the typical blackbody photon energy associated with an object of luminosity $L_x$ and effective emitting area $A \sim 1\ \mathrm{km}^2$ comparable to the polar cap regions of magnetic neutron stars,

$$\hbar\omega_{bb} \sim kT_{bb} \simeq k\left(L_x/\sigma_B A\right)^{1/4}\left(1+z_s\right)^{-1/2}$$
$$\simeq 9.9\left(L_x/10^{38}\ \text{ergs s}^{-1}\right)^{1/4}\left(A/10^{10}\ \text{cm}^2\right)^{-1/4}$$
$$\times\left(1+z_s\right)^{-1/2}\ \text{keV}, \tag{7.1.1}$$

where $z_s = [(1 - 2GM/Rc^2)^{-1/2} - 1]$ is the surface gravitational redshift of the star of mass $M$ and radius $R$. Some typical phase-averaged AXP spectra are shown in Figure 7.1.3. In general, at low energies, the photon spectrum is absorbed by the interstellar medium, in which heavy elements provide bound-bound and bound-free transitions whose cross section typically behaves toward lower energies as $(E/E_L)^{-3}$, where $0.1 \lesssim E_L \lesssim$ few keV is the line or edge energy. As a function of pulse phase, the continuum above a few keV often shows regular variations of the spectral index $\alpha$ and of the turnover energy $E_b$ (e.g. Pravdo *et al.*, 1977; Holt, and McCray 1982; White, Swank, and Holt, 1983).

Many AXPs show an emission line around 6.4 keV (see Fig. 7.1.3), ascribed to incompletely ionized iron fluorescence transitions arising in a relatively cooler plasma (Ohashi *et al.*, 1984). The plasma responsible for this is probably located in the accretion flow somewhat farther away from the immediate neighborhood of the neutron star, either near the so-called Alfvén surface or in the stream of matter coming from the companion (e.g. Holt and McCray, 1982; Nagase, 1989a). The equivalent width of this iron fluorescence feature is typically $\gtrsim 100$ eV, with a relatively narrow line width $\Delta E \lesssim 0.5$ keV.

At low energies ($\lesssim 1$–$2$ keV) most AXP spectra show the effects of absorption by metals in the interstellar gas, or sometimes in the accretion flow gas. This absorption causes a turnover at low energies, and the energy at which this occurs is variable when the gas optical depth changes, as is the case for occultation by the companion star or the accretion flow. An example is Vela X-1, which has a giant star companion that feeds the neutron star via a wind, the orbit having a radius $r_X \sim 1.7R_*$ (Nagase, 1989b). In this source random gas clumps, as well as the more regular eclipse ingress and egress, cause large variations in the absorption.

The X-ray luminosities of observed AXPs can be determined from the observed flux if one knows the distance and the beaming fraction of the radiated luminosity, as well as the spectral distribution. The distances are by far the largest uncertainty, being often known with

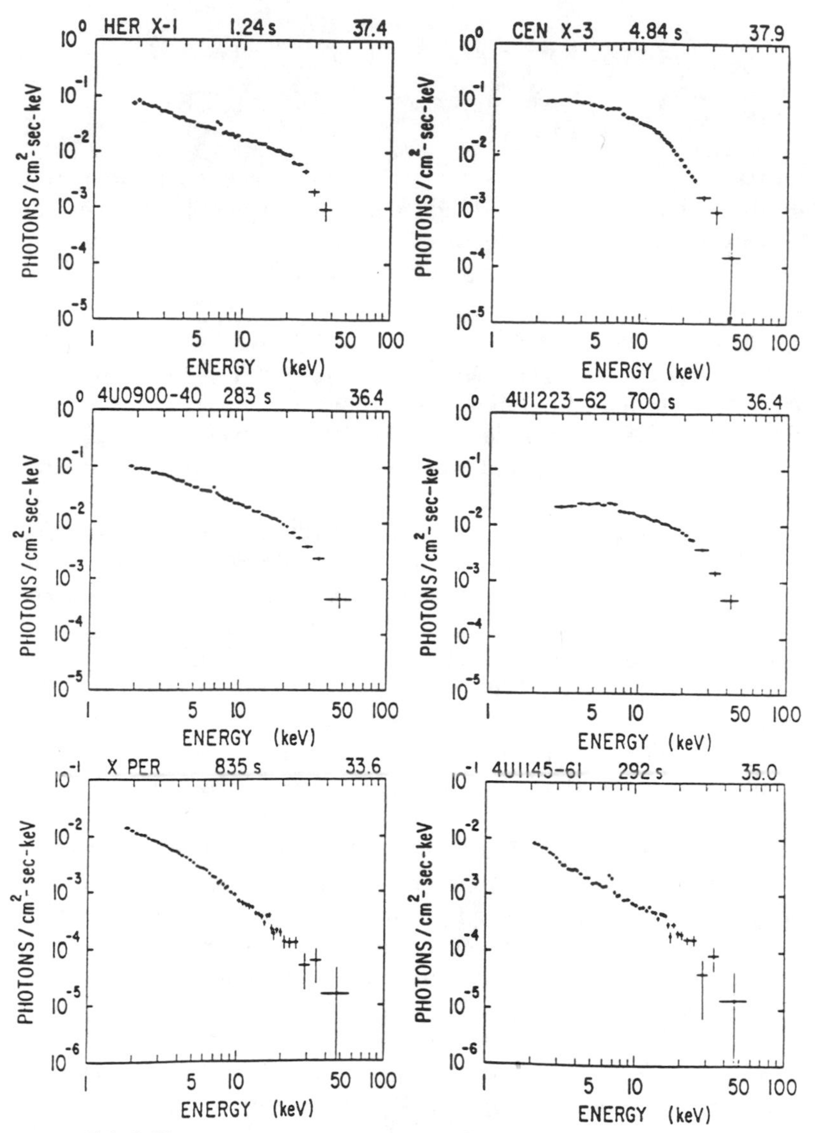


FIGURE 7.1.3 Various AXP spectra averaged over pulse phase. Notice the Fe line around $E \sim 6.7$ keV in some cases. The labeling is as in Fig. 7.1.2. From N. White, J. Swank, and S. S. Holt, 1983, *Ap. J.*, 270, 711.

errors as large as 50%, based on optical observations of the companion. The luminosity values so determined range between the approximate bounds

$$10^{34} \text{ ergs s}^{-1} \lesssim L_x \lesssim 10^{39} \text{ ergs s}^{-1} \sim 10 L_{\text{Ed}}, \tag{7.1.2}$$

where the upper limit occurs in only a few sources observed in the Magellanic Clouds, which appear to be intrinsically brighter than other galactic AXPs. The brighter galactic AXPs have luminosities which in general are somewhat below the Eddington luminosity

$$L_{\text{Ed}} = \frac{4\pi G M_* m_p c}{\sigma_T} = 1.26 \times 10^{38} \left( \frac{M_*}{M_\odot} \right) \text{ ergs s}^{-1}, \tag{7.1.3}$$

above which radiation pressure (calculated using the unmagnetized Thomson cross section $\sigma_T$, an isotropic radiation pattern, and spherically symmetric infall of a pure hydrogen ionized plasma) exceeds the gravitational attraction of the star. Sources which exceed the Eddington limit may be attributable to plasma which is rich in He or metals or an accretion and/or radiation pattern which is anisotropic, or they may be a case where the use of resonant magnetic opacities is necessary.

*e) Cyclotron Lines*

A number of AXPs have line features in the energy range $10 \lesssim E \lesssim 40$ keV which are interpreted as being due to free electron cyclotron transitions in the strong magnetic field $B \gtrsim 10^{12}$ G of the neutron star. The first and most prominent example with these features is Her X-1 (Trümper *et al.*, 1978; Voges *et al.*, 1982), showing a $15\sigma$ feature which could be interpreted as either an emission feature at $\sim 50$ keV or an absorption-like feature at $\sim 38$ keV. Various attempts at alternative explanations of these line features have proved untenable. Thus, an atomic line would have to come from an element like $\text{Pt}^{+77}$, or would have to be a nuclear line from, e.g. $^{241}\text{Am}$, the abundances of which are negligible compared to other astrophysically expected elements. The only reasonable explanation is that this line is caused by the cyclotron process. The energy of the ground cyclotron harmonic is given by

$$E_c = \hbar\omega_c (1 + z_s)^{-1} = 11.6 (B/10^{12} \text{ G}) (1 + z_s)^{-1} \text{ keV}, \tag{7.1.4}$$

where $\omega_c = eB/m_e c$ is the cyclotron frequency and $(1 + z_s) = (1 - 2GM/Rc^2)^{-1/2}$ is the gravitational redshift correction, so that the Her X-1 observations imply a magnetic field of the order $3 \times 10^{12}$ G under the absorption line interpretation. This detection was also confirmed with other instruments (Gruber *et al.*, 1980; Scheepmaker *et al.*, 1981; Tueller *et al.*, 1984). More recently, the cyclotron line in Her X-1 has been observed with the LAC detector aboard the *Ginga* satellite (Mihara *et al.*, 1990).

A similar cyclotron interpretation is ascribed to a line feature found in 4U 0115 + 63 (Wheaton *et al.*, 1979) at $\sim 20$ keV. A later analysis of this object by White, Swank, and Holt (1983) indicated two lines at energies of $\sim 11$ keV and $\sim 23$ keV, which, interpreted as first and second cyclotron harmonics, give a $B \sim 1.1 \times 10^{10}$ G. More recently, observations with the *Ginga* satellite have produced a cyclotron line detection in 4U 1538 − 52 at $\sim 20$ keV, or $B \sim 2 \times 10^{12}$ G (Clark *et al.*, 1990), and in X0331 + 53 (Makishima *et al.*, 1991) at $\sim 30$ keV, or $B \sim 3 \times 10^{12}$ G. The spectrum of X0331 + 53 is shown in Figure 7.1.4. Even when only one harmonic is seen, the *Ginga* data strongly indicate that a second harmonic must be included in the fitting procedure in order to reproduce the steep falloff above the first harmonic, e.g. in Her X-1, 4U 1538 − 52, and X 0331 + 53. In other *Ginga* observations, features thought to be cyclotron lines were found in 1E 2259 + 568 (Shinoda *et al.*, 1988; Koyama *et al.*, 1989) indicating $B \sim 5 \times 10^{11}$ G, in V0332 + 53 indicating a field of $\sim 2.5 \times 10^{12}$ G (Makino *et al.*, 1989), and in 4U 1907 + 09 indicating $B \sim 1.7 \times 10^{12}$ G (Makishima *et al.*, 1990). The cyclotron line energy and line profile also show a regular variation with pulse phase $\Phi$ (Voges *et al.*, 1982; Clark *et al.*, 1990; Nagase, 1989a). This phenomenon can be understood in terms of the properties of magnetized scattering atmospheres viewed at different angles depending on the phase (see § 7.6).

### *f) Secular Period Changes and Mass Exchange*

The spin period $P$ detected from individual pulsars changes over the years, the absolute value of the relative change $\dot{P}/P$ varying between $\sim 10^{-2}$ yr$^{-1}$ and $\sim 10^{-5}$ yr$^{-1}$ depending on the object. Generally there is a long-term overall trend for a negative $\dot{P}/P$, i.e. for a spin-up or decreasing of the period, indicating that overall angular momentum is being gained by the neutron star, provided by the accreting matter. However, there are objects which (occasionally for years) show instead a spin-down rather than a spin-up, while some objects (Her X-1, Cen

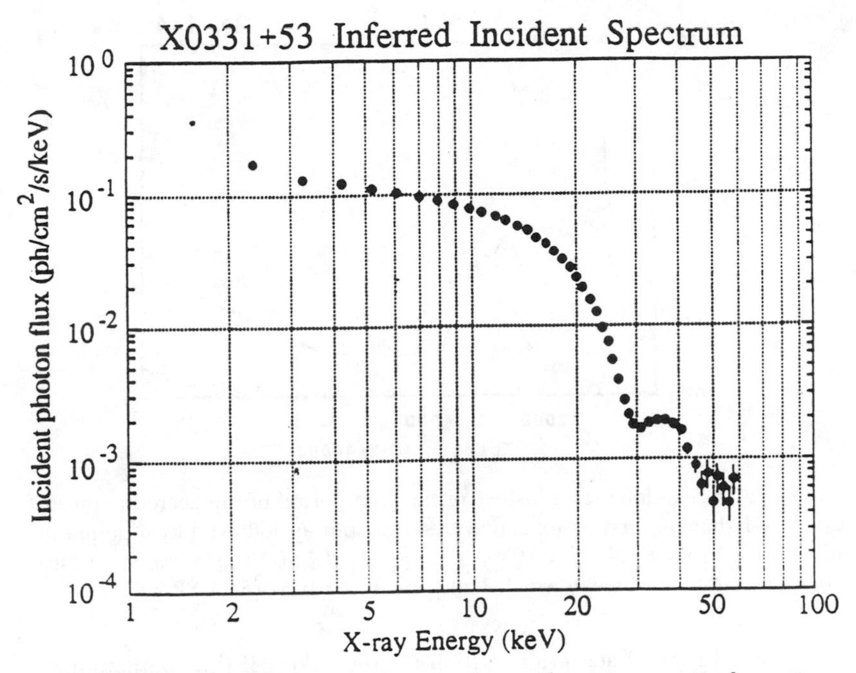


FIGURE 7.1.4 X-ray spectrum of the accreting pulsar X0331 + 53 from *Ginga* observations showing a power law spectrum with a turnover and a line feature at 30 keV interpreted as a cyclotron absorption line. The size of the dots indicates the instrumental uncertainty. From K. Makishima *et al.*, 1991, *Ap. J.* (*Letters*), 365, L59.

X-3) show an alternation of ups and downs superposed on an overall spin-up pattern. Extensive material on these period changes accumulated over the years, supplemented by surveys from the *Hakucho*, *Tenma*, *EXOSAT*, and *Ginga* satellites, provides coverage over time scales of up to 10–15 years for about 30 AXPs (Nagase, 1989b). The data for one of these are shown in Figure 7.1.5. The long-term secular period variations are believed to be caused by the accretion torques caused by the fact that the accreting material has some angular momentum (e.g. Joss and Rappaport, 1984; Nagase, 1989a). These period changes provide valuable information on the interaction between the stellar magnetic field and the flow of material approaching it. The torques should be different for disk-fed accretion, as expected if the binary companion is overflowing its Roche lobe (which is more likely in short orbital period, low-mass companion systems) or for a

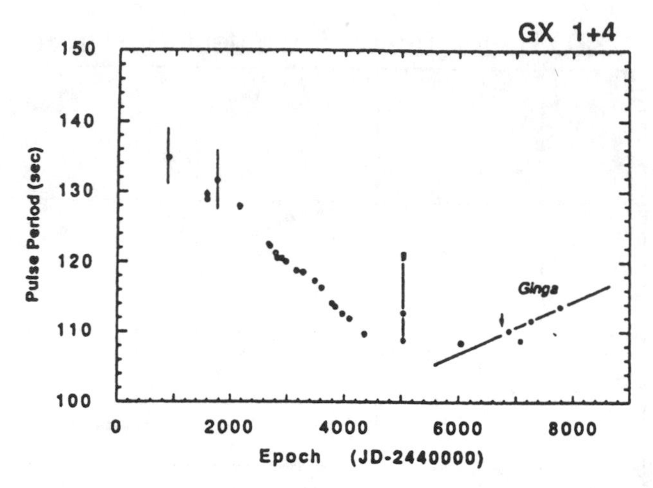


FIGURE 7.1.5 The long-term history of the spin period of the accreting pulsar GX 1 + 4 showing first an extended trend of spin-up followed by a period of spin-down. From F. Nagase, 1989, in *Proc. 23rd ESLAB Symposium on Two Topics in X-Ray Astronomy*, ed. J. Hunt and B. Battrick (ESA SP-296), p. 45.

stellar wind type of accretion, which is more likely if the companion is a massive early-type object (e.g Henrichs, 1983). This allows one to test both the structure of the magnetosphere-external flow interaction and the binary stellar evolution models that have been developed (e.g. van den Heuvel, 1988).

There are also short time scale variations, ranging from less than a day to months, whose character is more stochastic. These may be due to internal torques caused by the interaction between the core and crust of the neutron star. The analysis of these pulse period variations has as its goal an understanding of the internal dynamics of neutron stars (e.g. Boynton *et al.*, 1984; Deeter *et al.*, 1989; Lamb, 1988). This may be the only way to obtain information about the interior structure: relative size of crust and core, degree of superfluidity, mass distribution, moment of inertia, etc. (e.g. Pines, 1985, 1987). The rotational evolution of accreting pulsars under the influence of such torques is further discussed in § 7.2*d* and Chapter 11.3.

## 7.2 Accretion Flow and Magnetosphere Models

### *a) Distant Accretion Flow*

Models of accretion flow are typically simplest at distances sufficiently far from the magnetized neutron star so that the gravity of the latter is still dominant but stellar magnetospheric stresses may be neglected. Early work on this subject, soon after accreting pulsars were reported, was done by Pringle and Rees (1972), Davidson and Ostriker (1973), and Lamb, Pethick, and Pines (1973); and important milestones are represented by the works of Arons and Lea (1976a, b), Elsner and Lamb (1976, 1977), Gosh and Lamb (1978, 1979a, b), and Arons and Lea (1980). The subject has been reviewed by Henrichs (1983) and more recently by Nagase (1989a). An excellent discussion of the plasma physics, magnetohydrodynamic (MHD) processes, and astrophysical considerations related to the interaction of the stellar field with the accretion flow is given by Lamb (1988).

The basic type of accretion flow is determined by the characteristic distance at which matter becomes gravitationally captured by the neutron star and by the mean specific angular momentum $l \equiv \dot{J}/\dot{M}$ that is being carried by the matter at that distance. Depending on the type of binary stellar companion (or its absence) there are two distinct possibilities (see van den Heuvel, 1988, for a recent review).

*Disk accretion* can occur when the neutron star has a low-mass companion which does not lose much mass in the form of a stellar wind. In this case, mass exchange will occur after the atmosphere or the envelope of the companion fills its own Roche lobe and matter starts overflowing into the Roche lobe of the neutron star. Expressions for the size of the Roche lobe in binary systems are given by Joss and Rappaport (1984). The rate of mass transfer $\dot{M}$ will be determined by the stellar evolution of the companion, by gravitational radiation losses, and possibly by some other mechanisms (see Chap. 11.1). The mean specific angular momentum of the matter captured by the neutron star is

$$l_0 = \eta_0 a^2 \Omega_{\mathrm{orb}}, \tag{7.2.1}$$

where $a$ is the binary separation, $\Omega_{\mathrm{orb}}$ is the orbital angular velocity of the system, and $\eta_0$ is a dimensionless constant of order unity. This specific angular momentum is much larger than that of matter in Keplerian rotation near the neutron star, or even at larger distances from the neutron star. The Coriolis force causes the matter flowing in through the Roche cusp to swing sideways in the orbital plane and to start orbiting the neutron star, and the viscosity of the gas between neighboring streams causes a slow inward spiraling, resulting in an accretion disk far away from the neutron star. The physics of accretion disks has been discussed in some detail by e.g. Shakura and Sunyaev (1973), Pringle (1981), and Frank, King, and Raine (1985).

*Quasi-spherical accretion* can arise if the neutron star is isolated (without a companion) or if the companion is not filling its own Roche lobe or is of an early enough type that it produces a substantial wind which provides the accreted matter. For a star accreting from a wind, the approximate distance inside which matter is gravitationally captured is given by the "accretion radius" (Bondi and Hoyle, 1944; Henrichs, 1983):

$$r_a = \zeta \frac{2GM_x}{\left(v_{\rm rel}^2 + c_s^2\right)} \simeq 2.65 \times 10^{10} \times m_x\left[\left(v_{\rm rel}^2 + c_s^2\right)/1000 \text{ Km s}^{-1}\right]^{-1} \text{cm}, \tag{7.2.2}$$

where $\zeta \sim 1$ is a geometrical constant, $c_s$ is the sound speed of the gas, $v_{\rm rel} = (v_w^2 + v_{\rm orb}^2)^{1/2}$ is the relative velocity between the neutron star and the gas, $v_w$ is the wind velocity at $r_a$, and $v_{\rm orb}$ is the orbital velocity of the star. This expression contains the special case of a single (isolated) neutron star accreting from the interstellar medium ($v_{\rm rel} = 0$). The accretion rate, assuming the gas is collisional, i.e. in the fluid regime, is given by the Bondi rate,

$$\dot{M} = 4\pi\rho_0 v_{\rm rel} r_a^2 = 16\pi\zeta^2\rho_0 G^2 M_x^2 v_0^{-3} \text{ gm cm}^{-3}, \tag{7.2.3}$$

where $\rho_0$ is the gas mass density and $v_0 \equiv \sqrt{v_{\rm rel}^2 + c_s^2}$, both taken at the accretion radius $r_a$. The specific angular momentum of the matter captured in the wind case is rather lower, compared to the Roche lobe overflow case,

$$l_w \simeq \epsilon r_a v_{\rm rel} = \eta_w r_a^2 \Omega_{\rm orb}, \tag{7.2.4}$$

where $\epsilon$ and $\eta_w$ are dimensionless constants $\lesssim 1$. Various analytic estimates of $\eta_w$ have ranged from $\sim 0$ to being positive and $\sim 1$. For a constant-velocity homogeneous wind, Shapiro and Lightman (1976) (see also Davidson and Ostriker, 1973, and Illarionov and Sunyaev, 1975) found that small density variations with a typical gradient of dimension $\epsilon_\rho \simeq (v_{\rm orb}/v_w)(r_a/a)$ would give a positive $\eta_w \sim 1$. Results of later analyses (Davies and Pringle, 1980; Wang, 1981; Soker and Livio, 1984; Ho and Arons, 1987; Ho, 1988) have ranged from $\eta_w \sim 0$ to $\eta_w \sim \pm 1$. Numerical 2-D and 3-D calculations have so far proved inconclusive (see discussion and references in Lamb, 1988). It is possible that, especially in strongly inhomogeneous winds, there could be enough specific angular momentum captured to form a small disk, whose orientation and sense of rotation would be stochastically variable. It appears more likely, however, that inside radii $r \ll r_a$ the flow becomes more or less spherically symmetrical, within time scales comparable to the free-fall time. If the relative velocity is large, one also expects over length scales comparable to $r_a$ an accretion wake or accretion cylinder in the downstream portion of the wind, before the flow becomes more spherically symmetrical.

*b) Characteristic Dimensions of the Magnetosphere*

The distant accretion flow is important in good part because it provides the boundary conditions, or initial conditions, for the material that eventually falls onto the star to give rise to the accretion column. This is because, at a certain distance loosely called the "magnetospheric radius" $r_m$, the matter inflow (either disk or quasi-spherical) is stopped by the growing magnetic stresses of the stellar field, which it must at least partially penetrate. The matter which has diffused onto the stellar field lines is channeled by these toward the stellar surface (although in some cases matter may approach the surface in a more complicated manner; see below). If and when the matter latches onto the field lines, barring the uncertain effect of exchange or other instabilities, it will remain attached to the same field lines on which it started out at $r_m$ so that the geometry of the accretion column (the region near the star with matter-loaded field lines, typically near the polar caps) is determined by the details of the loading at $r_m$. The accretion at distances beyond $r_m$ may have roughly spherical symmetry, e.g. if matter is captured from the stellar wind of a companion star, since in this case the net angular momentum of the accreted matter is very low. Alternatively, the flow beyond $r_m$ may be in the form of an accretion disk if the accreted matter has enough angular momentum, as would be the case when matter from a companion star is accreted through Roche lobe overflow. In the spherical accretion case the magnetospheric radius is given approximately by equating the matter stresses to the magnetospheric stresses. For radial infall this is $\rho v_r^2 \simeq B^2/8\pi$, which leads to the value (Lamb, Pethick, and Pines, 1973)

$$r_{m,s} \simeq 2.9 \times 10^8 B_{*,12}^{4/7} R_6^{10/7} m^{1/7} L_{37}^{-2/7}, \qquad (7.2.5a)$$

where $B_{*,12}$ is the magnetic field at the stellar surface in units of $10^{12}$ G, $R_6$ is the stellar radius in units of 10 km, $m$ is stellar mass in solar mass units, and $L_{37}$ is the accretion-induced stellar luminosity in units of $10^{37}$ ergs $s^{-1}$. In the case of disk accretion, the same formula remains approximately valid, or somewhat more accurately one can use the modified criterion $\rho v_r v_\phi \simeq B^2/4\pi$ (Lamb and Pethick, 1974). Using the values of $v_r$, $v_\phi$ for a thin $\alpha$-disk in the so-called regime (b), where scattering is the main opacity and gas pressure dominates (Shakura and Sunyaev, 1973), one gets the following expression for the magnetospheric radius from disk accretion (Király and Mészáros, 1988):

$$r_{m,d} \simeq 1.8 \times 10^8 B_{*,12}^{40/61} R_6^{104/61} m^{-25/61} L_{37}^{-16/61} \alpha^{-2/61}, \tag{7.2.5b}$$

where $\alpha \leq 1$ is the disk viscosity parameter. Alternatively, one could use for the magnetospheric radius the criterion that $\gamma B_p^2 r^2 = \dot{M} v_K$, where $\gamma \equiv (B_\phi/B_p)(\Delta r/r)$ and $\Delta r$ is the width of the transition zone where the dipole field lines start mixing with disk material. In this case, for regime (b) of Shakura and Sunyaev (1973) one gets (Lamb, 1988)

$$r_{m,d} \simeq 2 \times 10^8 \gamma^{2/7} B_{12}^{4/7} R_6^{10/7} m^{1/7} L_{37}^{-2/7}. \tag{7.2.5c}$$

The radius $r_m$ gives the inner radius of the magnetosphere. These quantities (7.25) have been defined using an aligned ($\vec{\mu} \parallel \vec{\Omega}_{\rm orb}$) dipole. Oblique dipole-disk configurations are more complicated (Aly, 1980, 1986; Riffert, 1980), but the order of magnitude and functional dependence of the magnetospheric radii should still be approximately that given in equations (7.2.5).

Another important dynamical quantity to which the previously defined radii have to be compared is the corotation radius, or centrifugal radius, $r_c$ given by the balance between gravity and the centrifugal force acting on matter which is in corotation with the star (i.e. attached to a rigid part of the magnetosphere),

$$r_c = \left(GM/\Omega_s^2\right)^{1/3} = 1.5 \times 10^8 P^{2/3} m^{1/3} \text{ cm}. \tag{7.2.6}$$

Obviously, if $r_m \lesssim r_c$, matter is not dynamically inhibited from falling onto the star inside $r_m$, but in the opposite case, matter may instead be thrown out by the centrifugal force (the so-called propeller effect: Illarionov and Sunyaev, 1975). One may define in general a "fastness" parameter $\omega_s$ (Elsner and Lamb, 1977) for accreting magnetic neutron stars given by

$$\omega_s \equiv \left(\frac{\Omega_s}{\Omega_K(r_m)}\right) = \left(\frac{r_m}{r_c}\right)^{3/2}. \tag{7.2.7}$$

The stellar rotation is unimportant for $\omega_s \ll 1$. For observed X-ray sources (which therefore presumably were able to accrete, in order to produce X-rays) the inferred values of $\omega_s$ range between $\sim 10^{-3}$ and $\sim 1$.

*c) Matter Entry into the Magnetosphere*

The mechanism by which the matter from the external accretion flow enters the magnetosphere depends largely on the type of external flow, e.g. whether it is quasi-radial or disk-like, as when accreting from a stellar wind or from an overflowing Roche lobe, respectively. It depends also on the relative velocity of the magnetospheric field at $r_m$ and the accretion flow just outside $r_m$, and also on the plasma conditions valid in the accretion flow (for instance, the degree to which the disk or flow may be considered diamagnetic, i.e. such that it excludes magnetic fields from itself). It is clear that, in order to achieve transfer from the exterior disk or flow to the stellar surface, there must be some departure from pure diamagnetism in order for the matter to connect to the inner magnetospheric, closed field lines that connect to the star.

*Disk accretion.* In the disk accretion case there are several mechanisms which can couple the disk to the magnetosphere, that is, provide transfer of both matter and angular momentum (e.g Lamb, 1988; Stella and Rosner, 1984; Ghosh and Lamb, 1978). (*i*) Given some relative velocity between the disk and the magnetosphere at $r_m$, the interface should be Kelvin-Helmholz unstable, and penetration of the flow into the magnetosphere may occur if the unstable modes of wavelength $\lambda$ can grow to an amplitude $\gtrsim h$, the semithickness of the disk. The growth times for these long-wavelength modes can be estimated from the MHD linear dispersion relations to be $\sim 10^{-5}$ of the disk radial inward drift time. (*ii*) Turbulent diffusion of the magnetospheric field into the disk may occur as a result of entrainment of the stellar field by convective or turbulent cells in the disk if their energy density near $r_m$ exceeds that of the dipole field. A mixing length type of estimate indicates time scales for this process of the order $\sim 10^{-3}$ of the radial drift time. (*iii*) Reconnection of the magnetospheric field with magnetic field loops flaring out of the disk may occur as a consequence of convective and turbulent motions which distort the predominantly azimuthal disk field. Estimates for this time scale are of the order $\sim 10^{-3}$ the radial drift time. As a result of this reconnection, some of the dipole field lines would become connected directly to the disk, and provide a coupling of the two. The typical length scale over which this coupling or transfer occurs is uncertain, and this is the quantity parametrized as $\Delta r$ (with $\Delta r \lesssim r$) that enters into the determination of equation (7.2.5c). For $\Delta r / r \ll 1$, the material will tend to fall onto a localized portion of the magnetic polar caps of the neutron star.

*Quasi-spherical accretion.* In the case of quasi-spherical accretion just outside the magnetospheric radius $r_m$ (given by eq. 7.2.1), the matter is expected to undergo a shock transition and settle down subsonically in a magnetopause covering most or all of $4\pi$. The basic question is whether subsequently there are preferred points for the entry of the matter into the magnetosphere. One important consideration is the cooling of the shock-heated material at the magnetopause, which may be caused by bremsstrahlung if not much accretion has occurred to produce significant X-ray luminosity from the star, or by inverse Compton cooling if the star is radiating significant amounts of X-rays. The bremsstrahlung time scale is much slower than the free-fall time scale, whereas the inverse Compton cooling time is much shorter. Another important consideration is whether and to what degree the magnetopause is unstable. The Kelvin-Helmholtz instability is a prime candidate if there are relative tangential motions between the gas inside and outside the magnetopause, and this would cause large-scale mixing of material, which would eventually distort the magnetosphere. Another obvious candidate is the Rayleigh-Taylor instability, since the gas in the magnetopause is subject to the gravitational pull of the star. Both of these could create large-scale inflows below the magnetosphere with characteristic velocities of the order of the free-fall velocity. Other mechanisms (reconnection, loss cone entry through the polar cusps, microscopic diffusion) appear to occur on much slower time scales than those of the two MHD instabilities (Elsner and Lamb, 1984).

The geometry of the matter infall from the magnetopause toward the star will depend on the relevant cooling time scales, and more importantly on the wavelength of the fastest growing unstable modes. For the Rayleigh-Taylor instability, which should dominate if there is no relative sideways motion, the length scale of the unstable modes is hard to ascertain. In one approach it is relatively long wavelengths which appear to be important (Elsner and Lamb, 1976), whereas in another approach it appears that only the short wavelengths can exist (Arons and Lea, 1976b). The long-wavelength instabilities would lead to rapid diffusion of plasma into the magnetosphere, leading to large-scale infall of blobs of matter over a large solid angle, whereas the short-wavelength instabilities would promote slow diffusion onto field lines, which might result in infall of matter more concentrated toward the polar caps at the stellar surface. Numerical studies of these instabilities have not resolved the issue (Wang, Nepveu, and Roberston, 1984). In either case there is some tendency for the matter to enter preferentially in the regions near (but not exactly at) the polar cusps of the magnetopause above the polar caps (Arons and Lea, 1980), especially if the cusp neighborhoods are preferentially illuminated by the X-rays from the star. However, if the luminosity is much larger than about $10^{36}$ ergs $s^{-1}$, the infall may occur over a very large area of the star (Elsner and Lamb, 1984).

### *d) Accretion Torques*

The accretion torque $\vec{K}_s$ on the star is given by the rate of change of the angular momentum $\vec{J}_s$ of the star, and can be represented formally as an integral over a closed surface S:

$$\vec{K}_s = \dot{\vec{J}}_s = -\int_S \left(\vec{r} \times \vec{\Pi}\right) \cdot \hat{n}\, dS, \tag{7.2.8}$$

where $\vec{\Pi}$ is the momentum flux density tensor and $\hat{n}$ is a unit vector normal to $S$ directed outward. The sign and magnitude of the torque in the case of stellar wind accretion are rather uncertain, since the sign and magnitude of the specific angular momentum of the captured matter are not well known (see discussion in § 7.2*b*). In the case of disk accretion, if the matter is moving at more or less Keplerian velocity and one assumes the value of the inner edge of the disk to be known (e.g. equal to $r_m$), the material torque is given by (Pringle and Rees, 1972)

$$K_0 \equiv \left(GMr_m\right)^{1/2} \dot{M}. \tag{7.2.9}$$

There are also other torques, in particular electromagnetic ones, and their evaluation hinges on the disk and magnetosphere interaction picture. The pitch angle $B_\phi / B_p$ and magnitude of the stellar field lines and the dimension of the disk-magnetosphere transition region $\Delta r$ play a large role in the determination of both the material and electromagnetic torques (Lamb, Pethick, and Pines, 1973). The most specific studies of the disk torques so far are those of Ghosh and Lamb (1978, 1979a, b). Their analytic and numerical results for an aligned rotator ($\vec{\mu} \| \vec{\Omega}_{\rm orb}$) can be parametrized in terms of a dimensionless function $n$ of the dimensionless fastness parameter $\omega_s$ defined in equation (7.2.7) so that the net total torque on the star can be written as

$$K_s \simeq n(\omega_s) K_0, \tag{7.2.10}$$

where the function $n(\omega_s)$ can be fitted to the numerical models by the approximate expression

$$n(\omega_s) \simeq 1.4 \left( \frac{1 - \omega_s / \omega_c}{1 - \omega_s} \right). \tag{7.2.11}$$

The value of the critical frequency is $\omega_c \sim 0.8$–$0.9$ (Lamb, 1988), instead of the value of 0.35 originally given by Ghosh and Lamb (1978). There are so far no detailed torque calculations for nonaligned rotators, but it is generally assumed that the order of magnitude and the functional dependence should be fairly close to those in the aligned rotator. The above torque is $\sim 1.4$ for slow rotators spinning in the same sense as the disk ($\omega_s \ll 1$), and it changes sign above the critical frequency $\omega_c$. Above frequencies $\omega_{\rm max} \sim 0.95$ no stable solutions were found for the torque within the framework of this model. Thus, the torque provides for a net spin-up at low rotation rates, but this becomes a spin-down for fast rotators with $\omega_s \gtrsim \omega_c$, and chaotic torques may be expected if $\omega_s \gg \omega_c$.

The corresponding changes in the period $P$ of the neutron star may be expressed in terms of its first time derivative $\dot{P}$. For quasi-spherical flows this can be written as

$$-\dot{P} = f_1(MR/I)\, l(L, \ldots) P^2 L, \tag{7.2.12}$$

where $MR/I$ is the combination of mass, radius, and moment of inertia, $L$ is the luminosity, $f_1$ is a dimensionless function, and the specific angular momentum $l \sim l_w$ may depend on the luminosity $L$ as well as other parameters. For disk flows, the period change can be written as

$$\begin{aligned} -\dot{P} &= n(\omega_s) f_2(\mu, M) P^2 L^{6/7} \\ &= 6 \times 10^{-5} n(\omega_s) \mu_{30}^{2/7} R_6^{6/7} m^{-3/7} I_{45}^{-1} \left( P L_{37}^{3/7} \right)^2 \ \mathrm{s/yr}, \end{aligned} \tag{7.2.13}$$

where $f_2$ is another dimensionless function of the mass and the magnetic moment $\mu$ (Ghosh and Lamb, 1979b). The order of magnitude of these period derivatives and the possible changes of sign implicit in $n(\omega_s)$ are in the right range for explaining many of the observed pulse period histories described in § 7.1. A plot of $\dot{P}$ against the $PL^{3/7}$ gives diagonal curves running from upper right to lower left, followed by a cutoff at low $P$, which when compared against the data on different accreting X-ray pulsars allows in principle the determination of their dipole magnetic moment. However, because the expression for $\dot{P}$ against $PL^{3/7}$ gives a curve, it is possible to fit two solutions to each object, a high-moment and a low-moment solution. Additional information, such as the availability of a cyclotron line measurement, can help to fix this. The apparent discrepancies between such cyclotron line field values and values inferred in some AXPs from this torque model have been discussed by Ghosh and Lamb (1979b), Anzer and Börner (1983), and Ruderman (1985).

## 7.3 The Accretion Column: Dynamics and Geometry

### *a) The Approach to the Polar Cap*

After the matter latches onto the dipole field lines at

$$r \sim r_m, \tag{7.3.1}$$

it moves along these toward the stellar surface until it is decelerated near the latter. Because the field lines that get loaded are concentrated toward the polar caps (Baan and Treves, 1972), the matter near the stellar surface fills a subset of the magnetic dipole field lines close to the poles, in a configuration which is more or less azimuthally symmetric about the pole. Near the star, this region is limited on the outside by a roughly column-shaped surface (a flaring column, due to the flaring of the field lines) which is referred to as the "accretion column," whose radius at the base is denoted $a$. If the accretion flow beyond $r_m$ is disk-like with axis parallel to the magnetic axis, this accretion column will be azimuthally symmetric and the matter will form a hollow funnel, because the matter at $r_m$ latches onto a limited range of filed lines located between $r_m$ and $r_m + \Delta$, where $\Delta$ is the width of the transition zone. That is, the matter will eventually settle onto the star, not over the whole of the polar cap of radius $a$ but on a ring of finite thickness $d$ and outer radius $a$ about the polar axis (Basko and Sunyaev, 1976a). If instabilities along the infall path exchange matter between loaded and unloaded lines, then the accretion column may be filled rather than hollow, at least in a rough time-averaged sense. This filled column may also occur for spherical accretion beyond $r_m$, when matter is able to enter the magnetosphere preferentially near the poles (Arons and Lea, 1980). On the other hand, if the accretion flow beyond $r_m$ is disk-like and the stellar magnetic axis does not coincide with the axis of symmetry of the accretion disk, the area of contact where the dipole magnetosphere rubs against the circular inner rim of the disk is not continuous along the latter, because the dipole configuration is more extended along the magnetic equator than along the magnetic pole. It rubs along two oppositely located regions of the disk inner rim. Thus, the hollow accretion funnel that reaches the star is actually not loaded along $2\pi$ of the magnetic azimuth but rather along two oppositely located finite arches spanning an azimuthal angle range $\Delta\varphi$, two curved walls or portions of the funnel, of outer radius $a$ and thickness $d$ (Basko and Sunyaev, 1976a). Again, exchange instabilities, if present, could alter this configuration. At any rate, the filled column, the hollow column, and the partial hollow column cover some of the possible configurations to be expected in a time-averaged picture. In a time-dependent picture, one needs in addition to consider that these configurations may have a stochastic blob structure, elongated along the field lines, with finite length spikes of material falling along the field lines within the confines of the previously mentioned funnel regions (e.g. Hameury *et al.*, 1980; Trümper *et al.*, 1982).

In order to calculate the spectrum of the radiation from the accretion column, it is necessary to know the density and temperature distribution of the gas within it. The gas which gets onto the field lines at $r_m$ is relatively cold,

$$T \lesssim \text{few keV} \tag{7.3.2}$$

(Sunyaev, 1976; Basko and Sunyaev, 1976b; McCray and Lamb, 1976), and as it travels toward the star it cools by radiating, this occurring especially rapidly for the field-perpendicular component of the electron momentum via synchrotron radiation. The parallel component gets radiated away through the inverse Compton effect so that the matter as it approaches the stellar surface is expected to be relatively cool, although there are no detailed calculations of what the value exactly is for this quasi freely falling approaching matter. However, as the matter comes near the surface, it is clear that it should be decelerated more or less suddenly, in either a collisionless shock or a radiation-dominated shock some distance above the surface, or, failing this, through Coulomb or nuclear collisions in the neutron star atmosphere, in either of which cases the temperature of the decelerated matter can be estimated. These possibilities also exist in the case of accretion onto an unmagnetized neutron star (Zel'dovich and Shakura, 1969; Shapiro and Salpeter, 1975), with the difference that in the presence of a strong magnetic field the cross sections for the two-body interaction or the collective plasma processes are more complicated. The collisionless shock and the Coulomb/nuclear deceleration cases can be expected mainly in the low-luminosity sources, where the radiation pressure is not sufficient to sustain a radiation shock. The critical luminosity for this to occur can be obtained by requiring that over a distance of order $a$ the radiation be able to impart a momentum sufficient to balance the ram pressure of the matter,

$$L_c \lesssim \frac{GM_* c}{\kappa_{\parallel}} \cdot \frac{a}{R_*}, \tag{7.3.3}$$

where $\kappa_{\parallel}$ is the scattering opacity along the magnetic field. For accretion over a polar cap of area $\pi a^2$ this criterion is equivalent to

$$\varepsilon_c \equiv |c/\kappa_{\parallel}\phi a| = \frac{l_{\rm ph}}{a} > \pi, \tag{7.3.4}$$

where $\phi$ is the mass accretion rate per unit area and $l_{\rm ph} = (c/v)\lambda_{\rm ph}$ is the length over which the momentum transfer occurs, $\lambda_{\rm ph}$ being the photon mean free path. This expression is applicable only in the optically thick case, and if for $\kappa_{\parallel}$ one uses the nonmagnetic Thomson cross section one gets a critical luminosity $L_{c,\,\rm nm} \simeq 10^{-2} L_{\rm Ed}$ (Basko and Sunyaev, 1976a). Morc accurately, $\kappa_{\parallel}$ should be some average over the frequency-dependent cross section including the resonances, weighted with the (self-consistent) radiation field through the atmosphere (Gnedin and Nagel, 1984; Zheleznyakov and Litvinchuk, 1987).

*b) Radiation Hydrodynamics in the Accretion Column*

The dynamic infall problem can be solved using either a radiation hydrodynamical approach or an individual-particle approach. In the hydrodynamical approach, the problem is simplified considerably if one introduces the diffusion approximation for the treatment of the radiation. The radiation field then enters through its angle-averaged energy density $J$ and the radiation flux vector $\vec{F}$, measured in the local rest frame of the matter. The discussion below and in the next two sections follows the general treatment of Kirk (1984) (see also Arons, Klein, and Lea, 1987). Expanding the transfer equation to lowest order in the plasma velocity $v/c$ and neglecting time derivatives

(Castor, 1972), one has two equations for the radiation quantities:

$$\vec{\nabla} \cdot \vec{F} + (\vec{v} \cdot \nabla)J + 4J(\vec{\nabla} \cdot \vec{v})/3 = C_E, \tag{7.3.5}$$

$$\vec{\nabla}J = \vec{C}_P, \tag{7.3.6}$$

where $C_E$ and $\vec{C}_P$ are the appropriate collision integrals for energy and momentum transfer between the radiation and the plasma. One has in addition the energy and momentum balance equations for the plasma,

$$(\vec{v} \cdot \vec{\nabla})p/(\gamma - 1) + (\vec{\nabla} \cdot \vec{v})\gamma p/(\gamma - 1) = -C_E, \tag{7.3.7}$$

$$\rho(\vec{v} \cdot \vec{\nabla})\vec{v} + \vec{\nabla}p - \rho\vec{g} = -\vec{C}_P, \tag{7.3.8}$$

with $\vec{g}$ the acceleration of gravity and $\gamma$ the ratio of specific heats. Heat conduction is not included in the plasma energy equation (7.3.7), but energy conduction by radiation does enter into the energy equation for radiation (7.3.5). The plasma equations (7.3.7) and (7.3.8) have a singular point at the gas sound speed $v_g = (\gamma p/\rho)^{1/2}$, whereas the radiation equations (7.3.5) and (7.3.6) possess a singular point only if $\vec{\nabla} \cdot \vec{F} = 0$. The characteristic velocity for the latter set of equations is the sound velocity in the radiation fluid, $v_r = (4J/9\rho)^{1/2}$. The free-fall velocity above the star is always well above both $v_g$ and $v_r$, while the final plasma velocity close to the surface must always become lower than both. The transition through $v_g$ is discontinuous and is called a collisionless shock if the corresponding dissipation is provided by collective plasma processes, or a collisional shock if (in the absence of collective effects) the dissipation is provided mainly by two-body particle collisions. The transition through $v_r$, on the other hand, is not a discontinuity if the diffusion approximation is used, but nonetheless this smoothed-out transition is called by analogy a radiative shock.

The collision integrals $C_E$ and $\vec{C}_P$ require an angle and frequency integration over the (magnetic) cross sections and the radiation field, which is in general anisotropic. These complicated integrals can be somewhat simplified using the diffusion approximation and assuming a distribution (Wien, for example) for the radiation. This is the approach of Riffert (1987), who uses a Rosseland mean type of integration, leading to explicit expressions for $C_E$ and for $\vec{C}_P$, the latter being expressed along the parallel and perpendicular directions as

$$\vec{C}_P = (C_{p\parallel}, C_{p\perp}) = (-\kappa_\parallel \rho F_\parallel/c, -\kappa_\perp \rho F_\perp/c). \tag{7.3.9}$$

The effective opacities $\kappa_{\parallel,\perp}$ are then functions only of the ratio of the cyclotron energy to the radiation temperature. Other, simpler prescriptions have been given by Wang and Frank (1981) and Braun and Yahel (1984).

The energy collision integral $C_E$ also needs to include separately the effect of resonant and nonresonant photons. In the diffusion approximation and assuming a Wien spectrum, expressions including both of these photon classes have been given by Riffert (1987), while Wang and Frank (1981) and Braun and Yahel give expressions based on the contribution of nonresonant photons only. The situation is much simpler in the high-luminosity case, if radiation pressure dominates over gas pressure (that is, if eq. 7.3.3 or 7.3.4 is satisfied) and the diffusion approximation can be used. In this case, an explicit expression for $C_E$ is not needed, as it can be eliminated from the system of equations. Combining the radiation and plasma momentum equations (7.3.7 and 7.3.8) and using $\vec{C}_P = -\kappa_T \rho \vec{F}/c$ for simplicity (e.g Basko and Sunyaev, 1976a), one obtains the following set of equations for the dependent variables $\rho$, $v$, $J$, and $\vec{F}$ (Kirk, 1984):

$$\vec{\nabla} \cdot \vec{F} + (\vec{v} \cdot \vec{\nabla}) J + 4J(\vec{\nabla} \cdot \vec{v})/3 = 0, \tag{7.3.10}$$

$$\vec{\nabla} J/3 = -\kappa \rho \vec{F}/c, \tag{7.3.11}$$

$$\rho\left[(\vec{v} \cdot \vec{\nabla})\vec{v} - \vec{g}\right] = \kappa \rho \vec{F}/c, \tag{7.3.12}$$

$$\vec{\nabla} \cdot (\rho \vec{v}) = 0, \tag{7.3.13}$$

where equation (7.3.12) applies only along the magnetic field direction and equation (7.3.13) is the equation of continuity. Equation (7.3.11) can be used to eliminate $\vec{F}$, and equation (7.3.12) can be integrated to eliminate $\rho$ in terms of the (given) accretion flux at the surface $\dot{M}$, or $\phi$. This is a two-dimensional description, the relevant quantities depending on the parallel ($\parallel$) and transverse ($\perp$) coordinates with respect to the magnetic field.

This basic system of equations can be further simplified if one performs an average over the transverse spatial coordinate, thus reducing the system to one spatial dimension (e.g. Inoue, 1975; Basko and Sunyaev, 1976a; Braun and Yahel, 1984). In cylinder symmetry, the equation of continuity integrates to $\rho v = \phi$, while in dipole geometry, close to the star the column opening and thickness are given by

$$a(r) = a(R) \cdot (r/R)^{3/2}, \quad d(r) = d(R) \cdot (r/R)^{3/2} \tag{7.3.14}$$

so that for $d \ll a$

$$\rho v r^3 = \text{constant} = \phi R^3. \tag{7.3.15}$$

In dipole geometry, separating $\vec{F}$ into $\parallel$, $\perp$ components, equation (7.3.10) becomes

$$\frac{1}{r^3}\frac{\partial}{\partial r}\left(\frac{r^3}{3}\frac{\partial J}{\partial r}\right) + v\frac{\partial J}{\partial r} + \frac{4J}{3r^3}\frac{\partial}{\partial r}(r^3 v) + \vec{\nabla}\cdot\vec{F}_\perp = 0. \tag{7.3.16}$$

Here the only term that depends on the perpendicular coordinate is $\vec{\nabla}\cdot\vec{F}_\perp$, which must be calculated from equation (7.3.11). Assuming that the energy sources are distributed uniformly across the column width or layer thickness, one can write

$$\vec{\nabla}\cdot\vec{F}_\perp = \text{constant} = A, \tag{7.3.17}$$

and applying at the outer (side) edge the boundary condition $(cJ/2) = |\vec{F}_\perp|$, one gets an expression for $J$ as a function of the perpendicular coordinate $x$,

$$J = 2A\left[d + 3\kappa\rho(d^2 - x^2)/4\right]/c. \tag{7.3.18}$$

Along the longitudinal coordinate $r$, taking the values of $J$ and $v$ at the center equation (7.3.16) becomes

$$\frac{1}{r^3}\frac{d}{dr}\left(\frac{r^3}{3}\frac{dJ}{dr}\right) + v\frac{dJ}{dr} + \frac{4J}{3r^3}\frac{d}{dr}(r^3 v) + \frac{8cJ}{3\kappa\rho d^2} = 0 \tag{7.3.19}$$

for a hollow column with walls of thickness $d$. For a filled column, one gets in spherical geometry the similar expression

$$\frac{\varepsilon_c a}{3R^2 r^2}\frac{d}{dr}\left(v r^4 \frac{dJ}{dr}\right) + v\frac{dJ}{dr} + \frac{4J}{3r^2}(r^2 v) + \frac{4\varepsilon_c v J}{3R} = 0, \tag{7.3.20}$$

where $a$ is the column radius at $r = R$. This is complemented by the equation of motion,

$$v\frac{dv}{dr} + \frac{GM}{r^2} - \frac{\kappa}{c}F_\parallel = 0, \tag{7.3.21}$$

and the parallel diffusion equation,

$$F_\parallel = \frac{\epsilon_c v r^2 a}{3R^2}\frac{dJ}{dr}, \tag{7.3.22}$$

where the equation of continuity has been used to eliminate $\rho$. This set of equations can be solved, yielding the radiation density and surface flux as a function of height along the column (e.g. Basko and Sunyaev, 1976a).

## 7.4 Negligible Radiation Pressure Models

Negligible radiation pressure obtains when $L < L_c$ (eq. 7.3.3), where care must be taken to include the effects of resonances. In this case, equations (7.3.7) and (7.3.8) must be used, with $\vec{C}_P \simeq 0$ and $C_E$ calculated as indicated above. The solution of these equations must contain either a collisionless or a collisional shock.

### *a) Collisionless Shock Model*

The low-luminosity collisionless shock accretion column is relevant if the instabilities necessary for such a shock are able to develop. There is as yet no quantitative study of such plasma instabilities for plasma motion along the field, the assumption being that something similar to the zero magnetic field two-stream instability might occur as the infalling ionized plasma encounters a stationary or slower moving plasma made up of decelerated or stationary material that must necessarily exist in or above the atmosphere. Whether this indeed is the case or whether the strong magnetic field stiffens up the plasma and reduces its tendency to become unstable is unclear at the moment. Assuming that such a collisionless shock does develop, the properties of the postshock accretion column were considered by Langer and Rappaport (1982). The shock transition is expected to be very thin, of the order of the preshock ($\sim$ free-fall) velocity over the plasma frequency, or possibly of the order of the gyroradius, although the latter is more appropriate for motion across the field. Behind the thin shock transition region both electrons and protons are heated to the adiabatic postshock temperature. If these two components are not coupled by collective effects behind the shock, they will have different temperatures, since they have different masses. Since collective effects are needed to make the collisionless shock, the assumption of the absence of collective effects behind the shock would require careful examination, but if correct, one would have behind the shock a two-temperature plasma. In addition, the electrons will be initially quasi-relativistic, and are subject to rapid radiative cooling by the synchrotron and the inverse Compton effect with X-ray photons from the surface below. The protons on the other hand would not radiate significantly and would lose their energy more slowly via Coulomb collisions with the electrons, a process which strives toward achieving an equalization of the two temperatures. However, to preserve charge neutrality, the bulk velocity of the electrons and ions must be the same. Thus, equations (7.3.7) and (7.3.8) must be expanded to three equations to describe separately the electron and proton components, and a term must be included describing the energy exchange between the two components. Under these conditions, the shock standoff distance is roughly comparable to the product of the postshock proton velocity and the magnetized proton-electron energy exchange time scale $v_{p,2} t_{pe}$. The proton temperature slowly drops toward the surface by giving up energy to the electrons, while the electrons have a height-dependent temperature given by the equilibrium between their heating by protons and cooling by radiation. An example of this behavior is shown in Figure 7.4.1, from Langer and Rappaport (1982).

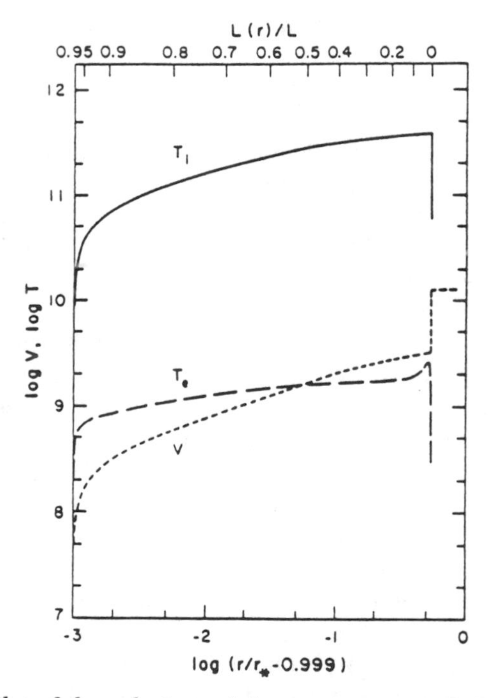


FIGURE 7.4.1 Plot of the velocity and the ion (proton) and electron temperature versus depth in a collisionless shock model. The shock is on the right, and the stellar surface is on the left. From S. H. Langer and S. Rappaport, 1982, *Ap. J.*, 257, 733.

The spectrum from such an accretion column is calculated by these authors in an approximate manner, by assuming that the continuum photons created by bremsstrahlung escape freely (since for negligible radiation pressure the column is optically thin in the continuum) while for the resonant photons, which are not optically thin, an escape probability is estimated on the basis of the time needed for these photons to diffuse in frequency out of the resonance.

*b) Coulomb Collision Decelerated Model*

In the absence of collective effects which could lead to a collisionless shock, and with negligible radiation pressure, the freely falling accreting matter which approaches the polar cap first interacts with the neutron star atmosphere through the moving (beam) and the quasi-stationary atmospheric electron components, possibly through a saturated two-stream instability or else via Coulomb collisions. These two components become indistinguishable within a very short distance. However, most of the inertia of the accretion flow is carried by the ion component, which interacts much less easily with the atmosphere due to its larger mass and thus tries to forge ahead downward past its own and the atmosphere's electrons. They cannot be stopped by the buildup of an electrostatic field, since a small return current is established in the plasma, with the electron gas slowly drifting down through the atmospheric ions. Thus, the only process capable of ultimately slowing down the beam ions is two-body collisions involving the ions.

The situation is basically described by a set of four equations (Zel'dovich and Shakura, 1969). The first two of these are the equations for the proton energy and momentum losses through the atmosphere,

$$\frac{dE}{dx} = Q_E, \tag{7.4.1}$$

$$\frac{dP}{dx} = Q_P, \tag{7.4.2}$$

where $E$ and $P$ are the accreting proton's average energy and momentum, and $Q_E$, $Q_P$ are the loss rates of these quantities to the atmosphere. In a single-particle approach, the connection between the (nonrelativistic) proton momentum and energy is $E = P^2/2m_p$, and early calculations of this deceleration mechanism (Basko and Sunyaev, 1975a, b; Pavlov and Yakovlev, 1976) used this description and assumed that the atmospheric electrons on which the protons were scattered could, to a first approximation, be taken to be cold. In a more detailed description, account must be taken of the fact that an initial small spread in proton momentum is amplified during the deceleration process, so that the average momentum and average energy are not connected by the above one-particle prescription, and the thermal dispersion of the atmospheric electrons must also be considered. This was done by Kirk and Galloway (1982), who calculated the evolution of the accreting proton momentum distribution during the deceleration process. This work was expanded and improved upon by Miller, Salpeter, and Wasserman (1987), Miller, Wasserman, and Salpeter (1989), and Pakey, Bussard, and Lamb (1989).

The other two equations needed for the two-body collisional deceleration problem describe the effects of the deceleration upon the atmosphere, and may be expressed as

$$C_E + Q_E \frac{\phi}{m_p} = 0, \tag{7.4.3}$$

$$\frac{dp}{dx} - \rho g = \frac{Q_P \phi}{m_p}, \tag{7.4.4}$$

where $\phi$ = constant is the mass flux and $m_p$ is the proton mass. These are one-dimensional versions of equations (7.3.5) and (7.3.6) specialized to a stationary atmosphere.

A system of equations similar to (7.4.1)–(7.4.4) but with the additional inclusion of a radiation pressure term in (7.4.4) was solved by Wang and Frank (1981), taking for the scattering a single, constant cross section given by the average value of the strong interaction *p-p* cross section. These were integrated as an initial value problem, starting with a given velocity and density at the top and calculating downwards until both the incoming proton velocity and the temperature of the background particles vanish at the bottom, since they were considered to be heated only by the incoming protons. A solution of a set of equations similar to (7.4.1)–(7.4.4) was also discussed by Braun and Yahel (1984) using the diffusion approximation and including a radiation pressure term in (7.4.4) but neglecting the ram pressure term $Q_P$ in (7.4.1). For $Q_E$ they used the earlier, single proton on cold electron deceleration formulae of Pavlov and Yakovlev (1976), which in this case depart substantially from the results of Kirk and Galloway (1982).

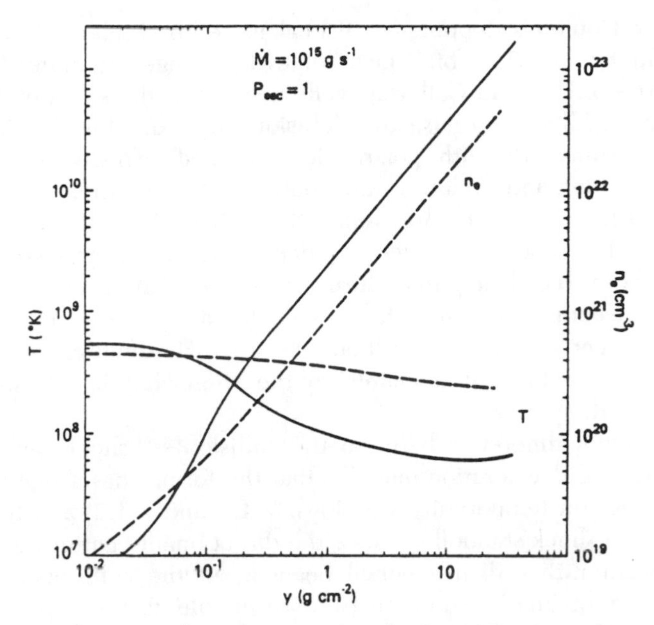


FIGURE 7.4.2 Density and temperature in a Coulomb-decelerated atmosphere plotted against the integrated density $y = -\int_{\infty}^{0} \rho(z)\, dz$ for a magnetic field $B = 5 \times 10^{12}$ G (*solid lines*) and $10^{13}$ G (*dashed lines*). From P. Mészáros, A. K. Harding, J. G. Kirk, and D. Galloway, 1983, *Ap. J.* (*Letters*), 266, L33.

A more complete version of equations (7.4.1)–(7.4.4) given by Mészáros *et al.* (1983) and Harding *et al.* (1984) used the numerical results of Kirk and Galloway (1982) for the distribution-averaged Coulomb and nuclear deceleration of protons, and a two-stream magnetized radiative transfer scheme for calculating the atmospheric losses to estimate $C_E$. The boundary conditions at the top were a given infall velocity and density as well as free radiation escape; at the bottom a complete reflection condition was assumed so that the bottom temperature would not tend to zero. The atmosphere density and temperature profile thus calculated is dynamically self-consistent, being in hydrostatic and energy balance with the radiation and the energy and momentum deposition from the infalling protons. The radiation spectrum and directionality were also calculated. An example of the density and temperature profile is given in Figure 7.4.2.

A more recent calculation of collision-decelerated self-consistent atmosphere models is that of Miller, Wasserman, and Salpeter (1989), using the Coulomb stopping lengths calculated by Miller, Wasserman, and Salpeter (1987), which give somewhat longer stopping lengths than those of Kirk and Galloway while neglecting the effect of nuclear collisions. Miller *et al.* use the diffusion approximation for treating radiation transport, with polarization-averaged expressions for the bremsstrahlung and cyclotron line cooling. For the latter they use an escape probability term and assume that a fixed fraction $10^{-1}$ of the cyclotron line cooling energy is redeposited in the atmosphere by the electron recoil acquired during resonant scatterings. The self-consistent collisional atmospheres calculated by these authors are somewhat cooler and denser than those of Harding *et al.* (1984), which may be attributed mainly to the somewhat longer stopping lengths used.

The main difference between the collisionless shock model and the collisional deceleration model is that the former has a significantly higher electron temperature (see Figs. 7.4.1 and 7.4.2) and that due to the large shock standoff distance the directionality pattern is that of a fan beam rather than a pencil beam as in the collisional model. These two qualitative differences may enable future observers to distinguish between the two models (e.g. White, Swank, and Holt, 1983).

## 7.5 Models with Radiation Pressure

### a) Qualitative Properties

The high-luminosity models $L \gtrsim L_c$ or $\varepsilon_c > \pi$ are of particular interest since they are obviously easier to observe experimentally (e.g. Her X-1). Theoretically they present the simplification that the diffusion approximation can be used, and with radiation pressure dominant over gas pressure $C_E$ does not need to be used explicitly. At the same time, the dependence on the choice of boundary conditions and the numerical difficulties associated with a diffusion solution in two dimensions become larger issues (e.g. Kirk 1984).

The pioneering treatment of this problem is due to Davidson (1973) (see also Maraschi, Reina, and Treves, 1974), who solved equations (7.3.10)–(7.3.13) in the approximation where nonmagnetic Compton exchange is the only significant photon-electron interaction and the gravity term $\vec{g}$ is neglected. This allows equation (7.3.12) to be integrated analytically, so the rest of the system reduces to a second-order partial differential equation which is integrated numerically. The boundary conditions used in this case are zero velocity and radiation flux at the stellar surface, zero radiation density $J$ along the sides and top of the column to represent free escape of radiation, and free-fall velocity at the top. These solutions show that the radiation pressure decelerates the matter and creates a mound-like pile of dense, subsonically settling gas below the radiation shock. This was the first qualitative indication of the type of geometry to be expected in a realistic accretion column, although in most subsequent studies the effect of gravity was typically retained, as it is important in the subsonic settling regime.

*b) One-dimensional Radiation-dominated Shocks*

Several of the subsequent studies including gravity and other effects were carried out using the one-dimensional approach described by equations (7.3.20)–(7.3.22) (Inoue, 1975; Basko and Sunyaev, 1976a). These equations require three boundary conditions, and the most straightforward (although not necessarily correct) procedure would seem to be again to set the gas velocity and the radiation flux equal to zero at the bottom ($r = R$) and the gas velocity equal to the free-fall velocity at the top ($r = r_1$),

$$v(r_1) = \sqrt{(2GM/r_1)}\,, \tag{7.5.1}$$

$$v(R) = 0, \tag{7.5.2}$$

$$F_{\parallel}(R) = 0. \tag{7.5.3}$$

However, equation (7.3.21) does not allow solutions with $v \to 0$, $F_{\parallel} \to 0$ at the lower boundary unless one also sets the gravity term there equal to zero. This is a problem typical of the equations dominated by radiation pressure: a pressure gradient must exist even at the bottom of the column, and if this can only be provided by radiation, there must be a finite flux there. Also, if the velocity tends to zero at the surface, from the continuity equation the density would have to tend to infinity there and the gas pressure could no longer be negligible. An alternative possibility is to consider a surface just above that where the velocity tends to zero and to set there (instead of 7.5.3) a condition including the advected energy,

$$(4/3)vJ + F_{\parallel} = 0. \tag{7.5.4}$$

This means that another condition must be used instead of (7.5.2), depending on what further approximations or modifications are made to the one-dimensional equations (7.3.20)–(7.3.22). A critical discussion has been given by Kirk (1984).

The simplest approximation in dealing with the one-dimensional equations is to neglect gravity, as was also done by Davidson (1973) in two dimensions (e.g. Basko and Sunyaev, 1976a). These authors do not use the condition of zero velocity at the surface, equation (7.5.2), but instead demand that $J \to 0$ at $r \gg R$. The solutions obtained for low luminosities, $\varepsilon_c > 1$, give a significant velocity at the stellar boundary, as expected since the radiation pressure would not be able to significantly decelerate the matter, while for $\varepsilon_c = 1$ a singular solution is found which has $v(R) = 0$. For higher luminosities, $\varepsilon_c < 1$, deceleration starts to occur in a narrow range $\Delta r$ at $r > R$ in a structure reminiscent of a radiation shock. In this zone, gravity can no longer be neglected, but on the other hand the inertia term $v(dv/dr)$ may be set approximately equal to zero in equation (7.3.21). This gives two first-order differential equations, the divergence of the radial flux being identically zero. In this second approximation, Basko and Sunyaev (1976a) choose a top boundary condition appropriate for the subsonic side of radiation-dominated shocks,

$$v = -v_{\rm ff}/7 = -(2GM/r)^{1/2}/7, \tag{7.5.5}$$

supplemented by another condition, which is needed since one does not know beforehand the position at which they would apply,

$$\frac{4}{3} vJ = \frac{\phi R^2}{2r^2}\left[v_{\rm ff}^2 - v^2\right] = \frac{\phi R^2 v_{\rm ff}^2}{2r^2} \tag{7.5.6}$$

(Zel'dovich and Raizer, 1972). Using as a remaining boundary condition equation (7.5.4), Basko and Sunyaev (1976) obtain two equations for $J$ and $v$ as a function of radius which give a solution to the problem under these conditions. However, the radius at which the shock front condition (7.5.6) could be fulfilled is finite only if $\varepsilon_c \gtrsim 1$, i.e. $L \lesssim L_c$, where one would not expect a radiation shock. This problem is associated with the use of the bottom boundary condition (7.5.4). In order to achieve a physically sensible solution, Basko and Sunyaev (1976a) introduced instead another boundary condition at the bottom, which stipulates that the radiation pressure at the bottom is just balanced by the magnetic field confining pressure. This implies a finite flux of energy into the star, or in the case where $B \gtrsim 2 \times 10^{13}$ G, when the temperature reached is sufficient for neutrino losses to become important, a nonradiative flux of energy out of the accretion zone. If this solution is stationary, it represents a radiation shock at a standoff distance which increases with the accretion rate (a behavior opposite to that in shocks dominated by gas pressure) and a subsonic settling zone beneath it, with the interesting property that the radiative losses from the sides tend to a maximum limiting value. The interior temperature and the lateral effective surface temperature are calculated by Basko and Sunyaev (1976a) as a function of radius for values near this limiting luminosity, within the framework of this model.

A different physical possibility is that radiation pressure becomes dominant only after deceleration through a collisionless shock has occurred. This was considered by Inoue (1975), who used similar boundary conditions as (7.5.1)–(7.5.3). A short distance below the collisionless shock, matter and radiation were assumed to come into equilibrium, and Inoue gave an approximate solution of the one-dimensional equations resulting in a settling column somewhat similar to that previously discussed. This possibility was also considered by Braun and Yahel (1984), who extended the one-dimensional equations to include the effect of a finite gas pressure, the possibility of a two-temperature situation, and the effect of Coulomb energy exchange between ions and electrons. These authors use a more realistic boundary condition on the radiation field than that of setting $J = 0$ at the free boundary, as assumed by Davidson (1973) and implicit in Basko and Sunyaev's (1976a) use of equation (7.5.1) for the top of the column. Braun and Yahel (1984) instead use a more appropriate Marshak boundary condition,

$$J = -\frac{2}{3\kappa\rho}\hat{n}\cdot\vec{\nabla}J. \tag{7.5.7}$$

This condition applies at the boundary between plasma and vacuum at the sides of the column, and Braun and Yahel (1984) assume this also to be valid at the top surface, where they place the collisionless shock. A more appropriate surface for this boundary condition might be at the position where the preshock plasma becomes optically thin, which at high accretion rates may be farther above the collisionless shock (Kirk, 1984). Under the previous assumption, Braun and Yahel concluded that a collisionless shock would not lead to stable solutions and that only a Coulomb-decelerated flow would be stable.

### *c) Two-dimensional Radiation-dominated Shock Solutions*

The first two-dimensional treatments (Davidson, 1973; Maraschi, Reina, and Treves, 1974) led to the concept of an accretion mound.

Some of the simplifications in these early treatments, such as the use of nonmagnetic cross sections and the neglect of gravity, were lifted by Wang and Frank (1982), who used a flux-limited form of the diffusion equations with magnetic parallel and perpendicular opacities valid at frequencies below the resonance. They solved numerically the equations (7.3.10)–(7.3.13) for an axisymmetric column and applied the boundary condition (7.5.7) at a layer of constant $r$ well away from the deceleration zone, which for the top does not in general coincide with the distance at which the flow just becomes optically thin (see previous discussion). At low accretion rates $\varepsilon_c > 1$ they found solutions in which the flow is not substantially decelerated, while for high accretion rates $\varepsilon_c \lesssim 1$ their solutions produce a mound of subsonic plasma qualitatively similar to that in Davidson (1973). For these higher accretion rates, a deceleration zone similar to the radiation shock appears which moves outward with increasing accretion rate, in qualitative agreement with Basko and Sunyaev (1976a). It is interesting that Wang and Frank obtain this behavior using the bottom boundary condition (7.5.4) at the stellar surface, whereas Basko and Sunyaev obtain it assuming that there is a finite energy flux into the star.

An analytical two-dimensional treatment combining a collisionless shock with an extended radiation-dominated sinking zone beneath it was investigated by Kirk (1985). In this case one expects a thin transition layer below the collisionless shock where the gas and the radiation have not yet achieved equilibrium and where gas pressure is important. However, in the sinking zone, equations (7.3.10)–(7.3.13) are valid, and these are solved in cylinder symmetry. Instead of using the Marshak boundary condition (7.5.7), which would be valid at some surface where the flow becomes optically thin between the collisionless shock and the sinking zone, jump conditions are derived which are applied between the preshock plasma and radiation and the equilibrium plasma and radiation in the sinking zone. At the bottom, the conditions are (7.5.4) and the demand of a particular functional dependence, in order to ensure the existence of stationary solutions. With the help of this last bottom condition, which introduces an arbitrary parameter $k$ that may vary between two extreme values, Kirk (1985) obtains a set of simple analytic solutions which describe in two dimensions the shape of the accretion mound, the gas and radiation temperature in the sinking zone, and the location of the collisionless shock. These solutions (see Fig. 7.5.1) are valid for the low-luminosity case, $\varepsilon_c > 1$.

The two-dimensional density and temperature structure in the

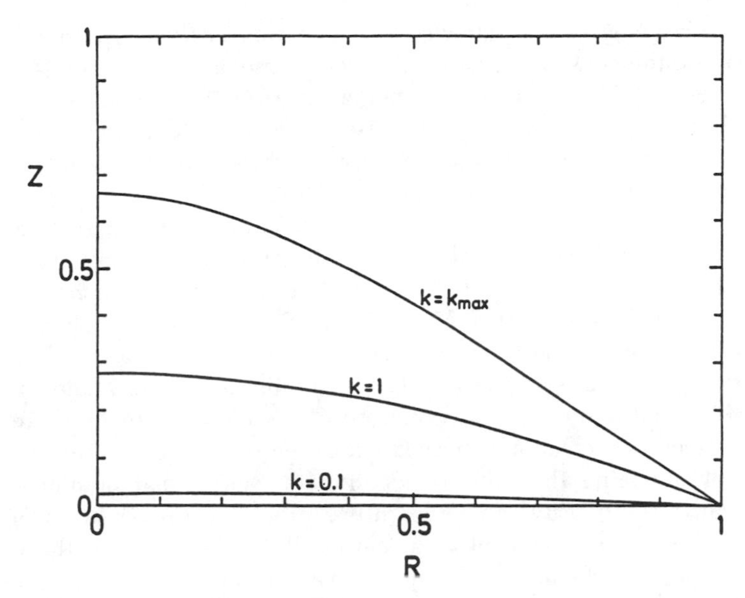


FIGURE 7.5.1 Shape of the accretion mound for the analytic solution of the low-luminosity collisionless shock, with a radiation-dominated region below it, for different values of the parameter $k$ related to the luminosity. From J. G. Kirk, 1985, *Astron. Ap.*, 142, 430.

case where the deceleration is caused mainly by the radiation pressure and the gravitational acceleration is neglected has been computed numerically by Herold, Wolf, and Ruder (1987). This allows the equation of motion to be integrated explicitly and its solution inserted into the energy equation, leading to an elliptical partial differential equation for $Q = v^2/v_\infty^2$, as a function of the dimensionless parameter $R_0 = \kappa\rho_\infty(v_\infty/c)R_{\rm col}$, where $\kappa$ is the effective mean opacity per unit mass and $R_{\rm col}$ is the cylindrical column radius. The resulting velocity and mass density profiles are functions of the height and the distance from the center of the column, $v(z, r)$, $\rho(z, r)$. Rebetzky *et al.* (1988, 1989) have used these density and velocity gradients, together with the assumption of a constant plasma temperature, a constant magnetic field, and negligible Comptonization effects, to solve the two-dimensional radiative transfer in the diffusion approximation, using magnetic Rosseland mean transverse and longitudinal diffusion coefficients and the Marshak boundary conditions. These authors find that advection plays a major role in the dynamics, the more so at frequencies near the resonance, where the scattering cross section is large. At these frequencies, the constant-density surfaces are much closer to the surface than they are for frequencies well away from the resonances. The conclusion is that, when bulk motions are included, cyclotron line photons can only escape from a narrow height around the column, just above the stellar surface.

The validity of the diffusion approximation in moving media is considered by Rebetzky *et al.* (1989), who find that it breaks down for velocities $\beta = v_\infty/c > 0.3$ or for accretion rates $\phi = \rho v > 2.3 \times 10^6$ g cm$^{-2}$, or $\dot{M} > 2 \times 10^{-9}\ M_\odot$ yr$^{-1}$ at $R_{col} = 1$ km. An alternative method of solving the transfer equation in moving media based on an expansion in scattering orders has been discussed by Maile *et al.* (1989). In all of the above calculations of the radiation hydrodynamics of the accretion column, coherent scattering was used as an expedient way of obtaining the diffusion coefficients and radiation moments, leading to the density and temperature profile. However, the Compton heating and cooling plays a major role in determining the temperature, and should be included. For a medium in hydrostatic equilibrium this has been done by Riffert (1987), using magnetized resonant cross sections. A similar treatment for a moving medium remains to be done.

A major effort in solving the self-consistent two-dimensional magnetic radiation hydrodynamic problem of the accretion column at the near critical accretion rates has been under way for some time (Arons, Klein, and Lea, 1986; Arons, 1987; Klein and Arons, 1989). This is a time-dependent numerical treatment where matter is assumed to fall along the dipole field lines but inhomogeneities in both the radial and transverse coordinates are considered. The radiation transfer is treated in the flux-limited diffusion approximation, using polarization-averaged resonant and nonresonant cross sections to derive parallel and transverse diffusion coefficients, and Compton heating and cooling is included. For the hydrodynamic problem, the question of the boundary conditions is given particular attention, especially at the top surface where the transition between optically thin and thick flow occurs. The numerical difficulties are considerable, particularly at high luminosities, where the radiation pressure becomes important, since in two dimensions this appears to lead to the formation of convective structures resembling photon "bubbles" which move up subsonically, while around them denser regions of matter continue to fall inward. The radiation field in these high-luminosity simulations is of the fan beam type, coming from the sides of the plasma-filled dipole surface. An

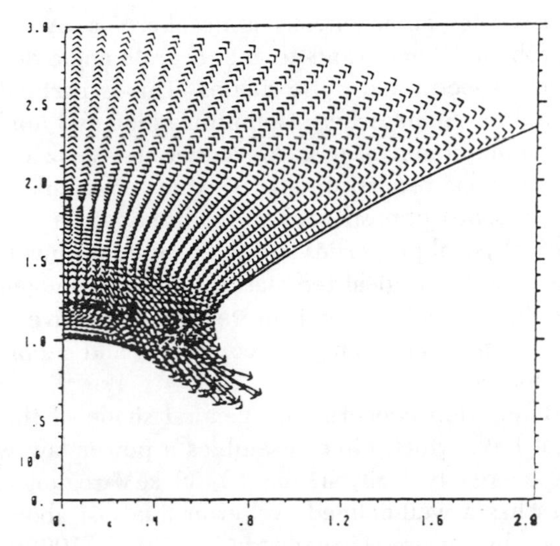

FIGURE 7.5.2 Radiation field in a dynamic infall calculation with $L_x = 5 \times 10^{37}$ ergs $s^{-1}$. The vectors show the direction of the photon flux, with the length of the vector representing the flux magnitude. From J. Arons, 1987 in *IAU Symp. 125, The Origin and Evolution of Neutron Stars*, ed. D. Helfand (Reidel, Dordrecht), p. 207. Reprinted by permission of Kluwer Academic Publishers.

example from one of the preliminary calculations (Arons, 1987) is shown in Figure 7.5.2. The radiation spectrum from such a configuration will of course be made up of the contributions from surface elements at varying temperatures extending from the stellar surface to several stellar radii above it, so that the spectrum calculation is complicated, and requires the development of approximate techniques. Spectral calculations so far have been limited to much more simplified accretion models, discussed in the next section.

## 7.6 Spectrum and Pulse Shape Models

### *a) Angle-averaged Spectrum*

Detailed spectra and pulse shapes of AXPs have been calculated for simplified plasma configurations, where the density, temperature, and optical depth are obtained from simplified static equivalents of the accreting atmosphere, or from approximate analytic or numerical dynamic models. This approach has been found expedient due to the complicated nature of both the transfer and the dynamics. The transfer in a static, homogeneous atmosphere is by itself a major numerical problem, if one wants to get accurate angle-dependent line and continuum spectra and pulse shapes. This is particularly true of the cyclotron line regime, where Comptonization and nonlinear transfer effects become dominant (e.g. Alexander and Mészáros, 1991a, b). In Chapter 6 various results of radiative transfer calculations for such simple homogeneous atmospheres were discussed in detail, concentrating on the physical properties and the techniques involved. It turns out that, despite their idealized nature, static homogeneous atmospheres are able to yield several interesting qualitative answers to a number of questions concerning the conditions and geometry of some typical AXP sources.

The first question concerns the general shape of the continuum above several keV, which often resembles a power law with a break above some energy typically in the 10–30 keV region and in some notable cases has a well-defined cyclotron line just above the break. As emphasized by Nagase (1989a) and Clark *et al.* (1990), when a line is present, its energy is typically $\sim 0.5$–$0.7$ of the cyclotron energy. As shown by Nagel (1981b), for any reasonable optical depth the line will be a negative feature, looking like an absorption line although in reality what happens is that photons are being scattered out of the line. The question is then to see whether with reasonable optical depths, temperatures, and densities one can get a flux comparable to that observed, with a ratio of continuum (below the break) to line flux similar to that observed. For the plasma parameters, one can take temperatures compatible with the accretion rate (eq. 7.1.1) (or higher if there is a shock above the surface), a density at least as large as the free-fall density $\rho = \dot{M}/v_{\rm ff}2A \sim 10^{-4}L_{38}$ g cm$^{-3}$ (or larger if the matter has gone through a shock), and an optical depth somewhere between the Thomson depth of free-falling material across the column width of $\sim 1$ km and the nuclear stopping length of infalling protons on atmospheric protons $\sim 55$ g cm$^{-2}$, i.e. $6L_{38} \lesssim \tau_T \lesssim 22$. In order to get a sufficiently realistic looking line profile and line depth in an angle-averaged (or phase-averaged) spectrum, as observed for instance in Her X-1, it is necessary to use cross sections which include at least thermal and vacuum effects (the latter so that both polarizations are strongly resonant). With these, the total flux, the shape of the continuum above $\sim 5$ keV, the break energy, and the line to continuum ratio of the angle-averaged spectrum can be reproduced with reasonable parameters (Mëszáros and Nagel, 1985a). This is shown in Figure 7.6.1. In the above calculation, a source of soft photons at energies

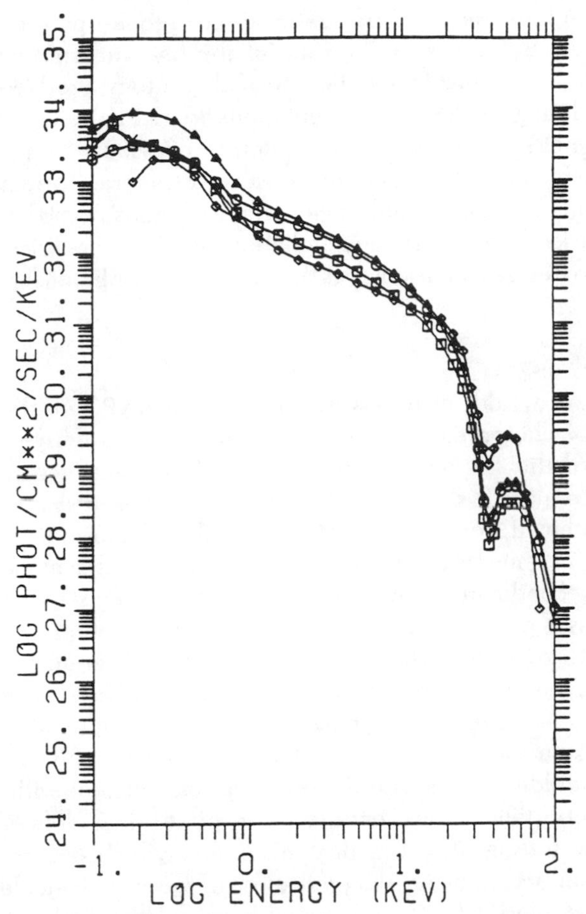


FIGURE 7.6.1 Angle-averaged spectrum of a static homogeneous atmosphere with a soft-photon source compared to the observations of Her X-1 (*diamonds*). Three models are shown with $\hbar\omega_c = 38$ keV, $kT = 7$ keV, and $\rho = 1.67 \times 10^{-4}$ g cm$^{-3}$. The triangles and circles are for a soft photon source at the outer (inner) boundary, no escape at the inner boundary, and $\tau_T = 6.6$, while the squares are for a soft photon source at the inner boundary, free escape from both outer and inner boundary, and twice the previous optical depth. From P. Mészáros and W. Nagel, 1985, *Ap. J.*, 298, 147.

$E_s \lesssim 1$ keV was assumed, in order to provide the low-energy portion of the spectrum. These soft photons are up-Comptonized by the hot electrons in the atmosphere to produce a quasi-power law shape between $E_s$ and $E \sim kT$. The ratio of the line and continuum below the break is dominated by the thermal photons produced through bremsstrahlung in the slab. Calculations (e.g. Figure 6.4.3 or 6.4.4) which do not include such a soft-photon source show a turnover below $E \lesssim kT$, which is not seen in most observed spectra. The source of the soft photons is not clear, although surface photons from the star or the Alfvén surface could produce radiation at these energies, and two-photon processes can also act as a source of soft photons.

### *b) Pulse Shapes*

There are a wealth of observational data on AXP pulse shapes as a function of photon energy, and the challenge is to derive from these data information about the physical conditions and geometry of the emission region. As first pointed out by Nagel (1981a), the pulse shapes obtained from a static cylinder configuration radiating as a fan beam are in general too broad to be acceptable as a model for most AXPs. A second and stronger argument in favor of a slab configuration emitting as a pencil beam follows from a consideration of the models required to reproduce the cyclotron line energy variation with pulse phase that is seen e.g. in Her X-1 (Mészáros and Nagel, 1985a). From the spectra as a function of angle discussed in Chapter 6 (Figs. 6.4.3 and 6.4.4) one sees that the cyclotron line energy, and the energy of its blue shoulder, varies with the observational angle $\theta$ with respect to the magnetic field. However, the pulse phase $\Phi$ is defined observationally by setting $\Phi = 0$ at flux maximum, which occurs when the emitting surface is being sampled closest to normal incidence. This occurs for $\theta$ closest to 0 in the pencil beam (slab), but for $\theta$ closest to $\pi/2$ in the fan beam (cylinder). As a result the energy of the line, or of the blue shoulder, varies inversely with pulse phase $\Phi$ in the pencil and fan beam models. In fact, only the pencil beam models are able to reproduce the observed behavior of decreasing line shoulder energy as the phase $\Phi$ increases away from zero (see Fig. 7.6.2). This behavior is in contrast to that of fan beam models, where the line blue shoulder energy would increase with increasing phase. Thus, one is led to infer that the beaming is of the pencil type. Notice that, strictly speaking, what is meant by pencil and fan beam in this case is that the radiation escapes predominantly along or perpendicular to the magnetic field

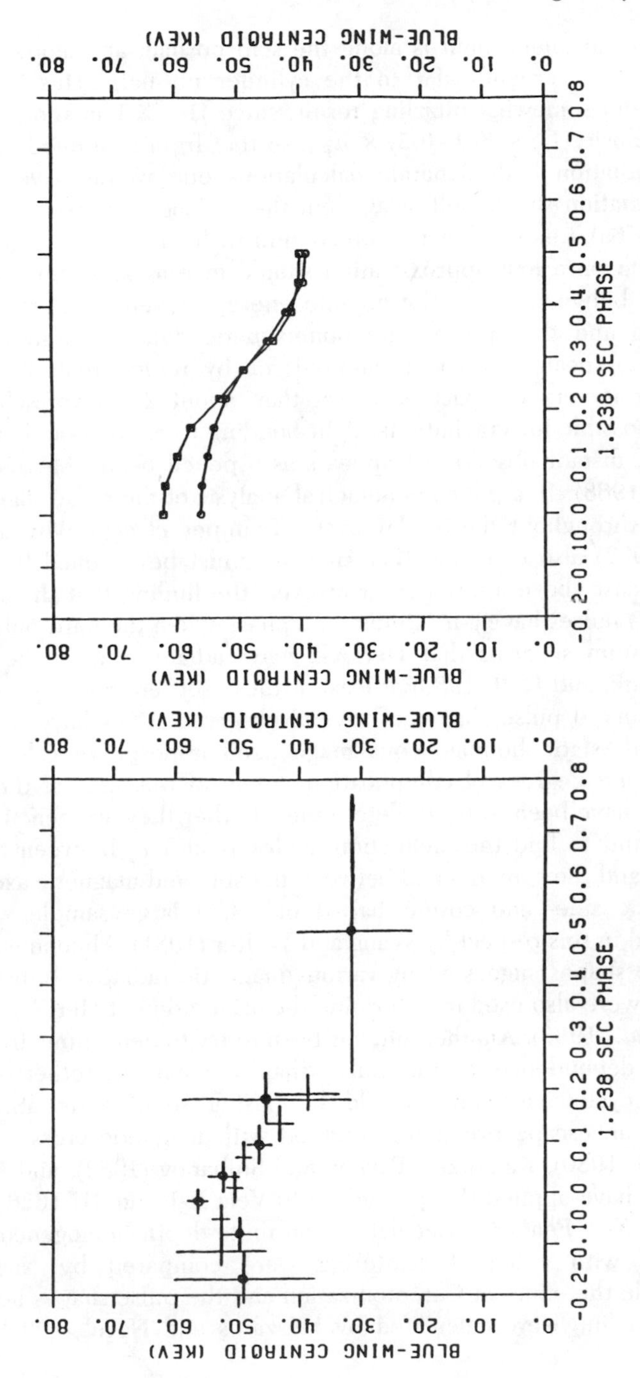


FIGURE 7.6.2 Energy of the blue shoulder of the cyclotron line in Her X-1 as a function of pulse phase Φ. *Left*: The curves show the results for two generic pencil beam models (from P. Mészáros and W. Nagel, 1985, *Ap. J.*, 298, 147). *Right*: The observed line blue shoulder energies from the MPI/AIT and MIT experiments (from W. Voges *et al.*, 1982, *Ap. J.*, 263, 803).

direction (the magnetic field is along the slab normal, and along the cylinder axis or perpendicular to the cylinder normal). This is an interesting and somewhat puzzling result, since Her X-1 is known to have a luminosity $L_x \sim (0.1-0.2) \times L_{Ed}$, so that from simplified nonmagnetic radiation hydrodynamic calculations one would have expected a radiation shock well away from the surface (e.g. Basko and Sunyaev, 1976a), i.e. a high accretion column with an extensive lateral surface which on a first approximation should give a fan beam. This discrepancy between the cyclotron line energy evidence favoring a pencil beam and the approximate nonmagnetic dynamic argument favoring a fan beam may be accounted for by noting that, if the neutron star is very compact (size less than about 2 Schwarzschild radii), it is possible for gravitational light bending to curve a fan beam so that, to a distant observer, it appears as a pencil beam (Mészáros and Riffert, 1988). In a phenomenological analysis of the pulse shapes of Her X-1 throughout the 35-day cycle, Trümper *et al.* (1986) and Kahabka (1987) also conclude that Her X-1 must be a pencil beam emitter, at least above a few keV. However, the finding that the soft ($E_x \lesssim 1$ keV) pulses have a maximum at a phase when the hard pulses are at minimum suggests that Her X-1 may radiate as a fan beam (White, Swank, and Holt, 1983) at least at these soft energies.

The observed pulse shapes of several observed AXPs have been modeled with static homogeneous magnetized atmosphere calculations of varying degrees of complication. Some of the aims of these comparisons have been to try to determine whether they are pencil or fan beams, and to find the inclination angles $i_1$ and $i_2$ between the line of sight and the spin axis and between the spin and magnetic axes. Using generic sine- and cosine-shaped pulses, a large sample was characterized in this respect by Wang and Welter (1981). Phenomenological pulse shapes suggested by various magnetic radiative transfer calculations were also used to determine the orientation of Her X-1 by Trümper *et al.* (1986). Another aim has been to try to determine, from the energy dependence of the pulse shapes, what the otherwise unobserved cyclotron energy would be. Using semiinfinite atmospheres and an escape probability method with magnetic cross sections, Kanno (1980), Kaminker, Pavlov, and Shibanov (1982), and Kii *et al.* (1986) have applied this procedure to Vela X-1 and 4U 1626 − 67. For Her X-1, Feautrier calculations on finite-depth homogeneous atmospheres with coherent scattering were compared by Nagel (1981a), while the effect of Comptonization and the pulse shapes near the cyclotron line were calculated by Mészáros and Nagel (1985b).

More recently, Clark *et al.* (1990) detected a cyclotron line in 4U 1538 − 52 and fitted the pulse shapes of Mészáros and Nagel (1985a) to the observed pulses, obtaining good qualitative agreement both in the continuum and in the line.

### c) *Effects of Mass Motions and Secondary Components*

The next step in modeling observed AXPs involves using spectra and pulse shapes calculated from models incorporating bulk motions and inhomogeneities. Only preliminary efforts have so far been made along these lines. The effects of inhomogeneities on the observed spectra and pulse shapes have been used in some specific models (e.g Langer and Rappaport, 1982; Harding *et al.*, 1984; Burnard, Klein, and Arons 1990; Burnard, Arons, and Klein, 1991). These models have not been compared in detail to observations, either because the transfer was not sufficiently detailed or because the temperature and density profiles used were uncertain. An inclusion of the effects of bulk motions requires a knowledge of the velocity profile, for which an accurate solution of the radiation hydrodynamic problem is required. However, various approximate or simplified velocity profiles have been incorporated in preliminary attempts to evaluate the interaction between mass motions and the angular dependence of the spectra. Thus, a self-consistent velocity profile obtained from a nonmagnetic diffusion hydrodynamic treatment was used by Rebetzky *et al.* (1989) to calculate two-dimensional angle-dependent spectra in the coherent scattering limit, for low accretion rates. At higher accretion rates, where they find the diffusion treatment breaks down, Maile *et al.* (1989) used a homogeneous constant bulk velocity equal to the final free-fall velocity at the surface to calculate the photon beam shapes. These calculations include the special relativistic Doppler shifts and aberration, which affect both the spectra and the pulse shapes. More interestingly, the convection and aberration of photons cause a significant fraction of the flux to be beamed downward onto the surface, where it is thermalized and ultimately reradiated. The effect of this is to create a halo of photons around the base of the column (Kraus *et al.*, 1989). The beaming of these blackbody (and therefore soft) photons reradiated by the surface will be at right angles to the main pulse of harder energies if the latter comes from the side surface of the column. Such a qualitative mechanism had been long suspected to provide the explanation for the phase difference observed between the very soft ($E \lesssim 1$ keV) and hard ($E \gtrsim 2$ keV) pulses.

The effect on the pulse shapes expected from a mound-shaped settling region was investigated by Burnard, Arons, and Klein (1991), using their inclined magnetic field inhomogeneous code (Burnard, Klein, and Arons 1988). The accretion mound was modeled as a set of five static inhomogeneous slabs whose density and temperature gradient is given by an analytic estimate and whose surface normals have increasing angles with respect to the magnetic dipolar axis (i.e. more convex than a spherical polar cap). This treatment uses the coherent scattering approximation so that Comptonization effects are ignored, which does not allow a study of the cyclotron line frequency region. It also prevents a consideration of the Comptonization of the mound photons in the radiation deceleration shock region which lies above the mound, which is expected to alter the spectrum considerably. These authors suggest that the static slab calculations of Mészáros and Nagel (1985a, b) involving a slab with $\tau_T \sim 6$–$8$ backlit by a soft-photon source (Fig. 7.6.1) actually represent the shock region Comptonizing the mound photons. As was also the case in that study, a satisfactory source of soft photons remains elusive, as the mound is expected to produce somewhat harder photons (several keV) than required for the Comptonizing slab model to reproduce the observations (0.1–0.3 keV). An interesting result from the Burnard *et al.* study is that they find a decrease in the degree of pulse modulation in the mound case, as opposed to the case of a flat slab model. This is understood in terms of the fact that for a mound even at large angles with respect to the magnetic axis one is still looking at a considerable fraction of the surface almost face-on, i.e the limb-darkening effects present in a flat slab are diminished. Another consequence is that the expected degree of polarization is diminished relative to that of a flat slab, since one is sampling photons arriving from a range of regions that have differing angles between the surface and the magnetic field direction. However, this effect does not appear to be severe, the calculation indicating values of the linear polarization degree up to 50%, not much below the values calculated by Mészáros *et al.* (1988) for the simple slab and column models.

A calculation of the special and general relativistic effects of a two-component model involving a settling region below plus scattering in an infalling plasma column above was carried out by Brainerd and Mészáros (1991). In this calculation the density in the column is given by the continuity equation, using a free-fall velocity profile modified by radiation pressure. The dipolar variation of the magnetic field is taken into account, while the transverse density variation is averaged out. The photon spectrum from the base of the column, where the matter has already stopped and thermalized, is assumed to be given by one of the static magnetized atmospheres calculated by Mészaros and Nagel (1985b). The scattering of this radiation by the infalling column electrons is calculated with a Monte Carlo code, some of the photons escaping and others being beamed down onto the surface, where they thermalize and are reradiated. The blackbody temperature around the base depends on the flux of incident photons, and ultimately on the accretion rate. The unscattered, scattered, and reradiated photons move along the Schwarzschild metric characteristics to reach the observer. The total spectrum shows a larger flux at low energies than expected from the thermalized static lower portion, because of the soft reradiated flux. An interesting feature is that each cyclotron harmonic seems to appear twice, once at the zero-velocity resonant frequency and once at the Doppler-shifted frequency of the rest frame of the infalling matter. It remains to be seen whether in a more realistic velocity profile with a transverse density gradient this second Doppler-shifted cyclotron feature remains distinguishable. Another interesting feature, similar to that discussed above, is that the beaming is part pencil, part fan, depending on the photon energy and the accretion rate. In this case, however, the angular shape and energy distribution also depends, because of general relativistic effects, on the radius to mass ratio of the neutron star.

# 8. Rotation-powered Pulsars

## 8.1 Observational Overview

### *a) Introduction*

Rotation-powered pulsars (RPPs), which are usually detected via their radio emission, derive their luminosity from their rotational energy. The rotational energy decreases as the pulsar radiates, as evidenced by the long-term spin-down behavior of virtually all single pulsars that have been observed for a sufficiently long time. This is in contrast to accreting X-ray pulsars (AXPs), which, even though also rotating, derive most of their luminosity from the gravitational accretion energy of infalling gas that they capture, and in most cases show a secular spinning-up behavior. Historically, RPPs were the first type of neutron star observationally discovered (Hewish *et al.*, 1968), having been theoretically predicted much earlier (Landau, 1932; Baade and Zwicky, 1934a, b). The existence of a strong magnetic field in these objects had been suspected, on the basis of flux freezing arguments, and some of the observational properties of fast rotating, strongly magnetized neutron stars were predicted in remarkable manner by Pacini (1967) previous to the observational discovery. The regularly pulsating radio signals from a discrete radio source (PSR 1919 + 21) detected by Jocelyn Bell and Anthony Hewish were quickly interpreted as coming from a rotating neutron star, an interpretation which became unavoidable after all alternative possibilities had been disproved (e.g. Gold, 1968). The observational material and much of the theoretical work previous to 1976 have been discussed by Manchester and Taylor (1976), and a number of excellent reviews have appeared since then (Taylor and Stinebring, 1986; Lyne, 1987a, b; Manchester, 1987). Two recent books on the subject are those of Lyne and Graham-Smith (1990) and Michel (1991).

The total number of RPPs that have been observed as of 1990 is close to 500, which makes it the largest sample of neutron stars studied. All the RPPs so far identified (except possibly PSR 0540 − 69) emit detectable radio radiation, and all but 2 of these have been discovered via their radio emission. Two objects were discovered in the X-ray band (PSR 1509 − 58 and 0540 − 69; Seward and Harnden, 1982; Seward *et al.*, 1984). At least 2 are strong and pulsating $\gamma$-ray sources, the Crab (PSR 0531 + 21) and Vela (PSR 0833 − 45), both of which as well as 0540 − 69 also pulsate in the optical wave band. The Crab, 1509 − 58, and 0540 − 69 are also detected as pulsating X-ray sources, but no pulsed X-ray emission has been observed from the Vela pulsar. At hard X-ray energies and low $\gamma$-ray energies the Crab and Vela have been detected with balloon-borne scintillators, and at medium $\gamma$-ray energies by satellites such as *SAS 2* and *COS B*, using spark chamber detectors, plastic scintillation counters, and Cherenkov counters. The Crab and Vela pulsars have been positively identified as pulsating at energies $E \gtrsim 50$ MeV, with the same period as observed in the radio range. There are in addition about 20 or so unidentified *COS B* sources at these energies which, although not observed to pulsate, are strongly suspected of being pulsars. The 4 pulsars detected so far at high energies have unusually short periods and large period derivatives, and the 3 of them which are associated with an observable SN remnant (Crab, Vela, and 1509 − 58) have the shortest observed spin-down times $P/\dot{P}$. So far only a half-dozen pulsars have been detected near the centers of detectable SN remnants, which is related to the fact that the latter dissipate on a time scale about two orders of magnitude shorter than the typical RPP emission lifetime.

### *b) Galactic Distribution*

All but two of the RPPs discovered thus far are in our galaxy. Their spatial distribution is typical of population I objects distributed along the galactic plane, albeit with a somewhat larger mean height above the plane. The typical pulsar scale height is $H_z \sim 400$ pc, as opposed to $\sim 50$–100 pc for supernova remnants and massive stars, which are the presumed progenitors. This larger scale height is probably explained by their observed relatively large transverse peculiar velocities ($v_t \sim 170$ km s$^{-1}$ or $|v| \sim 200$ km s$^{-1}$), which would cause them to wander off the plane. The radial distribution in the plane is almost constant from the center out and starts to taper off at about the galactocentric radius of the sun, although some RPPs are inferred to be at distances as large as $\sim$ 15 kpc. The plane-projected luminosity function $\Phi(L)$ in units of [pulsars $kpc^{-2}$] is approximately constant below luminosities of $L \sim 1$ mJy $kpc^2$ and drops off as a power law above that. Most of the active pulsars in the galaxy (after correction for observational biases: Lyne, 1987b) are inferred to have luminosities in the range 0.3–3 mJy $kpc^2$, but most of the observed pulsars have about a hundred times higher luminosity. After correction for observational biases, there should be about $7 \times 10^4$ pulsars beaming toward earth, and assuming a circular beam shape and a beaming fraction $f \sim 0.2$, the total number of active pulsars in the galaxy with $L \gtrsim 0.3$ mJy $kpc^2$ is estimated to be $N \sim (3.5 \pm 1.5)10^5$ (Lyne, Manchester, and Taylor, 1985). Since the escape velocity from the galaxy is about 300 km $s^{-1}$, only 20–30% of pulsars can have escaped the galaxy. The rest would reach a maximum distance of several kpc in a time scale of 30–60 Myr. However, there is evidence that the pulsar (radio) lifetime is several Myr (see below) so that pulsars populating the halo of the galaxy would be mostly defunct. The total number of dead pulsars in the galaxy is thus estimated to be $N_T \sim 3 \times 10^8$–$3 \times 10^9$. The birthrate of pulsars needed to maintain the number estimated to be active in the galaxy is about $60^{+60}_{-30}$ per year (Lyne, Manchester, and Taylor, 1985).

*c) Secular Period Changes*

The distribution of observed RPP spin periods $P$ is dominated by objects in the range $0.2 \lesssim P \lesssim 2$, with a smattering of shorter-period objects down to about 1.6 ms and one object as high as 4.3 s. The median period is $P_m = 0.67$ s. The principal evolutionary effect detected in these objects is an overall lengthening of the spin period, expressed by its derivative $\dot{P} < 0$. The observed spin-down time scales $P/\dot{P}$ are typically of order several Myr. A plot of observed $\dot{P}$ versus $P$ is shown in Figure 8.1.1. The very young pulsars still associated with SN remnants have very short periods and large period derivatives, and are found in the upper left of this diagram. If the pulsar radiates according to the magnetic dipole law and if magnetic fields decay exponentially, as originally assumed (Gunn and Ostriker, 1970), objects in this diagram would move toward the lower right along a slope $-1$ line and then drop downward. However, the data suggesting field decay have been the subject of controversy, and currently the question of magnetic field decay is a subject of intense inquiry (e.g. Bhattacharya and Srinivasan, 1990); this question is discussed further in

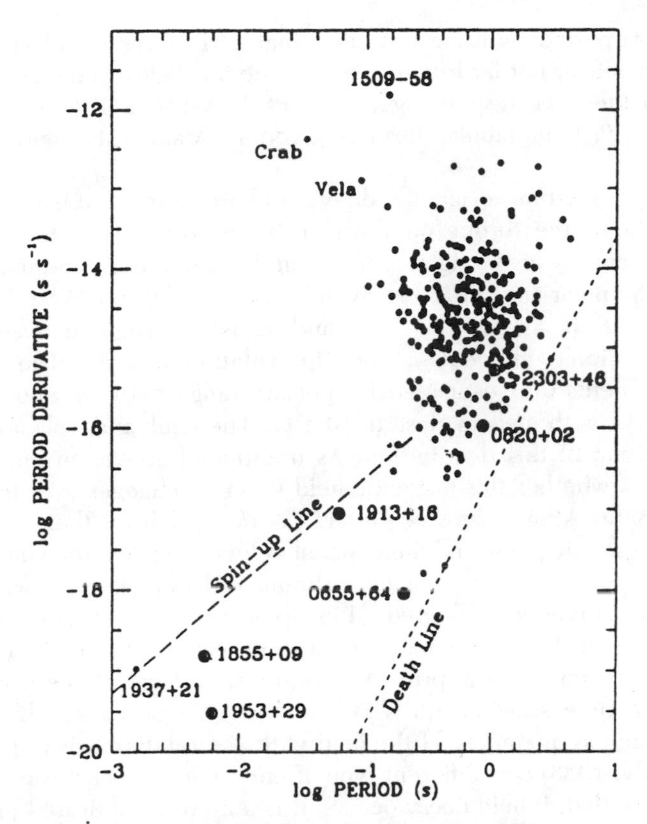


FIGURE 8.1.1 $\dot{P}$ versus $P$ distribution for 308 pulsars. Six binary pulsars are shown with a circle around the dot. The death line indicates the region below which pulsar emission appears to cease, while the spin-up line indicates the shortest period achievable by spin-up of a recycled pulsar in a binary, discussed in Chap. 11.3. From Taylor and Stinebring (1986). Reproduced, with permission, from the *Annual Review of Astronomy and Astrophysics*, vol. 24, © 1986 by Annual Reviews Inc.

Chapter 11.4. If pulsars spun down without changing their luminosity, their numbers would increase continuously toward long periods, which is not seen to be the case in Figure 8.1.1. In fact, the radio luminosity is found to decrease toward the lower right in Figure 8.1.1, with $L_R \propto P^{\alpha}\dot{P}^{\beta}$ with $\alpha = -1.04 \pm 0.15$, $\beta = 0.35 \pm 0.06$ (Proszinski and Przybcien, 1984), and no pulsars are detected to the right of and below the line labeled "death line," corresponding to $\dot{P} \propto P^3$. The supershort-period (millisecond) and binary RPPs are found either to the lower left or not far from the death line but below the line labeled "spin-up line," corresponding to $\dot{P} \propto P^{3/4}$. A list of RPP properties including $P, \dot{P}$ in tabular form is given in Manchester and Taylor (1981).

In the vacuum magnetic dipole radiator model (Ostriker and Gunn, 1969), the torque on the star is proportional to the flux of magnetic energy at the light cylinder, and consequently the magnetic field at the polar cap surface is given by $B_0 \sim 10^{12}(\sin \alpha)^{-1}(P\dot{P}_{-15})^{1/2}$ G, where $\dot{P}_{-15} = \dot{P}/10^{-15}$ s s$^{-1}$ and $\alpha$ is the angle between the magnetic dipole moment $\vec{\mu}$ and the rotation axis $\vec{\Omega}$. The stellar magnetic fields thus estimated for pulsars range between about $10^{13}$ and $10^{10}$ G, with median close to $10^{12}$ G. The Crab and Vela are near the high end of this distribution. As mentioned above, an important question is whether this magnetic field strength changes over time. If one plots the kinetic ages of pulsars (as derived from their distance from the galactic plane and their spatial velocity) against the characteristic spin-down age for a vacuum dipole radiator ($n = 3$), which is $P/2\dot{P}$, the curves for observed RPPs appear to level off (Fig. 8.1.2), indicating that the torque decays on a time scale of about $9 \times 10^6$ yr. This leveling off could in principle be due to a decay of the magnetic field on a time scale of $4.5 \times 10^6$ yr but may be caused by other effects, such as alignment of the field with the rotation axis (e.g. Blair and Candy, 1988) or a different type of energy loss mechanism. There is evidence that, if field decay occurs, it is a more complicated process than previously thought; some pulsars appear to have low ($10^8$–$10^9$ G) but very long-lived fields ($\gtrsim 10^9$ yr), the decay may not be simply exponential, and there may be mechanisms that generate or boost up the field strength (Chap. 11.4). These difficulties notwithstanding, the RPP radio luminosity is also inferred to decay on a time scale $\sim 5$ Myr, comparable to that of the torque decay (Lyne, Manchester, and Taylor, 1985). Pulsars appear to be moving toward the right in the $\dot{P}, P$ diagram and cross the death line on this time scale. Another interesting observation is that of a positive correlation (Cordes, 1987; Helfand and Tademaru, 1977) between the strength of the magnetic moment ($\propto B_0$, as derived from $P\dot{P}$) and the transverse spatial velocity of the pulsar.

A second derivative $\ddot{P}$ of the period has been measured in three pulsars so far (Crab, 1509 − 58, and 0540 − 69). The measured values of the braking index $n = \Omega\ddot{\Omega}/\dot{\Omega}^2$ are $2.509 \pm 0.001$, $2.83 \pm 0.03$, and $2.02 \pm 0.01$, respectively, the last one in X-rays (Nagase

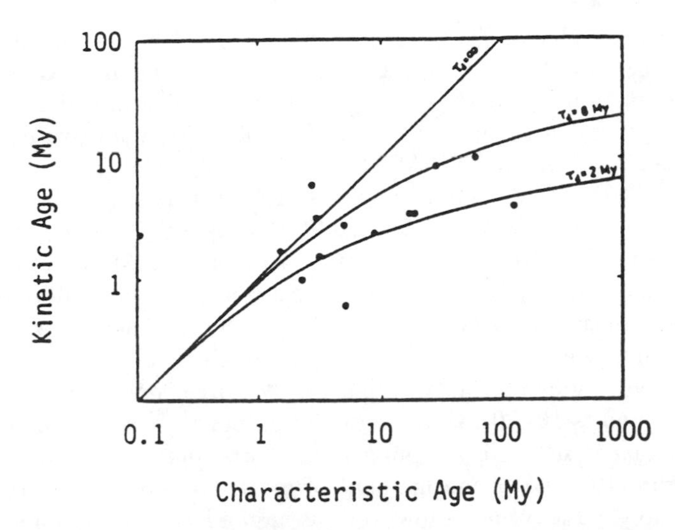


FIGURE 8.1.2 Relation between the kinetic and characteristic ages of pulsars. Assuming that the change in $P\dot{P}$ represents an evolution of the magnetic dipole moment, the curves indicate the path expected for the various field decay times (see text for alternative explanations). From A. G. Lyne, 1987, in *The Origin and Evolution of Neutron Stars*, ed. D. Helfand and J.-H. Huang (Reidel, Dordrecht), p. 63. Reprinted by permission of Kluwer Academic Publishers.

*et al.*, 1990). For a vacuum magnetic dipole radiator one expects the braking force to be $\propto \Omega^3$ so that $\dot{\Omega} = -K\Omega^n$, and the braking index is expected to be $n = 3$, somewhat larger than the observed values. This indicates that some other energy loss mechanism besides magnetic vacuum dipole radiation must be operating.

*d) Period Instabilities*

Besides the long-term spin-down, shorter time scale period instabilities such as glitches and noise are also detectable in some pulsars. Episodic period glitches are seen in a small number of RPPs in which $P$ and $\dot{P}$ abruptly change, showing a spin-up instead of the secular spin-down, and then relax toward the extrapolation of their former behavior over a time of days to months. Vela is the most frequent glitcher, having had eight at 2–3 yr intervals, with typical relative changes $\Delta\Omega/\Omega \sim 2 \times 10^{-6}$, $\Delta\dot{\Omega}/\dot{\Omega} \sim 8 \times 10^{-3}$, and the latest event (Hamilton *et al.*, 1989; Flanagan, 1990; McCulloch and Hamilton,

1990) giving values of $1.8 \times 10^{-6}$ and $2 \times 10^{-1}$–$4.4 \times 10^{2}$, respectively. The Crab has had three glitches, with relative changes $\sim 10^{-8}$ and $\sim 10^{-3}$, and PSR 1641 − 45 and 0525 + 21 have had one each. The largest glitch so far is that of PSR 0355 + 54, with two changes, the second of which had $\Delta\Omega/\Omega \sim 4.4 \times 10^{-6}$ and $\Delta\dot{\Omega}/\dot{\Omega} \sim 10^{-1}$ (Lyne, 1987b). Two other glitchers are also young pulsars, PSR 1823 − 13, with $\Delta\Omega/\Omega \sim 2.5 \times 10^{-6}$, $\Delta\dot{\Omega}/\dot{\Omega} \sim 0.06$, and PSR 1737 − 30, with $4.2 \times 10^{-7}$ and $2.6 \times 10^{-3}$, respectively (McKenna, 1988). The presence of glitches indicates the existence of two or more poorly coupled stellar components, e.g. the solid crust and a superfluid component (see Chap. 11.3). In the giant glitch of PSR 0355 + 54 and the latest glitch of the Vela pulsar, the naive conclusion would be that as much as 10–20% of the moment of inertia becomes decoupled, if what one is seeing is coupling to the superfluid core. This would imply that 10–20% of the moment of inertia resides in the crust, an improbably large value. However, as argued by Alpar, Pines, and Cheng (1990), the recovery time of $\lesssim 1$ day in the latest Vela glitch is too fast to represent coupling to the core and may instead result from the linear coupling of a portion of the pinned crustal superfluid (about $5 \times 10^{-3}$ of the total stellar moment of inertia) to the core superfluid.

Another interesting perturbation of the period is a component with noise-like characteristics which is seen in the power spectrum of pulsars with large $\dot{P}$, such as the Crab. This perturbation is typically white noise, but there is also a red component (increasing toward low frequencies). The $\dot{P}$ dependence of the noise (e.g. Cordes and Helfand, 1980) implies that the millisecond pulsars would be expected to be the stablest in this respect (very low $\dot{P}$), which is observed to be the case. The nature of the noise has received much attention because of the potential information it conveys about the internal structure of neutron stars (e.g. Pines, 1987).

### e) Binary and Millisecond Pulsars

Since the discovery of the first binary pulsar (PSR 1913 + 16: Hulse and Taylor, 1974), with a very short period of 59 ms, a number of other objects with similar or related characteristics have been found. As of 1990 the list of millisecond and/or binary RPPs comprises over thirty objects. A list is shown in Table 3 (after Backer and Kulkarni, 1990, and Backer, 1990) near the end of the book, giving the $P$, $\dot{P}$, derived $B_0$, binary period $P_{\rm orb}$ in days, binary eccentricity $e$, companion mass $M_c/M_\odot$, and binary orbit semimajor axis $a/R_\odot$ (Taylor,

1987; Manchester, 1987; Fruchter *et al.*, 1988; Ables *et al.*, 1988; Lyne *et al.*, 1988; Wolszcan *et al.*, 1988). Typically these objects have very low $\dot{P}$ and very low to low $P$, and hence lie in the lower left of the $P$, $\dot{P}$ diagram (Fig. 8.1.1), between the spin-up line and the death line. As expected from the very low $\dot{P}$, they have little or no timing noise, being remarkably stable in their spin-down (Cordes and Downs, 1985). Their very low $(P\dot{P})$ also implies in general a very low magnetic field. As a group, they have smaller than average galactic mean scale height ($\bar{z} \sim 165$ pc, as opposed to the pulsar mean of $\sim 400$ pc), which is related to a lower than average spatial velocity $v_{\rm pec}$. Their integrated pulse shapes and pulse polarization appear not to differ significantly from those of the rest of the RPP population (next subsection). It is remarkable that $\gtrsim 60\%$ of the RPPs with periods $P \lesssim 60$ ms are binary.

The wealth of interesting physical information from the binary systems is remarkable. The masses derived for the companion typically indicate an evolved low-mass star, although in two cases the companion is inferred to be a neutron star on the basis of its mass ($\sim 1.4\ M_{\odot}$) and other constraints. The best studied system, 1913 + 16, has allowed a measurement of the general relativistic advance of the periastron (4.2°/yr) and a measurement of the transverse Doppler and the gravitational redshift, yielding masses $M_p = 1.442 \pm 0.003$, $M_c = 1.386 \pm 0.003$, and orbital inclination angle $i = 47° \pm 3°$ (Taylor and Weisberg, 1989). The accurate $\dot{P}$ measurement also allows us to distinguish a contribution $\dot{P}_{\rm GR} = -2.38 \times 10^{12}$ s s$^{-1}$ due to general relativistic gravitational quadrupole radiation which agrees within 1% with the theoretical values. A particularly interesting case is PSR 1957 + 20, where the companion mass is extremely low, $M_c = 0.02$–$0.05\ M_{\odot}$, and the pulsar is regularly eclipsed by the companion (Fruchter *et al.*, 1988). The companion is in fact inferred to be losing mass, being ablated by the energy input from the pulsar wind.

### *f) Radio Pulse Shapes, Structure, and Polarization*

Although the individual consecutive pulse shapes received from an RPP can vary erratically in shape and intensity, when they are integrated and averaged over a large number of periods (typically $\gtrsim 10^3$) the integrated profiles are remarkably stable, allowing an extremely accurate determination of the phase of arrival. The timing accuracy obtained with these integrated profiles is comparable in many cases to that of the best atomic clocks. Generally the integrated pulse shapes have a low duty cycle, having a typical average equivalent width $\overline{W}_e \sim 10°$, where 360° is a full spin period (Lyne and Manchester, 1988). The integrated pulse shapes can be single, double, triple, or multiple peaked, often asymmetric, and in some cases with an interpulse about 180° separated from the main pulse (see Fig. 8.1.3). There is a slight dependence of the pulse shapes on frequency in that the components tend to separate or broaden slightly with decreasing frequency. In contrast to the integrated pulse shapes, the individual pulses are made up of one or more subpulses, which are generally narrower, $W_e \sim$ few°, falling somewhere in the window delineated by the integrated waveform. Their position inside this window is usually random, although in some cases there is an organized drifting behavior (see below). The intensity of subpulses varies greatly from one pulse to the next, often exhibiting periodic or quasi-periodic modulation (Taylor, Manchester, and Hugenin, 1975). On a shorter time scale (10–100 $\mu$s), micropulses are observed, with varying intensity, whose width is $\ll 1°$ (Manchester and Taylor, 1977). These micropulses imply an emission region $R \sim ct \lesssim 10^6$ cm, one of the arguments for a neutron star polar cap origin. The pulses and subpulses are believed to be due to rotation of a narrow beam of radiation emanating from the polar cap of a neutron star, while the micropulses are believed to be due to temporal modulation of this emission.

The linear polarization of the pulses is very high, often approaching 100% in the subpulses and up to 50% in the integrated pulses (Stinebring *et al.*, 1984), and in many cases there is also a smaller amount of circular polarization, $\lesssim 30\%$. The linear polarization is typically smoothly varying across the pulse width in a characteristic S-shaped curve (Fig. 8.1.4), changing sign at pulse center. This smooth S-curve is best explained by the polar cap model of pulsar emission (Radhakrishnan and Cooke, 1969), on the basis of curvature radiation from the open field lines emanating from the polar caps. The polarization sometimes shows a jump by 180%, suggesting that the dominant polarization may be either parallel or perpendicular to the plane of the magnetic field lines. In the subpulses the polarization is found to be either parallel or perpendicular to this plane, and the integrated pulses show the polarization which results from the sum of these, being dominated by the most frequently occurring polarization.

Both the pulse shapes and the smooth polarization changes are best explained in the hollow cone model of pulse shape, which may arise in the context of curvature radiation since the intensity is highest where the curvature is largest, i.e. in the outermost lines of the cap,

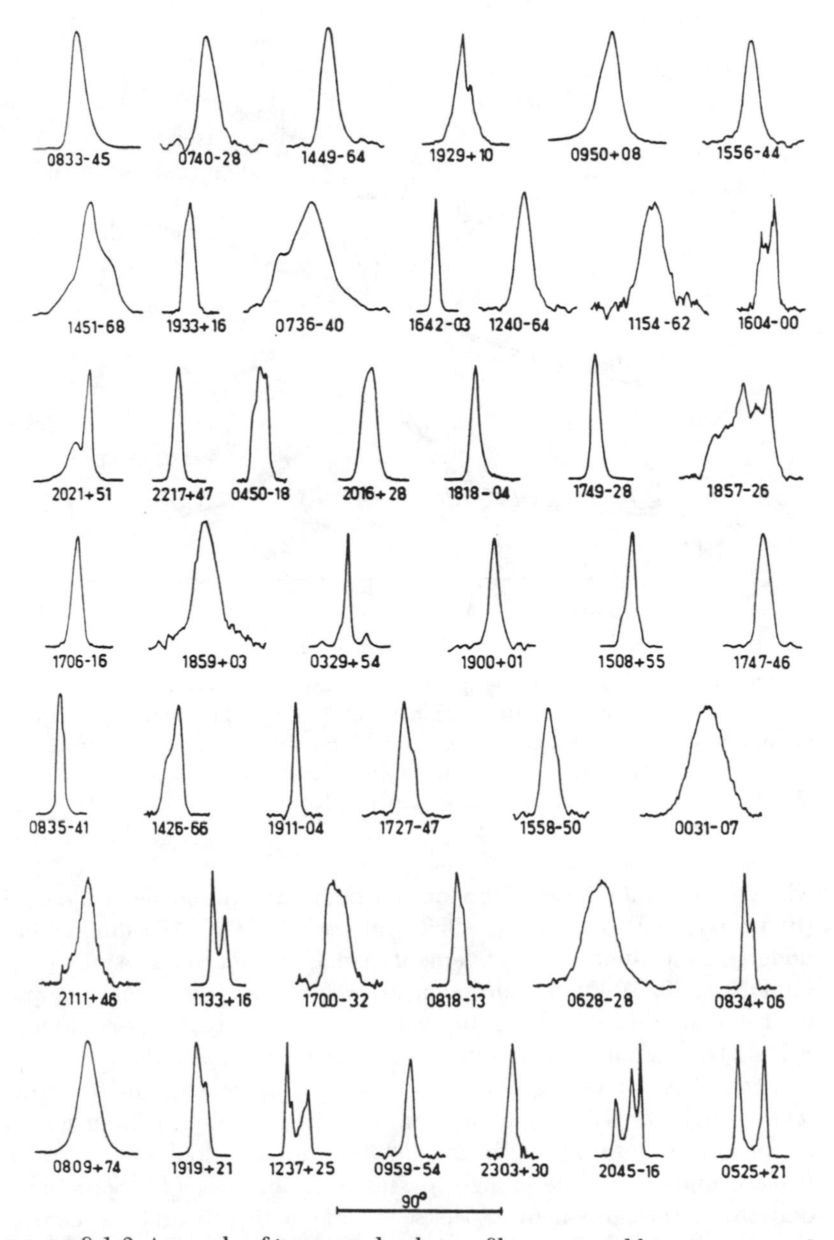


FIGURE 8.1.3 A sample of integrated pulse profiles measured between 400 and 650 MHz, drawn on the same longitude scale and arranged in order of increasing period. From R. N. Manchester and J. H. Taylor, *Pulsars*. Copyright © 1977 by W. H. Freeman and Co. Reprinted by permission.

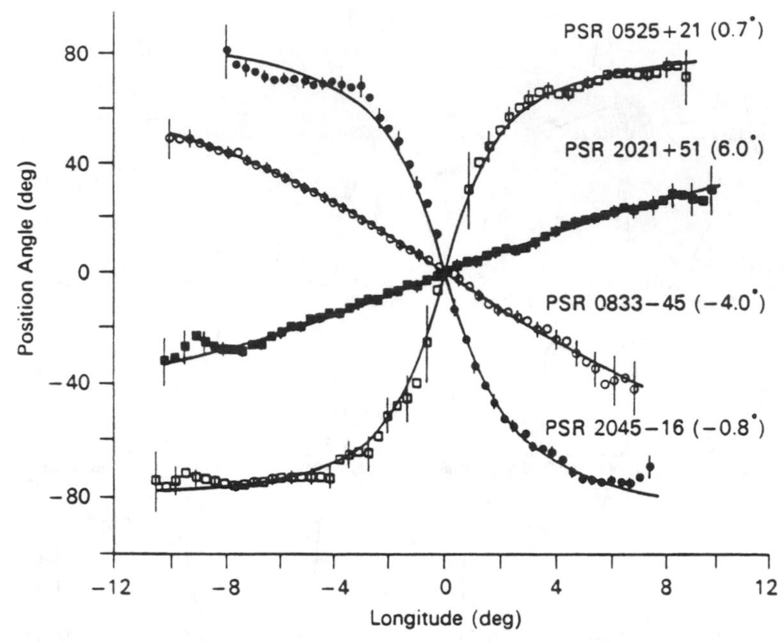


FIGURE 8.1.4 Variation of the polarization angle of the integrated pulse profiles of four pulsars with their best fit curve. The minimum angular separation of the line of sight to the magnetic axis is indicated in parentheses. From R. N. Manchester and J. H. Taylor, *Pulsars*. Copyright © 1977 by W. H. Freeman and Co. Reprinted by permission.

while the emission is polarized linearly parallel to the plane of the field (Rhadhakrishnan and Cooke, 1969; Komesaroff, 1970). This model has undergone a number of refinements since the earlier versions (e.g. Rankin, 1983, 1990). In particular, it is apparent that a core component is necessary to explain the variety of pulse shapes seen (Figure 8.1.5). Depending on where the line of sight intersects the polar cap, a single pulse is seen if one intersects the outer edge of the cone (conal single, or $S_d$), or a double pulse if the line of sight intersects it closer to the center but without touching the core (double, or $D$), or a triple profile if the line of sight passes over the pole ($T$). Sometimes only the core component ($S_t$) is seen, and both this and the central component of the $T$ exhibit significant circular polarization, of symmetrically alternating sense. There are also multiple ($M$) profiles, with a core and two sets of outriding conal components. The conal emission

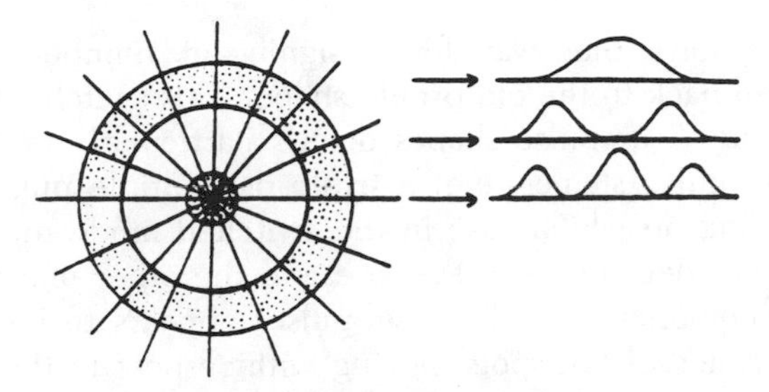

FIGURE 8.1.5 Schematic diagram of pulsar cap geometry showing the core and cone beam components and the influence of the line of sight impact parameter upon the pulse morphology. From A. G. Lyne, 1987, in *High Energy Phenomena around Collapsed Stars*, ed. F. Pacini (NATO ASI, Kluwer, Dordrecht), p. 121. Reprinted by permission of Kluwer Academic Publishers.

is believed to arise at distances of a few stellar radii above the surface, while the core component appears to come from very close to the surface (Rankin, 1990). The increasing conal component separation with decreasing frequency is understood in terms of the decreasing frequency of emission and increasing field line opening with height (Cordes, 1978). There are some indications that the integrated beam shape as given by Figure 8.1.5 may not be spherical but elongated in the altitude direction (Narayan and Vivekanand, 1983), based on the predominance of pulsars with a small polarization angle swing, which would imply a large impact parameter. Other authors (Lyne, 1987a; Barnard, 1988) argue that, since the emission is lumpy (random subpulses, etc.), a circular beam fit cannot be ruled out. The fact that the core component shows circular polarization which changes handedness near pulse center, and that this appears to be connected to the integrated pulses rather than to the subpulses, has been interpreted by Rhadakrishnan and Rankin (1990) as evidence for a geometric origin of this effect, requiring at least the core beam to be acircular to break the symmetry.

### *g) Pulse Nulling, Mode Change, and Subpulse Drift*

Pulse nulling is a phenomenon observed in many pulsars whereby the intensity drops by more than two orders of magnitude for a number of periods, which in some pulsars may be very many (e.g. thousands). The emission then switches back on and continues to behave as previously. In the mode change, the integrated profile changes shape abruptly and stays that way for a significant number of periods, reverting then back to the old profile shape. This switch between two different "modes" or pulse shapes occurs particularly in the case of "incomplete" pulse shapes, e.g. a triple or multiple pulse with one outrider missing, in which case in the switched mode the previously suppressed outrider comes to the fore and the other one disappears.

The phenomenon of drifting subpulses appears to be connected with the presence of hot spots moving with respect to the integrated pulse shape. There is appreciable systematic subpulse modulation in the conal components of *D*, *T*, and *M* pulsars, but a regular, progressive subpulse drifting seems to occur only in the conal single pulsars ($S_d$), i.e. when the line of sight grazes the outer rim of the cone. The drift generally occurs from the trailing edge toward the leading edge of the integrated pulse (Fig. 8.1.6). In the case of the regular drifting subpulses, a Fourier timing analysis yields besides the spin period $P = P_1$ a second periodicity $P_2$, the timing difference between the

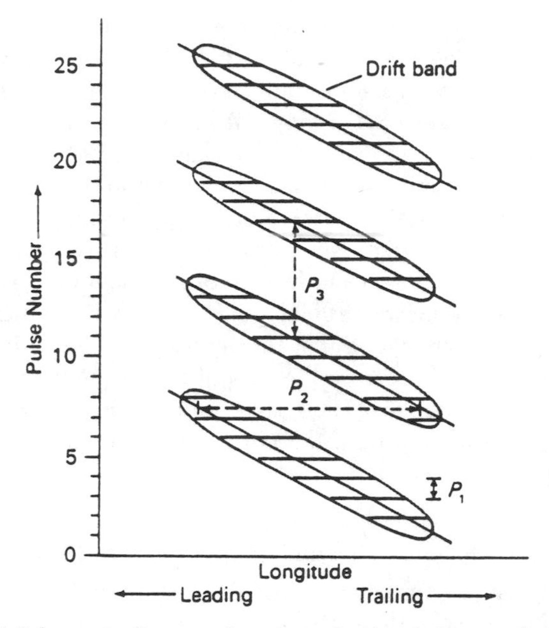


FIGURE 8.1.6 Schematic diagram showing subpulse drifting and identifying the three main periods. The main pulse period is $P_1$, the spacing between subpulses is $P_2$, and the spacing between bands is $P_3$. After R. N. Manchester and J. H. Taylor, *Pulsars*. Copyright © 1977 by W. H. Freeman and Co. Reprinted by permission.

leading and trailing edges of the integrated pulse, and a third period $P_3$, which is the separation between groups of marching subpulses or the time between a cycling of the subpulses between the trailing and leading edges of the integrated pulse. Pulse nulling sometimes occurs in the middle of a subpulse drifting sequence, and after nulling ceases it takes some time before settling down to an orderly drifting again.

### *h) Beam Radius and Magnetic Alignment*

A correlation appears to exist between the half-opening angle of the conal emission beam, $\rho$, and the period $P$ for pulsars within the same pulse shape family. For cone-dominated pulsars with both leading and trailing components present, Lyne and Manchester (1988) get a dependence of the form $\rho \simeq 6.5^\circ P^{-1/3}$, with $\rho$ in degrees, under the assumption of a perpendicular rotator. Rankin (1990) obtains separately for the $M$ objects $\rho \simeq 4.3^\circ P^{-1/2}$ and for the core single $S_t$ objects $\rho \simeq 6.1^\circ P^{-1/2}$. In general, she finds the core single $S_t$ objects to have the smallest characteristic ages, the triple to have intermediate ages, and the $S_d$, $D$, and $M$ to be relatively old. Young pulsars thus appear to possess a strong, dominant core beam, and as they age the conal component becomes more competitive at increasingly lower frequencies.

The inclination angle between the rotation and magnetic axes of pulsars can be obtained by fitting the pulse shapes with the magnetic polar cap model (Lyne and Manchester, 1988). These authors find that, except for the millisecond pulsars (which may have a more complicated history), the relative inclination angle appears to become smaller with increasing characteristic life $P/2\dot{P}$. This finding is very interesting, because in the usual vacuum magnetic dipole radiator model one expects the pulsars to align as they age, whereas in other models involving polar cap longitudinal currents a counteralignment is predicted, which depending on the current could overwhelm the alignment tendency (see § 8.2.$c$). If the characteristic time is a good measure of age, the previous pulse fitting results imply, however, an alignment time scale of $t_{\rm align} \simeq 10^7$ yr, comparable to the time scale for magnetic torque (or alternatively magnetic field strength) decay. This agreement certainly suggests a causal connection between the two phenomena.

### *i) High-Energy Spectra and Pulse Shapes*

Only a handful of RPPs have been observed at energies above the radio range, and of these only a few are pulsating. The ones so far firmly identified as pulsating at high energies, the Crab (0531 + 21) and Vela (0833 − 45), are among the very youngest, short-period objects. The remarkable thing is that these pulsars, which are also detected at high energies, are actually much more luminous in this range than in the radio, emitting most of their photon energy in the gamma-ray range.

The Crab pulsar was first shown to emit pulsed radiation in the optical range by Cocke, Disney, and Taylor (1969), in the X-ray range by Fritz *et al.* (1969), and in the gamma-ray range by Hillier *et al.* (1970). Subsequently, detailed gamma-ray pulse shapes and spectra were obtained with the *SAS 2* experiment (Thompson *et al.*, 1975) and the *COS B* experiment (Wills *et al.*, 1982). A composite of the pulse shapes of the Crab pulsar from radio through gamma-rays, using a summation of five *COS B* observations, is shown in Figure 8.1.7. The period of the high-energy pulses coincides within timing errors with that of the radio pulses, the phase of arrival of the high-energy pulses coinciding with those of the radio main pulse and interpulse, which are separated by a phase interval of $0.404 \pm 0.002$. In the radio range below 1 GHz there is a precursor pulse slightly ahead of the main pulse, for which there is no high-energy counterpart in the Crab. Observations at even higher energies up to the TeV range have also been reported, and are discussed in Chapter 10.1. Unlike at radio energies, the gamma-rays show a significant pulsed fraction, even between the pulse and interpulse, whose amount is energy dependent, and the relative strength of the pulse and interpulse is also energy dependent (Bignami and Hermsen, 1983). The amount of the pulsed fraction increases with increasing energy, the differential photon number spectrum of the pulsed component between 0.1 and 1.0 GeV being approximated with a power law of slope $-2.17$ (Hermsen, 1981). The maximum power emitted $E^2N(E)$ (including all bands from radio to gamma-rays) occurs around 50 keV (Bignami and Hermsen, 1983). Assuming that the emission occurs in two conical beams, the total luminosity between 50 MeV and 10 GeV is estimated to be (Buccheri, 1980) $L_\gamma \sim 2 \times 10^{35}$ ergs $s^{-1}$, which is about 0.04% of the total spin-down energy.

The Vela pulsar is the brightest object in the sky at $E \gtrsim 100$ MeV, with a flux $F \sim 1.3 \times 10^{-5}$ $cm^{-2}$ $s^{-1}$, as compared to $3.7 \times 10^{-6}$ $cm^{-2}$ $s^{-1}$ for the Crab (the distances are $\sim 0.5$ and $\sim 2.0$ kpc for Vela and the Crab, respectively), but unlike the latter, Vela has not so far been observed to pulsate in the X-ray range. Vela has also been reported at TeV energies (see discussion in Chap. 10.1). Pulsed

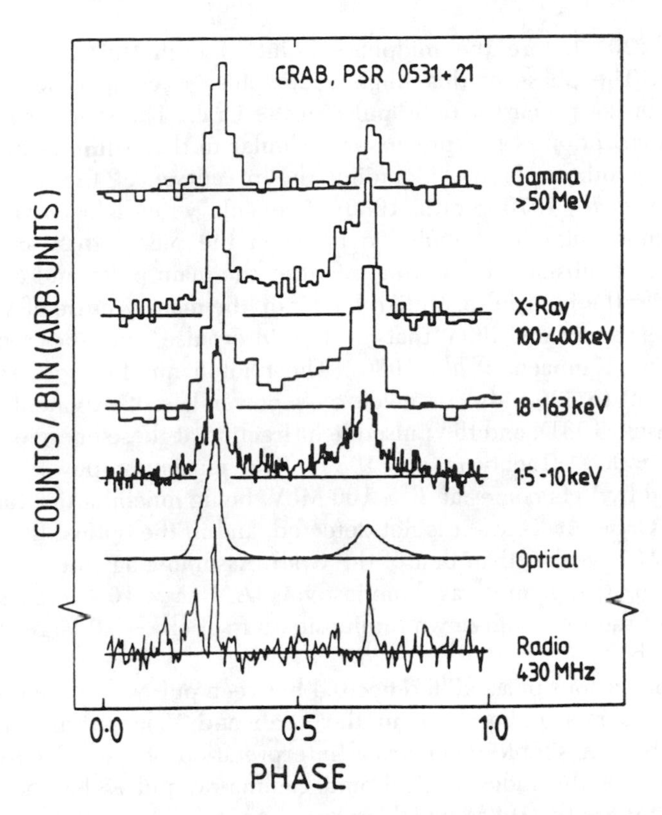


FIGURE 8.1.7 Crab pulsar light curves from 100 MeV (based on the *COS B* data of Wills *et al.*, 1980) through X-rays and optical to the radio range. From Bignami and Hermsen (1983). Reproduced, with permission, from the *Annual Review of Astronomy and Astrophysics*, vol. 21, © 1983 by Annual Reviews Inc.

optical emission from this source was first obtained by Wallace *et al.* (1977), and $\gamma$-ray pulsations were first detected with the *SAS 2* experiment by Thompson *et al.* (1975) and with *COS B* by Buccheri (1976) and co-workers. The light curves of Vela are significantly different from those of the Crab. While the phase separation of the gamma-ray pulse and interpulse is $0.420 \pm 0.007$, which is compatible with the separation of all pulses in the Crab, the optical main and interpulse of Vela are separated by about 0.25, with a midphase point coincident with that of the $\gamma$-ray pulses. The radio emission shows a single pulse which comes earlier than the gamma-ray main pulse, about 120° before the midphase point of both the $\gamma$ and optical pulses. The phase of this single radio pulse is compatible with the phase of the precursor radio pulse in the Crab. The shape, spectrum, and polarization of this precursor is similar to that of the radio pulses of many other pulsars, including the precursor of the Crab (e.g. Rankin *et al.*, 1970; Smith, 1986). The Vela $\gamma$-ray light curve has an even more significant emission between the pulses than the Crab, showing a substantial decaying tail after the main pulse and a smaller one after the interpulse. Unlike the Crab, the main $\gamma$ pulse of Vela has a softer spectrum than that of the interpulse and the region in between (Kanbach *et al.*, 1980). The photon number spectrum between 50 to 3200 MeV is given by a power law of exponent $-1.89$ (Hermsen, 1981), and the pulsed light fraction at these energies is very high, $\gtrsim 90\%$ (Lichti *et al.*, 1980). The maximum power $E^2N(E)$ emitted by Vela comes at $E > 100$ MeV, being much harder than that of the Crab. At X-rays it is not detected, and in the optical its flux is a factor $10^4$ weaker than that of the Crab. Assuming again two cones of emission, the gamma-ray luminosity is $L_\gamma = 4 \times 10^{34}$ ergs $s^{-1}$, or 0.6% of the total spin-down luminosity, a fraction $\sim 10$ larger than in the Crab.

The curious phase difference 0.4 between pulses and the presence of a precursor radio pulse in the Crab and Vela pulsars are very suggestive. A simple geometrical interpretation of the relative phase positions of the radio, optical, and gamma-ray pulses has been proposed by Smith (1986) which is based on a fan beam interpretation with the $\gamma$-ray emission occurring near the light cylinder and the radio emission coming from near the polar caps (Radhakrishnan and Cooke, 1969). An interesting observation that bears on the radiation mechanism and its location is the report of the possible detection of a very high degree of linear polarization in the $\gamma$-ray pulses of the Vela pulsar (Caraveo *et al.*, 1988). The observation used the *COS B* spark chamber data on Vela, amounting to over 3100 photons, and is based on a method proposed by Kozlenkov and Mitrofanov (1985) to measure $\gamma$-ray polarization using the relation between the plane of linear polarization and the position of the azimuthal plane formed by the $e^+e^-$ pair produced in the spark chamber. In this observation, the polarized photons are spread fairly broadly in azimuth ($\sim 30°$) but appear to vary significantly with pulse phase. In particular, the polarization of the main pulse and interpulse is weak at best, but the region between the main pulse and interpulse appears almost completely polarized. While the significance of the observation is not very high,

Caraveo *et al.* (1988) conclude tentatively that fan beam $\gamma$-ray emission near the light cylinder is probably most compatible with their data. However, models of both the high-energy and low-energy emission of RPPs are still in process. We turn next to a discussion of some of these models.

## 8.2 The Standard Magnetic Dipole Model

### *a) The Vacuum Magnetic Dipole Rotator*

We give here a simplified discussion of the structure of a rotating pulsar magnetosphere, outlining only the basic physical elements. Fuller discussions of this still incompletely solved problem are to be found e.g. in Michel (1982) and Beskin, Gurevich, and Istomin (1983, 1988).

In its simplest form, the neutron star may be considered an almost perfectly conducting sphere whose external magnetic field is, to a first approximation, of the dipole form. Let us first consider the case of a magnetic dipole moment $\vec{\mu}$ which is aligned with the rotation axis $\vec{\Omega}$, and let us assume for the moment that the external magnetosphere is empty of charges. In spherical coordinates, the nonrelativistic components of the magnetic dipole field are

$$B_r = \frac{2\mu}{r^3}\cos\theta\,, \quad B_\theta = \frac{\mu}{r^3}\sin\theta\,, \tag{8.2.1}$$

where the magnetic moment $\mu = B_0 R^3$ is given in terms of the surface field strength at the pole $B_0$ and the stellar radius $R$. Because of the assumed alignment, the external vacuum field is static. On the surface of the rotating conducting stellar surface, which moves with velocity $\vec{v} = \vec{\Omega} \times \vec{R}$, the electrical charges will be subjected to a Lorentz force $\vec{f} \sim (\vec{v}/c) \times \vec{B}$ which will cause them to move along the surface and rearrange themselves until a static distribution is reached for which the sum of the electric forces along the surface on the charges is zero. That is, one must have on the surface (on the rotating frame) $E'_\theta = 0$. This field is related to the external fields by

$$E'_\theta = E_\theta + \left(v_\varphi/c\right)B_r = 0\,, \tag{8.2.2}$$

where $v_\varphi = \Omega R \sin\theta$ and $B_r$ is given by the external vacuum magnetic field (8.2.1). Therefore, the external vacuum electric field has a $\theta$ component just outside the surface which is

$$E_\theta = -\frac{2\mu\Omega}{R^2}\sin\theta\cos\theta\,. \tag{8.2.3}$$

The dependence on $\sin\theta\cos\theta = \frac{1}{2}\sin 2\theta$ indicates a quadrupole field, and from the fact that the tangential component of the electrical field at the surface of a conductor must be continuous we see that the external (vacuum) electric field must be

$$E_r = \frac{9Q}{r^4}\left(\cos^2\theta - \frac{1}{3}\right), \quad E_\theta = \frac{6Q}{r^4}\cos\theta\sin\theta, \tag{8.2.4}$$

where $Q = \mu R^2\Omega/3c = B_0\Omega R^5/6c$ is the quadrupole moment. This external vacuum electric field can also be derived from an electrostatic potential, $\vec{E} = -\nabla\Phi$, which satisfies Laplace's equations $\nabla^2\Phi = 0$. Together with the boundary conditions at the stellar surface, the electrostatic potential that solves the Laplace equation and corresponds to the external electric field (8.2.4) is

$$\Phi = -\frac{Q}{r^3}\left(3\cos^2\theta - 1\right). \tag{8.2.5}$$

This system is a unipolar inductor, since sliding contacts between two points at different latitudes would measure a difference of potential (i.e. an e.m.f.) given by equation (8.2.5). Between one of the poles and the equator this maximum potential difference is $\Delta\Phi = B_0\Omega R^2/2c$, or a voltage drop $\Delta V = e\Delta\Phi/(1\ \text{eV}) \sim 3\times 10^{14} B_{12} R_6^2 P^{-1}$ V. The vacuum external electric field (8.2.4) has a nonzero component along the magnetic field lines, $\vec{E}\cdot\vec{B} \neq 0$, which at the surface has the value

$$\left(\vec{E}\cdot\vec{B}\right)_R = -\frac{\Omega R}{c} B_0^2 \cos^3\theta. \tag{8.2.6}$$

The electric field parallel to $\vec{B}$ at the surface is therefore

$$E_\parallel \simeq \frac{\Omega R}{c} B_0 \simeq 6\times 10^{10} B_{12} P^{-1}\ [\text{V cm}^{-1}], \tag{8.2.7}$$

where the rotation period $P = 2\pi/\Omega$ is in seconds and $B_{12}$ is the surface field in units of $10^{12}$ G. This electric force acting on the surface electrons and ions would be orders of magnitude stronger than the gravitational force and, provided the surface binding energies are not too large, would lead one to conclude that electric charges would be ripped off the surface and accelerated, filling the magnetosphere. Therefore, the magnetosphere cannot be empty; there must be a plasma distribution in it (Goldreich and Julian, 1969).

*b) The Rotating Pulsar Magnetosphere*

In a plasma-filled magnetosphere, assuming the conductivity to be infinite and neglecting particle inertia, the rotating version of Ohm's law implies that

$$\vec{E} + \frac{1}{c}\left(\vec{\Omega} \times \vec{r}\right) \times \vec{B} = 0 \tag{8.2.8}$$

is satisfied. The electric field induced in the plasma-filled magnetosphere is defined by this relation, and it has the property that the electric equipotential surfaces contain the magnetic field lines, i.e. the component of the electric field parallel to the magnetic field $E_{\parallel}$ vanishes,

$$\vec{E} \cdot \vec{B} = 0. \tag{8.2.9}$$

The presence of this electric field also determines the electric charge distribution in the magnetosphere,

$$\rho_c = \frac{1}{4\pi} \vec{\nabla} \cdot \vec{E} = -\frac{1}{2\pi c} \vec{\Omega} \cdot \vec{B}. \tag{8.2.10}$$

This result corresponds to a charge number density $n_c = 7 \times 10^{-2} B_z P^{-1}$ cm$^{-3}$, where $B_z$ is the $z$-component of the magnetic field in gauss and $P$ is the period in seconds, and is referred to as the Goldreich-Julian, or corotation, charge density. This charge distribution is made up of particles coming from the surface, and equation (8.2.9) implies that these charges are in strict corotation with the star. However, strict corotation cannot exist beyond the surface where the tangential velocity equals the speed of light. This cylindrical surface is called the light cylinder, its radius being

$$R_L = c/\Omega \simeq 5 \times 10^9 P \text{ [cm]}. \tag{8.2.11}$$

Strict corotation may exist along field lines which close at distances smaller than the light cylinder (see Fig. 8.2.1), whereas those field lines which, in the absence of rotation, would have closed at larger distances penetrate the light cylinder and become open field lines. The radius of the polar cap region which contains the open field lines is

$$R_p \simeq R \sin\theta_p = R\left(\Omega R/c\right)^{1/2}, \tag{8.2.12}$$

since for dipole field lines $\sin^2\theta / r =$ constant. Charged particles, which to a first approximation can only move along the magnetic field,

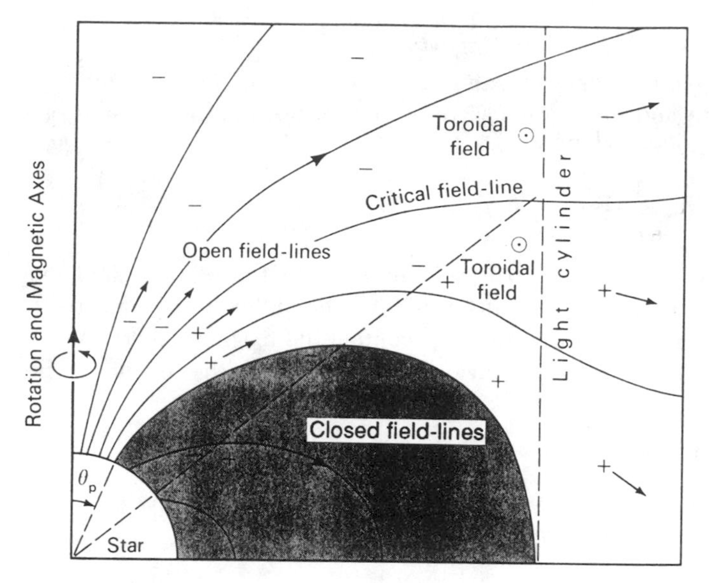


FIGURE 8.2.1 Magnetosphere of an aligned pulsar. The open-field lines are swept back to form a toroidal component after crossing the light cylinder. The closed-field lines encompass the corotating portion of the magnetosphere. The critical field line divides regions of positive and negative current flow, while the diagonal dashed line gives the locus of $B_z = 0$ where the sign of the Goldreich-Julian space charge changes. From R. N. Manchester and J. H. Taylor, *Pulsars*. Copyright © 1977 by W. H. Freeman and Co. Reprinted by permission.

can escape to infinity only along these open field lines. The currents associated with these escaping charges generate a toroidal magnetic field component, which is largest near the critical field line that separates the open from the closed region. For a magnetic field aligned with the rotation axis, as in Figure 8.2.1, the potential at the base of the open field lines near the axis is negative with respect to the exterior so that negative charges stream out. However, there cannot be a net charge outflow from the star so that the potential of the lines near the edge of the polar cap at $\theta_p$ must be positive with respect to the exterior, and positive charges stream out from the star along these lines which form an annulus on the outer part of the polar cap. There is an intermediately located set of critical field lines separating the regions of negative and positive outflow, which is where the potential on the surface of the star equals the potential of the exterior interstellar medium.

*c) Energetics and Evolution*

Near the light cylinder the poloidal and toroidal magnetic field components are expected to become comparable so that the field lines are bent back and penetrate the light cylinder at an angle of about a radian. Beyond the light cylinder the magnetic field is an outgoing wave (transverse) field, with an associated transverse electric field $E \sim B$ which carries a Poynting energy flux $S \sim cB^2/4\pi$. Thus, the light cylinder corresponds to the transition between the classical near field zone, where the fields are approximately static, and the wave zone at $R \gtrsim R_L$, where the fields are transverse and carry away energy. The rate at which the pulsar loses energy in this way is

$$\dot{E} \simeq -4\pi R_L^2 S_L \sim -\frac{B_0^2 R^6 \Omega^4}{c^3}, \tag{8.2.13}$$

where $S_L$ is the Poynting flux evaluated at $R_L$ and the magnetic field is evaluated there using the approximate dipole dependence $B_L \sim B_0(R/R_L)^3$. This energy loss rate agrees in order of magnitude with that predicted by the oblique vacuum magnetic dipole radiation formula $\dot{E} = -(2/3c^3)|\ddot{\vec{\mu}}|^2$, except that here it is obtained for the plasma-filled magnetosphere and an aligned rotator. The Maxwell stresses associated with the wave field also carry away angular momentum, exerting on the star a torque

$$K_s = -\frac{\kappa}{8c^3}\left(B_0 R^3\right)^2 \Omega^3, \tag{8.2.14}$$

where $\kappa$ is a constant of order unity depending on the exact field configuration. This torque agrees in order of magnitude with that which one would expect from the magnetic dipole radiation of a rotating oblique magnetic dipole in a vacuum,

$$K_{\text{md}} = -\frac{2}{3}\frac{\Omega^2}{c^3}\left(\vec{\mu} \times \vec{\Omega}\right) \times \vec{\mu} + \frac{1}{Rc^2}\left(\vec{\mu} \cdot \vec{\Omega}\right)\vec{\Omega} \times \vec{\mu} \tag{8.2.15}$$

(Davis and Goldstein, 1979; Michel and Goldwire, 1970; Goldreich, 1970), or $K_{\text{md}} \simeq -2\mu^2\Omega^3\sin^2\alpha/3c^3$, where $\alpha$ is the angle between $\vec{\mu}$ and $\vec{\Omega}$. This torque has a tendency to align the magnetic dipole axis with the rotation axis, and is of the right order of magnitude to explain the spin-down observed in pulsars (see Gunn and Ostriker, 1969; Manchester and Taylor, 1977). The predicted behavior for the period first derivative, from equation (8.2.13) and $E = I\Omega^2/2$, is $\dot{P} \propto P^{-1}$, or, more generally,

$$\dot{P} = AB_0 P^{-1} e^{-2t/t_D}, \tag{8.2.16}$$

where $A$ is a constant, $B_0$ is the initial surface field value, and $t_D$ is a characteristic decay time scale for pulsar emission to cease (via either field decay or alignment). The second derivative of the period, defined through the braking index $n = \Omega\ddot{\Omega}/\Omega^2$, would be, for a pure vacuum magnetic dipole radiator inclined at an angle $\alpha$, $n = 3 + 2\cot\alpha^2$, which is always larger than 3 and increases as the dipole aligns. However, in a more realistic model with currents and electromagnetic losses given by an approximation such as equation (8.2.13), one expects generally $n = 3$ for a dipole. The value of the surface magnetic field can be deduced from the fact that $\dot{E} = d/dt(I\Omega^2/2) \propto \dot{P}/P^3$, which equated to equation (8.2.13) gives

$$B \simeq 10^{12}(\sin\alpha)^{-1}\left(P\dot{P}_{-15}\right)^{1.2}\ \mathrm{G}, \tag{8.2.17}$$

where $\alpha$ is the angle between the magnetic and rotation axes. This decay is reached when the pulsars cross a line $\dot{P} \propto P^3$ for the vacuum magnetic dipole model, marked "death line" in Figure (8.1.2).

The standard model, as described above, has a number of inconsistencies which to this day have not been resolved completely. The most important among these are (*a*) the return current problem (where does the circuit utilizing the potential difference close?), (*b*) the fact that it envisages charges of one sign flowing through regions of the opposite charge sign (see Fig. 8.2.1), (*c*) the fact that the parallel field that is supposed to pull out the charges vanishes for the $\vec{E} \cdot \vec{B} = 0$ magnetosphere of charge density $\rho_c$, and (*d*) the fact that it is not possible to have everywhere the equilibrium charge density $\rho_c$ given by equation (8.2.10) streaming at velocities close to the light speed (see Michel, 1982). Nonetheless, the remarkable thing is that the more sophisticated (and still incomplete) treatments of oblique, self-consistent charged magnetospheres give a general picture which does not differ drastically from that described previously, and the value of the physical quantities involved agrees in order of magnitude with those discussed above. The consensus is that, while having its limitations, some or many of the elements in the "standard" model just described will be present in a yet to be found complete model.

## 8.3 Polar Cap Models

### *a) The Sturrock-Ruderman-Sutherland Model*

A natural extension of the standard model was proposed by Sturrock (1971), who investigated the consequences of the particle outflow from the star along the open field lines. Assuming that the magnetosphere has a charge density given by the Goldreich-Julian value $n_c \sim \Omega B_0/2\pi ce$ (eq. 8.2.10) which flows out at the speed of light from two polar caps of area $2\pi\Omega R^3/c$ (eq. 8.2.12), the primary particle flux from the pulsar is

$$\dot{N}_p \sim \frac{\Omega^2 B_0 R^3}{ec} \simeq 2 \times 10^{31} B_{12} P^{-2} \text{ s}^{-1}, \tag{8.3.1}$$

which for the Crab pulsar is $\sim 10^{34}$ s$^{-1}$. The accelerating potential responsible for this flow of particles is $\Phi \sim (2\Omega B_0/c)h^2$, since by definition $\rho_c = (\Omega B_0/2\pi c) = (1/4\pi)\nabla^2\Phi \sim (1/4\pi)\Phi/h^2$. Here $h$ is the typical height over which the particles are accelerated, which Sturrock takes to be comparable to the polar cap radius $h \sim R_p$ (eq. 8.2.12). This value is essentially equivalent to the potential difference between the center and the edge of the polar cap, as estimated from eqs. (8.2.5) and (8.2.12),

$$\Delta\Phi \sim \frac{\Omega B_0}{2c} h^2 \sim \frac{\Omega^2 R^3 B_0}{2c^2}, \tag{8.3.2}$$

and amounts to a potential drop of $\Delta V \sim 6 \times 10^{12} B_{12} P^{-2}$ V. Sturrock (1971) assumed that this potential was available to accelerate particles from the surface so that electrons would reach Lorentz factors $\gamma \gtrsim 10^7$. Their perpendicular energy would be quickly dissipated by synchrotron radiation, but the longitudinal energy acquired in the acceleration would be radiated away by curvature radiation in moving along the curved dipole field lines. The curvature photon energy (eq. 5.6.2) is $\hbar\omega_0 \simeq \frac{3}{2}\hbar\gamma_e^3 c/r_c \simeq 10^9 \gamma_7^3 r_8^{-1}$ eV, normalized to $\gamma \sim 10^7$ and to a typical dipole line curvature radius $r_c \sim (rc/\Omega)^{1/2} \sim 10^8$ cm. Photons of such energy, however, are subject to magnetic (one-photon) pair production (Chap. 5.5). The asymptotic attenuation length is $\kappa = 0.23(\alpha_F/\lambda\!\!\!^{-}_c)(B_\perp/B_Q)\exp[-4/3\chi]$, where $\alpha_F \simeq 1/137$, $\lambda\!\!\!^{-}_c$ is the Compton wavelength, and $\chi = (B_\perp/B_Q)(\hbar\omega/2m_e c^2)$. Setting $\kappa r \sim 1$ for any reasonable distance $r$, one gets $\chi^{-1} \sim 15$, or

$$\hbar\omega \gtrsim 4 \times 10^6 B_{\perp,12}^{-1} \text{ eV}, \tag{8.3.3}$$

as a criterion for pair production. Sturrock therefore envisaged that, for a sufficiently strong electric field, these secondary pairs would also be accelerated and create $\gamma$-rays leading to further pairs, resulting in a pair cascade. This would explain the high-energy pulses at $\gamma$-ray energies seen in the Crab and Vela pulsars, while bunching of the radiating particles and coherent radiation at lower frequencies from these pairs would lead to the observed radio emission. Pulsars stop radiating because, as they slow down, the potential difference $\Delta V \propto BP^{-2} \sim \dot{P}^{1/2}/P^{3/2}$ will reach a critical value below which no pairs can be produced. The corresponding death line will be $\dot{P} \propto P^3$ (see Fig. 8.1.2). Some of the problems with this simple picture are inherent in the standard model. For instance, the charges $n_c$ cannot everywhere move at the speed of light in an aligned dipole, leading to a lower potential drop (Michel, 1974), and potential variations across the cap can also reduce it (Tademaru, 1974). On the other hand, the presence of strong higher multipoles near the cap surface can offset this reduction (Henriksen and Norton, 1975). These models, as the standard model, use the assumption that the pulsar loses charges of both sign in equal amounts.

A major development of this model was carried out by Ruderman and Sutherland (1975), who investigated the effect of a loss of electrons while the ions are retained, based on the higher binding energy of the latter. They assume an axisymmetric but counteraligned rotator, and conclude that, in the regions where positive charges would normally have been ejected, vacuum gaps would instead form. In these vacuum gaps one has $\vec{E} \cdot \vec{B} \neq 0$, which also means the field lines in this region are not forced to corotate. For a gap height $h$, the accelerating potential available is again of the order of equation (8.3.2), and one expects the height to be limited by $h \lesssim R_p$. Since any positive charges present in the outer magnetosphere would flow out at a speed $\sim c$, the gap height must grow at the speed of light until it reaches this limiting value $h \sim R_p$. When the gap potential drop reaches a value of $\Delta V \sim 10^{12}$ V, a spark discharge occurs inside the gap, producing electrons and positrons which are accelerated in the $E_{\parallel} \neq 0$ field of the gap, reaching energies of order $\gamma_{e^+e^-} \sim 10^6$. Assuming a strong multipole structure near the surface so that the radius of curvature of the lines is $r_c \sim 10^6$ cm, the curvature radiation photons generated by these pairs will have energies $\hbar\omega_c \sim 10^9$ eV, themselves capable of producing pairs by the magnetic one-photon process (eq. 8.2.4), which would lead to a cascade. The difference from Sturrock's model is that here the acceleration occurs in the vacuum gap region only, where $\vec{E} \cdot \vec{B} \neq 0$. As they move beyond the height $h$, where $\vec{E} \cdot \vec{B} = 0$, the pairs are no longer accelerated and stream outward with a Lorentz factor $\gamma_s \sim 3$. The density of the streaming pairs is

$$n_s \sim n_c \gamma_{e^+e^-} \gamma_s^{-1} \sim 10^3 n_c \sim 2 \times 10^{13} P^{-1} \left( R_*/r \right)^3 \text{ cm}^{-3}, \tag{8.3.4}$$

where account has been taken of the dipole field divergence with radius. This far exceeds the Goldreich-Julian density $n_c$, so the plasma will be essentially charge neutral. Thus, two of the objections to the standard model are circumvented. The model is strongly dependent on the binding energy of ions (mainly Fe) exceeding the available potential, and these binding energies have been variously calculated (Hillebrandt and Müller, 1976; Flowers *et al.*, 1977; Kössl *et al.*, 1988) to range between about 2.5 keV and 50 keV in pulsar fields. The $\gamma$-radiation in this type of model is produced by the curvature radiation (e.g. Harding, 1981; Daugherty and Harding, 1982), while radio radiation occurs above the spark regions in the filled, streaming region where bunching by plasma instabilities would lead to coherent radio emission. An additional attractive feature of the Ruderman-Sutherland model is that corotation is not enforced in the spark gap region and in fact the sparking regions are expected to move around the cap, which serves as a model for the phenomenon of marching subpulses (e.g. Manchester and Taylor, 1977). Also, the recent phenomenology of radio pulse analysis (Lyne and Manchester, 1988; Radhakrishnan and Cooke, 1969) including pulse shapes and the swing of the polarization vector across phase is in general agreement with a polar cap emission model, as opposed to models where the emission occurs near the light cylinder.

*b) The Arons et al. Model*

For a pulsar with a Goldreich-Julian charge density streaming out at relativistic speeds which utilizes the full potential difference across the polar cap given by equation (8.3.2), the total particle loss rate is given by equation (8.3.1) and the particle energy per second produced by the pulsar is (see Goldreich and Julian, 1969; Ostriker and Gunn, 1969; Pacini, 1967)

$$L_p = e\,\Delta\Phi_{\max} \dot{N}_p = \frac{1}{4} \frac{\Omega^4 R^6 B_0^2}{c^3} \simeq 1.4 \times 10^{31} R_6^6 B_{12}^2 P^{-4} \text{ ergs s}^{-1}. \tag{8.3.5}$$

This luminosity is sufficient to explain the energy losses of all pulsars observed so far. This value of the potential drop was also assumed by Sturrock (1971), while the screening of the accelerating field by the Goldreich-Julian charges of the force-free magnetosphere was neglected. This inconsistency was lifted in the Ruderman-Sutherland model by introducing a vacuum gap, which also implies a reduction in the maximum luminosity possible. This is because the discharges occur when $\Delta\Phi \sim \Delta\Phi_{RS} \sim 10^{12}$ V, which can be significantly less than the maximum value across the polar cap (8.3.2) for short-period pulsars ($P \ll 1$ s). As a consequence the total particle luminosity drops below the value given by equation (8.3.5) by the factor $\Delta\Phi_{RS}/\Delta\Phi_{max}$. For pulsars of periods near a second, the reduction is small, but for fast rotating pulsars the particle luminosity becomes too small to explain the energy losses.

A significantly different approach to magnetospheric energy balance was developed by Arons and co-workers (Fawley, Scharlemann, and Arons, 1977; Scharlemann, Arons, and Fawley, 1978; Arons and Scharlemann, 1979; Arons, 1981, 1983), who concentrated on oblique rotators with charge densities differing from the Goldreich-Julian corotating density. They also addressed (see Scharlemann and Wagoner, 1973) the problem of the return current, envisaging this to be carried by auroral currents along the boundary of the last closed field line surface. As discussed, in a symmetric rotator it is not possible to have everywhere the force-free $\vec{E} \cdot \vec{B} = 0$ corotation charge density of equation (8.2.10) together with an outflow everywhere at speed $c$, because in this situation the field lines diverge away from the rotation axis everywhere. This can only be achieved exactly if the field lines are straight. Because of the dependence of the corotation charge density on $B_z$, a flow at speed $c$ in a curved magnetic field must lead to the appearance of a $\vec{E} \cdot \vec{B} \neq 0$, in which case it is crucial whether the sign of the force is such as to accelerate or decelerate the particles. In a symmetric rotator, the sign is inconsistent with an accelerating outflow. In an oblique rotator, one is led to the concept of favorably curved field lines (those that bend toward the axis of rotation and form a bundle of lines which is nearly cylindrical), as opposed to unfavorably curved lines (those which bend away from the rotation axis, leading to diverging bundles of field lines). Scharlemann, Arons, and Fawley (1978) pointed out that in the unfavorably curved field lines the space charge necessary to ensure continuity of flow at speed $c$ would increasingly exceed the density needed to ensure that the magnetosphere is force-free, and the ensuring buildup of space charge would eventually decelerate and halt the outflow. In the favorably curved lines, however, one expects an accelerating field, and it is only here that a consistent relativistic charge-separated outflow may be expected. Inside the favorably curved lines, the current continuity equation $J_{\parallel} = J_*(B/B_*)$ implies that the charge density grows in proportion to $B$ while the corotation charge density grows in proportion to $B_z$ with increasing altitude, so the condition $E_{\parallel} = 0$ can be achieved only at the stellar surface. Above this, the flux tube is unable to short out the vacuum $E_{\parallel}$ completely, which means that in the favorably curved lines the flux tube is starved of charge relative to the corotating charge density (Arons, 1981). This leads to a residuum of the vacuum $E_{\parallel}$, or "starvation" electric field, which is not shorted out. Under the assumption that $\Phi \sim 0$ elsewhere, the accelerating starvation potential in the favorably curved flux tube is (Scharlemann, Arons, and Fawley, 1978)

$$\begin{aligned}\Phi &= -\operatorname{sign}(\rho_c) f(\theta) \frac{\Omega R}{c} B_0 R \theta_c^3 \left[ \left( \frac{r}{R} \right)^{1/2} - 1 \right] \\ &\simeq -\operatorname{sign}(\rho_c) f(\theta) 10^{10.5} B_{12} R_6^{7/2} P^{-5/2} \left[ \left( \frac{r}{R} \right)^{1/2} - 1 \right] \text{V},\end{aligned} \tag{8.3.6}$$

where $\theta_c$ is the polar cap opening angle and $f(\theta)$ is a function of the angles. This is the potential before any pair creation occurs, and Arons *et al.* assume that, ignoring binding energy differences, it accelerates particles of the appropriate sign, i.e. electrons for $\vec{\Omega} \cdot \vec{B} > 0$. These accelerated particles will radiate $\gamma$-rays by curvature radiation, and for a wide range of parameters this will lead to a pair cascade. The pair formation front is rather abrupt because of the exponential nature of the one-photon magnetic pair creation mean free path (eq. 5.5.16), and above this front surface, which begins at height $H < R$, the pairs become so abundant that they modify the potential and create a $E_{\parallel} = 0$ flow. An approximate value for the height, assuming that it is constant across the tube, is given by (Arons, 1981): $H \sim 3 \times 10^5 B_{12}^{-1} R_6^{-3} P^{17/8}$ cm. However, if the region near the side boundaries of the favorably curved tube is a good conductor with a small starvation $E_{\parallel}$, pair formation is never important enough there, so that $H$ must curve upward near the side boundaries of the tube and the surface of the pair formation front becomes parallel to the field lines for distances $r \gtrsim 2R$. The acceleration region, where only primary

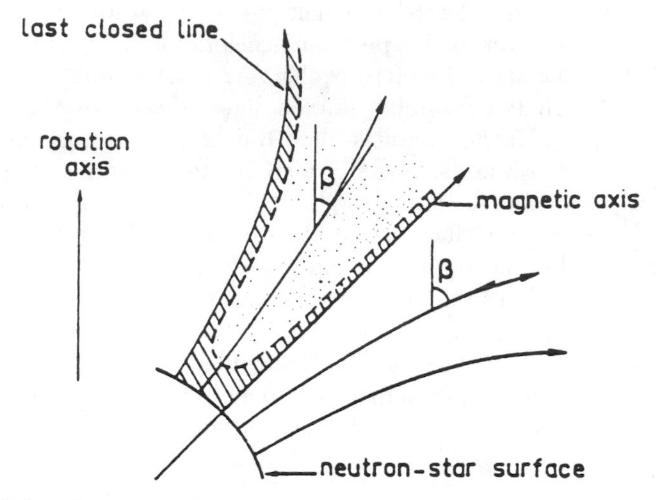


FIGURE 8.3.1 Slot gap model showing the pair discharge region on the upward (favorably) curved field lines. The excess space charge accumulation on downward-curved field lines is expected to choke off the flow, so only the upward-curving ones are expected to be able to maintain particle injection. From F. C. Michel, 1982, *Rev. Mod. Phys.*, 54, 1.

particles flow, has the shape of a slot gap (Fig. 8.3.1). The total particle luminosity is estimated from the density of the pair plasma created, which is a function of the field strength, the period, and the curvature of the field lines. In a dipole field, the pair luminosity is approximately (Arons, 1983)

$$\dot{N}_{e^+e^-} \sim 10^{36} B_{12}^{5/3} R_6^{8/3} P^{-7/6} \ \mathrm{s}^{-1}, \tag{8.3.7}$$

which for the Crab is $10^{39}$ $\mathrm{s}^{-1}$, an adequate number to explain the spin-up, as well as the luminosity, of the nebula around it, with a total number $\sim 10^{49}$ of pairs accumulated. On the other hand, this output may be too large for millisecond pulsars, unless one can explain the absence or nondetection of an energized nebulosity.

### *c) The Cheng, Ho, and Ruderman (CHR) Model*

The CHR model (Cheng, Ho, and Ruderman, 1985a, b) is based on the possibility of the existence of large outer magnetospheric gaps, the high-energy radiation being produced at distances extending out to the light cylinder in a fan beam. This is a significant departure from the polar cap models described above, whose emission has a pencil beam pattern. The possibility of the existence of outer magnetospheric gaps had been discussed by Holloway (1973) and Krausse-Polstroff and Michel (1985). In their model, CHR consider an oblique rotator with a magnetosphere which at least initially has a charge distribution similar to that of the standard model (Fig. 8.3.2). This figure is shown for $\rho_c > 0$, $\vec{\Omega} \cdot \vec{B} < 0$, but most of the discussion is the same except for a change of the sign of all charges in the opposite case of $\vec{\Omega} \cdot \vec{B} > 0$, $\rho_c < 0$. The assumption made by CHR is that the return current flows through the neutral sheet located where $\vec{\Omega} \cdot \vec{B} = 0$ (dashed line in Fig. 8.3.2) and that the negative charge of the regions on open field lines farther from the magnetic axis than this surface would tend to flow out through the light cylinder surface. This flow leaves behind a negative charge-depleted region, which acts as a positive charge with respect to the positive charges beyond the neutral sheet, repelling them and preventing their flow outward. This in turn leaves an empty gap which grows until it extends along the field lines between the neutral sheet and the light cylinder. The whole potential drop $\Delta V \simeq \Omega^2 B_0 R^2/c^2 \sim 6 \times 10^{12} B_{12} P^{-2}$ V of equation (8.3.2) would be achieved along $\vec{B}$, subject to modification by the effects of pair formation.

Pairs formed in the gap just above (beyond) the last closed field line surface will be accelerated and give rise to $\gamma$-ray beams which, because of the curvature of the lines, will tend to cross each other and produce new pairs via $\gamma + \gamma \rightarrow e^+e^-$ further away from the last closed surface. These pairs will fill the gap region extending up to the neutral sheet. The gap will thus be constrained between the last closed surface and another surface not too far above it so that the width $a$ between these two surfaces will be smaller than the breadth $b$ (in the azimuthal direction) and much smaller than the length $d$ along the magnetic field direction (see Fig. 8.3.3). It is in these large outer magnetospheric gaps with $\vec{E} \cdot \vec{B} \neq 0$, which for the Crab and Vela pulsar implies potential drops $\Delta V \sim 10^{15}$ V, that CHR envisage the primary particle acceleration to occur. The $\gamma$-ray emission from these and the subsequent pair generation provides charges for the rest of the region between the neutral sheet and the light cylinder, including the boundary layer change, which screens the electric charge deficit field of the gap, thus ensuring that $\vec{E} \cdot \vec{B} = 0$ elsewhere. The secondary and tertiary radiation from this plasma-filled, force-free region, however, is necessary to ensure the production of primary pairs in the gap (see below).

The gamma-rays and X-rays produced by the particles are beamed

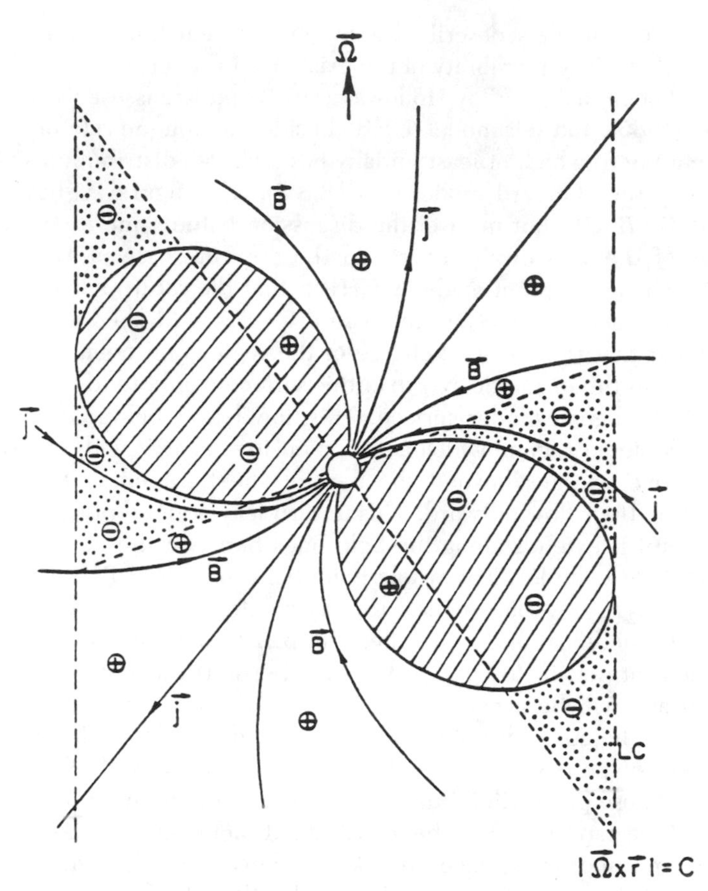


FIGURE 8.3.2 Schematic current flow pattern in the CHR model for a star with $\vec{\Omega} \cdot \vec{B} < 0$ above the polar cap. The charge is assumed to be given by the corotation value $\rho = \rho_c$, with $\rho_c = 0$ on the null surface $\vec{\Omega} \cdot \vec{B} = 0$. The current flow of positive charge away from the star is mainly in the form of a non-charge-separated plasma, which is assumed balanced by a return current through the null surface. From K. S. Cheng, C. Ho, and M. Ruderman, 1985, *Ap. J.*, 300, 500.

along the magnetic field, which leads to a fan beam pattern whose angular width depends on the width of the gap and the inclination angle between the magnetic and rotation axes. Because particles of both sign are accelerated and radiate, the radiation is beamed tangentially to the field in directions both toward and away from the rotation

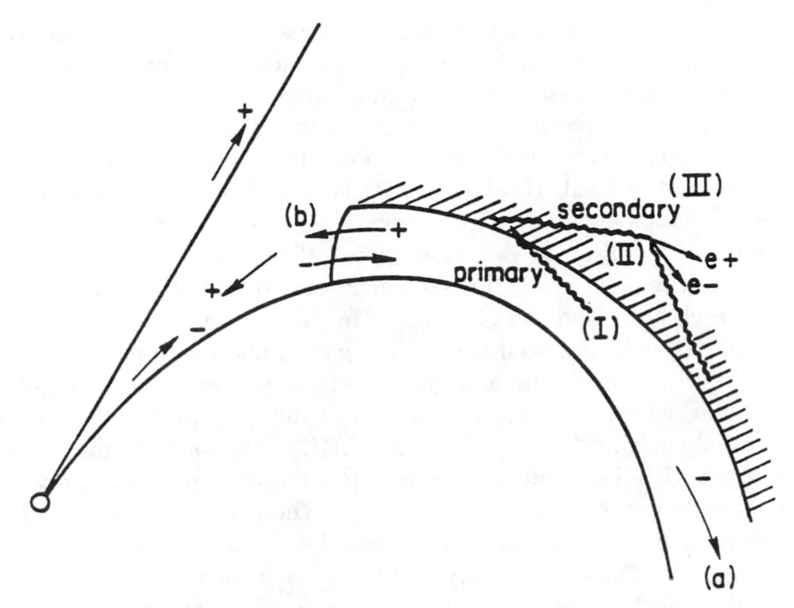


FIGURE 8.3.3 Outer-gap structure of the CHR model showing the relative positions of the primary (I), secondary (II), and tertiary (III) regions (see text). The current flow along $\vec{B}$ begins at (b) and extends to (a) for $\rho_c > 0$ above the polar cap. From K. S. Cheng, C. Ho, and M. Ruderman, 1985, *Ap. J.*, 300, 522.

axis. Thus, two pulses are expected, from the front and back regions (with respect to the observer) of the magnetosphere. Actually there can be four outer gaps, two long and two short, but one can argue that the short ones are thicker and harder to see if they are not quenched by the two longer, more powerful ones. Since these two regions are separated by a distance which is a significant fraction of the diameter of the light cylinder, the separation of the two pulses is a combination of the light travel difference, the relativistic aberration, and magnetic field line bending. Thus, in this model, the observed $\gamma$-ray pulse phase separation of $\sim 140°$ in the Crab is indicative of an average distance to the spin axis of $r_\perp \sim 0.6 R_L$, with the aberration effect leading to a cusped pulse profile.

The structure of the outer magnetosphere for a fast rotating pulsar consists of three different physical regions, labeled I, II, and III in Figure 8.3.3. In the gap itself (I), primary $e^+e^-$ are oppositely accelerated in the large $\vec{E} \cdot \vec{B}$ to extreme relativistic energies, until limited by energy losses to primary gamma-rays produced by curvature radiation (Crab) or inverse Compton scattering on infrared (IR) photons from the rest of the charge-separated magnetosphere (Vela). These $\gamma$-rays are partly converted in the gap, limiting its size, while a larger fraction goes into the next region (II). In region II, where there is only a very small residual $\vec{E} \cdot \vec{B}$, secondary pairs continue being created with sufficiently high energy to radiate secondary $\gamma$-rays and X-rays by synchrotron radiation. Since these are also produced by oppositely moving $e^+e^-$, this secondary radiation is also in beams that cross each other and can give rise to further, tertiary $e^+e^-$. The third region (III), which is well beyond the gap, is filled with tertiary $e^+e^-$, which are less energetic and give rise to softer radiation. This softer radiation, which is IR for parameters of the Vela pulsar, illuminates the whole magnetosphere, and in the CHR model for Vela, the density in gap region I is sufficient to give the required primary $\gamma$-rays by inverse Compton on the primary $e^+e^-$. These soft photons colliding with primary $\gamma$-rays are also instrumental in producing the $e^+e^-$ pairs in region II. Thus, the energy input into each region from the others has the effect of self-regulating the entire magnetospheric emission and allows a calculation of the power radiated in each region.

The high-energy spectrum in this model is given mainly by the secondary photons, the primary spectrum being used up largely in making particles. Depending on the abundance of tertiary electrons producing IR photons which can interact with primary $\sim 10$ MeV photons to make pairs, the gap energetics can be of the Crab or Vela type. The type is controlled by the value of the parameter (Ho, 1990)

$$\Sigma = \frac{\mu^2 \Omega^5}{e^4 m_e^3 c^{11}}, \tag{8.3.8}$$

where $\mu$ is the magnetic dipole moment. For $\Sigma \gtrsim 1$, a Crab-type gap is produced, whereas for $\Sigma \lesssim 1$, the model yields a Vela-type gap. The gap structure for the very fast pulsars such as the Crab ($P = 0.033$ s) consists of primary electrons in region I of energy $\gtrsim 10$ MeV losing energy primarily to curvature radiation, since not enough tertiary IR photons are present for the competing inverse Compton losses on these photons to dominate. The curvature radiation photons are radiated at energies $\sim$ GeV. These curvature photons enter region II and collide there with counterstreaming X-rays of keV energies to produce secondary pairs with energies $\sim$ several GeV. The secondary electrons emit synchrotron radiation in the energy range $\lesssim 1$ MeV, and inverse Compton photons in the range $\gtrsim 1$ MeV. Thus, the $\gamma$-rays are produced by inverse Compton scattering of secondary X-rays on secondary $e^+e^-$, and the spectrum from optical to GeV energies comes from both the synchrotron and inverse Compton contributions. The overall spectrum for a Crab-type gap can be approximated as a piecewise continuous power law of the form (Cheng, Ho, and Ruderman, 1985b; Ho, 1989)

$$N(\omega) \propto \omega^{-q},$$

$$\text{where } q \simeq \begin{cases} 0.66, & \hbar\omega/mc^2 \lesssim 3\times 10^{-6}; \\ 1.25, & 3\times 10^{-6} \lesssim \hbar\omega/mc^2 \lesssim 3\times 10^{-3}; \\ 1.8, & 3\times 10^{-3} \lesssim \hbar\omega/mc^2 \lesssim 1; \\ 2.2, & \hbar\omega/mc^2 \gtrsim 1. \end{cases} \tag{8.3.9}$$

For a somewhat slower pulsar such as Vela ($P = 0.089$ s) the gap structure differs from the above in that the primary electrons of energy $\sim 10$ MeV now lose their energy primarily through inverse Compton scattering on IR photons produced by tertiary electrons. This produces primary photons in the $\sim 10$ MeV range, which again collide with IR photons to produce secondary pairs of energy $\gtrsim$ TeV. These secondary pairs emit synchrotron radiation at $\lesssim$ several GeV. The secondary photons in the MeV range produce a tertiary population of tertiary pairs with Lorentz factors $\gamma \sim 1$ which are magnetically mirrored as they stream inward by the converging magnetic field lines. Near the inner mirroring point they radiate tertiary IR photons, which propagate throughout the gap region and act as the seed photons for the inverse Compton scattering of primary electrons. The self-consistent photon spectrum produced by synchrotron radiation of a Vela-type CHR gap is of the form

$$N(\omega) \sim \omega^{-3/2} \ln(3\ \text{GeV}/\hbar\omega), \qquad 100\ \text{keV} \lesssim \hbar\omega \lesssim 3\ \text{GeV}, \tag{8.3.10a}$$

$$N(\omega) \sim \omega^{-2/3}, \qquad \hbar\omega \lesssim 100\ \text{keV}. \tag{8.3.10b}$$

Although most of the energy of the secondary pairs is lost to producing the above secondary synchrotron spectrum, CHR estimate that a fraction $\sim 10^{-2}$ of the energy would be lost to inverse Compton boosting of the soft IR photons. The inverse Compton cutoff for these secondary photons is $\sim 10^{12}$ eV, much higher than the 3 GeV cutoff for secondary synchrotron photons. If $\rho_c < 0$ in Vela, negative electrons are sent down toward the polar caps and positive ions may be ejected. The latter do not suffer significant radiative losses and may attain the full potential drop energy of $\sim 10^{15}$ eV/nucleon. If there is a sufficiently thick target ($\sim$ 50–100 g cm$^{-2}$), they may produce PeV $\gamma$-rays via proton-collision-induced pion decay or, if the X-ray photon density is sufficiently high, via the photopion decay. The optical-IR radiation is tertiary photons from time-integrated (thick-target) synchrotron, with a small positive slope below $\Omega_{IR}$ and a flat slope above. The total power radiated in a fan beam at all energies is estimated by CHR to be $P_1(\text{Vela}) \simeq 3 \times 10^{35}$ ergs s$^{-1} \simeq 3 \times 10^{-2} I\Omega\dot{\Omega}$, which is compatible with observations and represents a few percent of the total spin-down energy. The total pair outflow is inferred to be $\dot{N}_{e^+e^-} \sim 10^{36}$ s$^{-1}$.

As a pulsar slows down, it is expected to progress from a Crab-type gap to a Vela-type gap, unless it was initially as the latter. As it continues to slow down, the Vela-type gap turns off when the production of tertiary electrons becomes insufficient, which is controlled by the parameter (Ho, 1990)

$$\Xi = \frac{e^{15}\mu^7\Omega^{17}}{\hbar m_e^{10} c^{38}} . \tag{8.3.11}$$

A self-sustained Vela-type gap can exist only for $\Xi \gtrsim 1$, and it gets quenched when $\Xi \lesssim 1$. Pulsars which are in the latter category would not be expected to produce optical, X-ray, or $\gamma$-ray radiation.

The radio emission in the CHR model is not expected to be caused by the outer gap $e^+e^-$ but rather to arise from a polar cap mechanism which could be related to ion flow. Since the pulse and interpulse from the Crab are in phase with the $\gamma$- and X-ray pulses, consistency requires the pencil-beamed radio emission, or at least part of it, to be converted into fan-beamed emission. CHR suggest that this conversion may be achieved by scattering on the downward moving secondary $e^+e^-$ that give rise to the high-energy radiation.

### *d) The Beskin, Gurevich, and Istomin (BGI) Model*

The success of the previous models has been remarkable despite the fact that a number of conceptual and formal uncertainties remained unresolved. Among these is the fact that for many of the energetic and dynamic estimates the vacuum magnetic dipole is used, although one also uses the fact that the magnetosphere contains a plasma whose density is approximately the corotation density $\rho_c$ of equation (8.2.10). Another complication which is usually waived is that associated with the inclination $\alpha$ of the dipole axis with respect to the rotation axis, and the effect of this on the magnetic field structure and on the relation between the potential difference available and the plasma current that flows out of the polar caps. In a series of papers, Beskin, Gurevich, and Istomin (1983, 1984, 1986, 1988) (BGI) have addressed a number of these questions and developed a largely self-consistent, although still approximate, model for an inclined rotator with plasma generation and longitudinal currents emanating from the polar caps. In this BGI model, a semianalytic description of the inclined pulsar magnetosphere is provided which builds upon some of the results of Michel (1982) and Mestel and Wang (1979, 1982). The value of the longitudinal current along the field lines is related to the value of the available potential difference $\Delta\Phi$ via a matching relation

$$\beta_0(\alpha) = \beta_M(\alpha)\left[1 - \sqrt{1 - i_0^2/i_M^2(\alpha)}\,\right], \tag{8.3.12}$$

where $\beta_0 = \Delta\Phi/\Delta\Phi_M$ is the dimensionless potential difference available, $\Delta\Phi_M$ being the maximum value (8.3.5), and $i_0 = j_\parallel/2j_c$ with $j_c = c\rho_c$ the corotation current defined through the corotation density (8.2.10). The dependence on the inclination angle $\alpha$ between the magnetic and rotation axes is calculated numerically and is roughly $\beta_M(\alpha) \sim \cos^a\alpha$, $i_m(\alpha) \sim \sin^k\alpha$, with $a \sim k \sim 1$. The polar cap radius is approximately $R_p(\alpha) \simeq R(f_*\Omega R/c)^{1/2}$, and $f_*(\alpha)$ varies from 1.59 at $\alpha = 0$ to 1.94 at $\alpha = \pi/2$.

In the longitudinal current polar cap models discussed in this section (Sturrock, 1971; Ruderman and Sutherland, 1975; Arons, 1983), the net surface torque is

$$K_{\rm bgi} = \frac{1}{c}\int \vec{r} \times \left(\vec{I}_s \times \vec{B}_0\right) dS = -g_*(\alpha)\frac{B_0^2\Omega^3R^6}{c^3}i_0\frac{\vec{\mu}}{\mu}, \tag{8.3.13}$$

where $\vec{I}_s$ is the surface current, $B_0$ is the surface field, $\mu = B_0r^3$ is the magnetic moment, and $g_*(\alpha) = f_*^2(\alpha)/8 \sim 0.4$. The second equality of (8.3.13) is specific to the BGI formulation and uses the fact that, of the potential and solenoidal components of the surface current, only the potential component contributes to the retarding moment $K$. This torque depends on the value of the dimensionless longitudinal current $i_0$, and for a maximal current $i_0 \to i_M$ the magnitude of the torque is comparable to that of the vacuum magnetic dipole radiator (8.2.14).

An important point (Heintzmann, 1981; BGI, 1983, 1986) is that, unlike the vacuum magnetic dipole case in which there is a tendency toward alignment ($\alpha \to 0$), the longitudinal current torque is of a sense as to increase the values of the angle, $\alpha \to \pi/2$. The evolution is given by

$$\dot{\Omega} = g_*(\alpha)\frac{B_0^2\Omega^4R^6}{c^3I}i_0\cos\alpha,$$

$$\dot{\alpha} = g_*(\alpha)\frac{B_0^2\Omega^4R^6}{c^3I}i_0\sin\alpha,$$

$$\dot{E}_{\text{rot}} = -W = \vec{K}\cdot\vec{\Omega} = I\Omega\dot{\Omega}, \tag{8.3.14}$$

where $E_{\text{rot}}$ is the rotational energy, $I$ is the stellar moment of inertia, and $W$ is the total energy loss. The energy loss is made up of a particle energy loss $W_p$ and a wave plus electromagnetic field energy loss $W_{\text{em}}$,

$$W = W_p + W_{\text{em}} = g_*(\alpha)\frac{B_0^2\Omega^4R^6}{c^3}i_0\left[\beta_0 + (\cos\alpha - \beta_0)\right], \tag{8.3.15}$$

where the first and second terms in square brackets correspond to $W_p$ and $W_{\text{em}}$, respectively. The total energy loss for the parallel current model has a factor $g_*(\alpha)\cos\alpha$, while the total loss from the vacuum magnetic dipole radiator has a factor $\sin^2\alpha/6$, other things being equal. The former vanishes when the dipole counteraligns ($\alpha \to \pi/2$), while the latter vanishes when the dipole aligns ($\alpha \to 0$). As emphasized by Good and Ng (1985) and Blair and Candy (1988), there are sufficient uncertainties involved that it would not be surprising if both counteraligning and aligning torques were present to some degree in a pulsar, competing with each other.

The BGI model, when applied specifically to a particle generation mechanism such as that of Sturrock (1971) or Ruderman and Sutherland (1975), has some specific observational predictions. Because of the reliance on curvature radiation which is zero near the magnetic axis, it reproduces the hollow cone emission model. In terms of the observational parameter

$$Q = 2P^{11/10}\dot{P}_{-15}^{-4/10}, \tag{8.3.16}$$

where $\dot{P}_{-15} = \dot{P}/10^{-15}$ s s$^{-1}$, one has for the case $Q \ll 1$ that the dimensionless current and potential are $i_0 \simeq i_M(\alpha)\sqrt{2\beta_0/\beta_M(\alpha)} = i_M(\alpha)\sqrt{2}\,Q$ and $\beta_0 \simeq \beta_M(\alpha)Q^2\cos^{2/5}\alpha$. The inner radius $r_i < R_p$

within which radiation and particles are not produced and the magnetic field strength as a function of $P$, $\dot{P}$ are, for $\beta_0 \ll \beta_M$ or $Q < 1$,

$$r_i \simeq Q^{7/9} R_p \cos^{7/45}\alpha,$$
$$B_{12} \simeq \dot{P}_{-15}^{7/10} P^{-1/20} \cos^{-1.1}\alpha, \qquad (8.3.17)$$

while, for $\beta_0 \to \beta_M$ or $Q > 1$,

$$r_i \to R_p,$$
$$B_{12} \simeq P^{15/8} \cos^{-1}\alpha. \qquad (8.3.18)$$

During its evolution, the pulsar born with $Q \ll 1$ loses rotational energy, and its first and second period derivatives evolve according to

$$\dot{P} = c_0 B_{12}^{10/7} P^{1/14} \cos^{2d}\alpha,$$
$$n \equiv \frac{\Omega\ddot{\Omega}}{\Omega^2} = 2 + 2d\tan^2\alpha, \qquad (8.3.19)$$

where $d \sim 0.7$–$0.8$. Unlike the vacuum magnetic dipole, where $\dot{P} \propto P^{-1}$ until it drops due to field decay or alignment, here $\dot{P}$ is almost independent of $P$, moving almost horizontally toward larger $P$ until it drops. The pulsar emission in this model starts to decay after $Q \gtrsim 2$, or after a characteristic pulsar lifetime

$$t_{\rm bgi} \simeq 3 P_0 \sin^{-1}\alpha_0 B_{12}^{-10/7} \text{ Myr}, \qquad (8.3.20)$$

where $P_0$, $\alpha_0$ are the initial period and inclination at birth. Thus, the death line for this model in a $\dot{P}$, $P$ diagram is a line $\dot{P} \propto P^{2.75}$ close to the $\dot{P} \propto P^3$ of the vacuum magnetic dipole. As $\alpha$ increases during its lifetime, a pulsar uses an increasing fraction of its total energy loss to produce particles. Short-initial-period pulsars are able to reach close enough to $\alpha \sim \pi/2$ before reaching the death line, and these may be observed as pulsars with an interpulse, where the opposite pole also comes into view. Pulsars with larger $P_0$ reach the death line before nearing $\alpha \sim \pi/2$. The statistics of pulsars with and without interpulses appears in reasonable agreement with the predictions of the relative numbers. Also, the radio luminosity in the longitudinal current model is expected to be approximately constant as long as $Q < 1$ and to drop for $Q > 1$, which seems to be in rough agreement with the observational data (Taylor and Stinebring, 1986). A more detailed comparison of the data with model predictions remains to be done.

# 9. Gamma-Ray Bursters

## 9.1 Observational Overview

### *a) Introduction*

Gamma-ray-bursters (GRBs) were discovered over 15 years ago with the CsI scintillation counters of the *Vela* satellites (Klebesadel, Strong, and Olson, 1973), which were sensitive to photons between 0.2 and 1.5 MeV. These satellites were designed to monitor energetic gamma-ray transient events produced by nuclear explosive devices, in order to verify compliance with the Nuclear Test Ban Treaty. This involved several satellites simultaneously above the earth in the very high orbit, so as to give a complete time coverage. The first such events that were detected looked somewhat suspiciously not unlike nuclear explosions, but after some analysis it was realized they could not possibly be coming from man-made devices on or near the earth. The simultaneous detection by two or three satellites allowed a measurement of the difference in time of arrivals of the signals, showing that these were arriving from directions away from both the earth and the sun, i.e. they were cosmic gamma-ray signals. Thus, the sophisticated *Vela* monitoring system designed mainly for defense purposes ultimately became famous for having achieved an unexpected and extremely significant purely scientific discovery. This is one of the examples where high technology experiments designed for "mundane," applied purposes occasionally lead to serendipitous advances in astrophysics. The cosmic signals thus discovered appeared as sudden burst-like events, lasting typically $\lesssim$ tens of seconds and distributed approximately uniformly in direction, and during the time that an individual event was in progress its luminosity would dwarf the contributions from all other quasi-steady $\gamma$-ray sources in the universe. In the years since, a large amount of information has been gathered about these objects (e.g Hurley, 1986a; Evans and Laros, 1986; Mazets *et al.*, 1988; Higdon and Lingenfelter, 1990).

The study of GRBs is carried out from space, since $\gamma$-rays are absorbed by the earth's atmosphere, occasionally from high-altitude balloons but more typically from satellites, since long exposure times are needed. The detectors used in this low to medium gamma-ray energy range extending from $\gtrsim$ 30 keV to about 50 MeV are often of the scintillator type, e.g. NaI(T) with passive shields, such as used on the KONUS experiments aboard the *Venera* spacecraft (Mazets *et al.*, 1981) or the GRS detector aboard *SMM* (Forrest *et al.*, 1980). These detectors typically have a modest energy resolution, $\Delta E/E \gtrsim 0.1$–$0.2$. Much higher resolution is possible, e.g. with Ge crystal detectors, such as used on *ISEE 3* (Teegarden and Cline, 1981). The GBD detector aboard *Ginga* consists of an X-ray proportional counter (PC) and a low-energy $\gamma$-ray scintillation counter (SC) (e.g. Murakami *et al.*, 1988), which together cover the range from $\sim$ 1 keV to $\sim$ 400 keV. The interpretation of the data coming from these and a number of other missions has led a majority of researchers to the view that GRBs must originate in strongly magnetized neutron stars (Mazets *et al.*, 1981; Woosley, 1982; Lamb, 1982), based on the extremely short variability time scales, the existence of periods in a few sources, and the presence of lines in a large number of cases interpreted as cyclotron and pair annihilation lines. In what follows we briefly review some of the main observational results.

### *b) Detected Fluxes and Temporal Structure*

The detection of GRBs occurs when their suddenly increasing $\gamma$-ray flux exceeds the triggering value preset for the detector. The energy flux levels observed when they reach maximum are typically in the range

$$10^{-7} \lesssim F \lesssim 10^{-4} \text{ ergs cm}^{-2}\text{ s}^{-1}, \qquad (9.1.1)$$

having started from a previous state where their $\gamma$-ray flux, if any, is somewhere below the typical detection threshold of $F_{\rm min} \sim 10^{-7}$ ergs cm$^{-2}$ s$^{-1}$. This minimum threshold is lower, about $10^{-8}$ ergs cm$^{-2}$ s$^{-1}$, for the BATSE detector flown on the *GRO* satellite. From burst statistics and occasional repetition rates, the preburst quiescent (or nondetection) period is estimated to be typically $\gtrsim$ 1 yr, although in some cases it is months. The GRB outburst lasts above the detection threshold for a time interval $\Delta t$ typically in the range 1–10 s, so the time-integrated flux, called the fluence $S$, is typically in the range $10^{-6} \lesssim S \lesssim 10^{-3}$. During the burst the flux level fluctuates strongly over time scales as short as $10^{-2}$ s. In some cases the total duration $\Delta t$ is as short as $10^{-1}$ s and in others as long as $\sim 1000$ s. In one case (GBS 0520 − 66, the peculiar March 5, 1979, event) the impulsive rise time was of the order of $10^{-4}$ s and, after decaying in about $\sim 0.1$ s, there followed a much lower flux level decay lasting $\lesssim 100$ s during which the flux varied regularly with a period of 8 s. There are a few other cases where evidence for a regular periodicity in the range 1–10 s has been reported (e.g. Hurley, 1986a, and references therein; Kouveliotou, 1988).

Roughly speaking, GRB time profiles are of three main types: (i) very brief bursts, (ii) bursts with a double temporal maximum or quasi-periodic time structure, which may be brief or extended, and (iii) long and irregular bursts. A somewhat more detailed classification (Desai, 1986) based on the GRB catalog obtained with the KONUS experiment (Mazets *et al.*, 1981) is as follows:

1) Weak bursts with insufficient statistics for characterization, 21%.
2) *e*-folding rise and decay times of $\sim$ 500–1000 ms, 15%.
3) *e*-folding rise and decay times of $\lesssim 500$ ms with secondary features, 3.6%.
4) Symmetrical rise and decay times, 6.5%.
5) Slow rise and decay times ($\sim$ s), with or without flat tops, 5%.
6) Bursts with a pair of narrow ($\sim$ 100 ms) spikes, 10%.
7) Paired pulses, 2.9%.
8) Complex multipeak profiles, long duration ($\gtrsim 10$ s), 18%.
9) Bursts with a single spike ($\sim$ 100 ms), 18%.
10) March 5, 1979, burst, rise time $\lesssim 0.2$ ms, decay time $\gtrsim 100$ ms, 0.7%.

An example of a burst of type (4) is given in Figure 9.1.1a, an example of type (6) in Figure 9.1.1b, and the example of type (10) in Figure 9.1.1c. Typically, these are broad-band time profiles. However, the exact profile is usually dependent on the particular energy observed. The detection and recording is straightforward only in those spacecraft (like *HEAO 1*) where all data were continuously telemetered. In most spacecraft, however, data recording starts only after a trigger criterion has been fullfilled (after a certain flux level is reached in some energy channel varying on a certain time scale, etc.), which differs between

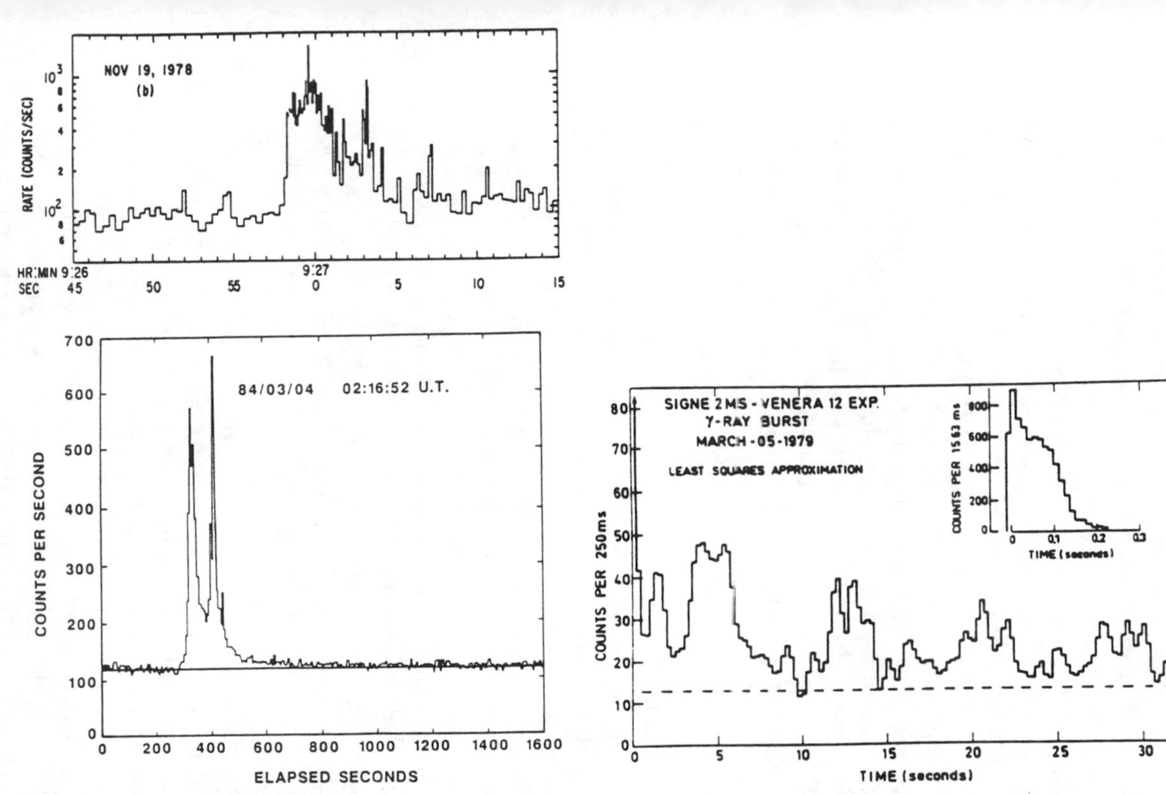


FIGURE 9.1.1 Three different GRB time profiles. (a, *top left*) An example of a burst with a rise time $\lesssim 0.2$ s and a slower decay over tens of seconds. This is the November 19, 1979, burst, which also presents evidence of high-energy spectral features measured by *ISEE 3* (from Teegarden and Cline, 1980). (b, *bottom left*) A two-peak burst with a long low-intensity tail of hard emission measured by the *PVO* experiment (from Evans and Laros, 1987). (c, *right*) The unusual March 5, 1979, burst as measured by the *SIGNE* experiment, showing the extremely short rise time of $\lesssim 0.2$ ms reaching a peak intensity 10 times larger than any other burst, followed by a fast 100 ms decay and a slower decay with an 8 s quasi periodicity. The main figure shows the 130–205 keV band, while the inset is integrated over the 130–723 keV band. From K. Hurley, 1986, in *AIP Conf. Proc. 141, Gamma-Ray Bursts*, ed. E. Liang and V. Petrosian (AIP, New York), p. 3.

spacecraft. A fuller discussion of these issues and of the flux and temporal structures of GRBs is given by Wood, Desai, and Schaefer (1986).

The repetition time scale of GRBs is highly uncertain, since no classical GRB has been observed to appear from exactly the same direction (within observational uncertainties) at a different time. The only exceptions are a small class (three members so far, including the March 5, 1979, event) of unusual-spectrum soft gamma repeaters (SGRs). For classical GRBs, which are the great majority, limits have been derived for the possible repetition time scale. A lower limit of 1–8 yr has been obtained by Atteia *et al.* (1987) by assuming all GRBs to have the same luminosity and comparing the frequency of close coincidences with that expected from Monte Carlo simulations. An upper limit can be derived by assuming that all the neutron stars in the galaxy are available to produce GRBs (e.g. Jennings, 1982) which is $\sim 3 \times 10^6$ yr for a disk distribution and $\sim 10^8$ yr for a halo distribution. This is evidently a model-dependent upper limit, which, however, is useful since it does appear that GRBs are neutron stars.

### *c) Counts and Spatial Distribution*

With the present instrumentation, GRBs are detected at the rate of $\lesssim 10^2$ yr$^{-1}$ with fluences $S \gtrsim 10^{-6}$ ergs cm$^{-2}$, which could perhaps be increased to a few $10^3$ yr$^{-1}$ at fluences $S \gtrsim 10^{-8}$ ergs cm$^{-2}$ (Meegan, Fishman, and Wilson, 1985). As emphasized by Hartmann and Woosley (1988) and Hurley (1986b), with $\gtrsim 500$ sources detected and at this rate of detection, $\gamma$-ray bursters will soon become the most common form of emission of observable neutron star sources, if that is what they indeed are. This is to be compared with $\sim 100$ bright galactic X-ray sources (Joss and Rappaport, 1984), $\sim 35$ X-ray bursters (Matsuoka *et al.*, 1985), $\sim 30$ accreting X-ray pulsars (Nagase, 1989a), and $\sim 500$ radio pulsars, the discovery rate for all of them being far slower.

The overwhelming majority of the detections are of so-called classical GRB events, as defined by their spectrum and time properties. Their spatial distribution appears isotropic, so far. The statistics for their isotropic distribution are not as good as one might hope, since $\gamma$-ray detectors are generally not direction sensitive, and directions have been established only for a subset of GRBs which have been detected by two or more spacecraft simultaneously, allowing a direction determination via a light front arrival argument. The error boxes (three or more satellites) or error circles (two satellites) are generally not too small, e.g. degrees. This situation should greatly improve with the accumulation of data from detectors with coded mask aperture or rotating modulation collimators, such as the WATCH instrument on *Granat* (Lund, 1990), which can achieve angular resolutions of $\gtrsim 10'$ by itself. A plot of the GRB angular distribution in galactic coordinates obtained from simultaneous observations by multiple spacecraft on 180 GRBs is shown in Figure 9.1.2. The current data give no detectable hint of either a dipole or a higher multipole to the angular distribution (Hartmann, Epstein, and Woosley, 1989).

The most frequently used tool for investigating the depth distribution of such objects where the distance is not a priori known is the so-called log $N$–log $S$ relationship adopted from the study of unidentified radio sources. In this case, it measures the number $N(> S)$ of sources detected at a fluence level greater than $S$. In general, if the sources are distributed isotropically and homogeneously in depth down to fluences less than the detection threshold $S_{\min}$, their number within $r$ is proportional to $r^3$, while the brightness of a source is proportional to $r^{-2}$, so that one would expect a distribution $N(> S) \propto S^{-3/2}$. This would also be the case for an isotropic homogeneous distribution unbounded in depth. On the other hand, if it were

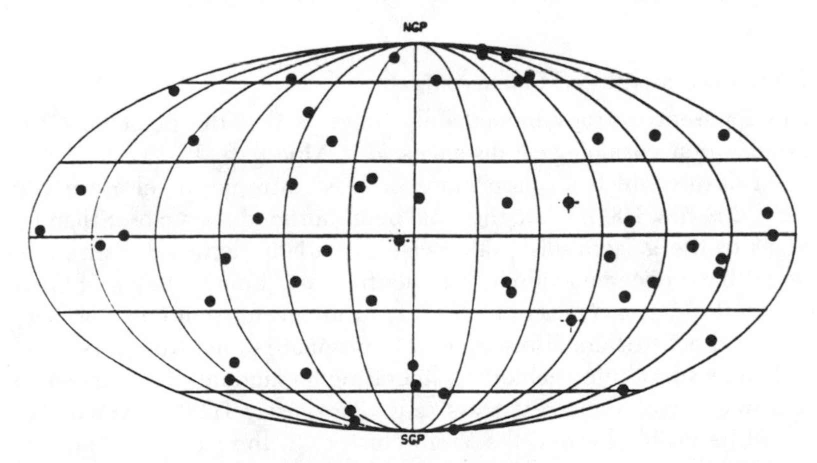


FIGURE 9.1.2 Angular distribution of GRBs in galactic coordinates, with the SGRs marked with crosses. Reprinted with permission from K. Hurley, 1990, *Adv. Space Res.*, 10, 179; © 1990 Pergamon Press PLC. Based on J. Atteia *et al.*, 1987, *Ap. J.* (*Supp.*), 64, 305.

bounded, there would be fewer sources at the low end of the fluence $S$. This might be expected, for example, if the GRBs were distributed along the galactic disk and the distance to which they were detectable was much larger than the disk half-thickness. In that case, the distribution of $N(> S)$ would flatten below the $-3/2$ power law for the sources fainter than the value $S$ characteristic of those sources at a distance comparable to the half-thickness of the galaxy (of course, with good angular information one would also see a departure from isotropy). In early analyses based on the KONUS catalog (Mazets *et al.*, 1981), it was evident that, while at high $S$ the $-3/2$ power law seemed well satisfied, at the lowest fluences one started seeing a flattening of the log $N$–log $S$ distribution. It has also been argued that instead of $S$ one should use $C$, the maximum count rate in a particular channel (e.g. Klebesadel, Fenimore, and Laros, 1984). In this case, the evidence for a flattening at the lowest $S$ disappears (e.g. Epstein and Hurley, 1988), down to the flux detection threshold (see Fig. 9.1.3). Another type of test performed on GRB distributions is the $V/V_{max}$ test (Higdon and Schmidt, 1990; Hartmann, Epstein, and Woosley, 1990), which unlike the log $N$–log $S$ method is independent of detector response problems and to a large degree also of selection biases. Results from this test indicate a possible departure from isotropy, but at the moment this is not significant, the results being compatible with isotropy (cf. also GRO 1991 preliminary results).

### *d) Distance Range and Counterparts*

The apparent isotropy immediately suggests that the detected GRBs either are at cosmological distances, $d \gtrsim$ Mpc (e.g. Paczyński, 1986), like galaxies which are also more or less isotropic, or else are very near, $d \lesssim$ few 100 pc, like the local neighborhood stars closer than the edges of the galactic disk; otherwise a spatially flattened distribution would have become evident. The neutron star interpretation of GRBs favors the latter (galactic) choice, although in some extragalactic models the GRBs are also supposed to originate in neutron stars which collide or spiral into each other, liberating an amount of energy which is a fraction of their rest mass (e.g. Paczyński, 1990a), which then would be visible beyond the Virgo cluster. In the galactic interpretation, taking into account the possible evolution of a neutron star population in space (including motion away from the plane) and variations in the intrinsic luminosity, the distances expected may range over 150 pc–2 kpc (Hartmann, Epstein, and Woosley, 1990; Paczyński, 1990b).

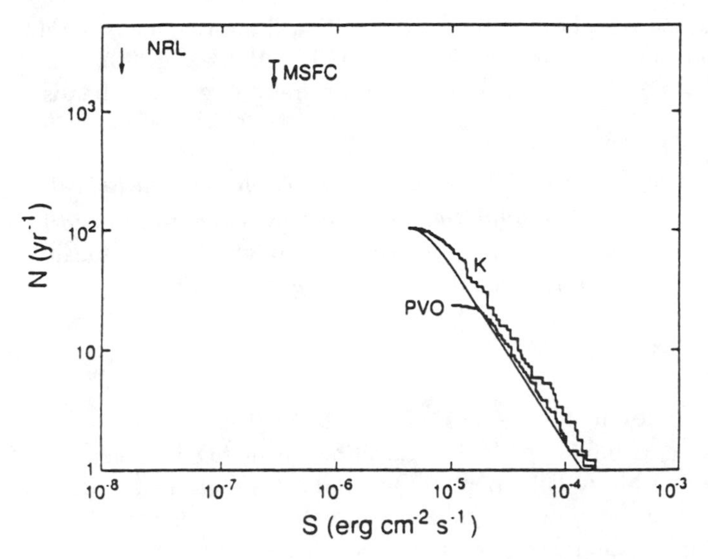


FIGURE 9.1.3 The brightness distribution (log $N$–log $S$) of GRBs. The abscissa is the effective peak flux $S$, and the ordinate is the burst rate per $4\pi$ steradians whose effective peak flux exceeds $S$. The data from the *PVO* experiment and the *Venera 11* KONUS catalog (K) are shown, while the smooth line is the expected rate from the *Venera 11* data for a homogeneous and isotropic distribution of monoluminosity sources, after corrections for observing biases. The MSFC and NRL data are upper limits from related studies. From R. I. Epstein and K. Hurley, 1988, *Astron. Letters and Comm.*, 27, 229. Reprinted with permission of Gordon and Breach Science Publishers.

The single greatest obstacle to a distance determination is that, with the single exception of the March 5, 1979, SGR source, whose positional error box coincides with the supernova remnant N49 in the Large Magellanic Cloud (e.g. Hurley, Cline, and Epstein, 1986), no GRB has been identified with an astronomical object known or studied at other wavelengths. This also puts limits on the type of companion, if any, the GRB may have. For main-sequence stars this limit is $\gtrsim 10$ kpc (Schaeffer *et al.*, 1983; Motch *et al.*, 1985), while for white dwarfs it is $\gtrsim 4$ kpc (Pederson *et al.*, 1983), based on an optical apparent magnitude limit $m_v \gtrsim 23$. Searches at X-ray energies for counterparts have also been unsuccessful, which puts a limit on the surface temperature $T \lesssim 10^6$ K and possible accretion rate $\dot{M} \lesssim 10^{-15}\ M_\odot\ \mathrm{yr}^{-1}$ of the quiescent (nonbursting) neutron stars (e.g. Pizicchini *et al.*, 1986; Grindlay *et al.*, 1983; Boer *et al.*, 1991). The accretion limits, which would be relevant for thermonuclear models, translate into distances

$d \gtrsim 2$ kpc. It is hoped that these limits will improve considerably with the advent of missions such as *HETE* (Ricker, 1990), which will simultaneously monitor GRBs over several energy wave bands.

### *e) Spectra: Classical Gamma Bursters and Soft Gamma Repeaters*

The classical GRBs, which make up the overwhelming majority of all GRBs, have a rather hard time-integrated spectrum, typically of the form of a broken power law in the energy spectrum

$$F_E \propto E^{-\alpha}, \tag{9.1.2}$$

whose index in the X-ray region is $\alpha_X \sim 0$, and a $\gamma$-ray power law index $\alpha_\gamma \sim 0.5$–$2$, with the transition occurring between about 100 keV and 1 MeV. This spectral form indicates that most of the power is emitted in the $\gamma$-ray range of the spectrum, as seen in Figure 9.1.4, which plots the power (or energy per decade) $E^2N(E)$ as a function of $E$ (e.g. Zdziarski, 1987). The spectra in this figure are for successive time intervals within the bursts and show a characteristic typical of

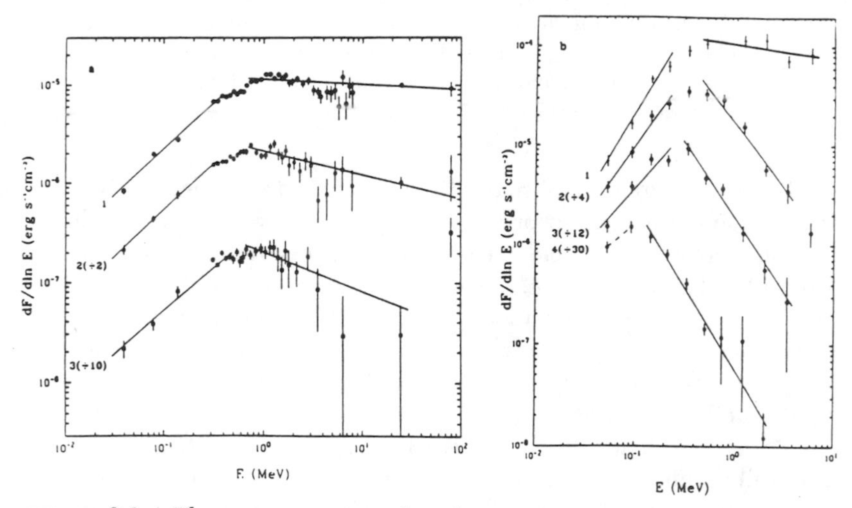


FIGURE 9.1.4 The spectra expressed as the power per decade of two classical GRBs: (a) GB840805 in three 16 s intervals, and (b) GB830801b in four 0.5 s intervals. From A. Zdziarski, 1987, in *13th Texas Symp. Relativistic Astrophysics*, ed. M. Ulmer (World Scientific, Singapore), p. 553.

most classical bursts, namely a gradual softening of the spectrum as time progresses. This is a feature of many GRBs (e.g. Norris *et al.*, 1986). Sometimes the burst is observed to begin sooner and last longer at lower, X-ray, energies (e.g. Laros *et al.*, 1984). A soft X-ray precursor (1–7 keV) has been observed for at least one GRB seen by *Ginga* (Murakami *et al.*, 1990), while about 25% of the *Ginga* GRBs appear to have X-ray tails that last for more than 50 s (Yoshida *et al.*, 1989). These X-ray precursors and tails can be fitted by blackbody curves with $T \sim 1.5$–$1.8$ keV, corresponding to a source size $\sim 7 \times 10^5 (d/1 \text{ kpc})$ cm typical of a neutron star polar cap.

An important property of classical GRBs, which is apparent from an inspection of spectra such as those in Figure 9.1.4, is the *X-ray paucity* constraint: the ratio of X-rays to $\gamma$-rays is extremely low, typically

$$\frac{L_X(3\text{–}10 \text{ keV})}{L_\gamma(> 100 \text{ keV})} \sim 0.02 \tag{9.1.3}$$

(Laros *et al.*, 1982). These spectra also show another characteristic of classical GRBs, which is the extremely hard high-energy power law tail extending in some cases up to 100 MeV. An analysis of the statistics of GRBs detected with the gamma-ray spectrometer aboard the *SMM* spacecraft shows in fact no clear indication of high-energy cutoff and indicates that a substantial fraction of all GRBs detected by this instrument extend above $\sim 10$ MeV (Matz *et al.*, 1985), with some up to $\sim 100$ MeV (Share *et al.*, 1986). A number distribution of GRBs with spectra extending to energies $> E$ is shown in Figure 9.1.5 (Matz *et al.*, 1985).

The special category of SGR bursters appears to be different from the classical GRBs discussed above (Mazets *et al.*, 1982a, b; Epstein and Hurley, 1988). The spectra of these objects are significantly softer than those of the great majority of classical GRBs, with most of the SGR's energy emitted below about 40 keV, which is in fact not too different from the spectra of accreting X-ray pulsars. An example is shown in Figure 9.1.6. This category is made up so far of only three objects, which besides their spectra are also characterized by being extremely brief bursts in their luminous phase and by showing *repeated* bursts coming from the same object or at any rate the same direction, as far as one can tell from the error boxes. One of these is GBS 0520 − 66 (the source of the spectacular March 5, 1979, burst, which has repeated at a lower level at least 15 times), GBS 1900 + 14

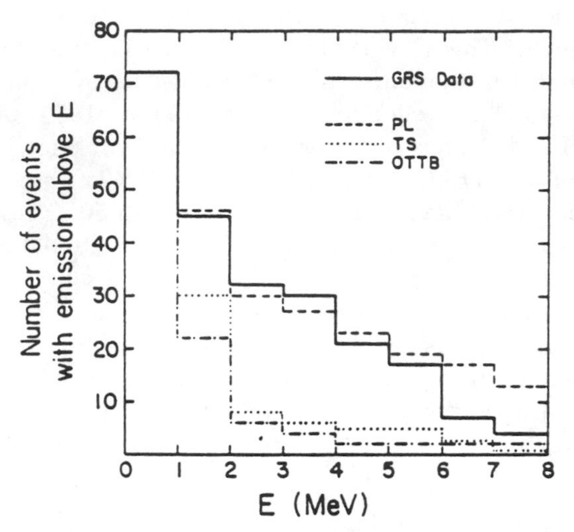


FIGURE 9.1.5 Number of bursters detected above a certain energy $N(>E)$ versus $E$, from the data of the GRS detector on the *SMM* spacecraft. The three broken lines show the number of events expected if the spectrum were a power law (PL), thermal synchrotron (TS), and optically thin bremsstrahlung (OTTB). From S. M. Matz *et al.*, 1985, *Ap. J.* (*Letters*), 288, L37.

(3 bursts), and SGR 1806 – 20 (the source of the January 7, 1979, burst and at least 110 subsequent events).

### *f) Spectral Line Features in GRBs*

Absorption features in the range 20–60 keV have been reported in 20–30% of the classical bursts observed with the KONUS detector (Mazets *et al.*, 1981, 1982a, b), and one case was reported from *HEAO A4* (Hueter, 1984). The width of these lines was relatively small, $\delta E/E \lesssim 0.3$–$0.5$. The interpretation of these features proved controversial because of the possibility of instrumental effects and other factors. However, the original and most straightforward interpretation was that these would be cyclotron lines in a magnetic field $B \gtrsim 10^{12}$ G (Mazets *et al.*, 1981). One of the problems appeared to be that this field might be too high to allow the escape of very high energy photons seen by Matz *et al.* (1985), and another that the gas producing them would have to be cold compared to that responsible for the higher-energy continuum. Also, second and higher harmonics

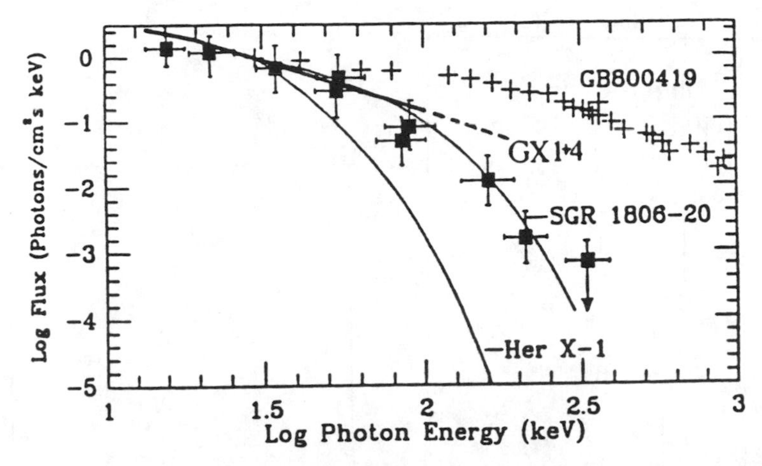


FIGURE 9.1.6 Average spectrum of the soft gamma repeater SGR 1806 − 20 (*filled squares*) fitted by an exponential, compared to the spectrum of the classical GRB GB800419 and the spectra of the X-ray pulsars Her X-1 and GX 1 + 4. From D. Hartmann and S. Woosley, 1988, in *Multiwavelength Astrophysics*, ed. F. Córdova (Cambridge University Press), p. 189.

would be expected, which were not seen in the observations mentioned.

This uncertainty disappeared to a large degree with the observation on two different bursts (GB 870303 and GB 880205), by the *Ginga* satellite (Murakami *et al.*, 1988) of *two* absorption features in each event, at energies of $\sim 20$ and $\sim 40$ keV, which would correspond to a first and second harmonic for a field of $B \sim 1.7 \times 10^{12}$ G. This was further confirmed with the detection of two harmonics at a different set of energies (26 and 47 keV) in another event, GB 890929 (Yoshida *et al.*, 1990). While the possibility of instrumental effects was seriously considered, the need for two narrow lines and possibly a third broader one to fit the spectra appears necessary (Fenimore *et al.*, 1988; Yoshida *et al.*, 1990). An example is shown in Figure 9.1.7, giving the raw count spectrum of GB 880205 at three different times and the fitted spectrum (Murakami *et al.*, 1988, 1989; Fenimore *et al.*, 1988). The analysis indicates that the line energies shift toward lower values as the burst progresses (Murakami, 1989). The continuum on which these lines are superimposed is a power law extending down to soft energies $\lesssim 5$ keV before it becomes absorbed, and up to the detector upper sensitivity of $\sim 375$ keV at the other end, showing a

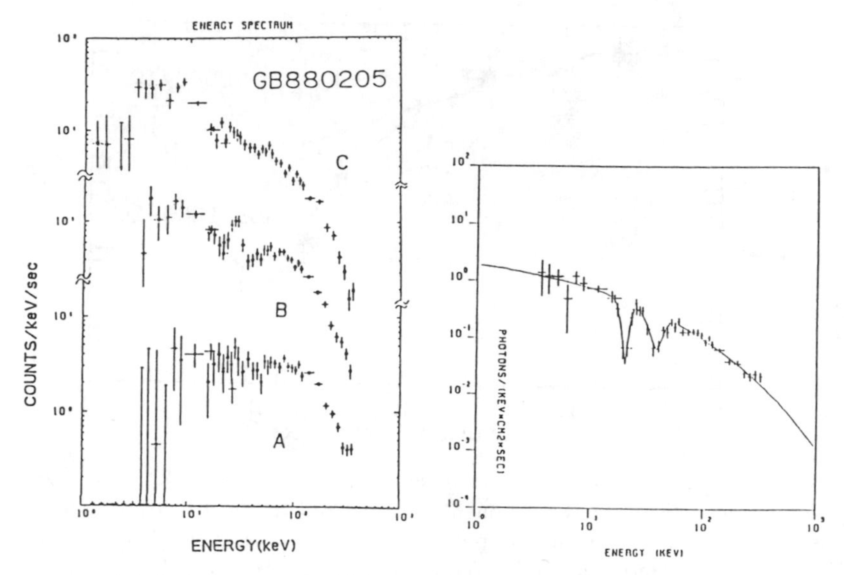


FIGURE 9.1.7 Spectrum of GB880205 observed with *Ginga* showing two and possibly three cyclotron harmonics. *Left*: Raw counts for three time intervals. *Right*: Fitted spectrum. From Murakami *et al.* (1988). Reprinted by permission from *Nature*, vol. 335, p. 234. Copyright © 1988 Macmillan Magazines Ltd.

definite softening in time. Because of the typical burst time structure and the simultaneous detection at high energies by the *same* spacecraft, there appears no doubt that this is the X-ray tail of otherwise normal GRB events, of which a total of 25 were observed by *Ginga* up to 1988. It is worth noting that several of these have been observed simultaneously with other detectors aboard the *PVO* and *Phobos* missions. The reality of the lines is supported by the fact that, out of about twenty bursts detected with sufficient statistics to determine spectra, only three have these line features, while the others do not (Murakami, 1989; Yoshida *et al.*, 1990).

Other features that have been reported in GRBs are broad emission lines at energies between 350 and 450 keV, which have been interpreted as normal two-photon $e^+e^-$ annihilation lines gravitationally redshifted near the surface of a neutron star (Mazets *et al.*, 1981, 1982a; see also Hurley, 1986a, b), and lines near 750 keV in the November 19, 1979, burst (Teegarden and Cline, 1981), possibly interpreted as nuclear lines or as due to redshifted one-photon annihilation in the presence of a strong magnetic field.

*g) Evidence for a Neutron Star*

The luminosity of a typical GRB is a function of the (unknown) distance, which can be written as

$$L \simeq L_{\mathrm{Ed}}\left(\frac{F}{10^{-4}}\right)\left(\frac{d}{100\ \mathrm{pc}}\right)^{2}\left(\frac{M}{M_{\odot}}\right)^{-1}, \tag{9.1.4}$$

where $L_{\mathrm{Ed}} \simeq 10^{38}(M/M_{\odot})$ ergs s$^{-1}$ and $d$ is the distance of the source. This is a typical neutron star maximum luminosity for a distribution of objects at distances $\lesssim$ few 100 pc which would be needed in order not to violate the log $N$–log $S$ slope of $-3/2$ (if they were at the alternative extragalactic distances, their luminosities would have to be enormous). The observations by *Ginga* (Murakami, 1990) of a large number of GRBs with a low-level, extended cooling period during which the spectrum becomes softer and concentrated in X-rays indicate a typical terminal X-ray flux of $\sim 10^{-8}$ ergs s$^{-1}$ and a blackbody spectrum with $T_{\mathrm{eff}} \sim 1.2$–$1.6$ keV. For a distance $d \sim 1$ kpc this implies a ratio $R/d \sim 1$, where $d$ is in kpc and $R$, the linear size of the emitting surface, is in km, or $R \sim 1$ km, compatible with a neutron star polar cap. Also, the typical variability time scale $\gtrsim$ few ms implies total sizes $R \lesssim 10^{8}$ cm, or for GB 790305 ($\lesssim 0.2$ ms) a size $R \lesssim 10^{6.7}$ cm, indicating the need for a neutron star. Finally, while keeping in mind the discussion about possible problems with hard MeV photons coming from a high field region, the spectral evidence mentioned in the previous section for cyclotron lines requiring a magnetic field $B \gtrsim 10^{12}$ G also indicates a neutron star.

## 9.2 Gamma-Ray Burster Models and Energetics

*a) Introduction*

The theoretical understanding of GRBs is still rather tentative, as compared to the case of accreting pulsars or rotation-powered pulsars described previously. Here we shall discuss the most widely considered models, in which these objects are taken to be magnetized neutron stars. This narrowing of the field of possible candidate sources to a few basic models based on magnetized neutron stars is a significant advance over the situation prevalent one year after the discovery of this type of sources (Klebesadel, Strong, and Olson, 1973), when the field of possible contending objects included not only neutron stars but also black holes, white dwarfs, and other stellar sources (Ruderman, 1975). The view that GRBs should be magnetized neutron stars, which became widely accepted with certain reservations since about 1979, had been based mainly on the early evidence for cyclotron lines, which became very strong with the analysis of the more recent *Ginga* data after 1988.

In what follows we address the question of how some of the current models attempt to explain the burst energy production, the burst time scales, and the features that cause this energy to appear predominantly at $\gamma$-ray energies rather than at X-ray energies. The specific models discussed here are the thermonuclear burst, the sudden accretion, the starquake, the flare, and the dormant pulsar models. As seen from this list of models, the exact source of the energy is not yet agreed upon. Neither is there as yet clear agreement as to why other neutron star sources are X-ray sources, including many accreting magnetized neutron stars (e.g. AXPs), whereas GRBs emit mainly $\gamma$-rays, with a typical ratio $L_X/L_\gamma \lesssim 0.02$ (see Imamura and Epstein, 1987). It should therefore be stressed that the problem of the cause of GRBs is still unresolved and that the models described here may well be ephemeral. Nonetheless, these objects demand some explanation, however preliminary, and it is in this spirit that we shall discuss them.

### *b) Thermonuclear Models*

In thermonuclear models the burst energy arises from the sudden ignition of thermonuclear fuel that has been slowly accumulated by accretion over the polar cap of a magnetized neutron star (Woosley and Taam, 1976; Maraschi and Cavaliere, 1977). The burst is produced by a helium flash triggered by an initial hydrogen flash, which occurs only if the temperature of the bottom of the accretion column is sufficiently low or if the accretion rate is sufficiently small over a small area of magnetized polar cap (e.g. Hameury and Lasota, 1986). At higher accretion rates over a magnetized polar cap of similar size such as is inferred for accreting pulsars, these instabilities do not arise, which is consistent with the fact that AXPs are not seen to burst. On the other hand, for high accretion rates but very low magnetic fields, where accretion occurs over the whole surface, thermonuclear flashes lead to X-ray bursts (e.g. Wallace, Woosley, and Weaver, 1982). The accretion rate needed for the magnetized cap accretion to lead to a thermonuclear burst depends on the assumed internal temperature of the star. For $T_i < 3 \times 10^7$ K the rate needed is $\dot{M} \sim 10^{-15}$–$10^{-16}$ $M_{\odot}$ yr$^{-1}$ (Hameury *et al.*, 1982), which leads to a H flash by electron capture on protons after a high enough density has been reached, typically requiring the accumulation of $\sim 10^{23}$ g. This heats the material above the $3\alpha$ reaction threshold, which results in a He thermonuclear detonation. On the other hand (Woosley and Wallace, 1982), if the internal temperature is higher than $3 \times 10^7$ K, a H flash can occur earlier, for accumulations as low as $\sim 10^{19}$ g, and the energy thus liberated will not be sufficient to raise the temperature to that required to ignite He so that after the initial flash the accreted H will burn stably into He. After the accumulated He reaches a mass $\gtrsim 10^{20}$–$10^{21}$ g at $\rho \sim 10^7$–$10^8$ g cm$^{-3}$, the $3\alpha$ reaction becomes possible, in the form of a runaway. The exact He temperature is determined by the accretion rate, leading either to a He deflagration or detonation for $\dot{M} \sim 10^{13}$–$10^{14}$ $M_{\odot}$ yr$^{-1}$. These low accretion rates are marginally compatible with X-ray limits on a steady luminosity, but the optical limits put a strain on the type of binary companion that could be supplying this gas. Alternatively, a single neutron star could accrete at this rate from the interstellar medium provided it were moving at relatively low speeds, $v \lesssim 30$ km s$^{-1}$.

The total burst energies, rise times, duration, and possible repetition rates appear to fit reasonably well those observed in typical GRBs (e.g Woosley, 1982). Of course, the thermonuclear energy deposition occurs at large optical depths in the polar cap which by itself would not lead to $\gamma$-ray emission but rather to X-ray emission at some typical energy $\hbar\omega \sim kT_{\mathrm{eff}} \sim$ few keV. At the same time, this large optical depth implies for the thermonuclear GRB a longer-duration, low-luminosity X-ray afterglow, which seems to agree with observations and is interpreted as a supporting argument for this model (Hartmann and Woosley, 1988; Murakami *et al.*, 1990). Two further predictions of the model are that the burst energy should be proportional to the burst recurrence time and that the burst recurrence time should be inversely proportional to the steady accretion luminosity, these being features that are thus far not well established.

The conversion of the energy into gamma-rays requires an additional mechanism for transmitting the bulk of the energy with little dissipation into a lower opacity region, where presumably it can convert more efficiently into $\gamma$-radiation. Two such possible transmission mechanisms (Hameury and Lasota, 1986) are magnetoconvection and Alfvén waves. The energy thus carried to the magnetosphere is presumed to lead to large-amplitude magnetic field disturbances, reconnection, and/or vacuum breakdown, which produces and accelerates $e^+e^-$ pairs that (being now in the lower-density outer magnetosphere) are able to produce $\gamma$-rays, as discussed below. The lower-luminosity X-rays would be produced by reprocessing in the stellar surface (Hartmann, Woosley, and Arons, 1988; Hameury and Lasota, 1989).

### *c) The Sudden Accretion Model*

There are two versions of the sudden accretion model, the comet capture model (Harwit and Salpeter, 1973; Tremaine and Zitkow, 1986), and the episodic gas accretion model (Lamb, Lamb, and Pines, 1973; Anzer, Börner, and Mészáros, 1976). Early versions of the comet impact model had some difficulties, e.g. the collision rates appeared too infrequent, the spectrum of the disrupted matter reaching the star would be thermalized to X-rays, and the typical time scales would correspond to free fall at the surface, which is too short for average burst durations (Howard, Wilson, and Baron, 1981; Colgate and Petschek, 1981). More recently, Pineault and Poisson (1989) and Pineault (1990) considered the possibility of comet capture from comet clouds around other stars than the capturing degenerate star and concluded that a significant increase in the capture rate could result, so that at least some GRBs may result from this mechanism. Colgate (1990) addressed the problem of the spectrum and time scale with a qualitative model in which the comet interacts with the outer magnetosphere of the neutron star at larger radii. The ensuing magnetic field disturbance leads to tearing mode instabilities and the generation of electric fields in which electrons are accelerated to relativistic energies and radiate nonthermal $\gamma$-rays, the typical dynamic time scale at such larger radii being in the right range for GRB time scales ($\gtrsim 1$–$10$ s).

The episodic gas accretion model is not as vulnerable to statistics problems, but it must accommodate the very low observational limits obtained on X-rays from a quiescent steady accretion rate. In its latest incarnation this model envisages the slow accumulation of matter in an accretion disk which piles up against the magnetospheric stresses at the Alfvén radius of a magnetized neutron star until a runaway instability dumps this reservoir of energy onto the neutron star (Kafka and Mayer, 1984; Michel, 1985; Epstein, 1985). The energies, burst rise times, and repetition rates in these models may be compatible with the observations. The difficulty is in obtaining a $\gamma$-ray spectrum with little X-ray emission, since in an accretion model the matter is expected eventually to reach the surface. One possibility is that thermalization is difficult to achieve if the magnetic field is very strong, but it is hard to see how this would be so extreme as to produce less than 2% X-rays. The observation interpreted as cyclotron lines at a few tens of keV (Murakami *et al.*, 1988; Yoshida *et al.*, 1990) in several GRBs indicates a field strength which is not significantly different from that of X-ray pulsars, although of course in an impulsive accretion event thermalization would be less complete than in the steady-state situation prevalent in the X-ray pulsars.

*d) Neutron Starquakes*

The first neutron starquake models assumed that the burst energy is provided by a relatively small and sudden change in the rotation rate of the neutron star (Pacini and Ruderman, 1974). In its original version, this event was supposed to be somewhat similar to the stellar quakes leading to glitches in radio pulsars such as Vela, except with a much larger energy. The early models for these glitches involved crustquakes, in which the pulsar is assumed to slowly lose rotational energy until its rigid crust is unable to withstand the strain associated with a diminishing centrifugal support and readjusts itself, leading to a small change in radius which represents a liberation of energy equal to the reduction in the gravitational energy. The problem with this original version of crustquakes is that the total energy available is insufficient to explain even the strongest glitches of the Vela pulsar, so pulsar glitches are currently explained within the framework of vortex unpinning theory (see Chaps. 8 and 11.3). For pulsars, the rotational slowdown leads to a strain between the pinned vortices and the crust until the former become unpinned, leading to a sudden exchange of angular momentum with the crust. However, if the stress of the pinned vortices on the crust leads to fracture or if the angular velocity difference is sufficiently large for dissipative effects to occur in the flow of superfluid neutrons through crustal nuclei, enough energy ($E \sim 10^{42}$–$10^{43}$ ergs) may be liberated to power a GRB (Epstein, 1988). An unconventional alternative to the angular momentum transfer catastrophe as a trigger for crustquakes proposed early on by Bisnovatyi-Kogan (1974, 1990) involved fission of high atomic number nuclei in the lower crust leading to an undersurface explosion. Another mechanism for inducing a crustquake could be the sudden release of nuclear energy from the transition of metastable nuclei from accreted matter which is compressed beneath the stellar surface (Blaes *et al.*, 1990).

The total energy requirements are more easily met by corequake models, which involve more of the neutron star mass and moment of inertia. The first proposal along these lines (Tsygan, 1975) involved the deformation of a solid core causing vibrations which couple to the exterior via the magnetic field lines. Later versions considered the possibility of a nuclear phase transition (e.g. Migdal, Chernoustan, and Mishustin, 1979) in the core which liberates a fraction of the total binding energy (Ramaty *et al.*, 1980; see also Hameury and Lasota, 1986). In this type of model the conversion of the mechanical motion of the readjusting crust into $\gamma$-radiation is generally expected to occur when the Alfvén waves induced by the crust motions become of large enough amplitude to accelerate electrons and pairs produced by magnetospheric breakdown in the large induced electric fields. These pairs would then produce the $\gamma$-ray spectrum by some combination of synchrotron, annihilation, and inverse Compton radiation, as in Ramaty *et al.* (1980) (see also Melia, 1988, 1990). As shown by Blaes *et al.* (1989), the efficiency of the crust-magnetosphere coupling is greatly increased in the presence of a magnetic field so that a large transmission coefficient may be expected. Also, torsional oscillations produced by the quake would not be damped by gravitational radiation so that the energy storage time may be comparable to the observed GRB duration (Miller *et al.*, 1990; Blaes *et al.*, 1989).

While starquake models appear as a likely possibility, a general problem with all versions is that they may overproduce X-rays since the efficiency of the transport of energy from the interior to the magnetosphere is unlikely to be very high. Another concern with corequake models based on phase transitions is that they are expected to occur only once in the lifetime of a neutron star. A similar problem arises in the nuclear energy release models leading to a crustquake, where the accumulation time needed for accretion over the whole surface is of the order of the Hubble time, unless accumulation and release are confined to a smaller region of the star. In glitch models, one of the questions is that, if vortex unpinning occurs in fast rotating stars, as observed to be the case with pulsars, it is not clear why pulsars are not observed to burst; alternatively, if they are slow rotators, it is not clear whether glitches can occur in such circumstances.

### *e) Flare Models*

There are several versions of the flare model, some of which to some degree overlap with the thermonuclear, quake, or dormant pulsar models. Thus, thermonuclear models were proposed to cause magnetic field line reconnection and particle acceleration in a flare-like manner by Mitrofanov and Ostryakov (1981), Woosley and Wallace (1982), and Bonazzola *et al.* (1984). Similarly, in the starquake models the acceleration of relativistic particles is thought to occur via Alfvén wave nonlinearities and field line reconnection (Ramaty *et al.*, 1980; Mitrofanov, 1984; Blaes *et al.*, 1989). In the model of Sturrock (1986) an otherwise quiescent or low-luminosity pulsar undergoes a flare-like event in its magnetosphere during which magnetic energy is liberated by the reconnection of field lines, leading to the production of pairs, which are accelerated and produce $\gamma$-rays in the electric field associated with the sudden time variation of the magnetic field. In the model of Liang (1987), a two-component model is envisaged in which a magnetic field loop high above the stellar surface undergoes a flare, leading to pair acceleration and $\gamma$-rays. The much smaller X-ray flux would be produced on the surface of the neutron star either by reprocessing of the impinging $\gamma$-rays or by radiating some of the original energy that gave rise to the magnetic field perturbation. Melia and Fatuzzo (1989) and Melia (1990a, b) have proposed that the equilibrium magnetospheric charge distribution of pulsars may be unstable to the development of plasma oscillations, and the energy of the oscillating relativistic particles would account for burst-like $\gamma$-ray production via the inverse Compton scattering of soft photons from the stellar surface. Sturrock, Harding, and Daugherty (1989) proposed that particle acceleration occurs in a pulsar magnetosphere via the electric field induced by an infalling body, causing an $e^+e^-$ cascade. These authors calculated the evolution of these cascades and obtain a radiation spectrum of a power law type extending to high energies which is in qualitative agreement with observations. This model is essentially a variant of the dormant pulsar model.

### *f) Dormant Pulsar Models*

Dormant pulsar models (Ruderman and Cheng, 1988) are based on the observation that some of the very-short-period rotation-powered pulsars emit most of their luminosity in $\gamma$-rays (e.g. the Vela pulsar) and a smaller amount of radio, optical, or X-ray photons. The emission mechanism (Cheng, Ho, and Ruderman, 1985a, b) is based on the presence of outer magnetospheric gaps, where pairs are created via crossed beams of synchrotron radiation and are accelerated in the electric fields of the gap, becoming the source of $\gamma$-rays and further pairs. A consideration of the evolutionary consequences of the unavoidable period lengthening for Vela-like pulsars is that these sources tend to put more and more of their luminosity into the $\gamma$-ray band, until they eventually quench themselves; there follows a relatively rapid alignment of the magnetic axis with the rotation axis. The number of such extinguished pulsars whose magnetic field still remains strong, assuming a field decay time scale of about $10^7$ years, is about $10^5$ in the galaxy, or more if the decay is slowed down, a number which is appropriate for explaining GRB statistics. These dormant pulsars can be reignited at random intervals by a sudden perturbation, e.g. a flash of X-rays from a thermonuclear burst going through the magnetosphere, or a passing comet or asteroid. The spectrum of the ensuing magnetospheric emission would resemble that of an extreme Vela-like pulsar, which could explain the gross features of the $\gamma$ spectrum and the low X-ray emissivity. The energy source in this case is the rotation of the neutron star, the gamma-ray luminosity being typically $L_\gamma \sim 10^{35}$–$10^{36}$ ergs s$^{-1}$. One of the problems with this model is in explaining the narrow cyclotron lines observed in some GRBs, since these appear at high field strengths and in absorption. The lines would require scattering or transmission of the continuum spectrum (originating in the outer gaps near the light cylinder) by gas near the polar caps of the neutron star, which has a very small cross section.

## 9.3 Spectrum Formation in GRBs

### *a) Introduction*

A variety of radiation mechanisms have been invoked to explain the spectrum of GRBs (for recent reviews, see Zdziarski, 1987, and Lamb, 1988). Initially a number of studies considered static homogeneous models using asymptotic analytic expressions for nonmagnetic radiative cross sections while some authors used similar radiation physics within the framework of expanding fireball models. Pair formation and annihilation played a large role in these models. The realization that strong magnetic fields were involved led to the inclusion of more complicated magnetic process in the calculations. For the most part, these processes have been studied within the simplest possible type of models, e.g. one-zone, optically thin, static models. In the last few years, multizone static models have also been considered, with detailed angle and frequency-dependent magnetic cross sections. An increasing trend is to include the effect of inhomogeneities in the plasma distribution, and the effects of mass motions and time variability. Below we discuss some of these radiation mechanisms and the spectral calculations performed with them.

*b) Bremsstrahlung and Blackbody Radiation*

The bremsstrahlung mechanism was invoked early on as an explanation of the low-energy ($\lesssim 1$ MeV) spectrum of GRBs (Gilman *et al.*, 1980; Mazets *et al.*, 1981). In an optically thin medium, this mechanism produces a spectrum

$$I_\omega \propto e^{-\hbar\omega/kT}. \tag{9.3.1}$$

At these lower energies, the bremsstrahlung fits are indeed rather good with $kT \sim 0.1-1$ MeV. The main concern with this mechanism was that it is rather inefficient, requiring high particle densities (since $I \propto n_e^2$), which would lead to high Compton optical depths that would have smeared out the lines unless a particular geometry was adopted (e.g Ramaty *et al.*, 1980; Lamb, 1982). The other problem is that it is not able to reproduce the high-energy power law spectra at energies 1–10 MeV and above detected with the *SMM* spacecraft (Matz *et al.* 1985). The spectrum problem could be alleviated somewhat by using nonthermal bremsstrahlung from a power law distribution of electrons, which would produce a power law spectrum (e.g. Harding, Petrosian, and Teegarden, 1986). However, the efficiency problem seems unavoidable.

At the other extreme, some authors have investigated fireball models in which the pair opacity is extremely large so that the spectrum is essentially blackbody. Since such pair fireballs would expand, the typical spectrum would be blueshifted (e.g. Rees, 1984). This type of model could avoid the problem of becoming pair opaque for radii which were too large to explain the time variability constraint and the hypothesis that the $\sim 0.4$ MeV lines reported were gravitationally redshifted $e^+e^-$ lines (Zdziarski, 1984), although a redshift might have been simulated by transfer effects (Rees, 1982). Other versions of the fireball model have been developed for the case where GRBs are assumed to be at cosmological distances (Goodman, 1986; Paczyński, 1986). The spectrum, however, would be essentially thermal in shape, and the observed power law spectrum at high energies would require special explanations.

*c) Synchrotron Mechanism*

For a thermal distribution of electrons, the low efficiency of the bremsstrahlung mechanism can in principle be avoided in the presence of a magnetic field by using the much more efficient thermal synchrotron mechanism. Including quantum effects, the spectrum at

$\hbar\omega \ll kT$ is given by

$$I_\omega \propto \omega \exp\left[-\frac{\hbar\omega}{2kT} - \left(\frac{9(\hbar\omega/m_e c^2)}{2(B/B_Q)(kT/m_e c^2)^2 \sin\theta}\right)^{1/3}\right] \quad (9.3.2)$$

(e.g. Pavlov and Golenetskii, 1986). In general, good fits can be obtained with this mechanism for energies $\omega \lesssim$ few MeV (Liang, Jernigan, and Rodrigues, 1983). If the low-energy features (Mazets *et al.*, 1981) are interpreted as the first few cyclotron harmonics, the magnetic field strength can be inferred. Deducing the temperature from the fits, an upper limit on the luminosity and on the distance can be obtained. However, the fits fall below most of the *SMM* spectra at $\hbar\omega \gtrsim 1$ MeV (Pavlov and Golenetskii, 1986) unless $\hbar\omega \gg kT$, in which case $L_x/L_\gamma \ll 1$ would also be violated (Brainerd and Lamb, 1987). While the mechanism is efficient, this very efficiency requires maintaining the electrons excited to a high transverse temperature. Collisions are not enough to do this, and other mechanisms must be invoked (e.g. collective effects or excitation by inverse Compton scattering with $\gamma$-rays: e.g. Hameury *et al.*, 1985).

A nonthermal synchrotron mechanism would be more plausible in view of the high typical particle energies necessary to explain power law emission above 5–10 MeV. Ramaty *et al.* (1980) and Ramaty, Lingenfelter, and Bussard (1981) considered thick-target synchrotron following monoenergetic particle injection at $\gamma_0 \gg 1$. In this version the spectrum is the integrated emission from the cooling electron distribution as it loses energy through radiation, and the spectrum has the form (Bussard, 1984)

$$I_\omega \propto \omega^{-1/2} \exp\left[-(\hbar\omega/m_e c^2)(3/2)\gamma_0^2(B/B_Q)\right]. \quad (9.3.3)$$

While this spectrum is able to fit the $\gamma$-ray portion of the spectrum, its low-energy behavior at X-ray energies gives $\alpha_X \sim -0.5$, which does not agree with the typical observed X-ray behavior $\alpha_X \sim 0$. Thick-target synchrotron spectra for an electron power law injection have been investigated by Bussard (1986a) and Brainerd and Lamb (1987). These spectra fit the high-energy *SMM* spectra quite well but also have difficulties at low energies reproducing $\alpha_X \sim 0$. In Sturrock's (1986) flare model, pairs accelerated by the electric field radiate synchrotron emission with a spectrum of index $\alpha \sim 1/2$ below the pair threshold ($\hbar\omega B_Q/m_e c^2 B \sim 0.2$), and $\alpha \sim 2$ above that threshold. This fits some bursts while leaving a number of other bursts where

$\alpha \sim 0.5$–$1.5$ unexplained. The Vela-type dormant pulsar model of Ruderman and Cheng (1988), on the other hand, produces a spectrum with $\alpha_X \sim 0$ and $\alpha_\gamma \sim 0.5$–$1$ between 50 keV and 2 MeV, which is better able to meet the observations.

*d) Inverse Compton Models*

A thermal Comptonization model gives good fits for many bursts at energies below about 1 MeV (Fenimore *et al.*, 1982), typical examples suggesting values of the Thomson optical depth $\tau_T \sim 1$–$3$ and temperature $kT \sim 150$–$250$ keV, with a spectral index at low energies $\alpha_X \sim 0$ and a Wien hump at energies near 0.4 MeV which might be used to explain the KONUS high-energy emission lines otherwise attributed to pair annihilation (Mazets *et al.*, 1980). In this model the hot electrons are assumed to be in a hot corona above the neutron star surface, while the soft photons which are upscattered by the corona come from the stellar surface. In general, in order to avoid too many $\gamma$-rays impinging upon the stellar surface which would be reradiated as X-rays one needs to assume that the corona is at distances $R \sim 3$–$10$ $R_{NS}$. At energies above $\sim 1$ MeV, the fits from this model fall well below the *SMM* spectra so that they would have to be complemented with a second high-energy component.

Nonthermal inverse Comptonization models were proposed by Zdziarski and Lamb (1987), who assume a power law electron injection which repeatedly upscatters soft photons. In order for this to occur the hot electrons must have an optical depth $\gtrsim 1$, and consequently it requires an electron luminosity $L_e \gg L_{soft}$; otherwise the hot electrons would rapidly cool, reducing the optical thickness of the remaining hot electrons. A self-consistent solution of the electron and photon kinetic equations yields a broken power law photon spectrum, with $\alpha_X$ depending on $L_{soft}/L_e$ and $\alpha_\gamma$ depending on the electron power law index. The break occurs around $\hbar\omega \sim m_e c^2$. This is able to explain the X-ray and $\gamma$-ray spectra of some burst sources. For a large compactness parameter $L_e \sigma_T / R m_e c^3$ of the injected electrons, where $R$ is the size of the source, the cooled $e^+e^-$ pairs produced by photons above pair threshold will accumulate in an optically thick thermal bath (Zdziarski, Coppi, and Lamb, 1990). An example is shown in Figure 9.3.1. In this type of model a value $\alpha_X \sim 0$ of the energy spectrum (as opposed to the power per decade) can be explained if the electron injection occurs at $R \gg R_{NS}$ and $L_{soft} \ll L_e$, given an appropriate electron power law.

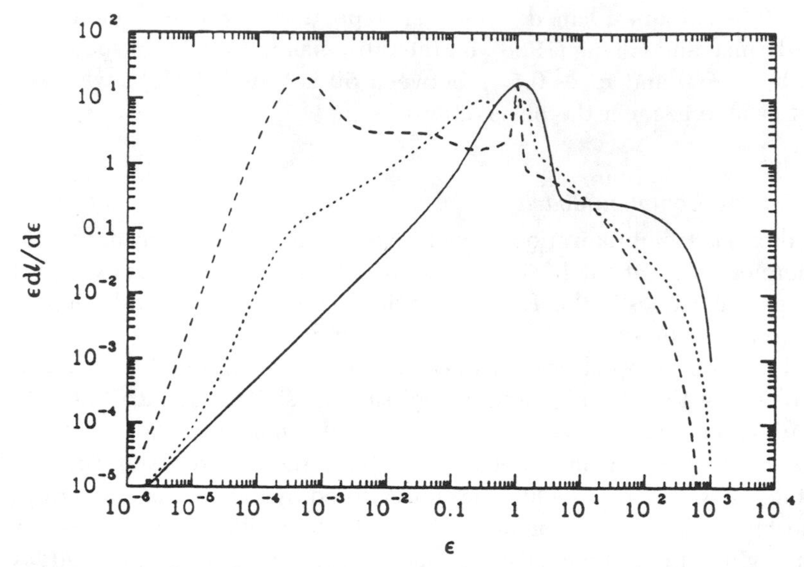


FIGURE 9.3.1 The GRB power per decade spectrum from a two-component nonthermal Compton model, where relativistic pairs with a power law ($\Gamma = 2.5$) energy spectrum are injected into a background of soft photons and thermal pairs. Solid curves have no soft injection, all photons coming from bremsstrahlung, dotted curves have $L_s = 10^{-2}\ L_e$, and dashed curves have $L_s = L_e$. In this figure the compactness parameter $L_e \sigma_T / R m_e c^3 = 30$. From A. Zdziarski, P. S. Coppi, and D. Q. Lamb, 1990, *Ap. J.*, 357, 149.

### e) X-Ray Cyclotron Line Models in GRBs

The low-energy (20–60 keV) single absorption features present in about 30% of the bursts in the KONUS catalog of GRBs were interpreted by Mazets *et al.* (1980, 1981) as being due most probably to the cyclotron process. On the basis of accreting pulsar model calculations, it was apparent that these absorption-like features would have to be due to cyclotron scattering, i.e. to resonant Compton scattering which depletes the photons in the cyclotron resonance. One problem experienced early on was that the continuum spectrum extended well above the single absorption feature, whereas one would have expected at least a second harmonic to be present and as strong as the first one (Bussard and Lamb, 1982). For a long time the only other confirming observation for a low-energy feature in a GRB source was from *HEAO 1* (Hueter and Gruber, 1982; Hueter, 1984). Because of the paucity of confirming data, doubts about alternative interpretations of these features were entertained for a considerable time. For instance, one alternative considered was that perhaps two continua with different cutoffs conspired to mimic an absorption-like feature. More recently, however, high-significance detections were made with the gamma-ray burst detector aboard the *Ginga* satellite in the GB880205, GB870303, and GB890929 events (Murakami *et al.*, 1988; Fenimore *et al.*, 1988, Yoshida *et al.*, 1990), showing clearly two approximately equal-strength absorption features, and in one case also a third, weaker feature, with the approximately equidistant spacing characteristic of cyclotron harmonics. The strength and shape of the features are somewhat dependent on the model assumed to deconvolve the raw data, but with these new data the evidence in favor of a cyclotron interpretation is thought now to be extremely strong.

The width of the *Ginga* lines indicates a temperature or electron energy dispersion $\sim$ 5–7 keV, rather cooler than the typical temperature or energy of the continuum, which is $\gtrsim$ 150–250 keV. The possibility of a two-component model, in which the gamma-rays and X-rays arise from different temperature regions, had been considered previously (e.g. Hameury *et al.*, 1985; Brainerd and Lamb, 1987; Hartmann, Woosley, and Arons, 1988; Melia, 1988). The X-rays, for instance, could arise from the cooler stellar surface or an accretion disk as it reprocesses gamma-rays produced in an optically thin, hotter region. Without going into such details, the *Ginga* observations suggested a phenomenological model where a power law continuum which extends to gamma-ray energies goes through a thin sheet of cooler magnetized gas (Fenimore *et al.*, 1988). Harding and Preece (1989) pointed out that in order to get a second harmonic as deep or deeper than the ground harmonic it would be necessary to include resonant scattering which leaves the electron in an excited state (Bussard and Lamb, 1982). This is because the first harmonic at $\omega \sim \omega_c$, which has a larger coefficient, is refilled with photons at $\omega \sim \omega_c$ produced by scatterings up to second harmonic, $\sim 2\,\omega_c$, which deexcite preferentially via two successive decays of $\omega \sim \omega_c$. Harding and Preece also quantitatively explained the ratio of the second to the third harmonic on the basis of a comparison of the excitation cross sections. However, to obtain quantitative results about the ratio of the first (ground) to second harmonic requires a radiative transfer calculation. Preliminary results of a transfer calculation were reported by Lamb *et al.* (1988), and detailed results were presented by Wang *et al.* (1989) and Alexander and Mészáros (1989). These calculations

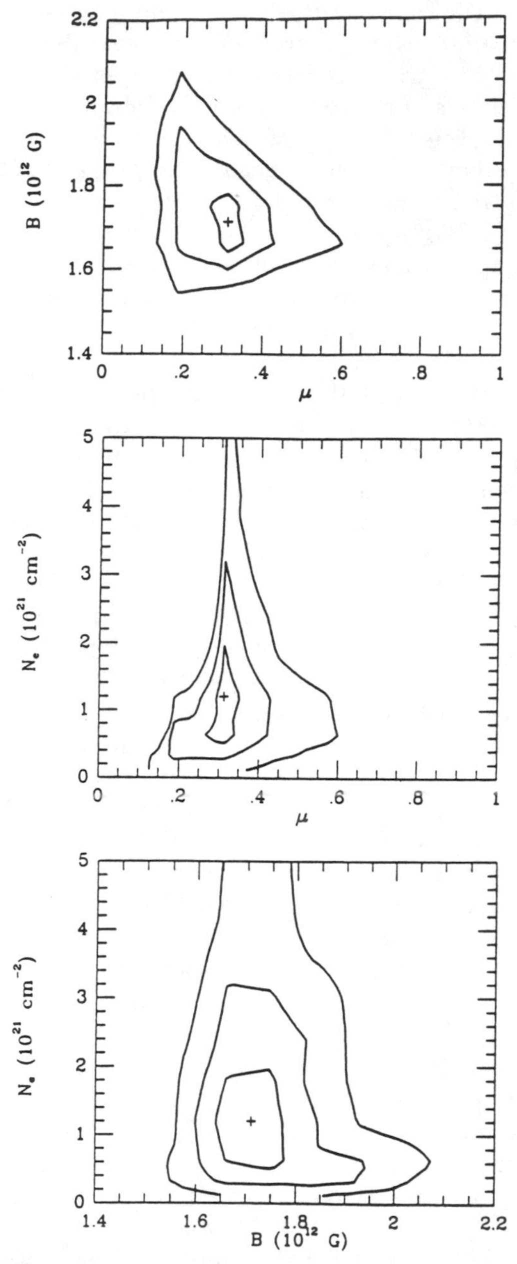


FIGURE 9.3.2 The transmission model cyclotron line fits for GB880205 showing the 68.3%, 95.4%, and 99.7% confidence levels for various computed quantities versus observations. (a, *top*) $B$ versus $\mu$, (b, *middle*) $N_e$ versus $\mu$, (c, *bottom*) $N_e$ versus $B$, where $N_e$ is the column density and $\mu$ is the cosine of the angle of observation assuming isotropic injection. From J. C. Wang *et al.*, 1989, *Phys. Rev. Letters*, 63, 1550.

were based on the phenomenolgical cool sheet transmission model of Fenimore *et al.* (1988), the aim being to reproduce the observed line ratios, line widths, and line shapes and to deduce therefrom the physical parameters (magnetic field strength, temperature, optical depth, etc.) of the X-ray line region. Wang *et al.* (1989) used semirelativistic cross sections and a Monte Carlo transfer code with multiple angle scattering (Fig. 9.3.2). Their $\chi^2$ best fits on the data indicate a temperature $T \sim 5.3$ keV, $\tau_T \sim 0.8 \times 10^{-3} \langle \cos\theta \rangle \sim 0.3$. In a different calculation, Alexander and Mészáros (1989) used the fully relativistic cross sections including vacuum polarization and a Feautrier code using the two-stream approximation (scattering from $\theta$ into $\pi - \theta$). Without going into a detailed $\chi^2$ fit, their best fit temperature is compatible with that of Wang *et al.* (1989), while the optical depth they obtain is larger. This difference is not expected to be due to the difference in the cross sections, since the semirelativistic expressions should be a good approximation at these temperatures; it may instead be due to the different angular treatment of the radiation and the fact that the latter calculation takes into account polarization, which by itself accounts for a factor of 3 difference with respect to an unpolarized treatment. This calculation shows some of the details of the fully relativistic quantum cross sections and indicates that a large degree of polarization is to be expected in the lines, especially the second and third harmonics, which have a lower optical depth and have not suffered as much refilling with photons from other resonances (see Fig. 9.3.3).

An interesting result in the context of the illuminated sheet model is that the relatively cool temperature of the gas which is producing the cyclotron scattering lines can be computed self-consistently if (as the fits indicate) the transmitting region is optically thin to Thomson scattering but optically thick to cyclotron scattering (Lamb, Wang, and Wasserman, 1990). In this case the temperature of the sheet will be dominated by inverse Compton cooling against the cyclotron photons (not the continuum), and in a steady state this temperature is $kT_e \sim (1/4)\hbar\omega \simeq (1/4)\hbar\omega_c \sim 5$ keV, as also obtained from the fits.

The previous models are still rather simple, assuming static conditions and leaving out any effects related to the high-energy portion of the spectrum (where most of the burst energy resides) except for using this spectrum as an input for the line calculations. It remains to incorporate these, or alternative elements, into a coherent picture of both the high-energy and the low-energy spectra which correctly reproduces both the spectra and the $L_X/L_\gamma$ ratio. A preliminary

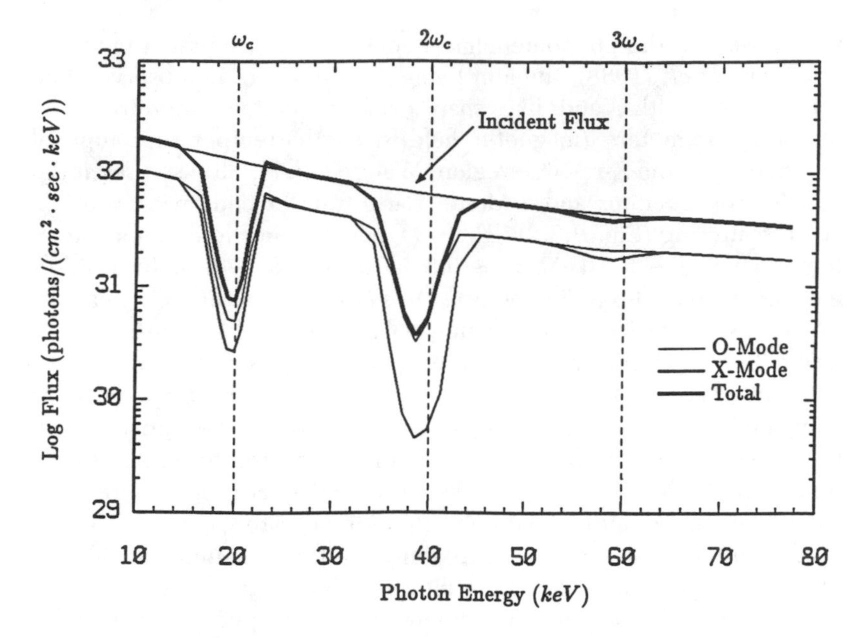


FIGURE 9.3.3 Cyclotron spectrum for the transmission model using the QED opacities, showing the total spectrum and the two normal polarization components. From S. Alexander and P. Mészáros, 1991, *Ap. J.*, 372, 554.

attempt in this direction is that of Hameury and Lasota (1989). Interestingly, it appears that narrow cyclotron lines may also occur in a relativistic wind such as might be expected following a burst on the surface of the neutron star (Miller *et al.*, 1990). These authors use a simplified treatment of the transfer involving only the higher harmonic lines, since in these the line profile may be calculated simply by assuming that photons disappear once they are scattered (actually they become first harmonic photons). They calculate the wind velocity in planar geometry from the resonant radiation pressure on the electrons, assuming an input high-energy spectrum, the wind density from continuity, and the wind temperature from Compton equilibrium, and find that, since the gas density and magnetic field fall off outward away from the surface, most of the line absorption occurs close to the stellar surface. The typical observable line broadening $\Delta E_{\rm FWHM}/E$ that they find in this simplified model for the second and higher harmonics is about 0.2, below the observed value of about 0.4, being dominated not by the wind velocity variation but by the wind temperature dispersion and the magnetic field variation. This finding is encouraging, and it remains to be seen whether a fuller investigation involving resonant transfer effects in the first harmonic as well as departures from planar symmetry can confirm these results.

*f) Multi-MeV Gamma-Ray Propagation in a Strong Magnetic Field*

The fact that many GRBs in the *SMM* and SIGNE catalogs show significant emission above 10 MeV (Matz *et al.*, 1985) was difficult at first to reconcile with the presence of a strong magnetic field $B \gtrsim 10^{12}$ G (Matz *et al.*, 1985; Mitrofanov *et al.*, 1986). This is because the one-photon pair creating process (e.g. Chap. 5.5) is very efficient in converting photons above the threshold $\hbar\omega > 2m_e c^2/\sin\theta$ into pairs (where $\theta$ is the angle between the photon and the magnetic field). The straightforward conclusion was that, if the field indeed was high, the radiation above 1 MeV would be able to escape only along a rather narrow bundle which hugs the magnetic field lines around the magnetic polar axis, and that this strong beaming would decrease the probability of detecting these sources at high energies. This conclusion would have been contrary to the statistics from *SMM* (Matz *et al.*, 1985), which indicated a very high proportion ($\gtrsim 50\%$) of GRBs having spectra extending above 5 MeV.

One way of avoiding this constraint is to assume that the GRBs have low or no magnetic field, $B_* \lesssim 4 \times 10^{11}$–$10^{12}$ G, but this assumption would run contrary to the KONUS and *Ginga* statistics (Mazets *et al.*, 1981; Yoshida *et al.*, 1990) indicating that about 30% of the bursters have cyclotron features. Another way out is to postulate that the high-energy gamma-rays are produced far from the stellar surface, where the magnetic field is weaker, e.g. $R_\gamma \gtrsim 3$–$10\ R_{NS}$, as also required to avoid the X-ray paucity constraint if the $\gamma$-rays are emitted isotropically (Imamura and Epstein, 1987). More recent calculations of the propagation of $\gamma$-rays emitted from the stellar surface through a strong magnetic field, however, indicate that the earlier estimates may have exaggerated the narrowness of the escape beam shape and the difficulty of the detectability.

Riffert, Mészáros, and Bagoly (1989) performed a detailed calculation of the $\gamma$-ray propagation in a strong field dipole magnetosphere with magnetic one-photon pair creation including general relativistic effects. The latter are important, since photons propagate not radially but along curved paths so that photons starting from the polar caps at the stellar surface bend in the same direction as the field lines, thus

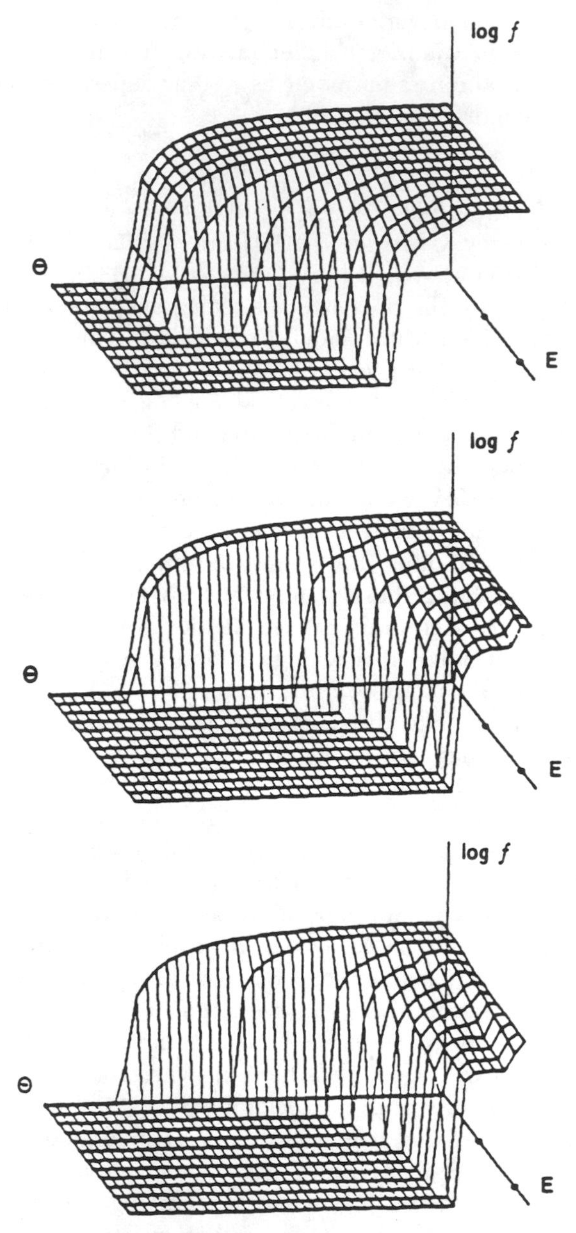


FIGURE 9.3.4 Gravitationally broadened gamma-ray escape beams at energies above 1 MeV for an assumed isotropic injection at the polar cap surface into a dipole magnetic field. The stellar radius is 2.5 Schwarzschild radii and the polar cap opening angle is 15°, for magnetic field strengths (a, *top*) $B = 10^{-2} B_Q$, (b, *middle*) $B = 5 \times 10^{-2} B_Q$, and (c, *bottom*) $B = 10^{-1} B_Q$. From H. Riffert, P. Mészáros, and Z. Bagoly, 1988, *Ap. J.*, 340, 443.

reducing the relative angle between photons and field lines and raising the threshold for pair production. As a result, the escape beam shapes at all energies are broadened by the gravitational light bending, as seen in Figure 9.3.4. That the broader beam shapes facilitate the detection, since there is a larger chance that they will point at the detector, was verified explicitly by simulating the detection statistics (Mészáros, Riffert, and Bagoly, 1989) using the theoretical escape beam shapes calculated by Riffert, Mészáros, and Bagoly (1989) folded over the detector response of the GRS detector aboard *SMM* used by Matz *et al.* (1985). In these simulated theoretical detection rates various magnetic dipole field strengths, polar cap opening angles, and power law input spectral indices were used. The results show that the *SMM* statistics are compatible with a population of GRBs made up of neutron stars with a range of magnetic fields, at least 30% of which have $B_* \gtrsim 4 \times 10^{12}$ G at the surface.

Another circumstance which helps energetic gamma-rays emitted from the surface to avoid excessive absorption by one-photon pair creation is to have the radiation strongly beamed. For instance, if the particles responsible for producing the $\gamma$-rays are being ejected at relativistic velocities (Ho, Epstein, and Fenimore, 1990), the $\gamma$-rays will be strongly beamed in a narrow cone about the magnetic field lines. A calculation of the integrated opacity along radial paths (neglecting general relativistic effects) for beamed radiation emitted from the whole surface of the neutron star indicates that a large fraction of the high-energy radiation can escape without absorption, since the photons start out by heading along the optimal escape route.

### *g) The X-Ray Paucity in GRBs*

The other important constraint provided by GRB observations is their low flux (see eq. 9.1.3 and Fig. 9.1.4) in the range of X-ray detectors, $E \lesssim 100$ keV, manifested by a slope $\alpha_X \sim 0$ for the energy spectrum $EN(E)$, or a power per decade slope $\sim +1$ up to a break energy of a few hundreds of keV. One of the early ideas for explaining this was that the radiation arises from particles which are strongly outward beamed, and therefore the $\gamma$-rays do not impinge upon the surface, where they would have been thermalized into X-rays. A variant of this has been proposed by Ho and Epstein (1989a) in which relativistic electrons or $e^+e^-$ pairs streaming out from a neutron star scatter the blackbody X-rays ($E \sim$ few keV) produced on the stellar surface during a burst. The electrons encounter fewer and more collimated photons as they move away from the star, and they effectively boost most of the photons to high energies. The cooling spectrum produced by this nonthermal inverse Compton mechanism in the nonmagnetic limit is X-ray poor compared to what one would obtain from a uniform isotropic flux of primary photons. The spectra obtained in these calculations are approximately flat in photon number up to some energy $E_{br}$ depending on the electron Lorentz factor and a negative power law above. In terms of power per decade, $E^2N(E)$, they have positive slope below the break and negative above, in qualitative agreement with observed spectra (e.g. Fig. 9.1.4), provided one chooses appropriately the electron injection energy spectrum and the soft photon energy. This process could be relevant for low magnetic field bursters ($B \lesssim 10^{11}$ G).

For the higher magnetic field bursters, such as those producing the cyclotron lines at $E_c \sim 20$ keV (see § 9.1*e*), any mechanism involving scattering must take into account the strong energy dependence of the Compton cross section introduced by the cyclotron resonances (see Chaps. 4.2 and 5.1). One possible mechanism has been proposed by Brainerd (1989). In his calculations, the resonant scattering from Landau level 1 to the ground level 0 is significantly larger than the more commonly considered 0 to 0 cross section, $\sigma_{10} \gg \sigma_{00}$, at field values $B \gtrsim 10^{12}$ G, and in addition the $\sigma_{10}$ cross section diverges as the photon energy decreases below the cyclotron energy, whereas $\sigma_{00}$ decreases $\propto (\omega/\omega_c)^2$ and the continuum cross section $\sim \sigma_T$ remains constant. Due to the high radiation density in GRBs at energies in excess of the typical ground cyclotron energy, one might expect that the population of level 1 would be sufficiently high that the opacity $\tau_{10} > \tau_{00}$, if $\gtrsim 0.1\%$ of the electrons are in the first excited level. Brainerd concludes that low-energy blackbody photons from the stellar surface would be preferentially scattered by excited electrons, a process which exchanges the soft photon (of energy below the cyclotron energy) for a cyclotron photon (energy a few tens of keV) while deexciting the electron. The divergence of the $\sigma_{10}$ cross section does not appear in the calculations of Geprägs *et al.* (1990), which include higher-order corrections in $\alpha_F$, so that the dependence of this suppression mechanism on the divergence needs to be considered.

Another mechanism which seeks to explain the positive slope of the power per decade up to a certain break energy is a magnetic version of the previously discussed scattering from beamed relativistic electrons (Dermer, 1989, 1990; Ho and Epstein, 1989b). In this case, the scattering of the soft surface photons occurs preferentially when

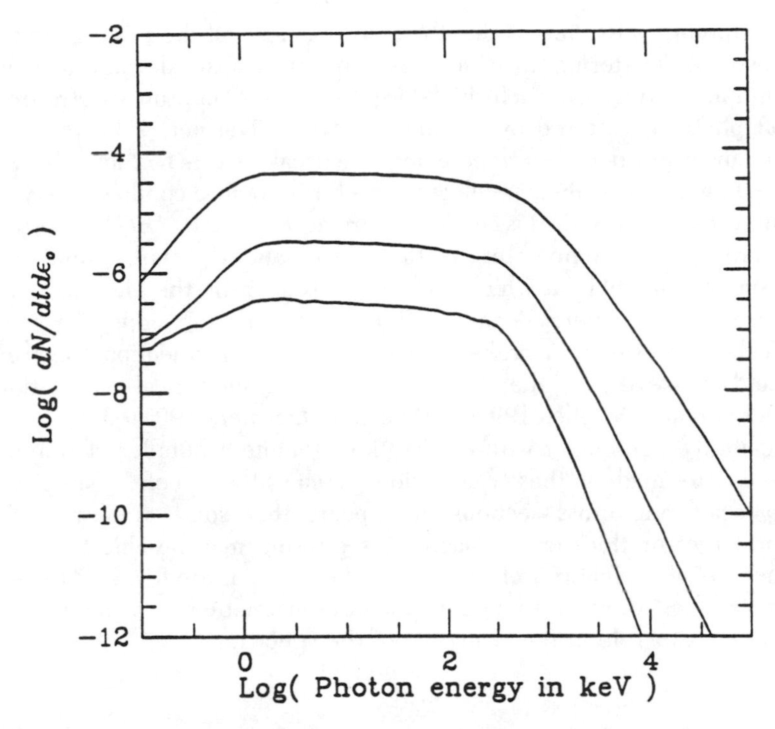


FIGURE 9.3.5 Angle-integrated thin-target spectrum $N(E)$ from resonant inverse Compton cooling of blackbody photons scattering off a power law distribution of beamed electrons in a magnetic field $B = 1.8 \times 10^{12}$ G. The three curves from the top are for electron steady state power laws $\gamma^{-1}$, $\gamma^{-2}$, $\gamma^{-3}$. From C. Ho and R. Epstein, 1989, in *Proc. GRO Workshop*, ed. W. N. Johnson (NASA Goddard, Greenbelt, MD), pp. 4–436.

they are resonant in the rest frame of the moving electrons. The resonance condition depends both on the electron Lorentz factor and on the angle between the photons and the electron velocity. For electrons beamed along a uniform magnetic field, thin-target Compton spectra (Dermer, 1989; Ho and Epstein, 1989b) give a photon number slope $\alpha \sim 0$ between the soft photon energy and the break energy, and $\alpha \sim (p - 2)$ above the break, where $p$ is the slope of the electron number distribution (see Fig. 9.3.5). In terms of power per decade, this value gives a positive slope of 2 below the break, which is adequate for some bursts but may be too steep for the average value, which is closer to 1.

Another possibility is that the electrons lose all their energy in the process of scattering, in which case one must also take into account multiple scatterings. Such thick-target cooling Compton spectra from soft photons scattered by beamed electrons (Dermer, 1990) give for the power per decade a slope below the break of $1 - \Gamma/2$ and above it $2 - \Gamma$, where the injection electron spectrum before cooling is $\propto \gamma^{-\Gamma}$. The break energy in this model occurs at $E_{\mathrm{br}} \sim 2E_c^2/kT(1 - \beta\mu_{\mathrm{min}})$, where $T$ is the temperature of the surface and $\mu_{\mathrm{min}}$ is the minimum cosine of the angle at which the photon is seen by the electron. This can fit the low-energy X-rays if $\Gamma \ll 2$, but then the slope above the break is too steep. A broken power law electron injection spectrum could, however, fit the data. More recent numerical calculations (Dermer and Vitello, 1990; Vitello and Dermer, 1991) show a fair qualitative agreement with observations. While a number of simplifications are made in these calculations, such as the use of the simplified $\sigma_{00}$ scattering cross sections, it appears that some variant of the thin-target or thick-target magnetic scattering may be able to reproduce the gross features of the observed spectra in the few keV to $\sim 1$ MeV range. Above this range the effect of magnetic pair formation will play a major role in the formation of the spectrum.

# 10. Super-High-Energy Gamma-Ray Sources

## 10.1 Observational Overview

### *a) VHE and UHE Techniques*

The reports that a number of well-known neutron star sources are strong emitters of radiation at energies above $10^{11}$ eV has generated uncommonly high interest and continues to inspire the considerable efforts of a large number of experimental groups. As of early 1991, the status of most of these observations remains controversial. Unlike in the experiments at X-ray and sub-GeV $\gamma$-ray energies, the energy range involved here is usually observed with ground-level detectors. Because of the difficulty in obtaining and interpreting data in this range, we go here into some details about the experimental setups. We then review the observations of individual sources, and at the end of the section we discuss some of the major difficulties encountered in the data analysis.

The VHE (very-high-energy) and UHE (ultra-high-energy) ranges are conventionally defined to correspond to TeV and PeV energies, where 1 TeV = 1000 GeV = $10^{12}$ eV, while 1 PeV = 1000 TeV = $10^{15}$ eV. Astronomical observations at these extreme energies have evolved out of cosmic ray studies and require specialized techniques, which to this day remain fraught with difficulties. An excellent review of recent and older work is given by Weekes (1988, 1990), and some recent advances and problems are summarized by Protheroe (1987), Bonnet-Bidaud and Chardin (1988), and Fegan (1990). The dominant difficulty in this energy range is the low photon flux. Whereas at medium gamma-ray energies or less (below $\sim 1$ GeV) satellite experiments like *SAS 2* and *COS B* have discovered quite a number of point sources (e.g. Ramana Murthy and Wolfendale, 1986), at energies above this the expected flux of photons per unit areas and unit time from any astronomical source becomes very small. For instance, extrapolating the Crab pulsar's spectrum up to 1000 GeV, one expects a flux of only 16 $m^{-2}$ $yr^{-1}$ so that a balloon or satellite experiment becomes impracticable. Whereas at $E \sim 10^4$–$10^6$ eV a typical RPP or AXP flux might be $\sim 10^{-3}$ $cm^{-2}$ $s^{-1}$ and a typical GRB flux is $\sim 1$ $cm^{-2}$ $s^{-1}$, a typical $10^{12}$ eV source flux is expected to be $\lesssim 10^{-11}$ $cm^{-2}$ $s^{-1}$, or at $10^{15}$ eV it is $\lesssim 10^{-14}$ $cm^{-2}$ $s^{-1}$.

Fortunately, at TeV and PeV energies the earth's atmosphere itself acts as a sort of detector: a very-high-energy photon or very-high-energy primary cosmic ray particle initiates in the atmosphere a cascade of $e^+e^-$ pairs, muons, hadrons, and neutrinos which reach the earth's surface in very large showers in a cone about the direction of the original photon or particle that entered the top of the atmosphere. For this reason, it is feasible to perform ground-based experiments using much larger-area or larger-volume detectors than would be possible in space, by detecting the secondary radiation. Unfortunately, cosmic ray particles (nuclei) also produce showers, which represent a large and irreducible background which arrives uniformly in time and essentially isotropically, aside from statistical fluctuations. Thus, any celestial point sources must be detected against this large isotropic background, which makes observation difficult. However, the sensitivity to a cosmic source can be improved by a factor of 5 to 10 if the source has a known time variability, e.g. is pulsating.

There are currently three basic types of experimental techniques for detecting celestial sources at these energies.

i) Atmospheric Cherenkov detectors, for primaries at $0.1 \lesssim E \lesssim 100$ TeV, which are in the VHE range. These detectors measure the optical light produced by the secondary $e^+e^-$ pair showers created by a VHE primary, which is produced in the upper atmosphere. The sky background (diffuse starlight, air glow, etc.) determines the threshold for detecting the showers, while the subsequent identification of a celestial source is limited by the isotropic shower background produced by cosmic rays. This technique can be used only on moonless, cloudless nights and is more suited to observing steady sources, or sources with bursts of minutes to hours, than to long-term monitoring of sources varying over periods of days. The total duty cycle of usable time is relatively low, about 5–10%. Observations are carried out either at mountain altitudes, where the low-energy threshold is smaller ($\simeq 0.3$ TeV), or at sea level, where the threshold for detection is higher due to the larger atmospheric depth ($\simeq 1$ TeV). The light is collected by mirrors and focused onto fast time resolution photomultiplier tubes. The direction of the source is determined by light front arrival techniques, which can be done with spatially separated mirror arrays and require nanosecond or better timing. Alternatively, steerable mirrors or mirror arrays can be pointed at a suspected source and compared to an off-source position. The angular resolution of typically $\Delta\theta \simeq 1\text{–}2°$ is limited by the Cherenkov angle of emission to values $\gtrsim 0.5°$.

ii) Extensive air shower (EAS) arrays in the UHE range 0.1 PeV $\lesssim E \lesssim$ 100 PeV. These are arrays of spatially separated scintillators located at sea level or mountain altitudes which have nanosecond time resolution electronics and use the light front arrival technique to determine the direction of incidence of the primary particle. These arrays have a long history as detectors for charged cosmic ray particles. The secondaries are spread in a cone about the direction of the primary, and a source appears as a spike in position over the isotropic EAS produced by cosmic rays, during drift scan observations. Two criteria have been used to attempt to distinguish between $\gamma$-ray-induced and cosmic-ray-induced showers. One of these is the use of the "age parameter," measured by the lateral extent and flatness of the shower, which if large has usually been interpreted to be due to photon-induced cascades that are supposed to attentuate much faster than nucleon-induced cascades. The other criterion involves the muon richness of a cascade, which is supposed to be smaller for photon-induced than for nucleon-induced cascades. However, both of these criteria are derived from accelerator experiments at energies $\lesssim$ 100 GeV, and there are strong suspicions that they may not be reliable at these higher cosmic energies (e.g. Drees *et al.*, 1988; see also § 10.1*e* below).

iii) Underground particle detectors. These devices, usually located in deep underground mines in order to filter out the charged-particle component of the cosmic ray background, were originally intended to measure neutrinos arising from the possible decay of protons. They consist typically of a large volume of water which acts as a Cherenkov detector, the primary cosmic particles (whether neutrinos or another weakly interacting species that penetrates that deep) producing secondary charged particles (e.g. charged muons) which radiate Cherenkov light. The latter is measured by photomultiplier tubes located around the boundaries of the water container, facing inward, which give a stereoscopic view of the muon track allowing measurement of the characteristics of the original primary particle. Most events are caused by downward-directed muons, whose direction and rate indicate a secondary cosmic ray muon origin. A small fraction of events are due to upward-directed muons coming from below the detector's horizon. Most (or possibly all) of these are from atmospheric neutrino interactions initiated by cosmic rays. These cosmic rays act as an irreducible background to any possible neutrino (or weakly interacting particle) flux from compact cosmic sources, which are also expected to produce upward-directed muons. The identification of events from compact sources is in principle possible if there is an excess of events above the background whose direction points at the cosmic source or if there is a distinct time signature.

*b) VHE Emission from Radio Pulsars*

Several RPPs have been reported as copious sources of $\gamma$-rays in the VHE, or TeV, range. The theoretical interpretation of TeV radiation from RPPs has a longer tradition and is perhaps easier than for other types of sources, given that potential drops of order $\gtrsim 10^{12}$ V are expected (e.g. Cheng, Ho, and Ruderman, 1985a), and that $\gamma$-rays at 1–100 MeV have been previously detected and theoretically modeled (see Chap. 8.2). In the PeV range, the claimed positive detections of an RPP (the Crab pulsar) are few and in need of verification, especially since they are DC level (unpulsed) detections (Dzikowski *et al.*, 1983; Boone *et al.*, 1984) and since other groups have reported (nonsimultaneous) nondetections. By contrast, the TeV emission from several RPP sources, including the Crab and Vela pulsars, has been reported by numerous groups, some of them simultaneously, and may appear reasonable on theoretical grounds. On the other hand, the number of nondetections (often unreported) is also high, and this fact has to be kept in mind.

i) PSR 0531 + 21: The Crab pulsar and nebula. A positive $3\sigma$ DC level detection was first reported by Fazio *et al.* (1972) using the 10 m dish Cherenkov detector of the Whipple Observatory at $E \gtrsim 10^{12}$ eV. This was a result averaged over three years of observation, giving an average flux $F \sim 4 \times 10^{-11}$ cm$^{-2}$ s$^{-1}$. Concentrating on three 60 day periods 60 days after radio glitch events, the detection level reported was $5\sigma$ and $F \sim 1.2 \times 10^{-10}$ cm$^{-2}$ s$^{-1}$, also consistent with steady emission. Previously, Chudakov *et al.* (1965) had established an upper limit of $\lesssim 5 \times 10^{-11}$ cm$^{-2}$ s$^{-1}$ at $E \gtrsim 5 \times 10^{12}$ eV. The emission therefore might be sporadic. Pulsed emission at the $P_s = 33$ ms period was reported in Jennings *et al.* (1974), Grindlay *et al.* (1976), and Ericksson *et al.* (1976). Gupta *et al.* (1978) reported observing two

VHE pulses separated by a phase interval $\Delta\phi \sim 0.42$, similar to the radio, optical, X-ray, and 100 MeV *COS B* pulse separation. Reports indicate that the emission may behave in a burst-like manner on 15 minute time scales (Bhat *et al.*, 1986b), being variable over scales of seconds and days (Gibson *et al.*, 1982; Dothwaite *et al.*, 1984c). The pulsed detection significances quoted are in some cases high, but most detections show no significant pulsed signal. The integral spectrum upper limit is $F(>E) \lesssim 1.7E_{\mathrm{GeV}}^{-1.17}$, and if the emission is assumed to be isotropic the time-averaged luminosity above 1 TeV is $L_{\mathrm{VHE}} \sim 10^{34}$ ergs $\mathrm{s}^{-1}$. More recently, the Crab Nebula has been observed with the Whipple VHE Cherenkov telescope (Weekes *et al.*, 1989), using a novel technique for distinguishing photon showers from those initiated by cosmic rays. This technique selects those showers which are elongated and have an orientation characteristic of photon showers as deduced from simulations, parametrized by an "azwidth" parameter. The small azwidth events correspond to photon events, and the difference of ON-source minus OFF-source observations as a function of this parameter is shown in Figure 10.1.1. With this azwidth selection criterion, Weekes *et al.* (1989) obtain for the Crab Nebula a DC signal with a $9\sigma$ significance but find no strong evidence for a periodic component pulsed at the period of the Crab pulsar, with an upper limit for the latter of less than 25% of the DC signal. The latest results on this experiment are reported in Lang *et al.* (1990).

At PeV energies, DC detections have been reported for the same day of observation, February 23, 1989, although not with high significance, by two widely separated groups, the Baksan group (Alexeenko *et al.*, 1989) and the Tata group (Rao *et al.*, 1989). In the Tata observations, the muon content of the showers was similar to that in cosmic ray showers, which is one of the difficulties in understanding the PeV emission from this and other sources as well.

ii) PSR 0833 − 45: The Vela pulsar. Pulsed TeV observations at the $4\sigma$ level and a flux $F \sim 10^{-11}$ $\mathrm{cm}^{-2}$ $\mathrm{s}^{-1}$ were reported by Grindlay *et al.* (1975b) at a period coincident with the X-ray period $P = 89$ ms, using observations from the Narrabri (Smithsonian-Sidney) group. Bhat *et al.* (1980) reported two peaks separated by $\Delta\phi \sim 0.42$ at the $4.4\sigma$ and $2.2\sigma$ levels, which agrees with the 100 MeV *COS B* peak separation, but not with the optical ($\Delta\phi = 0.24$) or radio (single-peak) data. This behavior has been confirmed by Vishwanath (1982), Gupta (1983), and Bhat *et al.* (1986c), although in general the significance is not high. The emission is variable, like that of the Crab, but the total flux seems to be three orders of magnitude smaller than the

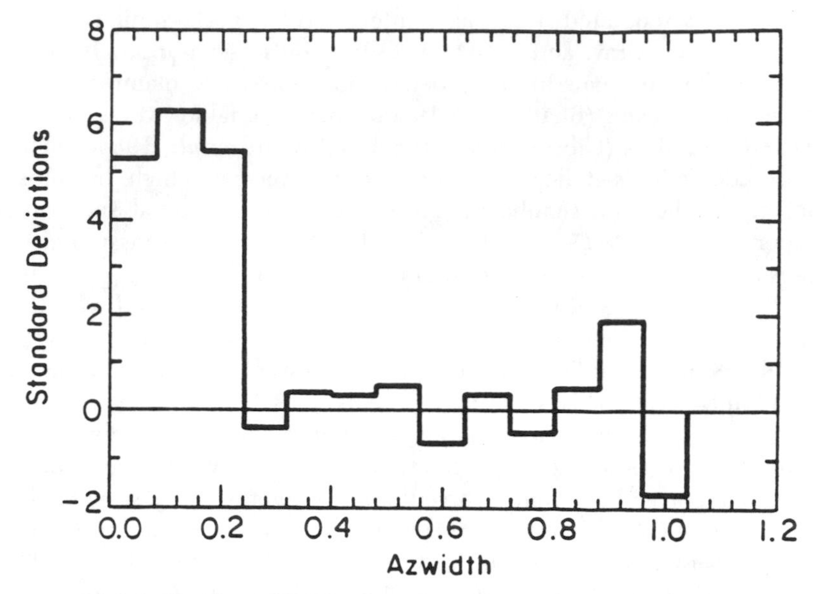


FIGURE 10.1.1 Number of VHE events from the Crab Nebula as a function of the "azwidth" parameter, the small values corresponding to photon events. From T. C. Weekes *et al.*, 1989, *Ap. J.*, 342, 379.

extrapolated *COS B* flux, being $F(>E) \lesssim 1.6 \times 10^{-8} E_{\mathrm{GeV}}^{-0.89}$ cm$^{-2}$ s$^{-1}$. If assumed isotropic, the time-averaged TeV luminosity at $E \gtrsim 10$ TeV is $L \sim 10^{32}$ ergs s$^{-1}$.

iii) Other fast pulsars. Pulsed TeV detections have also been reported from several other short-period pulsars, which together with the Crab and Vela pulsars lie in a particular region of the $P\dot{P}$ diagram. Indeed, in order to emit a significant amount of TeV radiation, outer-gap pulsar models must have a large value of $B^2/P$, and in general a short period is favored. Among such objects, PSR 1937 + 21 (1.5 ms) has been reported by Chadwick *et al.* (1985d), PSR 1953 + 29 (6 ms) by Chadwick *et al.* (1985a), and PSR 1509 − 58 (150 ms) by de Jager *et al.* (1987). The binary pulsar PSR 1855 + 09 has also been reported, at a marginal significance level, by de Jager *et al.* (1990).

### *c) VHE-UHE Emission from Accreting X-Ray Pulsars*

A number of AXPs have been reported at TeV and PeV energies, based on flux excess at the corresponding celestial position or on signal periodicities corresponding to the spin pulse, orbital, or other long-term time signatures. As in the case of RPPs, the number of nondetections is also considerable, so the emission appears sporadic. While the magnetic field in these sources is as large as in RPPs, the periods are usually much longer and an induced electric field mechanism may not appear as natural as in RPPs (see Chap. 8.2), so alternative mechanisms have been proposed, e.g. shock acceleration of protons followed by $\gamma$-ray emission. In general, the AXPs that have been reported to emit VHE-UHE radiation seem to be near corotation, i.e. the X-ray $\dot{P}/P$ is small and changes sign, alternating between spin-up and spin-down (Weekes, 1986).

i) Her X-1. This object has X-ray periodicities $P_s = 1.24$ s, $P_{\rm orb} = 1.7$ days, $P_{\rm long} = 35$ days. At TeV energies, Dothwaite *et al.* (1984c) reported a $3\sigma$ detection pulsed at the spin period $P_s$, around orbital phase $\phi_{\rm orb} \sim 0.76$, with $F \sim 1.2 \times 10^{-9}$ cm$^{-2}$ s$^{-1}$. Gorham *et al.* (1986, 1987) and Resvanis *et al.* (1987) reported a large number of detections, typically lasting up to 40 minutes, during the X-ray main ON and small ON of the 35 day cycle, modulated by $P_s$ and $P_{\rm orb}$, in some cases as much as 1 hour after the neutron star went into eclipse behind the companion star. A simultaneous observation was also made by two different groups (Whipple and Haleakala) on April 4, 1984, modulated at $P_s$ (Chadwick *et al.*, 1987b). A fascinating result (Resvanis *et al.*, 1988; Lamb *et al.*, 1988) is the report by the Haleakala and Whipple groups that the TeV period of Her X-1 is actually 0.16% shorter than the X-ray period $P_s = 1.24$ s. The significance of these results is enhanced by the fact that similar results were obtained by the Los Alamos group (see below). The power spectrum and pulse shape at TeV energies are shown in Figure 10.1.2. The measured period seems to be getting shorter, the values reported in 1983–84 averaging near 1.238 s and those in 1986–87 averaging near 1.236 s. The time-averaged luminosity in TeV photons is $L \sim 2 \times 10^{35}$ ergs s$^{-1}$, with a duty cycle (VHE time/total observation time) of about 7%.

At PeV energies, Her X-1 detections have been reported from the Fly's Eye experiment (Baltrusaitis *et al.*, 1985b) in a 40 minute observation pulsed at $P_s$. However, Chadwick *et al.* (1985a) did not find a simultaneous TeV signal. The Los Alamos group reported both a DC signal during 10 days of observing in July 1986 and a pulsed signal with $P_s$ shorter by 0.16% during another 10 day period, while Mitsui *et al.* (1987) observed burst-like PeV activity during July 1985 and July 1986 (referred to in Protheroe, 1987). A scatter diagram of TeV and

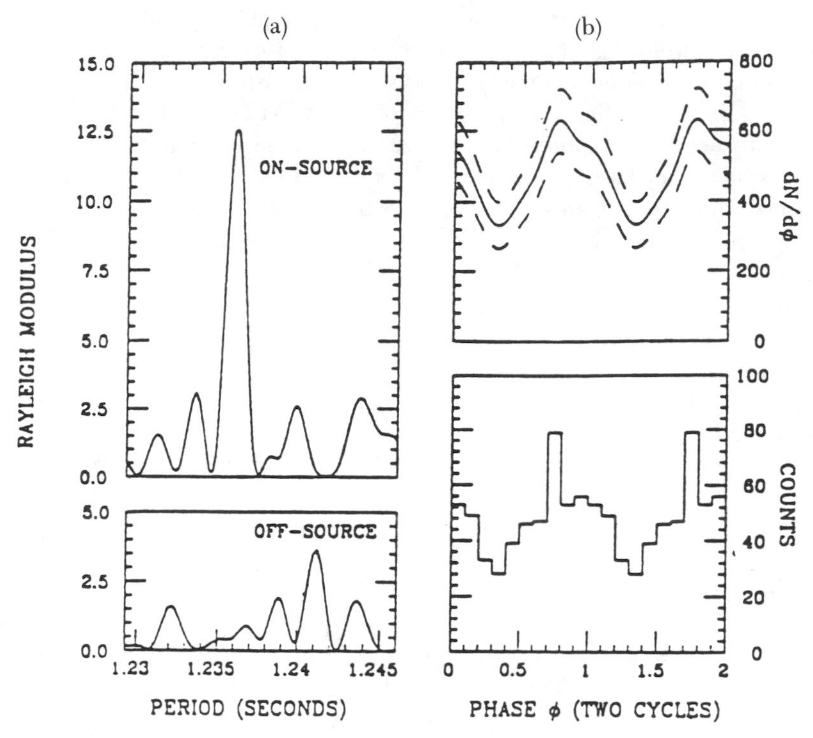


FIGURE 10.1.2 TeV emission from Her X-1. *Left panels*: Power spectrum near 1.24 s. *Right panels*: Pulse shape of Her X-1, the lower section being the raw data and the upper section representing the same data reduced by the normal kernel density estimator method. From L. Resvanis *et al.*, 1988, *Ap. J.* (*Letters*), 328, L9.

PeV observations of Her X-1 indicating the position in the 1.7 day to 35 day phase space (Gorham *et al.*, 1987) is shown in Figure 10.1.3. In this diagram, one sees a marked preference for emission during either the main X-ray ON or the small X-ray ON. However, it is also apparent that the source is sometimes observed up to 1 hour after eclipse by the companion star. As with some other sources, the rate of detection seems to be declining in time so that the experiments as of 1990 (Weekes, 1990) do not conclusively show a signal associated with Her X-1.

ii) Vela X-1. The X-ray periodicities are $P_s$ = 283 s, $P_{orb}$ = 8.96 days, and this source is believed to be accreting from a wind. At TeV energies a pulsed detection at $P_s$ was reported from the Potchefstroom

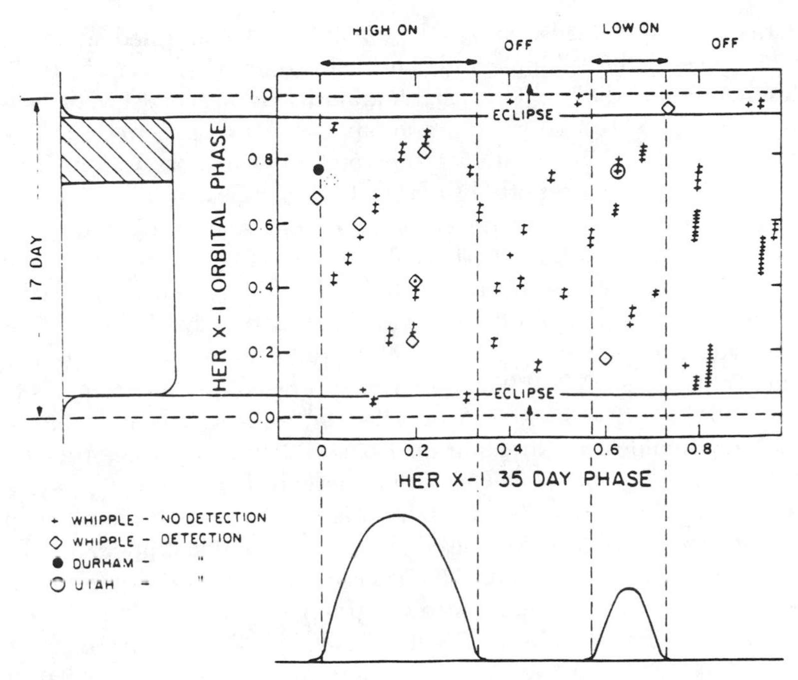


FIGURE 10.1.3 Distribution of Her X-1 events in the 1.7 day and 35 day phase space. From P. W. Gorham *et al.*, 1987, in *Very High Energy Gamma-Ray Astronomy*, ed. K. Turver (Reidel, Dordrecht), p. 125. Reprinted by permission of Kluwer Academic Publishers.

group (North *et al.*, 1987), enhanced shortly after eclipsing of the source, and similar results were reported by the Durham group (see below). The first reports on this source, at PeV energies, from the Adelaide group (Protheroe *et al.*, 1984), were based on EAS observations in 1979–81 binned over the orbital period, which showed a signal at phase $\phi_{orb} \sim 0.63$. On the other hand, Suga *et al.* (1985) reported $P_{orb}$ modulation at $\phi_{orb} \sim 0.5$ on the basis of 1967 observations, and North *et al.* (1987) reported a signal at $\phi_{orb} \sim 0.13$. A photon luminosity during bursts of $L \sim 1.6 \times 10^{34}$ ergs s$^{-1}$ per decade above $E \sim 3 \times 10^{15}$ eV is inferred, with a burst duty cycle of about 8% of the total time of observation.

iii) 4U 0115 + 63. With X-ray periodicities $P_s = 3.16$ s, $P_{orb} = 24.3$ days, this is a transient source in a very eccentric orbit. This object was reported as a transient TeV source pulsed at $P_s$ by the

Durham group (Chadwick *et al.*, 1985b), and confirmed at lower significance by the Whipple and Haleakala groups (Lamb *et al.*, 1987; Resvanis *et al.*, 1987c). The VHE luminosity during burst events was about $L \sim 2 \times 10^{35}$ ergs $s^{-1}$ and appeared to be active at TeV energies during 1% to 10% of the total time of observation. This source has not been reported so far at PeV energies.

iv) LMC X-4. This extragalactic transient source, located at 50 kpc in the LMC, has X-ray periods $P_s = 13.5$ s, $P_{orb} = 1.408$ days, and $P_{long} = 30.5$ days. It has been reported at PeV energies only, so far, by the Adelaide group (Protheroe and Clay, 1985) by binning EAS observations of several years over the orbital period, which gave a signal at $\phi_{orb} \sim 0.90–0.95$. The time-averaged photon luminosity above $10^{16}$ eV is estimated to be $L \sim 10^{38}$ ergs $s^{-1}$, comparable to the X-ray luminosity. No confirming observation has yet been made.

v) Cen X-3. This object has X-ray periods $P_s = 4.8$ s, $P_{orb} = 2.1$ days. It has been reported at TeV energies by the Durham group (Carraminana *et al.*, 1988), pulsed at $P_s$, with the signal preferentially observed at $\phi_{orb} \sim 0.7–0.8$. This source has also been reported at TeV energies by the Potchefstroom group (de Jager *et al.*, 1990). Notice that, while the absolute value of the X-ray $\dot{P}/P$ of this source is not as low as that of the other four sources (e.g. Nagase, 1989a), it does show frequent episodes of spin-up and spin-down.

### *d) Cygnus X-3 and Other Objects*

i) Cygnus X-3. The X-ray source Cyg X-3 has received most of the attention of the VHE-UHE community, numerous detections (as well as a number of nondetections) having been reported. This X-ray binary source has a low-mass companion, and the X-ray period is interpreted as being orbital, $P_{orb} = 4.8$ hr. The compact source is most probably either an accretion-powered or a rotation-powered pulsar, or even (with less probability) something else. The emission at all wavelengths has so far refused to show pulsations, although there have been reports of TeV pulsations (see below). The source seems to be enveloped in a cloud of gas which masks the inner properties. Thus, the X-ray emission shows smooth 4.8 hr sinusoidal variations, which may be interpreted as due to scattering of the neutron star emission by an asymmetric cloud, the asymmetry being introduced by the presence of the binary companion star. It also shows infrared radiation, modulated on similar time scales, as well as radio radiation, which shows strong flaring activity on time scales of years. It is also a powerful *SAS 2* source above 30 MeV energies (Lamb *et al.*, 1977; Fichtel, Thompson, and Lamb, 1987), although *COS B* did not detect it in the 50 MeV–5 GeV range (Hermsen *et al.*, 1987). A general review of this object at all wavelengths has been given e.g. by Molnar (1985) and Bonnet-Bidaud and Chardin (1988).

At TeV energies, Cyg X-3 was first reported to show a DC excess of events by the Crimea group (Vladimirsky *et al.*, 1973) a week after an intense radio flare, as shown in Figure 10.1.4. They also detected an excess after another giant radio flare in 1980. A modulation with the 4.8 hr period was reported by Neshpor *et al.* (1979) and discussed by Stepanian (1982). The 4.8 hr modulation at TeV energies was later detected by a number of groups (Bhat *et al.*, 1987; Chadwick *et al.*, 1987a; Resvanis *et al.*, 1987), which saw this source after the radio flare of October 10, 1980, the TeV emission occurring at $\phi_{orb} \sim 0.65$, and also by Fegan *et al.* (1987). The Durham group reported evidence for a 12.6 ms pulsation period in TeV observations of Cyg X-3 after two separate radio flares in September 1983 and October–November 1986 (Chadwick *et al.*, 1985a, 1987a). The Haleakala and Whipple groups could neither confirm nor negate this result in observations that detected a DC effect during October 1986 (Resvanis *et al.*, 1987a; Fegan *et al.*, 1987). Their results, however, are not incompatible with those of the Durham group. On the other hand, a somewhat

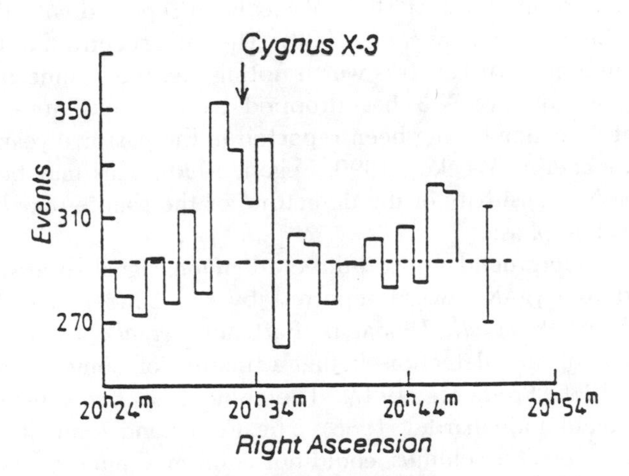


FIGURE 10.1.4 First DC detection of Cyg X-3 at TeV energies. After B. M. Vladimirsky *et al.*, 1973, *Proc. 13th ICRC* (Denver), 1, 456; from R. J. Protheroe, 1987, *Proc. 20th ICRC* (Moscow).

different period of 9.22 ms was reported for Cyg X-3 by the Crimea group (Ziskin *et al.*, 1987). More recently the Durham group presented new observations with the Mark IV Cherenkov telescope at La Palma and a reanalysis of their previous data (Brazier *et al.*, 1990) finding a period of 12.5962 ms at a probability level of $1.7 \times 10^{-6}$ and a spin-down period derivative between 1981 and 1989 of $\dot{P} = (1.9 \pm 0.3) \times 10^{-14}$ s s$^{-1}$. They find the TeV emission centered around phase 0.66 of the 4.8 hr period, although the TeV emission phase appears to shift very slightly in a regular manner with respect to the X-ray 4.8 hr maximum position. The time-averaged flux is $F \sim 5 \times 10^{-12}$ cm$^{-2}$ s$^{-1}$, while the flux around the X-ray maxima is $F \sim 2 \times 10^{-10}$ cm$^{-2}$ s$^{-1}$ and occasionally up to five times that value.

At UHE energies, Cyg X-3 was first detected by the Kiel group (Samorski and Stamm, 1983), who found a $4.4\sigma$ DC signal with an EAS array. The $\sigma$ value was reported to be even larger when binning over $P_{orb} = 4.8$ hr was introduced, being largest around $\phi_{orb} \sim 0.35$. The flux $F \gtrsim 7 \times 10^{-14}$ cm$^{-2}$ s$^{-1}$ implies a luminosity $L \gtrsim 2 \times 10^{37}$ ergs s$^{-1}$ at $E > 2 \times 10^{15}$ eV. Less significant confirming detections were reported by Lloyd-Evans *et al.* (1983) and Morello *et al.* (1983). Other orbitally modulated detections followed (Kifune *et al.*, 1987; Bhat *et al.*, 1986; Alexeenko *et al.*, 1987; Baltrusaitis *et al.*, 1985a), mostly peaked around $\phi_{orb} \sim 0.6$–$0.65$, although in some cases the signal appears stronger at $\phi_{orb} \sim 0.25$ (Eames *et al.*, 1987). A detection at energies in excess of $10^{18}$ eV has been reported with the Fly's Eye experiment (Cassiday *et al.*, 1990). The total spectrum of Cyg X-3 is shown in Figure 10.1.5. It is worth noting that the significance level of detections of Cyg X-3 has dropped as time has gone on, no significant detection having been reported in the past five years at any gamma-ray energy (Weekes, 1990; Fegan, 1990). This may be due to the increasing sensitivity of the detectors, or the source may be in an overall decline phase.

Deep underground upward-directed muon signal excesses (DC) attributed to Cyg X-3 were reported by the Soudan and NUSEX groups (Marshak *et al.*, 1985a, b; Battistoni *et al.*, 1985a, b), the significance of the detection being a matter of some uncertainty (Chardin, 1986; Protheroe, 1987). The subject has been reviewed by Bonnet-Bidaud and Chardin (1988). The Fréjus and Kamioka groups, using larger detector volumes, could not confirm a muon excess in the direction of Cyg X-3 during the same period (Berger *et al.*, 1986; Oyama *et al.*, 1986). The IMB group (Bionta *et al.*, 1987) also reported a positive DC signal, while the Baksan group and the HPW

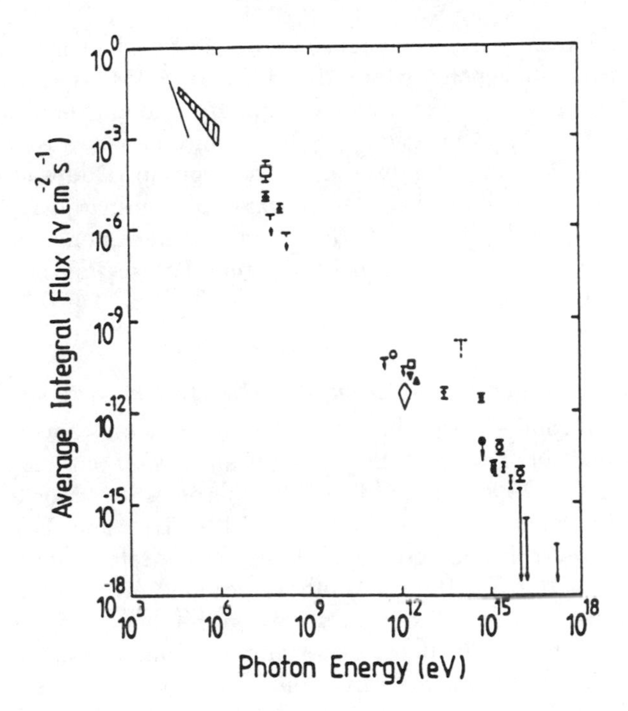


FIGURE 10.1.5 Composite spectrum of Cyg X-3 from keV to multi-PeV energies. From T. C. Weekes, 1988, *Phys. Rep.*, 160, 1; based on C. L. Bhat *et al.*, 1986, *Ap. J.*, 306, 587.

group (Chudakov *et al.*, 1985; Aprile *et al.*, 1985) reported no significant excess. New results may be forthcoming from underground muon detectors being readied at Los Alamos, Utah, Japan, and elsewhere.

ii) Other VHE-UHE sources. Several other sources have been reported to emit sporadically at these energies. Among them the intense *COS B* source Geminga may be a neutron star, and pulsations at 100 MeV and X-rays have been reported with $P_s$ = 60 s (Bignami *et al.*, 1984; Buccheri *et al.*, 1985). At TeV energies Zyskin and Mukhanov (1985) and Kaul *et al.* (1985) reported Geminga to show 60 s pulsations, whereas Helmken and Weekes (1979), Cawley *et al.* (1985), and Bhat *et al.* (1986b) did not see any pulsations. The TeV pulsation status of this source appears uncertain at present, while at PeV energies it has not so far been detected.

The black hole candidate Cyg X-1 has been reported at PeV energies by the MSU group, with a 3.5$\sigma$ DC effect during 1985–86 but no significant excess during 1984–85 (referred to in Protheroe, 1987). Also, a magnetic white dwarf binary, AM Her, has been reported at TeV energies by the Gulmarg group (referred to in Protheroe, 1987). The galaxy M31 (Andromeda) was reported marginally at TeV energies by the Durham group (Dothwaite *et al.*, 1984), while the radio galaxy Cen A was reported marginally at TeV energies by Grindlay *et al.* (1975b). None of these non-neutron star sources have so far been confirmed at either TeV or PeV energies by other groups.

### *e) Problems with the Interpretation of VHE-UHE Observations*

There are a number of problems and disagreements concerning the interpretation of TeV and PeV measurements of cosmic sources (Weekes, 1988, 1990; Fegan, 1990; Protheroe, 1987; Morse, 1987). One of the main concerns is whether the TeV and PeV events measured are really caused by photons or by some other type of particle (e.g. Halzen, 1990). Another, perhaps even more serious, concern involves a complex of questions on the degree of significance attributed to the observations, uncertainties about the statistics used to infer a DC excess in the direction of a source, the treatment of time-modulated components, and the evaluation of the background. We discuss some of these details below.

i) The age parameter of showers. This parameter characterizes the lateral extent of a shower, being larger for cascades that develop at higher altitudes. Generally, $\gamma$-ray induced showers are expected to develop at higher altitudes, at 9/7 radiation lengths, and therefore should have a large age parameter $s$, while nucleons typically penetrate deeper before initiating a cascade and have a smaller $s$. The Kiel and Adelaide groups used this criterion successfully to detect signals from Cyg X-3 and Vela X-1, while the Haverah Park and Adelaide groups did not use it in their detection of Cyg X-3 and LMC X-4, respectively. Numerical simulations of $\gamma$-ray and proton-induced cascade developments and the corresponding signal detected in a typical EAS array (Hillas, 1987) show that the lateral extent and $s$ parameters of these two types of showers may be fairly similar within observational uncertainties. Nonetheless, most of the claimed positive detections were obtained using an age cut (large $s$) criterion, which indicates that shower modeling or the observations are not entirely consistent.

ii) Muon content of showers. Gamma-ray initiated showers are expected to be relatively muon-poor (~ 0.1) relative to nucleon-induced showers (Stanev *et al.*, 1985; Edwards *et al.*, 1985), at least within the standard model. However, the data from surface showers have been contradictory: the Kiel and Los Alamos EAS results indicate a ratio of muons from the Cyg X-3 and Her X-1 events which is similar to that in cosmic ray showers, while the Akeno, Tien Shan, and Chacaltaya groups detected Cyg X-3 and Vela X-1 by looking only at the muon-poor showers. The underground muon signals are also problematic, because of their large scattering angle at $\pm 5°$ in the direction of the source, requiring an abnormal transverse momentum. If some of the cosmic source showers are indeed muon-rich or the underground muon signals are indeed too broad, then either a new type of particle is being measured instead of the suspected $\gamma$-rays (Baym *et al.*, 1985; Morse, 1987; Halzen *et al.*, 1987) or one may be dealing with known particles (e.g neutrinos) which interact in an unconventional way (Gaisser and Stanev, 1987; Domokos and Nussinov, 1987). Or they may be $\gamma$-rays which interact in nonstandard ways at these high energies (e.g. Drees, *et al.*, 1988).

iii) Cherenkov imaging. TeV $\gamma$-ray induced showers are expected to be narrow and regular shaped, while nucleon showers are broad and irregular in shape. This property is being exploited currently with the Whipple camera Cherenkov detector, set up in a high-angular-resolution mode ($\lesssim 0.5°$), in order to image various cosmic sources (Weekes, 1990). The idea is to increase the signal to noise ratio by making an image shape/size cut. Preliminary results using this technique confirm the Crab Nebula detections, but fail to detect Her X-1 or any other pulsar or binary source (Weekes *et al.*, 1989; Lang, 1990). Given the time variability and sporadic nature of this type of emission, more data using this technique will need to be accumulated before definitive conclusions can be reached.

iv) Statistics. There is a general problem with all VHE-UHE measurements in that the number of photons (or events) is fairly small. The flux levels measured by different groups are often different, and the background rate, especially at TeV energies, where atmospheric conditions are influential, is variable between locations and epochs. Bonnet-Bidaud and Chardin (1988) have made a strong case for being skeptical of many of the measurements. As far as the different flux levels by different groups observing at different epochs are concerned, the fact that the emission appears to be strongly sporadic provides at least part of the explanation. That the total number of detections is very large, from a large number of groups, has led to a cautious consensus among cosmic ray physicists (e.g. Protheroe, 1987; Weekes, 1988, 1990; Fegan, 1990) that at least some of the signals from cosmic sources are real.

## 10.2 Models of VHE-UHE Gamma-Ray Sources

### *a) The Beam-Dump Model of VHE-UHE Gamma-Ray Generation*

The very-high-energy (VHE) gamma-rays observed from discrete sources in the TeV energy range ($10^{12}$ eV) may be produced by electrons or positrons accelerated to relativistic energies ($\gamma_{\pm} \gtrsim 10^{6-7}$) by inverse Compton collisions with hard X-ray photons. An example of this has been calculated for short-period rotation-powered pulsar models with an outer magnetospheric gap (Cheng, Ho, and Ruderman, 1985b) to explain the Crab or Vela pulsar TeV emission. However, to explain emission at ultrahigh energies (in the PeV range, $10^{15}$ eV) in general requires some other kind of charged particle, since the electron radiative energy losses would make it difficult to accelerate them to the appropriate energies. Since protons have the same charge but their higher mass greatly reduces their radiative losses, they can in principle be accelerated to much higher energies, provided an appropriate mechanism can be found. The conversion of these superenergetic proton beams into $\gamma$-rays can occur through nuclear interactions if an appropriate column density of target material is traversed by the proton beam, where it can dump its energy. This proton beam-dump model (Vestrand and Eichler, 1979; Hillas, 1984) is currently the most widely considered mechanism for the interpretation of the PeV events, and perhaps also of the TeV events, reported in celestial sources. The target particles may be other protons in the reaction

$$p + p \rightarrow p + p + n_1(\pi^+ + \pi^-) + n_2\pi^0, \quad \pi^0 \rightarrow 2\gamma, \tag{10.2.1}$$

which has a threshold energy of about 290 MeV (~ the mass of the pion) and a cross section $\sigma_{pp} \sim 2.7 \times 10^{-26}$ cm$^2$ for $E_p \gtrsim 2$ GeV with multiplicity $m_{\pm} \sim 2E_p^{0.25}$, $m_0 \sim 0.5m_{\pm}$ (e.g. Lang, 1980; Stecker, 1973). The grammage that the proton must traverse in order to produce significant $\gamma$-rays is ~ 50–200 g cm$^{-2}$, while for much higher grammages the $\gamma$-rays get absorbed and the energy emerges predominantly in the form of neutrinos, as shown in Figure 10.2.1.

An alternative process for producing VHE-UHE gamma-rays is to have the proton going through a sufficiently dense soft-photon field that can produce $\gamma$-rays via the photopion process,

$$p + X \rightarrow p + \pi \rightarrow 2\gamma. \tag{10.2.2}$$

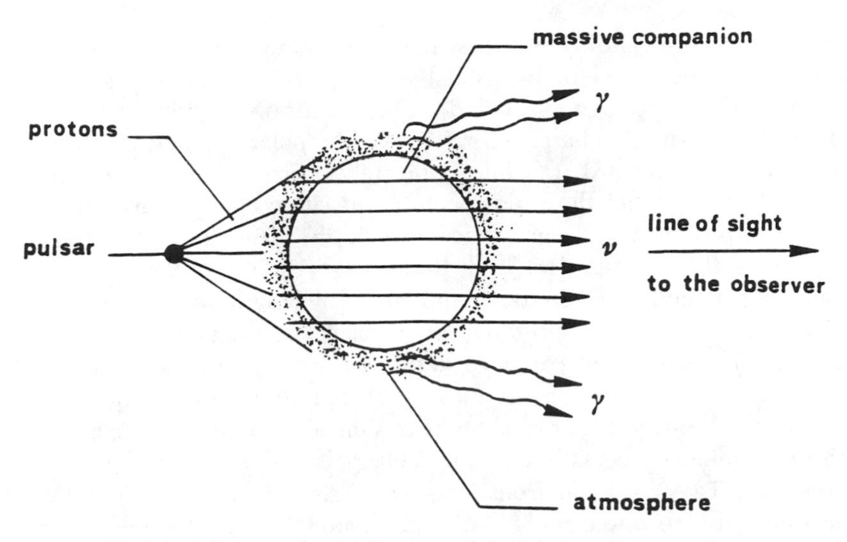


FIGURE 10.2.1 Schematic beam-dump model of the production of gamma-rays and neutrinos in a neutron star binary system. From V. S. Berezinski, C. Castagnoli, and P. Galeotti, 1986, *Ap. J.*, 301, 235.

The cross section peaks at a photon energy (in the proton frame) of about 300 MeV, where $\sigma_{\gamma,p} \sim 4 \times 10^{-28}$ cm$^2$, where the multiplicity is $m_\pi \sim 0.2$ (Stecker, 1968).

The beam-dump model addresses the production of γ-rays in the presence of a given flux of energetic baryons, typically protons. The acceleration and spectrum of the protons themselves are addressed in the models discussed below.

### *b) Induced Electric Field Acceleration Models*

Induced electric field acceleration models assume the source to be a rotation-powered pulsar (and some of the reported sources indeed are, whereas others are accretion-powered), or they rely on a pulsar analogue mechanism. In a rotation-powered pulsar, the rotating magnetic field induces an electric field $\vec{E} \sim c^{-1}\vec{v} \times \vec{B}$ (when it does not get shorted out by the presence of plasma, which would lead to $\vec{E} \cdot \vec{B} = 0$), leading to a maximum potential drop of order

$$\Delta V \sim 10^{13} B_{12} P^{-2} \text{ V} \tag{10.2.3}$$

(e.g. eq. 8.2.16). For $P \lesssim 0.1$ s, if the magnetosphere is such that protons are expected to be available along these open fields, they could reach energies $\Delta V \gtrsim 10^{15} B_{12}$ eV. For a pulsar with $\vec{\Omega} \cdot \vec{B} > 0$, the Goldreich-Julian charge $\rho_c < 0$ above the polar cap and electrons are sent down toward it, while oppositely charged ions or protons, which largely cancel these downcoming negative charges, are pulled upward from the polar cap surface, e.g. in the Arons (1983) slot gap model or the Cheng, Ho, and Ruderman (1985b) outer-magnetospheric-gap model. Upon traversing the full length of the gap, for short-period pulsars the protons would reach PeV energies. The main problem in an isolated (single) rotation-powered pulsar is in finding appropriate targets to convert these protons into $\gamma$-rays. In the Crab pulsar, the density of X-ray photons is sufficient for this to occur via the photopion process, but in the Vela pulsar this is not the case. However, TeV radiation from Vela may arise via inverse Compton scattering by secondary $e^+e^-$ of soft photons, while TeV radiation from the Crab may arise from synchrotron radiation (e.g. Sturrock, 1971; Arons, 1983; Cheng, Ho, and Ruderman, 1985b).

In accreting sources, the problem with this mechanism is to ensure that it does not get quenched by the infalling plasma, shorting out the field. One way of handling this is to use the accretion disk analogue of the pulsar model, which was originally invoked in the context of extragalactic radio sources (Lovelace, 1976). In the context of a binary accreting magnetized neutron star, Chanmugam and Brecher (1985) consider the typical disk accreting pulsar scenario (see Chap. 7.2), where the disk inner radius $r_1$ is given by the magnetospheric radius $r_m \sim 1.5 \times 10^8 B_{12}^{4/7} R_6^{10/7} m^{1/7} L_{38}^{-2/7}$ cm (eq. 7.2.5b); for simplicity we have used the spherical magnetospheric radius, $m$ is the stellar mass in solar units, while $L_{38}$ is the total accretion luminosity. The magnetic field in the disk is expected to have a perpendicular component $B_z$ due to ballooning out of the predominantly azimuthal field, and the maximum value of this field at $r_m$ is roughly comparable to the stellar dipole field strength at that radius, $B_z \sim B_*(R/R_m)^3$. The maximum velocity mismatch between the stellar dipole field and the accretion disk at $r_m$ occurs if $\Omega r_m \ll v_k(r_m) = (GM/r_m)^{1/2}$. In this case the potential drop, which is given by $\Delta V \sim |\vec{B}_z(r_m) \times \Delta\vec{v}/c| \, r_m$, takes a maximum value

$$\Delta V \sim B_z(r_m)[GMr_m]^{1/2}/c \sim 10^{14} B_{12}^{-3/7} L_{38}^{5/7} \text{ V}. \tag{10.2.4}$$

As emphasized by Chanmugam and Brecher (1985) and Ruderman (1987), this mechanism for an accreting pulsar could produce protons of energy $\sim 10^{16}$ eV, as required to explain photon energies of $\sim 10^{15}$ eV, provided that the magnetic field is rather weak, $B \lesssim 10^8$ G for an accretion rate near the Eddington limit $L_{38} \sim 1$, since in this case $r_m \sim R$ and the Kepler velocity of the disk is large. In the case of Her X-1 and 4U 0115 + 63, the cyclotron line measurements argue for a surface field $B_{*12} \gtrsim 1$, so this mechanism probably does not apply here. Some ways around this difficulty were inspired by the fact that there is some discrepancy between the surface field as measured from cyclotron line measurements and as deduced from spin torque modeling (Ghosh and Lamb, 1979b). One possibility is that the surface field strength is dominated by short-range higher multipoles, with a much weaker long-range dipole field. Another possibility (Ruderman, 1987) is that the thermal dynamo mechanism (Blandford, Applegate, and Hernquist, 1983) produces a higher-strength multipole surface field outside those regions where a weaker dipole field communicates with the exterior accretion flow.

*c) Shock Acceleration Models*

The identification of a large number of VHE and UHE sources with high-luminosity accreting pulsars (Her X-1, 4U 0115 + 63, Vela X-1, LMC X-4, and Cen X-3) suggested the possibility that a completely different proton acceleration mechanism might be at work: the diffusive acceleration process by the first-order Fermi mechanism in a shock wave. This mechanism has been extensively investigated as a way of producing the diffuse cosmic ray background, using collisionless shocks in supernova remnants or terminal galactic shocks (see Blandford and Eichler, 1987, for a review). The protons are scattered back and forth between the upstream and downstream regions, which act as mutually converging mirrors. The scattering typically is done by field fluctuations present in the turbulent downstream region and by Alfvén waves induced by protons streaming upstream. In each scattering cycle the photon is boosted in energy by a factor $\Delta E/E \propto \Delta v/c$, where $v$ is the shock front velocity. The energy therefore exponentiates, and given an appropriate injection rate, the particles increase their energy until they escape from the acceleration region, developing in the process a power law energy distribution $n(E) \propto E^{-\alpha}$, with $\alpha \sim 2$–$2.5$, (e.g. Drury, 1983).

Collisionless shocks are invoked in some low-luminosity models to explain the X-ray emission of accreting pulsars, while in high-luminosity ($L \gtrsim 10^{-2}$–$10^{-1} L_{\mathrm{Ed}}$) accreting pulsars some authors have considered the possibility of there being a precursor collisionless shock above the radiation pressure dominated deceleration front (see Chap. 7.3*c*). Eichler and Vestrand (1985) and Kazanas and Ellison (1986) used the possible existence of such collisionless shocks in the accretion flow not far above the neutron star polar cap in order to accelerate protons up to PeV energies by the first-order Fermi mechanism. The acceleration time scale is given by $t_a \sim \gamma/\dot{\gamma}$, and the relative energy boost per cycle is $\Delta\gamma/\gamma \sim \beta$, while the cycle time is roughly the ratio of the diffusion length $\lambda_d \sim r_g/(B_1/B_0) \sim r_g/\beta_s$ to the instantaneous speed $\sim c$, where $r_g$ is the proton gyroradius and $\beta_s$ characterizes the velocity gradient across the shock. We have then for the acceleration time

$$t_a \simeq \frac{\xi r_g}{\beta_s^2 c} = \frac{\xi \gamma m_p c}{\beta_s^2 ZeB}, \tag{10.2.5}$$

where $\xi \gtrsim 1$ and $B$ is the average magnetic field at the shock. The energy loss time scale will be given by the combined nuclear interaction and synchrotron losses. The latter are typically dominant, so

$$t_{\rm sy} = \frac{4\pi m_p c}{\gamma \sigma_T B^2 Z^2} \left(\frac{m_p}{m_e}\right)^2, \tag{10.2.6}$$

where $\sigma_T$ is the Thomson cross section. The protons will diffuse out of the acceleration region when the diffusion length becomes comparable to the size of the acceleration region, $R_g/\beta_s \sim \pi^{-1} R_s$, where $R_s$ is the shock radius, and this combined with the requirement that the acceleration time be less than the loss time gives an upper limit on the proton Lorentz factor that can be achieved,

$$\gamma_m < \beta_s \left[\frac{1}{2\xi}\left(\frac{m_p}{m_e}\right)\frac{R_s}{Z^2 r_e}\right]^{1/3}, \tag{10.2.7}$$

where $r_e$ is the classical radius of the electron. This maximum energy assumes the acceleration time (10.2.5) for quasi-parallel shocks (with magnetic field close to parallel to the flow direction); however, for oblique or quasi-perpendicular shocks, the acceleration time could be more than an order of magnitude shorter (e.g. Drury, 1983) and therefore the maximum energies could be larger by a similar factor. Using equation (10.2.7) for a shock radius near the neutron star, $R_s \sim 10^6$ cm, and $\beta_s$ approximately the free-fall velocity, energies of about $10^{16}$ appear possible. However, several difficulties with such nearby shocks must be considered. One problem is that, in order for much of the energy to go into the highest-energy particles, the shock Mach number must be high, i.e. $\rho_s v_s^2 \gg B^2/8\pi$, which requires $B \lesssim 10^8 \beta_s^{1/2} L_{38}^{1/2} R_{s,6}^{-1}$ G (Eichler and Vestrand, 1985). These authors therefore argue that, in sources such as Her X-1 where cyclotron line measurements at X-ray energies indicate fields $B_* \gtrsim 10^{12}$ G, the shock may be further out, at distances of several tens of stellar radii, or if it is close to the star, the acceleration must occur in regions of lower magnetic field strength. Another problem if the shock occurs close to the surface is that the high-energy charged protons would escape diffusively, which would be difficult to reconcile with a sharp beam as inferred from the UHE pulse shapes. To avoid this difficulty, Kazanas and Ellison (1986) proposed that the escaping high-energy particles are neutrons formed by proton-proton collisions with 25% efficiency and by $^4$He photodissociation, which would not be affected by the magnetic field. There is also a problem if the protons produce UHE $\gamma$-rays before escaping the region by photopion or proton-proton collisions near the surface, since the magnetic field would be strong enough there to absorb the $\gamma$-rays by magnetic one-photon pair production.

A distant shock model may be the likeliest way of resolving the above difficulties. However, since the particle luminosities are high, comparable to the X-ray luminosities, which themselves are close to the Eddington value, one cannot postulate an accretion shock at $R_s \gg R_*$ since that would liberate only a fraction $R_*/R_s \ll 1$ of the total available gravitational energy. This could only be done if the accretion rate were much larger than critical, and that would have observable consequences for the X-ray emission, which does not seem to be the case. An alternative way is to assume a wind of particles from the star which undergoes a shock farther out, as suggested by Eichler (1986). A specific example based on a 12 ms period rotation-powered pulsar has been considered by Quenby and Lieu (1987) in connection with Cyg X-3.

A self-consistent shock acceleration model for an accreting pulsar must be able to explain the energetics, the origin of the flow of matter which is shocked, and the VHE-UHE spectrum while remaining consistent with the wealth of accumulated X-ray observations and the phenomenological models developed to explain the X-ray emission. Besides these constraints on the VHE-UHE emission mechanism, there is an observational fact which appears to hold a clue (Trümper, 1987): the accreting pulsars thus far reported to be VHE-UHE sources all seem to have episodes during which their period derivative changes sign, i.e. they alternate between periods of spinning up and spinning down, and for the most part they seem to have abnormally low $\dot{P}/P$, a fact usually interpreted to mean that the inner edge of the accretion flow is rotating at a speed close to that of the outer boundary of the magnetosphere touching it (Ghosh and Lamb, 1979a, b).

### *d) Corotating Jet Model*

A corotating jet model has been calculated by Király and Mészáros (1988) incorporating the above-mentioned theoretical and observational constraints. The key feature of this model is that the gravitational energy gets liberated near the stellar surface, and some of it is converted into a jet of outflowing material which undergoes a shock at sufficiently large distance. The cause of the outflow is that in high-luminosity accretion pulsars one may expect some of the matter to be blown out by the radiation pressure, especially since the cross sections are resonant. The accretion column may consist of a funnel (Basko and Sunyaev, 1976a) so that the inflowing and outflowing matter may both be axisymmetric without occupying the same solid angle. Time-dependent calculations of inhomogeneous, 2-D high-luminosity accretion columns (Arons *et al.*, 1986) also indicate plumes of ejected matter between regions of accreting matter. In the model of Király and Mészáros (1988), this ejected matter carries a significant fraction of the total accretion energy in two jets directed along the magnetic axis out to the magnetospheric radius. At this radius the stream of gas must undergo a shock, since here the accretion disk material has been backed up by the stresses of the stellar magnetic field, causing the accreted plasma to latch onto the stellar magnetospheric field lines. At this location the accreted gas spreads out over a large solid angle, forming the Alfvén surface (e.g. Chap 7.2). The inertia and transverse magnetic fields of the Alfvén shell are assumed to provide some of the back pressure needed to support such a shock, the rest being provided by the accretion flow beyond it. Proton acceleration is expected to occur at this (collisionless) shock. The corotation condition deduced from the X-ray observations of $\dot{P}/P$ is crucial in ensuring that the protons reach the highest PeV energies. Indeed, the protons must be scattered between the downstream and upstream regions several hundred times for these energies to be reached, and these two regions must be moving at approximately the same transverse (corotation) velocity so that the protons do not drift out of the acceleration region before reaching high energies (see Fig. 10.2.2). Taking the radius of

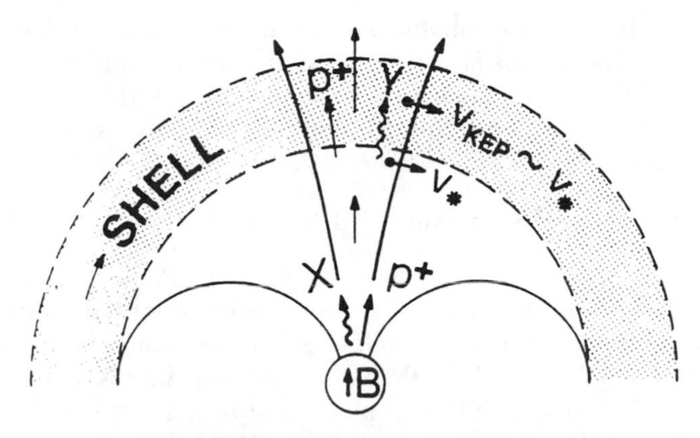


FIGURE 10.2.2 Shock acceleration in the polar cap jet model at the corotating accretion shell. From P. Király and P. Mészáros, 1988, *Ap. J.*, 333, 719.

the Alfvén shell to be the shock radius, if this is close to the corotation radius $R_{co} = 1.5 \times 10^8 m^{1/3} P^{2/3}$, we have $R_s \simeq \delta r_{co}$, where $\delta \sim 1$. The magnetic field at the surface of the star, if not known from cyclotron observations, can be estimated from this quasi-corotation condition by equating the magnetospheric radius of equations (7.2.5) to $\delta r_{co}$. Using for simplicity the spherical approximation for the magnetospheric radius and assuming a dipole dependence $B \propto r^{-3}$, the surface field is

$$B_{*,12} \simeq L_{38}^{1/2} P^{7/6} R_6^{-5/2} m^{1/3} \delta^{7/4}. \tag{10.2.8}$$

With the shock radius and the magnetic field at this radius thus determined, the proton acceleration time is, from equation (10.2.5),

$$t_a \sim 2.7 \times 10^{-2} L_{38}^{-1/2} P^{19/18} R_6^{-1/2} m^{7/9} \delta^{19/12} \xi^{2/3} \beta_s^{-1} Z^{-4/3}\ \mathrm{s}, \tag{10.2.9}$$

where for the proton Lorentz factor the maximum value of equation (10.2.7) is used. On the other hand, the photopion time scale on which these protons can convert into UHE photons while in the shock region is

$$t_{p\gamma} \sim \frac{(m_p/m_e)}{m_\pi \sigma_{p\gamma}} \frac{4\pi R_s^2}{\eta L_x} \frac{\epsilon}{(1+\tau)}$$

$$\simeq 1.7 \times 10^{-1} L_{x,38}^{-1} P^{4/3} m^{2/3} \delta^2 \eta^{-1} (1+\tau)^{-1} \epsilon_{\mathrm{keV}}, \tag{10.2.10}$$

where $\epsilon_{\rm keV}$ is the X-ray photon energy in keV, $m_\pi \sigma_{p\gamma} \sim 8 \times 10^{-29}$ cm$^2$, $\tau \sim 1$ is the Thomson optical depth of the shell, and $\eta \sim 2$–$4$ is a geometric factor. The radius $r_B$ beyond which UHE photons can escape without magnetic pair production from the stellar field is approximately (eq. 5.5.6)

$$r_B \simeq 2.8 \times 10^8 R_6 \epsilon_{15}^{1/3} (\sin\theta/0.17)^{1/3} B_{*,12}^{1/3} \text{ cm}, \tag{10.2.11}$$

where $\theta$ is some average angle between photon and field in the jet. In general, these radii are larger than the corotation radius. This effect therefore helps to produce a collimated $\gamma$-ray beam. In fitting this model to Her X-1, LMC X-4, 4U 0115 + 63, and Vela X-1, Király and Mészáros (1988) find that the acceleration times are $\lesssim$ the photopion time scale, and therefore one can expect UHE $\gamma$ pulses in phase with the X-ray pulses even when an accretion disk or binary companion dump target is not in the right position. This is required by Whipple observations, which saw VHE pulses in the middle of some of the X-ray ON periods which were very narrow and in phase with the X-ray pulses. On the other hand, since the photopion time scales are not much shorter than the acceleration time scale, one expects a comparable amount of energy to be carried out in the form of escaping high-energy protons, which then can interact via *pp* collisions to make UHE photons when the line of sight nearly grazes the disk or the binary companion. Typical spectra to be expected from e.g. Her X-1 and Cyg X-3 (assuming the latter is an AXP) within the framework of this model have been calculated by Rudak and Mészáros (1991).

*e) Binary Pulsar Shock Acceleration Model*

In the previous sections, the models consisted of either accreting binary X-ray pulsars or isolated rotation-powered pulsars. However, one must also address the possibility of acceleration in a binary radio (rotation-powered) pulsar system. The number of such systems is significant (see Chap. 8 and Table 3 near the end of the present book), and in some of them, e.g PSR 1957 + 20, the binary companion is so close that the particle wind from the pulsar is interacting with the companion atmosphere, causing it to lose mass in a wind. The interaction of the pulsar wind with the companion atmosphere or companion wind results in a shock (Phinney *et al.*, 1988; Ruderman, Shaham, and Tavani, 1989; Ruderman *et al.*, 1989; Cheng, 1989).

A model for the acceleration of protons at such a shock, as well as the ensuing VHE-UHE gamma-ray spectrum from the system, has been investigated by Harding and Gaisser (1990), and compared to current observations and upper limits on various systems. One of the distinguishing features of this type of shock in a pulsar wind is that, contrary to the case in some other models, it is expected to be quasi-perpendicular, i.e. with the magnetic field perpendicular to the flow direction. Using for simplicity the nonrelativistic shock acceleration time estimate $t_a \simeq (E/ceB)[\xi(\xi+1)/(\xi-1)]$, where $\xi$ is the compression ratio, and equating this to the diffusion time of particles out of the shock region, $t_d = r^2/D$, where $D \sim D_{\min} = r_L v/3$ is the diffusion coefficient ($r_L$ is the Larmor radius), Harding and Gaisser obtain a maximum proton energy. For the case where the pulsar wind interacts with the stellar wind or the magnetosphere of the companion, this maximum proton energy is

$$E_p^{\max} \simeq 10^7 \text{ TeV } B_{12} P_{\rm ms}^{-2} \left[ \frac{3(\xi-1)}{\xi(\xi-1)} \right]^{1/2} \times \begin{cases} 1, & r_s \ll a - r_*, \\ (a/r_s - 1), & r_s \simeq a - r_*, \end{cases} \tag{10.2.12}$$

where $a$, $r_s$, and $r_*$ are binary separation, shock radius, and companion stellar radius, $B_{12}$ is the pulsar surface field, and $P_{\rm ms}$ is the pulsar period in milliseconds. For the case where the pulsar wind interacts directly with the atmosphere of the companion, the maximum energy is an order of magnitude smaller. Notice that this maximum proton energy is proportional to the square root of the pulsar spin-down luminosity (eq. 8.2.13). Thus, in general, the maximum energy would be $\gtrsim 10$ TeV for $P_{\rm ms} B_{12}^{-1/2} \lesssim 900$ and $\gtrsim 10$ PeV for $P_{\rm ms} B_{12}^{-1/2} \lesssim 30$, which imply pulsar spin-down powers of $\gtrsim 6 \times 10^{31}$ ergs s$^{-1}$ and $\gtrsim 5 \times 10^{37}$ ergs s$^{-1}$. Taking into account the distance and the estimated binary separation, Harding and Gaisser estimate the fraction of pulsar energy interacting with the companion at a shock surface and the ensuing high-energy fluxes from a number of binary pulsars. The values are not very large, the largest flux ($\sim 9 \times 10^{-11}$ cm$^{-2}$ s$^{-1}$) being obtained for PSR 1957 + 20 for a particle energy index of 2 and an effective gamma-ray solid angle $\Delta\Omega \sim 1$. A TeV detection of this object, folded over the 9 hr orbit period, has been reported by von Ballmoos *et al.* (1989) at the phase corresponding to the L4 Lagrangian point and with a flux of $\sim 10^{-9}$ cm$^2$ s$^{-1}$, assuming from the phase width a solid angle $\Delta\Omega \sim 0.01$. Harding and Gaisser also apply this model to the case of Cyg X-3, adopting the values of $P = 12.6$ ms,

$\log \dot{P} = -13.5$, $P_{\rm orb} = 0.2$ days, $M_c = 0.5 M_\odot$, $a = 1.2 \times 10^{11}$ cm, and $d = 10$ kpc. With these values, and some additional assumptions about the duty cycle and spectrum, they obtain fluxes in agreement with the *SAS 2* measurements at 100 MeV, while at TeV energies the model fluxes can exceed the reported observed values. The magnetic field implied by the $P$, $\dot{P}$ values above is $B \sim 6 \times 10^{11}$ G, much larger than that of typical millisecond pulsars and incompatible with the recycled LMXB scenario (see Chap. 11.3), which requires fields $B \lesssim 10^9$ G. In this model, therefore, Cyg X-3 would have to be a young neutron star, younger than $\sim 7 \times 10^3$ yr.

*f) Magnetospheric Acceleration Models of Accreting Pulsars*

The fact that at least the TeV radiation of both rotation- and accretion-powered pulsars does not seem to be very different, if the observations are taken at face value, is thought provoking. A difference between these two classes of objects may be that PeV radiation seems to be present only in accreting pulsars, as of 1991. Nonetheless, it seems wise and worthwhile to investigate the extent to which a pulsar analogue of the magnetospheric acceleration by an induced electric field may be applicable to accreting pulsars. One example of this is the accretion disk analogue of the pulsar mechanism of Chanmugam and Brecher (1985), discussed above. A related, but even more pulsar-like, model is that of Cheng and Ruderman (1989). Their model assumes a magnetospheric structure similar to that of the (rotation-powered) pulsar model of Krausse-Polstorff and Michel (1985), which considers the fields and magnetospheric charges in the presence of a conducting disk (see Fig. 10.2.3). In this magnetosphere, the lines which do not intersect the disk corotate with the star, while those which connect to the disk corotate with the Keplerian disk velocity at the point where the line connects to it. In the open field line regions, the sum of the centrifugal, gravitational, and electrostatic forces on the negative charges $\vec{F}_-$ satisfies $\vec{F}_- \cdot \vec{B} = 0$, while the heavier protons and positive ions are subjected to much larger inertial and gravitational forces, leading to $\vec{F}_+ \cdot \vec{E} \neq 0$, which result in driving these positive charges out from this region of the magnetosphere so that only the negative remain. In the region above the disk, on the other hand, only $\vec{F}_+ \cdot \vec{B} = 0$, and here it is the positive charges that remain. Within these two charge-separated regions of opposite sign the electric field satisfies the force-free condition $\vec{E} \cdot \vec{B} \simeq 0$. Separating these two portions of the magnetosphere and centered on the neutral line $\vec{\Omega}_* \cdot \vec{B} = 0$ is an empty gap within which $\vec{E} \cdot \vec{B} \neq 0$ (see

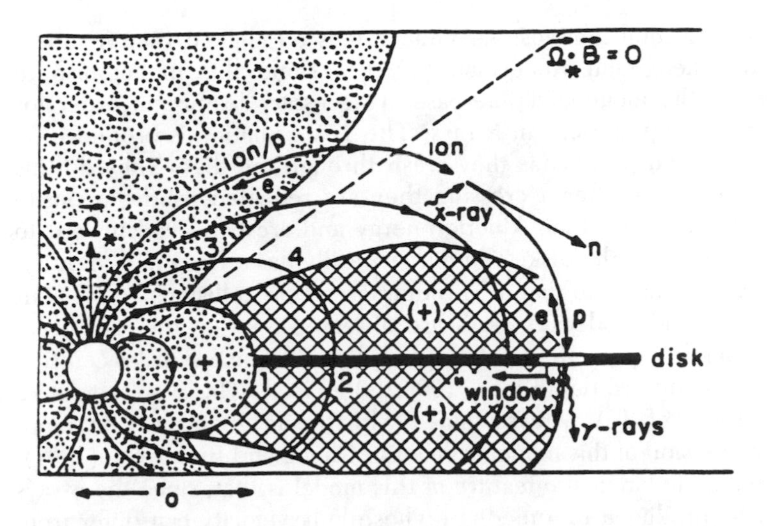


FIGURE 10.2.3 VHE gamma-ray generation in the corotating magnetosphere model, where the accelerated particles interact with the accretion disk. From K. S. Cheng and M. Ruderman, 1989, *Ap. J.* (*Letters*), 337, L77.

Chap. 8.3). It is this gap that serves as an accelerator in Cheng and Ruderman's accretion disk model. It also acts as a reflector, since a charged particle entering from either of the two magnetospheric portions into the gap is pushed back toward its magnetosphere. The maximum potential drop along a field line going from near the inner edge of the disk through the gap to the star, across the polar cap and back along $\vec{B}$ to the disk is again

$$\Delta V_{\max} \sim \frac{B_* R^3 \Omega_K(r_0)}{r_0 c} \sim 4 \times 10^{14} \xi^{-5/2} m^{1/7} R_6^{-4/7} L_{37}^{5/7} B_{*,12}^{-3/7} \text{ V}, \tag{10.2.13}$$

which is essentially the Chanmugam-Brecher (1985) value (eq. 8.2.2). Here $\xi = 2r_0/r_m \sim 1$ and the magnetospheric radius $r_m$ used is the spherical value of equation (7.2.4). A proton current flux directed from the star to the disk would be balanced by an upward flow of electrons from the disk or by a magnetospheric charge density, thus canceling any net charge density from the current flow and keeping $\vec{F} \cdot \vec{B} \sim 0$ everywhere except in the gap. The maximum particle power is of course limited to the fraction of the gravitational energy liberated at $r_0$, i.e. $r_0/R$ of the total, or about $10^{36}$ ergs s$^{-1}$ for an Eddington value of the total luminosity. While this is an aligned rotator model ($\vec{\Omega} \| B_*$), Cheng and Ruderman (1989) assume this model also to survive in the inclined dipole case, a necessary feature in order for them to show pulsations in X-rays. The TeV radiation is produced by the accelerated protons as they crash through the disk. They do this repeatedly, since after a crossing they are reflected by the opposite gap, until they lose their kinetic energy and are conveyed inward to the inner edge of the disk. The disk typically has a large path density, $\gg 50$–$200$ g cm$^{-2}$, so that $\pi^\circ$ production by *pp* collisions occurs, but the corresponding decay $\gamma$-rays are observed only on the rare occasions when a gap occurs in the radial direction in the disk accretion flow, whose optical depth is $\sim 100$–$200$ g cm$^{-2}$. This would create a pencil beam of TeV $\gamma$-rays aligned with the magnetic axis. A more detailed version of this model is given in Cheng and Ruderman (1991). An interesting kinematic feature of this model is that, since the $\gamma$-rays would be produced in a medium whose bulk velocity can differ from that of the frame in which the protons are accelerated, the gamma-rays are subject, in addition to the basic rotation period $P$, to small relative changes $\delta P$ associated with aberration, the radial motion of the disk "window," and the radial dependence of the $B$ field distortions. This may explain the $\delta P/P \sim 10^{-3}$ observed in Her X-1 (North *et al.*, 1987; Raubenheimer *et al.*, 1989). Other kinematic explanations for this period difference based on disk motions have been investigated by Slane and Fry (1989).

### *g) Acceleration in Accreting Magnetospheres*

A general discussion of the possible acceleration mechanisms and acceleration regions in accreting magnetospheres has been given by Katz and Smith (1988), avoiding any very specific commitment to particular models. They consider the possibility of particles being accelerated to their final energy $\gamma$ in $N$ repeated bounces (which includes the shock models and the disk electric field models as particular cases, with $N \sim 1$ in the latter). Considering the synchrotron loss limits on the maximum energy and the maximum radius of confinement $r \lesssim r_g$ within which particles have reached an energy $\gamma$ in a time $t \sim Nr/c$, the characteristic particle maximum energy is

$$\gamma_{\max} \sim \left( \frac{\mu m_p c^2}{N^2 e^3} \right)^{1/7} \sim 8 \times 10^7 \mu_{30}^{1/7} N^{-2/7}, \tag{10.2.14}$$

where $\mu = B_{\text{ast}} R^3$ is the stellar dipole moment. The characteristic radius at which this occurs is

$$r_{ch} \sim \left( \frac{e^5 \mu^3 N}{m_p^4 c^8} \right)^{1/7} \sim 10^8 \mu_{30}^{3/7} N^{1/7} \text{ cm}. \tag{10.2.15}$$

This radius is of the order of the Alfvén radius $r_A \sim r_m$ given by equations (7.2.4) and (7.2.5) and is in any case an upper limit to the size of the closed magnetosphere within which bounces may be expected to occur. The maximum observed values of $\gamma \sim 1\text{–}3 \times 10^7$ for the PeV events imply an upper limit on the number of bounces $N \lesssim 10^2\text{–}10^4$, but lower-energy events may be produced closer or farther than $r_{ch}$.

In order for bounces and acceleration to occur, there must be some kind of coupling between regular particle motions and some external force field in the magnetosphere. The three types of regular particle motion considered by Katz and Smith (1988) are the gyromotion, the bounce from end to end of the closed poloidal lines caused by mirroring, and drift motions, whose respective frequencies are given by

$$\omega_g = \frac{eB}{\gamma m_p c} \sim 10^{10} \mu_{30} \gamma^{-1} r_8^{-3} \text{ s}^{-1},$$

$$\omega_b \simeq \epsilon \omega_g \simeq c/r \simeq 3 \times 10^2 r_8^{-1} \text{ s}^{-1},$$

$$\omega_d \sim \epsilon^2 \omega_g \simeq \frac{\gamma m_p c^3}{r^2 eB} \simeq 10^{-5} \gamma r_8 \mu_{30}^{-1} \text{ s}^{-1}, \tag{10.2.16}$$

where $\epsilon = r_g/r = \gamma m_p c^2/(eBr) \simeq 3 \times 10^8 \gamma r_8^2 \mu_{30}^{-1}$. These frequencies may be compared to the characteristic Kepler frequency of the disk $\omega_K = (GM/r^3)^{1/2} \sim 1.4 \times 10^1 r_8^{-3/2} \text{ s}^{-1}$. Katz and Smith consider in particular the situation where the particles receive energy from fluctuating fields in the magnetosphere near the Alfvén surface caused by fluctuations in the accretion rate, slippage, etc., which ultimately derive their energy from the accretion gravitational potential energy. Their root mean square fluctuations are parametrized as

$$\langle (\Delta B)^2 \rangle = \eta^2 B_A^2, \tag{10.2.17}$$

where $\eta$ is a small dimensionless number and $B_A$ is the ordered magnetic field at the Alfvén radius. In order for this to couple efficiently with one of the regular motions of the particles, the field fluctuation frequency $\omega_f$ should be close to or resonant with one of the frequencies (10.2.14). Whereas in the shock acceleration model the particle energy gain per bounce is first order in $v_s/c$ (first-order Fermi mechanism), the particle acceleration from randomly fluctuating fields is a second-order Fermi mechanism (because in first order some fluctuations will give while others will take energy). The energy gain per bounce is then

$$\frac{\langle \Delta E \rangle}{E} \simeq \omega_f \left\langle \left( \frac{\Delta B}{B} \right)^2 \right\rangle, \tag{10.2.18}$$

leading to an energy gain time or acceleration time scale

$$t_a \sim \omega_f^{-1} \left\langle \left( \frac{\Delta B}{B} \right)^2 \right\rangle^{-1}, \tag{10.2.19}$$

where $\omega_f^{-1}$ is the characteristic time scale of the magnetic field fluctuations. If $\Delta B$ is not too sensitive to radius and $B \propto r^{-3}$, one has then for the number of bounces

$$N \simeq \frac{ct_a}{r} \simeq \left( \frac{c}{\omega_f r} \right) \left( \frac{1}{\eta^2} \right) \left( \frac{r_A}{r} \right)^6. \tag{10.2.20}$$

If the fluctuating field frequency $\omega_f \sim \omega_K$, this may be resonant with the bounce frequency $c/r$, and for $\eta \lesssim 1$, $r \gtrsim 0.5 r_A$, one may get to $N \gtrsim 10^2$, as required for approaching the highest energies. However, the fluctuation energy available at $R_A$ again represents only a fraction $r_A/R \simeq 10^{-2}$ of the Eddington luminosity. Another possibility considered by Katz and Smith (1988), which may tap a larger fraction of the total accretion energy, is that the accreting matter that reaches the neighborhood of the stellar surface produces disturbances there which could travel back upward as relativistic Alfvén waves. The power of these waves would be approximately conserved as they travel up the dipole field lines, whose solid angle increases $\propto r$, so the magnetic fluctuation amplitude grows as $\Delta B \propto r^{-3/2}$, and the acceleration time now becomes

$$t_a \simeq \omega_f^{-1} \eta^{-2} \left( \frac{r}{R_*} \right)^{-3}, \tag{10.2.21}$$

where $\eta$ is defined near the stellar radius $R_*$. A value of $\Delta B/B \sim 1$ at the Alfvén surface may be achieved with $\eta \gtrsim 10^{-4}$ at the star. This would require a fluctuating stress at the star which is $\sim 1\%$ of the hydrodynamic stresses, and with $\omega_f \simeq \omega_b \simeq c/r$ one has $N \sim 1$, With $\omega_f$ comparable to either $\omega_g$ or $\omega_b$, the value of $N$ may be of order unity, and the fraction of the total energy tapped may conceivably approach a sizable fraction of the total accretion energy.

# 11. Evolution of Neutron Stars

## 11.1 Stellar Evolution of Neutron Star Systems

### *a) Supernovae and Neutron Star Formation*

According to current calculations and extensive observational evidence, the most straightforward way to form neutron stars is in supernova (SN) explosions in massive stars ($M \gtrsim 8\text{–}10\ M_\odot$). This type of event, a supernova of type II (SN II), involves the collapse of the evolved core of the progenitor star. Core collapse occurs after all the thermonuclear fuel that is capable of exothermic reactions during the star's main sequence and later quasi-steady burning stages has been exhausted. This burning is accompanied by an increase in the core mass and the gradual change of its chemical composition toward increasingly heavy elements. In single stars of $10 \lesssim M \lesssim 60\text{–}100\ M_\odot$ the collapse occurs when the evolved CO core grows larger than the Chandrasekhar mass (eq. 1.2.6), triggered by the photodesintegration of the Fe-group elements; when the precursor star is within the more limited range $10 \lesssim M \lesssim 25$, collapse calculations typically predict a neutron star remnant (e.g. Woosley and Weaver, 1986; Baron and Cooperstein, 1991). The total energy liberated in this type of event, given roughly by the gravitational binding energy of the neutron star, is of order $E \sim 10^{53}$ ergs, most of which goes into neutrinos and kinetic energy of the ejected outer shell of matter. The optical spectrum of the nebula is expected to have substantial H lines from the ejected hydrogen-rich stellar envelope. After an intense and short initial intensity spike occurring as the radiation first breaks through the opaque outer layers, the optical light curve shows a distinctive leveling off, or plateau, caused by the hydrogen recombination front moving into the envelope, in good agreement with the observed characteristics of SN IIs. Because this type of explosion occurs in massive stars, SN IIs are associated with the young disk population (population I) of spiral galaxies and are not seen in elliptical galaxies. Thus, this mechanism can easily explain the presence of accreting pulsars distributed mostly along the disk or older pulsars that have wandered off the plane, but cannot so easily explain the presence of pulsars in globular clusters unless these pulsars are as old as the cluster itself.

A second and more controversial scenario for neutron star formation involves the accretion-induced collapse (AIC) of a white dwarf (WD), which may occur in some circumstances normally associated with supernovae of type I (SN Is). The latter arise when a thermonuclear instability on the surface of a WD produces a total energy which is somewhat lower than that of SN IIs. The observed SN Is are associated with the older galactic population II objects, and since the progenitor WD is hydrogen poor (e.g. a degenerate He, CO, or ONeMg object) the optical spectrum does not show appreciable hydrogen lines. The optical light curve is dominated by the exponential radioactive decay of $^{56}$Ni. In most calculations reproducing the observed characteristics of a SN I, the remnant appears to be entirely disrupted (Woosley and Weaver, 1986); however, for rapid accretion rates the WD core could collapse to a neutron star without a substantial optical display (Nomoto and Iben, 1985; Woosley and Weaver, 1985). For progenitor masses $3 \lesssim M \lesssim 8\ M_\odot$, the growth of a degenerate CO core leads in a single star to envelope expansion and a C deflagration which disrupts the remnant, while in a binary it leads to the loss of the envelope and to a CO WD, unless enough material is accreted from the companion to lead again to a C deflagration and remnant disruption. For masses $8 \lesssim M \lesssim 10 \pm 2\ M_\odot$, the growing core undergoes nondegenerate C ignition followed by growth of a degenerate ONeMg core. In single stars this proceeds until Ne ignition, and the ensuing collapse and rapid electron capture can lead to the formation of a neutron star remnant. In a binary it leads to an ONeMg WD, unless again accretion from the companion is rapid enough. For low accretion rates $\dot{M} \lesssim 10^{-9}\ M_\odot/\mathrm{yr}$ the nova-like thermonuclear flashes in the H-rich accreted material are violent enough that the accreted material is ejected, preventing growth and collapse of the WD. However, for $\dot{M} \gtrsim 10^{-9}$–$10^{-8}\ M_\odot/\mathrm{yr}$ the flashes appear weak enough that the accreted matter stays on the surface, leading to mass growth of the WD and eventual collapse to a neutron star (e.g. Canal, Isern, and Labay, 1990; Nomoto and Kondo, 1991). This can occur, for instance, when the system has gone through two mass exchange phases leading to a double WD system (Iben and Tutukov, 1984). After the orbit has shrunk sufficiently, the less massive WD forms an accretion disk around the more massive, until the latter exceeds the Chandrasekhar limit, leading to the formation of a neutron star remnant. The efficiency of the AIC neutron star formation scenario, i.e. what fraction of neutron stars arise from this as opposed to SN II events, is still undecided.

### *b) Observational Overview of Neutron Star Binary Evolution*

We review here briefly the observations of binaries containing neutron stars, also mentioning systems which are believed to have originated in the disruption of such binaries. The galactic population of X-ray binaries shows a fairly distinct separation between high-mass X-ray binaries (HMXBs), where the companion mass is $M_1 \gtrsim 8-10\ M_\odot$, and low-mass X-ray binaries (LMXBs), where the companion mass is $M_1 \lesssim 2\ M_\odot$ (van Paradijs, 1988). The HMXBs are generally galactic disk (population I) objects, the companion being a young massive main-sequence or giant star. Their distribution appears concentrated about the galactic equator, and many of them appear to have runaway velocities. The neutron star in almost all identified cases shows X-ray pulsations, this class of object being the common accreting X-ray pulsar (AXP). The pulsations indicate that the magnetic field is relatively high and nonaligned with the spin axis in these neutron stars. The X-ray spectrum is generally hard, as is typical of AXPs, extending typically to $E \gtrsim 20-40$ keV. By contrast, the LMXBs are distributed as old, population II objects, with no tendency to hug the disk and with a concentration toward the galactic center. A few LMXB systems contain AXPs with a hard spectrum (Her X-1, GX 1 + 4, 4U 1626 − 67, 1E 2259 + 59), but in the great majority of LMXBs the X-ray spectrum is rather soft, $E \sim 2-3$ keV, typically a blackbody occasionally complemented by a hard power law tail. Among the nonpulsating LMXB sources are the bright galactic bulge sources (most of which show quasi-periodic oscillations, or QPOs, and may have a small magnetic field), the X-ray bursters (XRBs), the globular cluster X-ray sources (GCXs), and the soft X-ray transient sources (SXTs), all of which almost certainly have a very low magnetic field, with the possible exception of the rapid burster. Among the low-mass binaries are also the binary rotation-powered pulsars (binary RPPs), which so far have been observed only at radio wavelengths. Some of these systems contain a low-mass He star, while others contain a second neutron star (e.g. Taylor, 1987; Backer and Kulkami, 1990; see also Table 3 near the end of this book).

In the HMXBs the optical companions generally have normal early-type stellar spectra, indicating that the disk is not a major optical contributor relative to the star, and the optical light curves show the nonuniform surface brightness distribution expected from tidally and rotationally distorted stars in a binary system. The HMXBs can be divided into the "standard" massive binaries and the Be X-ray binaries. The standard HMXBs are strong X-ray sources, usually have regular sinusoidal light curves, and often show eclipses, the companion being typically a massive giant with $M_1 \gtrsim 20\ M_\odot$ and orbital periods $P_{\phi 0} \sim 1.4$–$10$ days, which are near to filling their Roche lobe. The spectral type of the companion is earlier than B2, of luminosity class I–III (evolved), and the mass loss occurs either via atmospheric Roche overflow or via capture from the stellar wind, which can be very strong in early-type evolved stars, giving rise in both cases to a persistent strong X-ray emission. The Be binaries contain a main-sequence (III–V) early-type star with emission lincs, typically a Be star of mass $M_1 \sim 10$–$20\ M_\odot$ with orbital periods $P_0 \gtrsim 15$ days, sitting deep inside its Roche lobe and rarely eclipsing. The X-ray emission is variable, being due to outburst from the fast-rotating Be companion, which occasionally ejects a disk from its equator. While the orbits of the standard HMXBs are generally circular, the Be systems have a large eccentricity, which explains the variable emission, accretion occurring preferentially near the aphelion (Joss and Rappaport, 1984). A plot of the X-ray pulse period $P_s$ against the orbital period $P_0$ shows a different trend for the three types of HMXB pulsing binaries, the Be binaries, the standard wind accretors, and the standard Roche lobe overflowers, as shown in Figure 11.1.1 (Corbet, 1984; van Paradijs, 1988). A representative list of HMXB systems and their properties is given in Table 1 (from Nagase, 1989a, and Ögelman, 1988) near the end of the book (most of the known AXP systems are included).

In LMXBs the optical spectrum (when observable) is faint and abnormal for a stellar spectrum, indicating a large optical contribution from the accretion disk and from the accretion stream, similar to what is seen in cataclysmic variables. The optical light curve is disk-dominated and depends on the disk inclination angle $i$ with respect to the plane of the sky. For low $i$ it is approximately sinusoidal, while at high $i$ the curves often show dips caused by absorption from the outlying disk and stream material along the orbital equatorial plane. LMXBs can be divided into two classes.

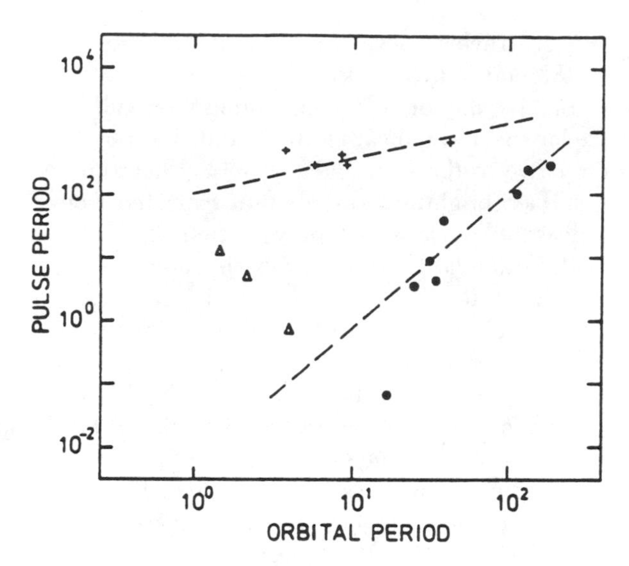


FIGURE 11.1.1 The relation between spin period versus orbital period in three different types of HMXBs: Be X-ray binaries (*filled circles*), non-Be wind-driven X-ray binaries (*crosses*), and non-Be X-ray binaries driven by Roche lobe overflow (*triangles*). From J. van Paradijs, 1988, in *Timing Neutron Stars*, ed. H. Ögelman and E. v. d. Heuvel (Kluwer, Dordrecht), p. 191; reprinted by permission of Kluwer Academic Publishers. Based on R. H. D. Corbet, 1984, *Astron. Ap.*, 141, 91.

i) *Type II or classical LMXBs*. Here the companion is a low-mass, low-luminosity main-sequence or dwarf star, typically (when seen) a red dwarf, of mass $M < 1\ M_\odot$. This class includes the X-ray burster sources, many of the bulge and transient sources, and at least one AXP, 4U 1626 − 67. Except for the last object, the X-ray spectrum is soft and indicative of a neutron star accreting over the whole surface, i.e. an unmagnetized or very low field neutron star. The orbital periods are typically $P_0 \sim$ hr, and because of the low mass eclipses are unlikely. The typical ages of the systems are $t \gtrsim 5 \times 10^9$ yr, and there are no clear indications so far for runaway system velocities, although this may be an observational problem due to the low luminosities.

ii) *Type I LMXBs*. In this class the companion is either a moderate-age main-sequence star or an evolved companion, which is typically detectable. Despite their type I name, these are not population I objects. The system ages are somewhat younger than the classical type II LMXB, $t \lesssim$ few $10^9$ yr, and the galactic distribution is characterized by significant runaway velocities. The orbital periods are $P_0 \sim$ 0.2–10 days, and eclipses are seen in several cases. In one case the companion is a $\sim 2\ M_\odot$ A$V$ (main-sequence) star and the companion is an AXP (Her X-1). In the other examples of this class, the companion is an evolved lower-mass star ($\lesssim 1\ M_\odot$). This class includes a number of bright QPO sources (e.g Cyg X-2, GX 1 + 4, Sco X-1). The magnetic field of the neutron star is evidently either high, in the AXP sources, or medium to low (but still important), in the QPO sources. A list of the characteristic properties of some LMXB sources whose orbital periods have been measured is shown in Table 2 (from van der Klis, 1989, van den Heuvel, 1988, and van Paradijs, 1983) near the end of this book. (For more details see Lewin, van Paradijs, and van der Klis, 1988, and van der Klis, 1989; see also Stella, White, and Priedhorsky, 1987, and Parmar *et al.*, 1991.) A list of general RPP properties is given in Manchester and Taylor (1977, 1981). Table 3 lists some current information concerning binary, millisecond, and globular cluster pulsars, after Backer and Kulkarni (1990) and Backer (1990). These lists are illustrative, more complete information being available in specific reviews or the references above.

*c) Mass Exchange Effects on Binary Evolution*

We consider here only the evolution of binaries which lead to or involve a neutron star when this involves either the direct core collapse of an initially massive star (a high-mass binary, or HMB) or the accretion-induced collapse of a WD in an old low-mass binary (LMB). When the mass exchange with a neutron star leads to significant X-ray emission, these binaries are referred to as HMXBs and LMXBs, although the "X" may apply only during part of the binary evolution. In general, the mass loss from the system which occurs in a supernova explosion is expected to lead to unbinding of the binary only if more than half the total mass of the system is ejected (Blaauw, 1961). In the case of the HMXB, disruption usually does not occur because the initially more massive star has become the less massive one (through exchange with its companion) by the time it collapses. In the case of the LMXB, the accretion-induced collapse of a WD, typically $\lesssim 0.1\ M_\odot$, is lost from the system (van den Heuvel, 1988) so that in a large fraction of cases the low-mass binary system can also avoid disruption.

Mass exchange in binary systems occurs either by capture of mass from the wind of the donor star or by capture of the mass overflow from a donor which fills its Roche lobe. The Roche lobe surfaces are the limiting gravitational equipotential surfaces of a binary system, in the form of two tear-shaped surfaces enclosing each star which meet at a point intermediate between the two stars (the interior Lagrangian point). Matter which is on this surface can move freely, leading to the possibility of mass exchange. An expression for the typical Roche lobe size is given by (e.g. Eggleton, 1983)

$$\frac{r_L}{a} = \frac{0.48\,q^{2/3}}{\left[0.6q^{2/3} + \ln\left(1 + q^{1/3}\right)\right]}, \tag{11.1.1}$$

where $q = M_*/M_c$ is the stellar mass ratio, $M_c$ being the companion and $a$ the binary separation. (For calculations of the radius of stars as a function of time at different stages of their nuclear evolution, see e.g. Trimble, 1982, Iben and Tutukov, 1984, and van den Heuvel, 1983, 1988.) Mass exchange by wind accretion is usually negligible for unevolved (main-sequence) stars, whose winds are rather weak, but it can be very significant in massive early-type stars or in the later stages of the expansion toward the giant region. In the case of Roche lobe overflow mass transfer, the exchange is referred to as type A, B, or C, depending on whether the mass donor fills its Roche lobe before having finished core hydrogen burning, after core hydrogen burning but before He burning (during its expansion), or after the second expansion following core He burning (Kippenhahn and Weigart, 1967).

The evolution of binary systems is characterized by three time scales: the dynamic time needed for a system to adjust into hydrostatic equilibrium, which is very short; the Kelvin-Helmholz or thermal time scale $t_{th} \propto E_{th}/L$, which is

$$t_{th} \propto 3 \times 10^7 \left(M/M_\odot\right)^{-2} \text{ yr}; \tag{11.1.2}$$

and the nuclear time scale, which for H-core burning at $T \sim 10^7$ K is

$$t_{nuc} \simeq \text{const.}\left(M/L\right) \simeq 10^{10}\left(M/M_\odot\right)^{-2.5} \text{ yr}, \tag{11.1.3}$$

where $M$ is the mass of the star. The hydrogen burning leads to a He core, which depending on the initial chemical composition leads to a He-core mass of $M_{He} \simeq 0.073\ (M_1/M_\odot)^{1.42}$, where $M_1$ is the original progenitor stellar mass (for $X = 0.70$, $Z = 0.03$: van Beveren, 1980).

### *d) Conservative and Nonconservative Binary Evolution*

The evolution of a binary system may occur between two broad limits: the conservative case, where mass and angular momentum are reapportioned but the total value remains constant within the system, and the nonconservative case, where mass and angular momentum are lost from the system.

i) *Conservative evolution*. This typically is a good approximation for massive ($M_1 \gtrsim 10\ M_\odot$) close binaries with mass ratios $q = M_2/M_1 \gtrsim 0.4$, where the more massive primary donor $M_1$ has a radiative envelope. In conservative systems, mass loss by the more massive star ($M_1$) leads to a decrease in the separation $a$ or the orbital period $P$ (since $P^2 \propto a^3$), while mass loss from the lighter star $M_2$ leads to an increase of $a$ and $P$. For circular orbits,

$$\frac{a}{a_0} = \left[\frac{(1+q^2)q_0}{(1+q_0^2)q}\right], \quad \frac{P}{P_0} = \left(\frac{a}{a_0}\right)^{3/2} \tag{11.1.4}$$

(van den Heuvel, 1983). A massive star smaller than its Roche lobe will slowly increase its size $R_1$ on a time scale $t_{\rm nuc}$, and as it fills $r_L$ the mass loss will disturb its dynamic and thermal equilibrium. The thermal equilibrium radius of the star $r_{\rm th}$ is more or less constant as the mass of the primary drops ($q = M_2/M_1$ increases), while the primary Roche lobe radius given by equation (11.1.1) initially decreases, reaches a minimum near $q \simeq 1.176$, and then starts growing again (van den Heuvel, 1988; Paczyński, 1970). During the time that $r_L < r_{\rm th}$, the initially more massive star $M_1$ is out of thermal equilibrium and loses mass (evolves) on a time scale $t_{\rm th}$. After the mass fraction has reversed, there comes a point where $r_{\rm th}$ drops below $r_L$ and the star continues evolving on a time scale $t_{\rm nuc}$. At this point, if the star has lost most of its envelope, it is well inside its Roche lobe and consists of an evolved (He and heavier) core, whereas if a remnant H envelope is still present and is in radiative equilibrium, the star fills the Roche lobe and through its slow expansion on $t_{\rm nuc}$ it continues to remain in contact with $r_L$, which expands with the star.

ii) *Nonconservative evolution*. This occurs either when the star has a radiative envelope but a small mass fraction $a = M_2/M_1 \lesssim 0.4$, or when the star has a convective envelope (in which case it is unstable to mass loss, on a semidynamic time scale $t_{\rm sd}$ intermediate between the dynamic and the thermal). In either of these cases, $M_2$ is unable to accept all of the mass and angular momentum shed from $M_1$, and some or most of this is lost from the system. The response of the intended recipient $M_2$ to the offering which it is unable to accommodate is to expand and fill its own Roche lobe $r_{L2}$, leading to a common envelope (CE) around both stars which is gradually lost from the system. In the limit where all the orbital energy change is used to expel the envelope, the binary parameters change according to (Tutukov and Yungelson, 1979; Iben and Tutukov, 1984; van den Heuvel, 1988)

$$\frac{G(M_{2f}+M_{2e})M_{2e}}{\lambda a_1 r_L} = \frac{GM_1 M_{2f}}{2a_2} - \frac{G(M_{2f}+M_{2e})M_1}{2a_1},$$

$$\frac{a_2}{a_1} = \frac{M_1 M_{2f}}{M_{2f}+M_{2e}} \frac{1}{[M_1 + (2M_{2e}/\lambda r_L)]}, \qquad (11.1.5)$$

where $r_L$ is dimensionless, $R_L a_1$ is the Roche lobe of $M_1$, $\lambda \leq 1$ is a stellar form parameter ($\lambda \sim 0.5$ for centrally condensed stars), $M_1$ is the recipient, $M_2$ is the donor, $M_{2f}$ is the donor final core mass, and $M_{2e}$ is the donor envelope mass. In the CE stage, the less evolved star moves through the envelope of the donor and frictional drag causes the orbit to shrink, the period to shorten, and the energy deposited by this frictional heating to blow the envelope away, which can lead to a very close low-mass binary, e.g. a cataclysmic variable (a WD plus an evolved core companion star) or a neutron star plus an evolved core companion star.

*e) Mass Transfer Mode, X-Ray Lifetimes, and Orbit Shrinking Processes*

In HMXBs, the mass donor is the massive star and therefore the orbit shrinks as the exchange progresses, as long as the mass fraction does not become inverted. In general, this leads to a gradual increase in the mass exchange rate, exchange becoming easier as the distance between the stars shrinks. In HMXBs, a steady long-term mass exchange via atmospheric Roche lobe overflow occurs typically when $M_1$ is still in the H-core burning (main-sequence) stage at the time it fills its Roche lobe (i.e. case A mass exchange). The systems satisfying this condition have typically $P_0 \lesssim 4$–$5$ days, and the mass transfer rate is approximately $\dot{M} \sim 10^{-9}$–$10^{-8}$ $M_\odot$/yr during $t \sim 10^5$ yr at a quasi-steady rate, which is in the right range to explain the steady X-ray emission of HMXBs. Another mechanism for steady transfer in HMXBs is wind accretion, for which a relatively strong wind $\dot{M} \gtrsim 10^{-6}$ $M_\odot$/yr is needed (since not all the mass is captured by the compact secondary). For stars $M_1 \lesssim 30$ $M_\odot$ the wind is strong only after termination of the H-core burning, in the blue supergiant stage, where for orbital periods $P \gtrsim 5$ days such a wind can be sustained for about $t \lesssim 2 \times 10^4$ yr (van den Heuvel, 1983). Some examples of such sources are 4U 0900 − 40 and GX 301-2. For larger masses $M \gtrsim 30$

$M_\odot$, the wind is already strong during the late H-core burning, but a significant amount of X-ray emission occurs only if the compact star can accrete a significant fraction of this, which implies that the neutron star should be in the low-velocity portion of the wind, near the optical star (since for wind accretion $\dot{M}_w \propto V_w^{-4}$; see Chap. 7.2). This requires a relatively short period system, $P_0 \lesssim 4$–$5$ days, and the exchange can proceed for about $t \sim$ few $10^6$ yr. Examples of such HMXB systems are 4U 1700 − 37 (Vela X-1) and 4U 1538 − 52. The general conclusion (van den Heuvel, 1983) is that short orbital period HMXBs ($P_0 \lesssim 4$–$5$ days) have an X-ray lifetime $t \sim 10^5$–$10^6$ yr, while longer orbital period HMXBs ($P_0 \gtrsim 5$ days) remain X-ray active for a shorter time $t \sim 10^4$ yr. As a result, one expects observationally to detect more shorter-period HMXB systems than longer-period ones.

In LMXBs, the winds are generally very weak and for appreciable X-ray emission a Roche lobe overflow is needed. In general, the companion donor star has the lower mass, $M_2 < M_{NS} \simeq 1.4\ M_\odot$, so that in systems with wide orbits where the transfer is driven by nuclear evolution the orbit separation increases with continued mass exchange, and the transfer is inherently stable (an exception is Her X-1, $M_2 \simeq 2\ M_\odot$). The duration of the steady mass exchange by atmospheric Roche lobe overflow provides a persistent X-ray emission for $t \sim$ few $10^6$ yr before the luminosities reach the Eddington limit, at which point smothering of the X-ray emission sets in (van den Heuvel, 1988).

Several mechanisms can drive the mass transfer. The most obvious one is the nuclear evolution of the donor star, which in the case of wind transfer determines when and how long the star will produce a wind of the appropriate strength for significant accretion to occur. In the case of atmospheric Roche lobe overflow, it determines how long the star will gradually (on $t_{nuc}$) lose its matter before it expands to become a giant and starts overflowing on the shorter $t_{th}$, dumping so much matter that the X-ray emission is smothered. The most frequent mechanism driving Roche lobe overflow is the shrinking of the orbit, which in some situations occurs faster than the star can grow. The mass transfer itself can do this (if $M_2 > M_1$) on a slow time scale, but there are other mechanisms for orbit shrinkage which can operate on shorter time scales, the most common being angular momentum losses. For instance, tidal forces can regulate the orbital evolution of some close HMXB systems. One noticeable effect of this is the circularization and synchronization of the orbit, which can be achieved typically in $t \sim$ few $10^6$ yr after the SN event that gave rise to the neutron star. Systems such as Cen X-3, SMC X-1, and Her X-1 (the last is a LMXB) have circular orbits, which supports the view that their ages are $t \gtrsim$ few $10^6$ yr (van den Heuvel, 1983). After circularization, tidal instability can set in if the ratio of orbital to spin angular momentum $J_0/J_s \leq 3$ (Savonije and Papaloizu, 1983). As a result of this instability, the neutron star spirals in on a time scale $t \sim 5 \times 10^5$ yr. Some objects which are close to satisfying this condition are LMC X-4, 4U 1700 − 37, and Cen X-3, which have very short orbital periods (Table 1). In fact, Cen X-3 has been observed to decrease its $P_0$ with a characteristic time scale $t \sim 4 \times 10^5$ yr (Kelley, 1981; van der Klis, 1983). The conclusion is that $\sim 5 \times 10^5$ yr ago these systems had periods $P_0 \gtrsim 3$–4 days.

Other mechanisms responsible for driving the mass transfer in LMXBs via the shrinking of the orbit by angular momentum losses include gravitational radiation, magnetic braking, and dynamic mass transfer instabilities (e.g. Rappaport, Verbunt, and Joss, 1983; Ruderman and Shaham, 1983; Verbunt and Rappaport, 1989). The gravitational radiation time scale is $t_{gr} \sim 2 \times 10^7[(M_1 + M_2)^{1/3}/M_1 M_2](P_0/1.6 \text{ hr})^{8/3}$ yr, important only for $P_0 \lesssim 10$ hr. The magnetic braking process is not fully understood, estimates for its time scale being $t_{mb} \sim 10^7$–$10^9$ yr (e.g. Savonije, 1983), and this is expected to operate on systems up to $P_0 \lesssim$ few days for evolved stars close to filling their Roche lobe. Nuclear evolution is the only mechanism for driving mass transfer that operates for $P_0 \gtrsim 0.5$ days, for both unevolved and evolved stars (Taam, 1983; Webbink, Rappaport, and Savonije, 1983). The dynamic mass transfer instability operates on a typical time scale $t \sim 2.3(H/R_s)t_{ev}$ for $(M_2/t_{ev}) \gg 10^{-9}$ $M_\odot$/yr, where $t_{ev}$ is the steady evolution time scale prior to onset of the instability, $H$ is the atmospheric scale height, and $R_s$ is a characteristic radius in the atmosphere at which the mass loss is occurring (Verbunt and Rappaport, 1989). More recently, motivated by observations of millisecond pulsars and the discovery of a binary pulsar that is ablating its small-mass companion (PSR 1957 + 20), as well as by the discrepancy between the inferred birthrates of these systems and that of their presumed LMXB progenitors, the role of high-energy radiation and particle winds has been investigated as the mechanism for self-exciting and accelerating the mass transfer from the companion of a pulsar (e.g. Ruderman *et al.*, 1989).

### *f) HMXB Evolutionary Scenarios*

For $M_1 \gtrsim 10$ $M_\odot$, the core mass is $M_{He} \gtrsim 2.8$–3 $M_\odot$ and the explosion occurs on $M_1$ when it has become the less massive of the two, so

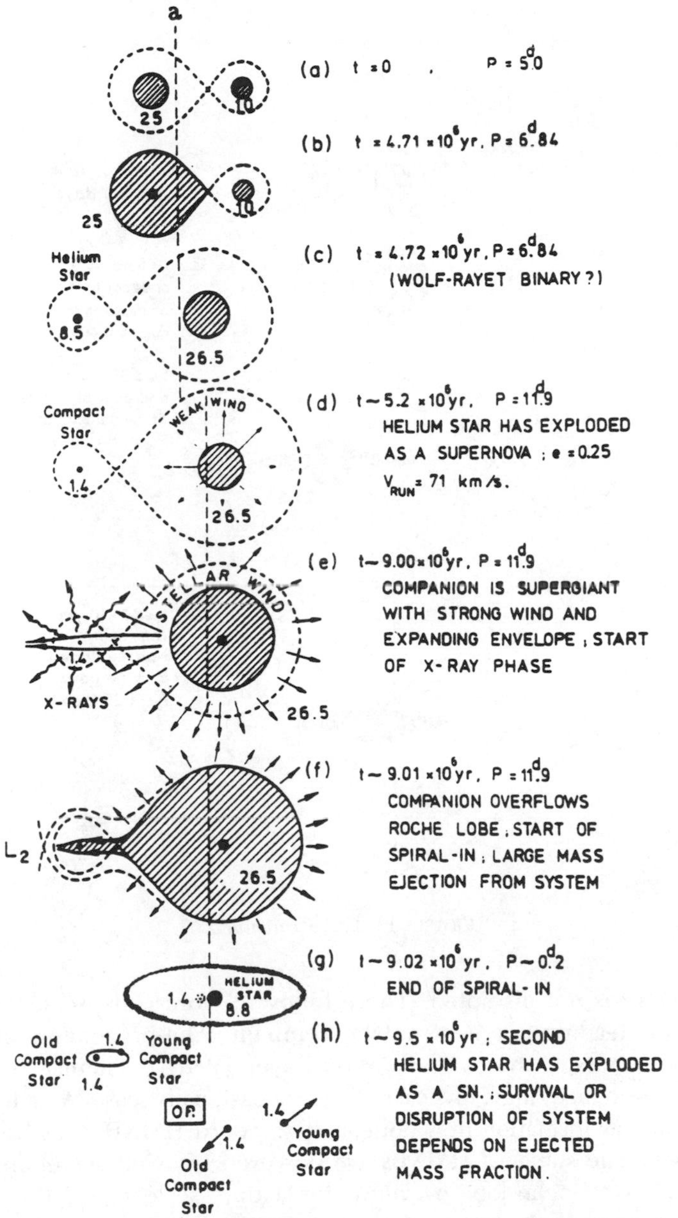


FIGURE 11.1.2 A particular evolutionary scenario of (a) a standard HMXB system and (b) (*next page*) a Be HMXB. Each state is labeled with the age of the system in years and the orbital period in days. From E. P. J. van den Heuvel, 1983, in *Accretion Driven Stellar X-Ray Sources*, ed. W. Lewin and E. v. d. Heuvel (Cambridge University Press), p. 303.

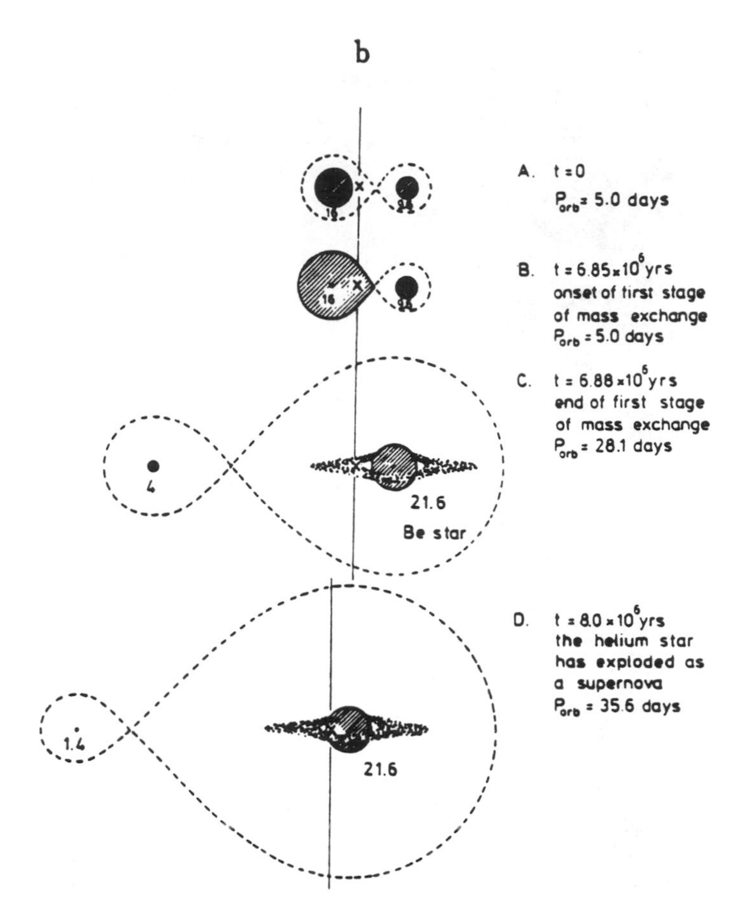


FIGURE 11.1.2 *Continued.*

the binary is not disrupted. There follows a core collapse supernova, generally leading to a neutron star remnant. The progenitors of most of these massive systems are Wolf-Rayet (WR) or similar binaries, which exchange mass by case B or occasionally case A, and after neutron star formation they appear as standard HMXB or as Be X-ray binaries. The standard HMXBs are powered by wind accretion or by atmospheric Roche lobe overflow, for a time scale $t_x \sim 10^4$–$10^5$ yr. The Be sources are powered by transient equatorial mass loss, typically for a time scale $t_x \sim 10^7$ yr. For a mass ratio $q = M_2/M_1 \gtrsim 0.4$ the initial evolution is conservative. Two possible conservative scenarios leading to these types of HMXBs are shown in Figure 11.1.2. The left-hand system is a typical standard HMXB with persistent X-ray emission, while the right-hand system is a Be X-ray binary with transient episodes of X-ray emission. These systems will evolve to a stage at which they no longer satisfy the conditions for conservative evolution, and a CE phase will follow. For instance, in Cen X-3, using the Tutukov-Yungelson equations (11.1.5) with $M_1 = 1.4\ M_{\odot}$, $M_{2e} = 12\ M_{\odot}$, $M_{2f} = 4\ M_{\odot}$, $a_1 = 17.8\ R_{\odot}$, one has $R_L = 0.6$ and $a_2/a_1 = 0.0043$, so $a_2 = 0.077\ R_{\odot}$ well inside the core of the He star $M_2$. Thus, one expects Cen X-3 to be in the CE phase, and the compact star will spiral in toward the center (Taam *et al.*, 1978). In general, the standard HMBX ($P_0 \gtrsim 40$ days) will spiral in; in order to avoid this, periods in excess of 100 days are needed. Two examples of the CE final evolution of a close and a wide (Be) HMXB are shown in Figure 11.1.3. In the left-hand case, after a spiral-in phase the secondary star is lost in a symmetrical way and a single, old neutron star results, with a low peculiar velocity. In the right-hand case, after an initial spiral-in phase the evolution leads either to a second spiral-in phase and a neutron star plus WD system (e.g. PSR 0655 + 64), to a second SN explosion and a system of a young plus an old neutron star (e.g. PSR 1913 + 16), or to a disrupted system of an old and a young neutron star with high peculiar velocities (e.g. a single millisecond pulsar, e.g. PSR 1937 + 21: Bonsema and van den Heuvel, 1985).

### g) LMXB Evolutionary Scenarios

Among the galactic bulge sources there are about 8 with $L_x \sim 10^{38}$ ergs s$^{-1}$, requiring $\dot{M} \sim 10^{-8}\ M_{\odot}$/yr, and 20–30 with $L_x \sim 10^{36}$–$10^{37}$ ergs s$^{-1}$, requiring a correspondingly lower $\dot{M}$. A study of the characteristic accretion rate and luminosity history of LMXBs evolving under the action of gravitational radiation plus magnetic braking (Rappaport, Verbunt and Joss, 1983) would predict a ratio of high- to low-luminosity sources $\sim 1/50$, which indicates that the evolution of these systems must be controlled by other processes. Calculations of the accretion rate and orbit evolution under the action of nuclear evolution (Webbink, Rappaport, and Savonije, 1983; Joss and Rappaport, 1983), on the other hand, can give $\dot{M} \sim 10^{-8}\ M_{\odot}$/ yr fairly steadily over the whole lifetime, which is compatible with the observed statistics of high- and low-luminosity objects. This implies that most of the LMXB companions must be subgiants, filling their Roche lobe.

In general, the mass and orbital parameters of many binary pulsars (Table 3) are somewhat similar to the final parameters resulting from

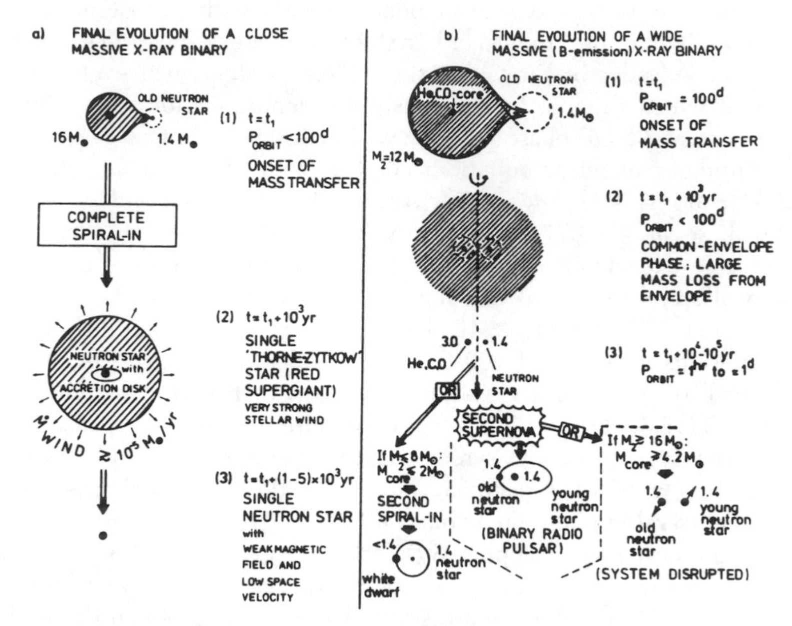


FIGURE 11.1.3 Final common envelope scenarios for HMXBs. (a) Short-period systems ($P_0 < 100$ days), leaving a single neutron star. (b) Long-period systems ($P_0 > 100$ days), which, depending on the mass of the secondary, leave a binary with a neutron star and the core of the companion, two neutron stars, or a disrupted system of two neutron stars. From E. P. J. van den Heuvel, 1988, in *Timing Neutron Stars*, ed. H. Ögelman and E. v. d. Heuvel (NATO ASI vol. 262, Kluwer, Dordrecht), p. 523. Reprinted by permission of Kluwer Academic Publishers.

LMXBs with evolved companions (e.g. neutron star plus He WD, $P_0 \sim 10$–$20$ days), especially under the assumption of nuclear evolution. However, the presence of some pulsars with $B \gtrsim 10^{11}$ G (PSR 0820 + 02) indicates that (unless field growth is allowed or the decay is not simply exponential) the neutron star must be young, $t_{NS} \lesssim 10^7$ yr (this being the typical field decay time scale: see § 11.4*b* and Chap. 8.1*c*) whereas the companion He star must have at least $t_{He} \gtrsim 5 \times 10^9$ yr (van den Heuvel, 1988). Such evidence argues for the presence of some relatively young neutron stars in old systems. The same need exists in Type I LMXB systems such as Her X-1. Since the companion is of low mass and would be disrupted if a SN explosion occurred, the only alternative seems to be the accretion induced collapse of a WD, which ejects $\lesssim 0.2\ M_\odot$. Thus, PSR 0820 + 02 could have formed from a WD accreting $10^{-8}\ M_\odot$/yr over $10^7$ yr. However, the evidence for magnetic field decay refers, strictly speaking, to a decay of the magnetic torque (e.g. Chap. 8.1*c*), and the data could in principle be compatible with a constant field, a more slowly decaying field, or a two-component model with a crustal field that decays in $t \sim 10^7$ yr and a core field $B \sim 5 \times 10^8$ G which decays on a much longer time scale (see § 11.4*b*). This would ease constraints on the youthfulness of the neutron star.

Type I LMXB systems and young runaway pulsars may start out initially as wide ($P_0 \gtrsim 100$ days), high-mass binary systems with mass ratio $q \lesssim 0.3$, which would evolve through a spiral-in CE phase. This would lead to a He star with $M \gtrsim 3\ M_\odot$ and a lower-mass unevolved companion; the He star would core collapse leading to a neutron star in an eccentric orbit about the unevolved companion with a significant peculiar velocity. The eccentric orbit would circularize by gravitational radiation or magnetic braking in a few million years, leading to a type I LMXB system such as Sco X-1 or Cyg X-2. Alternatively, and with higher probability, the core collapse SN explosion could lead to disruption of the binary, giving a high peculiar velocity runaway pulsar (van den Heuvel, 1983). In the case of the AXP Her X-1, the companion is a post-main-sequence star of 2 $M_\odot$, and the transfer must be occurring via atmospheric Roche lobe overflow from an evolved companion, providing $\sim 10^{-9}\ M_\odot$/yr for $10^4$–$10^5$ yr (afterward, the exchange would start occurring on a thermal time scale and the X-ray luminosity would be smothered). The neutron star must be young, since it pulsates and does so at $L < L_{\rm Ed}$. A likely scenario (Taam and van den Heuvel, 1986) is that this was a cataclysmic variable (CV) type of binary consisting of a low-mass star plus a WD, the latter having undergone accretion-induced collapse to a neutron star. The ejection of about 0.3 $M_\odot$ and a slight asymmetry of the SN event would impart to the system a runaway velocity $\gtrsim 125$ km s$^{-1}$. The AXP GX 1 + 4 is a LMXB with a red subgiant companion of $t \gtrsim 5 \times 10^9$ yr but requiring a high magnetic field $B \gtrsim 10^{11}$–$10^{12}$ G to pulsate, i.e. $t_{\rm NS} \lesssim 10^7$ yr. A scenario here would involve a long-period ($\gtrsim 10$ day) binary with a low-mass subgiant plus a WD, accreting at $\gtrsim 10^{-8}\ M_\odot$/yr over $10^6$–$10^8$ yr leading to a low-mass subgiant plus neutron star system.

The type II LMXBs include the globular cluster sources and the galactic bulge sources. The LMXB strong sources show a much higher incidence in globular clusters ($10^2$–$10^3$ times more frequent) than in the rest of the galaxy. This has led to the view that, in order to have so many more binaries, these may be formed by direct capture of a neutron star (Fabian *et al.*, 1975). Direct capture is possible since velocity dispersion in the core is only 10–20 km $s^{-1}$ and tidal dissipation can lead to capture (e.g. Verbunt, 1988). When the companion is a main-sequence star, $L_x \sim 10^{36}$ ergs $s^{-1}$; when the companion is a subgiant (class III), $L_x \sim 10^{38}$ ergs $s^{1}$. On the other hand, in the galactic bulge the velocity dispersion is too large ($\gtrsim 100$ km $s^{-1}$) for capture to occur. The progenitors of the galactic bulge sources are probably CV systems, where the neutron star formed via accretion-induced collapse of a WD.

The binary radio pulsar (RPP) systems and the single millisecond pulsars (MSPs), many of which are in globular clusters, are also believed to arise from type II LMXB systems. The very short periods in these RPPs may arise by spin-up due to the accreted material from the companion (e.g. Alpar *et al.*, 1982). The binary RPPs can be divided into two categories, high-mass binary pulsar (HMBPs) and low-mass binary pulsars (LMBPs), depending on the mass of the companion. The HMBPs generally have short orbital periods $P_0 \sim$ 0.1–10 days and eccentric orbits, while the LMBPs generally have longer periods $P_0 \sim$ 1–10 days and close to circular orbits.

i) The HMBP systems (e.g. PSR 1913 + 16 and 0655 + 64) require a substantial loss of angular momentum from the system, indicative of a common envelope phase. The progenitors are likely to be OB main-sequence binary systems which lead to core collapse supernovae (Taam and van den Heuvel, 1986). The more massive star becomes a neutron star first, and when the secondary has evolved away from the main sequence it starts to lose mass to the companion and becomes an X-ray producing LMXB. If the secondary is more massive than $\sim 8$ $M_\odot$, it also undergoes a SN explosion, which produces a second neutron star. The explosion usually sends the two neutron stars flying apart (providing a possible origin for single MSPs, like PSR 1937 + 21), but occasionally highly eccentric systems like PSR 1913 + 16 and 0655 + 64 can be formed.

ii) The LMBP systems such as PSR 1953 + 29 and PSR 0820 + 02 probably require a low-mass hydrogen-burning star to explain their large $P_0$. The original system may have consisted of a $\sim 1$ $M_\odot$ main-sequence star and a core-collapse neutron star. After $t_{nuc} \sim 5 \times 10^9$ yr the companion star moves off the main sequence and starts feeding the neutron star, which, being old, emits unpulsed X-rays (a LMXB), while the orbit gets larger and the spin period of the neutron star gets shorter, until one is left with a low-mass WD and a spun-up MSP in a wide orbit. One problem with this scenario is that the initial SN explosion that gave rise to the neutron star is likely to have disrupted the binary system. An alternative scenario involves a low-mass subgiant and a WD which underwent accretion-induced collapse. This scenario could also explain the relatively strong magnetic field of PSR 0820 + 02. In the case of the single 1.6 ms PSR 1937 + 21 or the 3.1 ms PSR 1821 − 24, this may have involved a core-collapse neutron star plus a WD of mass $\gtrsim 0.7\ M_{\odot}$, which following coalescence in $\sim 10^8$ yr underwent a spin-up at the expense of the disrupted WD (e.g. Michel, 1987; Verbunt *et al.*, 1987). The progenitor may have been similar to the wide Be systems. It could also be that a single MSP is left over after it has ablated its companion, as is seen to happen in the eclipsing binary PSR 1957 + 20 (Fruchter *et al.*, 1988).

A major problem with the general picture of binary pulsar formation in globular clusters is that it implies that millisecond pulsars have gone through an LMXB stage, but on the basis of their estimated numbers and active lifetimes, the birthrate of MSPs appears to be at least a factor $\sim 10^1$–$10^2$ larger than the birthrate of LMXBs (Bhattacharya and Srinivasan, 1986; Kulkarni and Narayan, 1988; Coté and Pylyser, 1989; Romani, 1990a). A possible solution of this would be if there are a large number of systems where the mass transfer does not lead to X-ray emission or if the X-ray emitting stage of LMXBs is shortened significantly, e.g. by accelerated evolution due to X-ray illumination (e.g. Ruderman *et al.*, 1989; Ray and Kluzniak, 1990). Uncertainties also arise concerning the neutron star formation mechanism itself. The AIC process is in principle reasonable, and a number of theoretical investigations indicate the need for it or otherwise rely on it (e.g. Bailyn and Grindlay, 1990; Ruderman, 1991), while some direct evolutionary calculations show that this process may be more likely than previously thought (Canal, Isern, and Labay, 1990; Nomoto and Kondo, 1991). Further detailed calculations may be needed in order to pin down quantitatively the efficiency of this process.

The nonpulsing LMXBs with a very low magnetic field (X-ray bursters and soft transient sources) do not impose the constraint of requiring a young neutron star under the simple field decay assumption, so their scenarios may involve either core collapse or accretion-induced collapse. In several XRB sources a low-mass companion has been detected and the mass transfer is thought to occur by atmospheric Roche lobe overflow. The low magnetic field is inferred from the lack of pulsations and the fact that the spectrum corresponds to a blackbody of 2–3 keV, which for accretion rates near the Eddington limit indicates accretion over the whole surface of a neutron star. The X-ray bursts are generally of type I, which are thermonuclear bursts with an energy release of $\sim 1$ MeV/nucleon, or occasionally type II bursts, which are accretion events releasing $\sim 100$ MeV/nucleon. The soft transient sources are powered by nonsteady accretion from a low-mass companion, and the lack of a strong field is deduced by the fact that, in at least two sources, Aql X-1 and Cen X-4, type I bursts were detected during the initial part of the outbursts, indicating accretion over the whole surface.

Some nonpulsating LMXB sources appear to require the presence of a small but nonnegligible magnetic field. This is the case with the rapid burster MXB 1730 − 335, which shows both type I and type II bursts. To explain the relationship between the luminosity of a burst and the preceding dead time, a reservoir mechanism is needed, which can be provided by a magnetic field $B \sim 10^9$–$10^{11}$ G (Lamb and Lamb, 1977). This view is strengthened by the fact that some transient AXPs such as GX 304-1, LMC X-4, and GX 301-2 show occasional type I bursts (Lewin, 1980; Hoffman *et al.*, 1978). A scenario for MXB 1730 − 335 (Taam and van den Heuvel, 1986) requires a relatively young neutron star in a binary in a globular cluster, which could be provided by a WD tidally captured by a low-mass star which undergoes accretion-induced collapse.

The QPO sources also require a small but nonnegligible field if one accepts the beat-frequency model (Lamb, 1988). An extensive review of QPO sources and properties is given by van der Klis (1989). In sources such as GX 5-1, Cyg X-2, and Sco X-1, which are bright galactic bulge sources, the beat-frequency model predicts spin rotation periods of $\sim 5$–$15$ ms and magnetic fields $B \sim 10^9$–$10^{10}$ G, while the orbital periods are $\sim 1$–$10$ days. The companion must be an evolved low-mass star, and the scenarios envisaged (e.g. Taam and van den Heuvel, 1986) could be similar to those of GX 1 + 4 and PSR 1953 + 29, namely, a LM subgiant plus a WD ($P_0 \sim 10$–$12$ days), where mass loss $\dot{M} \sim 10^{-8}$ $M_\odot$/yr for $10^7$–$10^8$ yr leads to the accretion-induced collapse of the WD.

## 11.2 Thermal Evolution of Neutron Stars

### *a) Cooling Mechanisms*

Neutron stars formed in core-collapse supernovae must reach internal temperatures in excess of $T_i \sim 10^{11}$ K (Woosley and Weaver, 1986).

The accretion-induced collapse of white dwarfs would lead to neutron stars of comparable or somewhat lower temperature (Nomoto, 1986). At these very high internal temperatures, the neutron star interior is mostly subject to neutrino energy losses, which are able to cool the star down to $10^9$–$10^{10}$ K within about a day. Subsequently, the star continues to cool, mainly by neutrino losses (Baym and Pethick, 1979; Shapiro and Teukolsky, 1983), until temperatures $T_i \sim 10^8$ K are reached, when photon cooling takes over.

The main neutrino cooling mechanism in the interior is the modified Urca process (Chiu and Salpeter, 1964),

$$n + n \to n + p + e^- + \bar{\nu}_e, \quad n + p + e^- \to n + n + \nu_e, \tag{11.2.1}$$

which is more efficient than the usual Urca reaction $n \to p + e^- + \bar{\nu}_e$, $e^- + p \to n + \nu_e$ in degenerate matter, where it is difficult to balance momentum in the reaction. In the modified process, this is achieved with a bystander particle which is used to absorb momentum. The neutrino luminosity from this process (Friman and Maxwell, 1979) is

$$L_\nu^{\text{Urca}} \sim 6 \times 10^{39} (M/M_\odot)(\rho/\rho_N)^{-1/3} T_9^8 \text{ ergs s}^{-1}, \tag{11.2.2}$$

where $\rho_N = 2.4 \times 10^{14}$ g cm$^{-3}$. Another possible process is nucleon pair bremsstrahlung, $n + n \to n + n + \nu + \bar{\nu}$, $n + p \to n + p + \nu + \bar{\nu}$, whose rates, however, are a factor $\sim 30$ below those of the modified Urca process (Friman and Maxwell, 1979). In the presence of superfluidity of the nucleons, the modified Urca and nucleon pair bremsstrahlung rates must be multiplied by a factor $\sim \exp(-\Delta/kT)$, where $\Delta$ is the superfluid gap energy, representing the reduction in the number of thermal excitations. In this situation, the process of neutrino pair bremsstrahlung in the crust becomes important,

$$e^- + (Z, A) \to e^- + (Z, A) + \nu + \bar{\nu}, \tag{11.2.3}$$

whose rate (Maxwell, 1979) is given by

$$L_\nu^{\text{brem}} \sim 5 \times 10^{39} (M_{\text{cr}}/M_\odot) T_9^6 \text{ ergs s}^{-1}, \tag{11.2.4}$$

where $M_{\text{cr}}$ is the mass of the crust.

In the presence of a pion condensate, the cooling can be significantly enhanced because the analogue of the unmodified Urca process can now occur, with conservation of energy and momentum. The excess momentum is taken up by processes in the condensed pion field (Bahcall and Wolf, 1965). The process can be represented as a quasi-particle $\beta$-decay,

$$N \to N' + e^- + \bar{\nu}_e, \tag{11.2.5}$$

where the quasi-particles $N$, $N'$ are linear combinations of neutron and proton states in the pion sea. The resulting luminosity is (Maxwell *et al.*, 1977)

$$L_\nu^\pi \sim 1.5 \times 10^{46}\theta^2 (M/M_\odot)(\rho/\rho_N)^{-1} T_9^6 \text{ ergs s}^{-1}, \tag{11.2.6}$$

where $\theta \sim 0.3$ is an angle that measures the degree of pion condensation. This rate is close to that obtained by Bahcall and Wolf (1965) for the cooling by decay of free pions, $\pi^- + n \rightarrow n + e^- + \bar{\nu}_e$, $\pi^- + n \rightarrow n + \mu^- + \bar{\nu}_\mu$ and their inverses. Its numerical value is significantly larger than that for the modified Urca process (eq. 11.2.2) and will be dominant at all temperatures of interest if a pion condensate exists.

If quark matter exists in the core of neutron stars, there could be large neutrino losses from the $\beta$-decay of degenerate relativistic quarks (Iwamoto, 1980; Burrows, 1980). The $u$ and $d$ quarks are lighter and are expected to be relativistic, while the heavier $s$ quark may be nonrelativistic. The relativistic rates, which will be more important, are in the simplest case of $\beta$-decay

$$d \rightarrow u + e^- + n\bar{u}_e, \quad u + e^- \rightarrow d + \nu_e. \tag{11.2.7}$$

The corresponding luminosity (Shapiro and Teukolsky, 1983) is

$$L_\nu^{\text{quark}} \sim 1.3 \times 10^{44} \, (M/M_\odot) T_9^6 \text{ ergs s}^{-1}. \tag{11.2.8}$$

The presence of $s$ quarks, which lead to reactions of the type $s \rightarrow u + e^- + \bar{\nu}_e$, $u + e^- \rightarrow s + \nu_e$, is not expected to alter this luminosity estimate significantly (Iwamoto, 1980). This rate is much larger than that for normal neutron star matter (eqs. 11.2.2, 11.2.4) and is comparable to cooling in the presence of a pion condensate (eq. 11.2.6).

### *b) Cooling Time Estimates*

The cooling time against a particular neutrino cooling mechanism can be estimated simply if one assumes the neutron star interior to be approximately isothermal, because of the very high thermal conductivity of the degenerate electron gas. This is not strictly correct (e.g. Nomoto and Tsuruta, 1987) but gives an order of magnitude estimate. The thermal energy of the star will reside almost entirely in the degenerate fermions, either neutrons or quarks, which will be $U \sim 6 \times 10^{47} (M/M_\odot)(\rho/\rho_N)^{-2/3} T_9^2$ ergs for neutrons. Equating $-dU/dt$ to the neutrino luminosities $L_\nu$ given previously, one obtains the following characteristic cooling times for the modified Urca and pion condensate neutrino cooling processes acting alone:

$$\Delta t_{\rm Urca} \simeq (1\ {\rm yr})(\rho/\rho_N)^{-1/3}[T_9^{-6}(f) - T_9^{-6}(i)], \tag{11.2.9}$$

$$\Delta t_\pi \simeq (20\ {\rm s})\theta^{-2}(\rho/\rho_N)^{1/3}[T_9^{-4}(f) - T_9^{-4}(i)], \tag{11.2.10}$$

where $f$ and $i$ denote the final and initial temperatures.

The photon luminosity $L_\gamma$ is additional to the neutrino luminosity and initially is negligible compared to the latter, but after the interior temperature has dropped below $\sim 10^8$ K it becomes the dominant energy loss. The crust being optically thick, this photon luminosity is simply

$$L_\gamma = 4\pi R^4 \sigma T_s^4 = 7 \times 10^{36} R_6^2 T_{s,7}^4, \tag{11.2.11}$$

where $\sigma$ is the Stefan-Boltzmann constant, $R_6$ is the stellar radius in units of $10^6$ cm, and $T_{s,7}$ is the effective surface temperature in units of $10^7$ K. The surface temperature is related to the interior temperature $T$ and can be approximated from numerical calculations through equation (11.2.16) so that the interior cooling time against photon losses can be estimated from equating $-dU/dt$ to $L_\gamma$, which gives

$$\Delta t_\gamma \simeq (2 \times 10^3\ {\rm yr})\alpha^2(M/M_\odot)^{1/3}[T_{s,7}^{-\frac{1}{2}}(f) - T_{s,7}^{-\frac{1}{2}}(i)] \tag{11.2.12}$$

in terms of the surface temperature $T_{s,7}$ with $\alpha$ defined in equation (11.2.16). The cooling curves for the interior temperature of neutron stars is schematically shown in Figure 11.2.1, assuming that each mechanism acts as if the others were not present. The previous estimates assume the interior to be perfectly isothermal. More accurate numerical integrations for nonisothermal stars were performed by Nomoto and Tsuruta (1987), who show that in the case of a stiff equation of state (e.g. the Pandharipande, Pines, and Smith, 1976, e.o.s.) it takes nearly $10^5$ yr before interior isothermality is reached, while for softer equations of state (e.g. the Friedman and Pandharipande, 1981, e.o.s.) it takes about 500–700 yr. Thus, the isothermal approximation is better at later times and for softer equations of state, but gives only approximate results for early times and stiff equations of state. The cooling time scales also assume the neutrons to be normal, whereas in a superfluid state the neutrino rates would be affected by the previously mentioned factor $\exp(-\Delta/kT)$, where $\Delta$ is the gap energy. Also, the heat capacity of the star increases

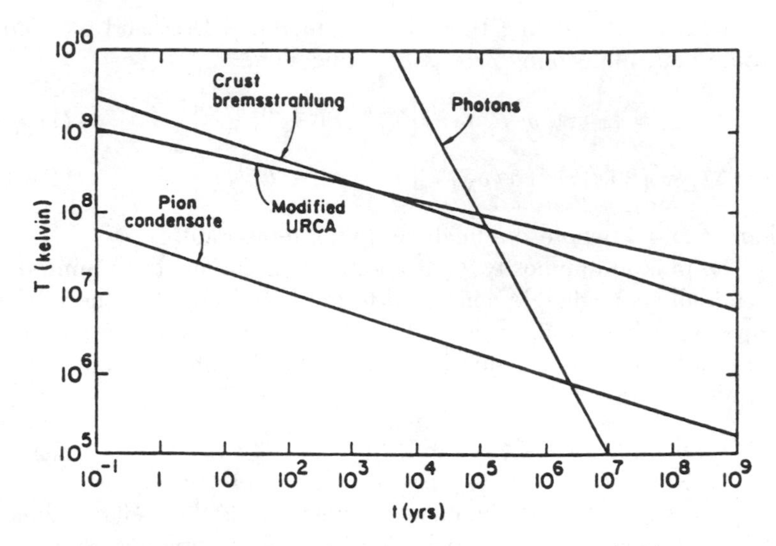


FIGURE 11.2.1 Schematic neutron star cooling curves showing interior temperature against time for various processes, computed as if each was acting alone. From Baym and Pethick (1979). Reproduced, with permission, from the *Annual Review of Astronomy and Astrophysics*, vol. 17, © 1979 by Annual Reviews Inc.

discontinuously, as the temperature drops below the transition temperature, and then decreases exponentially, so the cooling time scale increases above the transition temperature and decreases below it. The photon cooling also neglects magnetic effects (see below).

### *c) Numerical Models of Isolated Neutron Star Cooling*

The detailed calculations of the cooling of single neutron stars show, as expected, that the surface temperature of the neutron star is generally lower than that of the interior by a couple of orders of magnitude. The transition from the approximately isothermal interior to the envelope, where the temperature starts to drop rapidly, occurs approximately at the depth where the pressure of the electrons becomes nondegenerate. Approximate analytic integrations using simple radiative opacities are shown by Shapiro and Teukolsky (1983), and more exhaustive numerical and analytical calculations including conduction are given by Gudmundsson, Pethick, and Epstein (1983) and Hernquist (1984b), as well as Nomoto and Tsuruta (1987). The calculations solve the heat transport equation

$$F = \frac{16}{3}\frac{\sigma T^3}{\bar{\kappa}\rho}\frac{dT}{dz} = \frac{4}{3}\sigma\frac{dT^4}{d\tau}, \tag{11.2.13}$$

assuming the heat flux $F \equiv \sigma T_s^4$ to be constant through the envelope. The solution of equation (11.2.13) leads to a power law dependence on the total optical depth $\tau$,

$$\frac{T}{T_s} \simeq \left(\frac{3}{4}\tau\right)^{1/4}, \tag{11.2.14}$$

where the optical depth $d\tau = \kappa\, dz$ contains both radiative and conductive contributions,

$$\kappa^{-1} = \kappa_r^{-1} + \kappa_c^{-1}. \tag{11.2.15}$$

The outer regions are dominated by the radiative opacity, first by scattering and later by bound-free and free-free, while deeper down the conduction term dominates in general when the electrons become degenerate. The conduction occurs at the lower densities by electron-ion scattering, with electron-phonon and electron-impurity scattering taking over at increasing densities. These regimes are shown in Figure 11.2.2, which also indicates the melting line where the transition from liquid to solid occurs, taken from Gudmundsson, Pethick, and Epstein (1983). Extensive tabulations of these opacities in the nonmagnetic case are contained in the Los Alamos Opacity Library (Huebner *et al.*, 1977). The envelope opacity is constant in the scattering regime and starts to increase in the free-bound/free-free regime, where the Rosseland mean is used, decreasing again deeper down as degenerate electron conduction takes over. The region where the opacity switches from being dominated by radiation to being dominated by conduction is where the opacity is largest, offering the greatest resistance to the heat flow, so it is the value of the opacity in this "sensitivity region" (Gudmundsson, Pethick, and Epstein, 1983) that determines the characteristic value of the heat flow. The numerical results show that the interior temperature $T$ and the surface effective temperature $T_s$ are related as

$$\frac{T_s}{T} \sim 10^{-2}\alpha, \quad 0.1 \lesssim \alpha \lesssim 1, \tag{11.2.16}$$

where $\alpha$ is a dimensionless factor. Simple fits to the radiative and

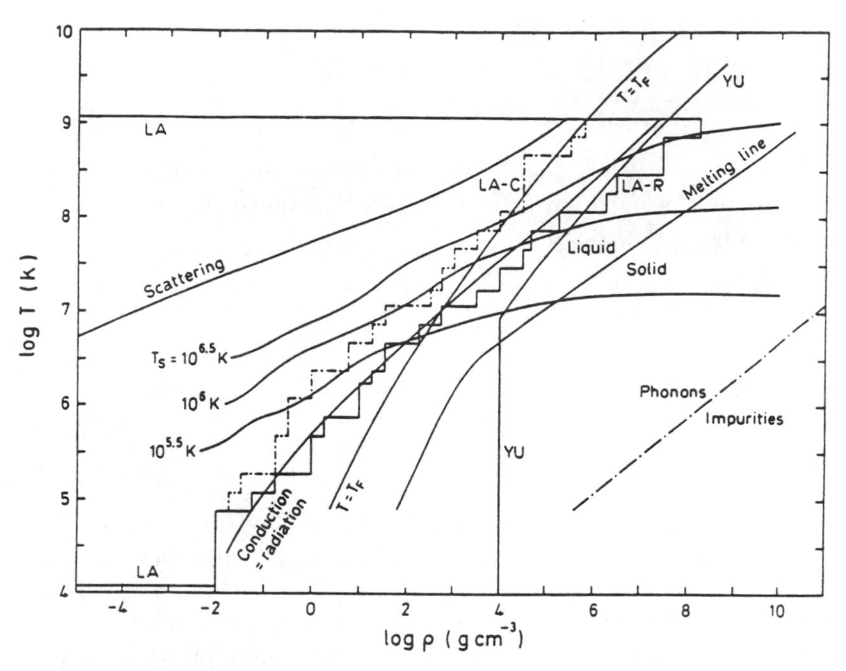


FIGURE 11.2.2 The dominant sources of opacity as a function of density and temperature, also showing the melting line and the Fermi temperature line. The LA-C, LA-R lines correspond to two regimes of Los Alamos opacity tables, while the YU curves correspond to the region of validity of the Yakovlev-Urpin opacities. Also shown are the temperature-density profiles for neutron star envelopes at three values of the surface temperature and a surface gravity of $10^{14}$ cm s$^{-2}$. From E. H. Gudmundsson, C. J. Pethick, and R. I. Epstein, 1983, *Ap. J.*, 272, 286.

conductive opacities allow an analytical integration of the heat transport equation (Ventura, 1988) leading to the analytic expression

$$\frac{T_s}{T} \simeq \left(\frac{\rho}{\rho_0}\right)^{-4/13}, \tag{11.2.17}$$

where $\rho_0 \simeq [0.1, 0.027] T_{s,6}^{5/4}$ g cm$^{-3}$ for a [$^4$He, $^{56}$Fe] envelope. Thus, for a $^{56}$Fe envelope, $T_s/T \simeq 1/113$, which occurs at an optical depth $\tau_r \simeq 2.2 \times 10^8$, where $\rho_r \simeq 1.3 \times 10^5$ g cm$^{-3}$ at the base of the radiative envelope.

The effect of a strong magnetic field is to reduce the radiative opacities below the values used in the previous expressions by a factor

$$\frac{\kappa_B}{\kappa} \simeq 41\left(\frac{kT}{\hbar\omega_c}\right)^2 \tag{11.2.18}$$

in the Rosseland mean opacity. This is because both the Thomson and free-free opacity of the extraordinary mode at $\hbar\omega \sim kT \ll \hbar\omega_c$ is down by a factor $(\omega/\omega_c)^2$ (e.g. Chap. 4). The conductive opacities are also modified (Yakovlev, 1982, 1984; Hernquist, 1984a, b) according to

$$\frac{\kappa_B}{\kappa} \sim \rho^{-2} \tag{11.2.19}$$

in the degenerate case, while in the nondegenerate case both are equal to the nonmagnetic value. When the Fermi energy surpasses the cyclotron ground energy $\hbar\omega_c$, quantum oscillations set in for the degenerate conduction, caused by the quantum nature of the Coulomb collision process (Ventura, 1973, 1988). However, because of various cancellations, these magnetic effects do not lead to a significant change in the surface temperatures calculated previously (Hernquist, 1984a), at least in an angle-averaged sense, for a one-dimensional transport.

The cooling calculations for single isolated neutron stars, which are the only ones that have been modeled with some confidence up to now, can be compared with observations of neutron stars whose ages are known (e.g. van Riper and Lamb, 1981; Gudmundsson, Pethick, and Epstein, 1983; Nomoto and Tsuruta, 1986, 1987). The age can be estimated if the gaseous remnant of the supernova envelope is observable, from dynamic considerations (e.g. its size, assuming an initial velocity or explosion energy, plus a model of the remnant expansion). The above-mentioned calculations assume that there was no additional heat input into the star after it was born. The calculations of Nomoto and Tsuruta (1986, 1987) solve the cooling numerically, without assuming isothermality, and compare the results for various neutron star equations of state to the X-ray observations of 12 isolated neutron stars for which temperature upper limits exist. This is shown in Figure 11.2.3. These calculations use standard neutrino luminosities and opacities, assuming superfluidity but neglecting pion condensation or quarks. As can be seen, within the theoretical uncertainties there is agreement with the observations for a number of sources, but the observed upper limits for Cas A, SN 1006, Tycho, and Vela fall below

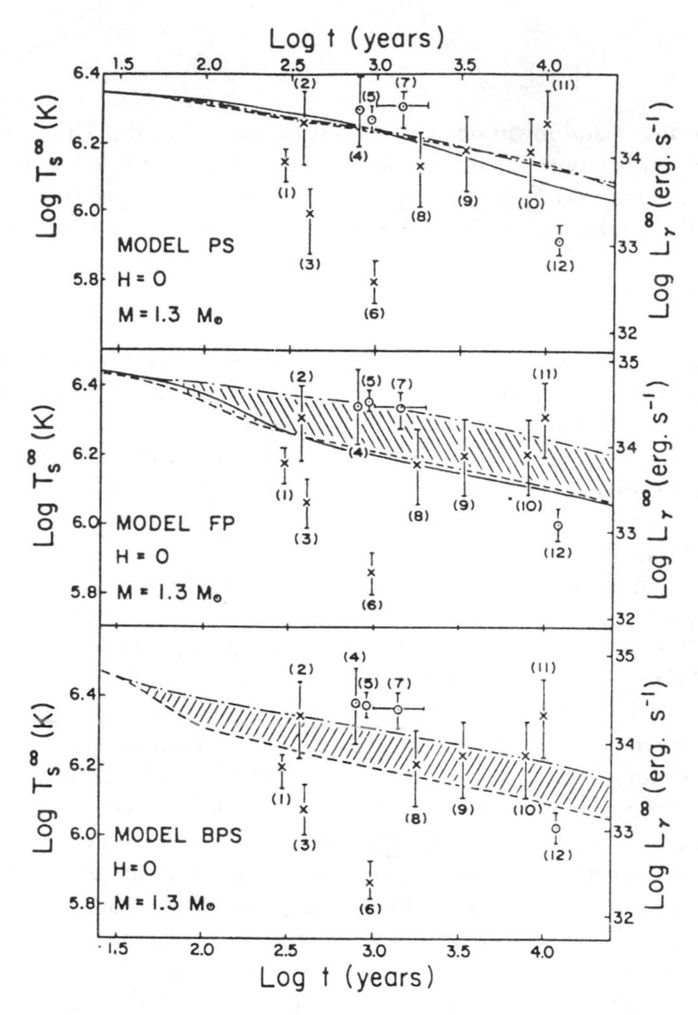


FIGURE 11.2.3 The surface temperature versus time calculated for various neutron star models, compared with upper limits set by *Einstein* observations. The circles and crosses refer to SNRs with and without detected point sources, while the numbers identify the SNRs. From K. Nomoto and S. Tsuruta, 1987, *Ap. J.*, 312, 711.

the standard cooling models. This may be an indication that the physics used must still be further refined, or it may be indicating that additional cooling could be present, possibly from a pion condensate. These upper limits could become even more stringent if additional heating sources are included. For instance, in isolated neutron stars vortex-creep frictional heating between the crust and the superfluid (Shibazaki and Lamb, 1989; Alpar, Cheng, and Pines, 1989) could contribute sufficient heating to modify the cooling profiles.

## 11.3 Rotational Evolution of Neutron Stars

### *a) Observational Evidence for Changes in the Spin Rate*

The initial spin rotation rate of neutron stars at birth is unknown. Among RPPs the youngest example with a confirmed period is the Crab pulsar, with $P = 0.033$ s, which is the smallest so far. More will be known about this question if a period is detected from a neutron star in the SN 1987a. There is, however, some indication that a fraction of radio pulsars should be born with larger periods in the range 0.2–0.5 s, or at any rate pulsed radio emission in a fraction of pulsars should be delayed until they have reached such rotation rates, in order to obtain agreement with the observed period distribution (e.g., Taylor and Stinebring, 1986). On the other hand, AXPs have periods mostly above 0.5 s and may have undergone considerable evolution in a binary system before being observed in this form. The spin rotation rate inferred for weakly magnetized beat-frequency models of QPOs is of order $\sim 0.01$–$0.05$ s, and only one GRB has shown strong evidence for rotation with a period $\sim 8$ s, this being the unusual March 5, 1979, burst.

The rotation rate of a neutron star will change in the course of time as the result of either external or internal torques. The external torques include radiative losses (either of photons, winds, or gravitational waves), and magnetic or tidal torques in the case of binary systems. In the case of AXPs, there is usually an overall spin-up trend, which is interpreted as being due to an accretion torque which on average tends to increase the stellar angular momentum. This behavior alternates in some objects with episodes of spin-down. An example is shown in Figure 7.1.5. In the case of RPPs, the behavior is overwhelmingly one of overall spin-down, which was early on interpreted as supporting evidence for a strong magnetic field, which leads to an external radiative torque tending to slow the rotation. However, in neutron stars one also expects internal torques. These can be

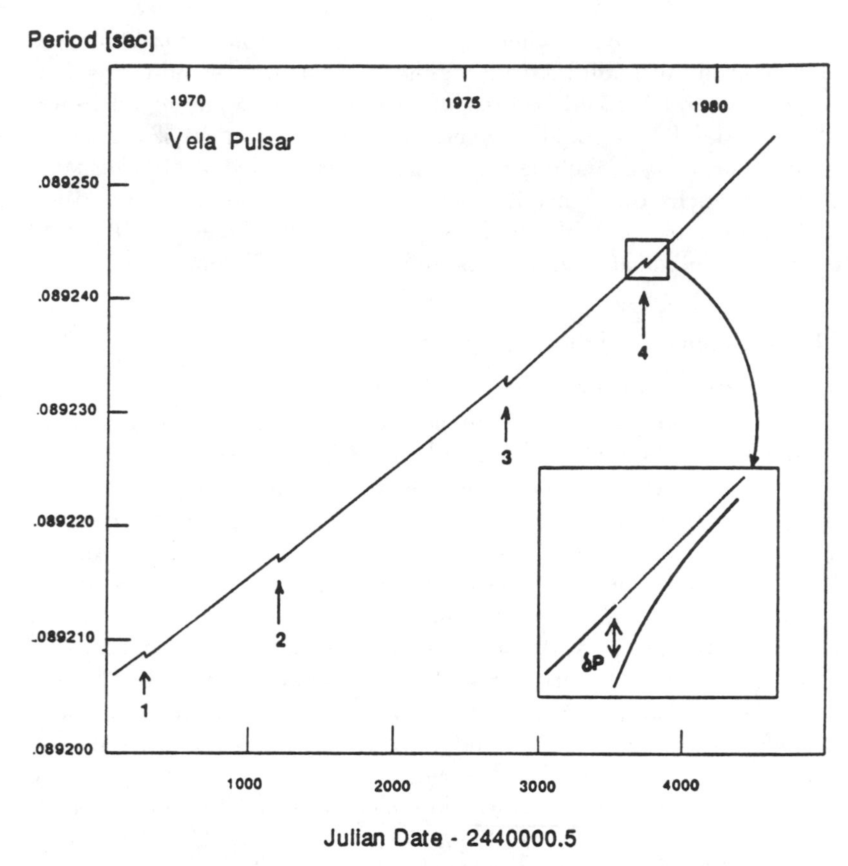


FIGURE 11.3.1 Schematic pulse period history of the Vela pulsar showing four discrete glitches as episodes of spin-up, superposed on an overall long-term spin-down behavior. The change in period of each glitch is $\Delta P/P \sim 10^{-6}$, and the recovery time is $\tau_r \sim$ 1–3 months. From G. S. Downs, 1981, *Ap. J.*, 249, 687.

produced by friction and dissipation inside the neutron star associated with the possibly different rotation rate of the solid crust, which connects to the external torques, and the superfluid neutrons in the core or the inner crust. Episodes of sudden spin-up and subsequent relaxation to a secular spin-down trend in some RPPs (glitches) are interpreted as being caused by internal torque readjustments (see Figure 11.3.1). In some AXP systems, there is also evidence of short time scale fluctuation and of possible precession periods (e.g. Boynton *et al.*, 1986; Deeter *et al.*, 1989), which are thought to be connected with internal torque coupling. There is also evidence for quasi-periodic oscillations in several AXPs (Cen X-3, 4U 1626 − 67, EXO 2030 + 375), and low-frequency red (power law) and white noise (flat) power spectra are observed in many more AXP systems (Nagase, 1989b).

*b) Internal Torques*

The superfluid neutrons in the inner crust and in the core partake in the rotation of the neutron star by forming quantized vortices whose density per unit area perpendicular to the bulk spin angular velocity $\vec{\Omega}$ is

$$n_v = \frac{2\Omega}{\kappa} = \frac{2\Omega m_N}{\pi \hbar}, \qquad (11.3.1)$$

where $\kappa = \pi\hbar / m_N \simeq 2 \times 10^{-13}$ cm$^2$ s$^{-1}$ is the vorticity quantum (Ruderman, 1976; Sauls, 1988). While a superfluid cannot perform a perfect rigid rotation in the usual sense, it can mimic it by forming an appropriate amount of such quantized vortices. Changes in the superfluid $\vec{\Omega}$ are achieved by changing the vortex density. In the case of spin-down the vortices must move out and annihilate near the boundary, while in spin-up the vortices move in and new ones are created near the outer boundary.

In the core, superfluid neutrons coexist with a smaller proportion (~ 5%) of superconducting protons and nonsuperfluid electrons, forming a spatially homogeneous medium, where the vortex lines are able to move freely under the action of the forces exerted on them. The neutrons circulating in the vortices will drag along protons, which form a type II superconductor, creating a strong magnetic field at the core of the vortices. As a result, the vortex lines and the rotation of the core superfluid are strongly coupled to the rotation rate of the electrons. The electrons couple to the protons and to the neutron star crust at the boundary of the core on a time scale which is of order

$$\tau_c \sim 10^2 P \left(\delta m_p / m_p\right)^2 s \qquad (11.3.2)$$

(Alpar and Sauls, 1988), where $\delta m_p$ is the difference between the effective and the bare mass of a proton in a neutron sea, whose value $\delta m_p / m_p \sim 0.1$–$0.5$, and $P$ is the stellar spin period in seconds. Because of the drag between the neutrons and the protons, one expects the entire core to be rotating rigidly with the crust, responding to changes in the crustal velocity on a time scale of order $\tau_c$ (Alpar *et al.*, 1984c). This is a departure from earlier models (e.g. Baym *et al.*, 1969) which assumed the core to be only weakly coupled to the crust. The analyses of macroglitches in the Crab and Vela pulsars (Downs, 1981; Demianski and Proszynski, 1983), of microglitches in the Crab pulsar (Boynton, 1981), and of the angular acceleration power spectrum in Her X-1 (Boynton and Deeter, 1979) and Vela X-1 (Boynton *et al.*, 1986) have shown that a weakly coupled core is unlikely. The observational evidence and the model analyses (c.f. Lamb *et al.*, 1978) leading to the current view that the core is rigidly rotating with the crust are reviewed by Lamb (1985).

The inner crust of the neutron star also contains a smaller fraction of superfluid neutrons, making up $\lesssim 10^{-2}$ of the moment of inertia of the star. This superfluid coexists with the nuclear crystal lattice, which provides a very inhomogeneous medium through which the superfluid neutron vortices move. For a sufficiently small velocity difference, the vortices in fact can get "pinned" to the nuclei, to the spaces between nuclei, or to lattice defects (Anderson and Itoh, 1975; Alpar, 1977). Thus, in regions where the vortex lines are pinned, the angular velocity of the superfluid remains tied to that of the crust. The rotational relaxation of the pinned superfluid in response to changes in the rotation rate of the crust occurs by thermal creep of the vortex lines through the pinning barriers at a finite temperature (Alpar *et al.*, 1984a). The observed glitch behavior of some RPPs (e.g. Vela) has been interpreted to be due to events of sudden unpinning, with the subsequent exponential resumption of spin-down reflecting the dynamical relaxation behavior of the pinned superfluid in the inner crust, and the observed changes in the spin-down rate have been interpreted to provide an estimate of the fraction of the moment of inertia resident in the crust (Pines and Alpar, 1985):

$$\frac{\Delta\dot{\Omega}}{\dot{\Omega}} \sim \frac{I_{\rm dec}}{I}. \tag{11.3.3}$$

Here $I$ is the total moment of inertia and $I_{\rm dec}$ is the inertia of the portion that becomes decoupled. The values of $\Delta\dot{\Omega}/\dot{\Omega} \sim 10^{-3}$–$10^{-2}$ from most of the observed glitches on a number of pulsars (see Chap. 8.1*d*) agree with the estimates of the fraction $I_i/I$ for the superfluid component of the crust. However, for the largest glitches observed in Vela and PSR 0355 + 54 (see Chap. 8.1*d*), which have $\Delta\dot{\Omega}/\dot{\Omega} \sim 10^{-1}$, the recovery time observed is $\sim$ 1 day, which is too short for crust-core coupling, and this may be caused by linear coupling of a portion of the pinned crustal superfluid to the core superfluid, as argued by Alpar, Pines, and Cheng (1990). The superfluid unpinning explanation of glitches is in contrast to the older starquake model of glitches, where due to the buildup of strain the star crust readjusted itself, leading to a change in the overall moment of inertia $\Delta\Omega/\Omega \sim \Delta I/I$, which encounters serious problems in explaining the order of magnitude of the average Vela glitches, in contrast to the superfluid unpinning estimate (11.3.3). The proposed interpretation of the 35 day pulse behavior of the AXP Her X-1 (Trümper *et al.*, 1986) in terms of free precession can also be understood in terms of the crust pinning model (e.g. Pines, 1987; Alpar and Ögelman, 1987). This type of information is invaluable in that it provides a much needed handle on the physical conditions in the interior of neutron stars.

*c) External Torques and Spin Evolution in AXPs*

The external torques in AXPs depend on the environment, on the type of accretion flow, and on the details of the magnetic field configuration of the neutron star. Typically, the dominant effect is provided by the accretion torques, whose order of magnitude is given by $K_0 = (GMr_m)^{1/2}\dot{M}$ (eq. 7.2.9), but the sign of the torque depends on whether the magnetospheric radius (7.2.5) is larger or smaller than the corotation radius (7.2.6). In terms of the dimensionless fastness parameter $\omega_s$, which is the ratio of the stellar angular velocity to the Keplerian angular velocity at the magnetospheric radius, Ghosh and Lamb (1978) define a dimensionless function $n(\omega_s) \simeq 1.4(1 - \omega_s/\omega_c)(1 - \omega_s)^{-1}$, which has positive or negative sign depending on whether $\omega_s$ is less or more than the critical $\omega_c \sim 0.8$–$0.9$ (Lamb, 1988). The accretion torque in this model therefore is

$$K_s \simeq n(\omega_s)(GMr_m)^{1/2}\dot{M}, \tag{11.3.4}$$

where $r_m$ is the magnetospheric radius. This gives a torque whose sign depends on $\omega_s$. For $\omega_s \ll 1$, $K_s \simeq 0.4K_0$ and there is a spin-up; for $\omega_c \leq \omega_s \leq 1$, there is a spin-down; and for $\omega_s > 1$, no accretion occurs, since matter is centrifugally ejected. The parameter $\omega_s$ itself depends on the magnetospheric radius, i.e. on the accretion rate, the magnetic field strength, and on the geometry of the latter (see Chap. 7.2*d*). The complications and uncertainties arise in the details of the field geometry near this magnetospheric boundary (see eq. 7.2.10).

Other models allow spin-down to occur by matter ejection even while accretion is occurring, or differ in assuming a diamagnetic disk which is not penetrated by the magnetic field (e.g. Anzer and Börner, 1983; Arons *et al.*, 1982). These models have the expected property that the torque switches sign when the magnetospheric radius exceeds a fraction of the corotation radius, and share many of the uncertainties. Despite such uncertainties, the torque (11.3.4) reproduces well the observations of $P$ and $\dot{P}$ of a number of AXPs which may be accreting via a disk, and even some that are accreting from a wind. The pulse period of a $\omega_s \ll 1$ AXP which is spinning up is given from equation (11.3.4) as $-\dot{P} \propto \dot{M}P^2$ in general, and for disk systems (e.g. Ghosh and Lamb, 1978) as $-\dot{P} \propto n(\omega_s)(PL^{6/7})^2$ (eq. 7.2.13). The evolutionary tracks of AXPs in the spin-up phase are shown schematically in Figure 11.3.2. As accretion proceeds, the AXPs dribble down toward shorter periods until the torque drops due to the $n(\omega_s)$ factor as the fastness parameter $\omega_s \rightarrow \omega_c$ (which may be followed by a spin-down episode, not shown). While this picture works well for a number of

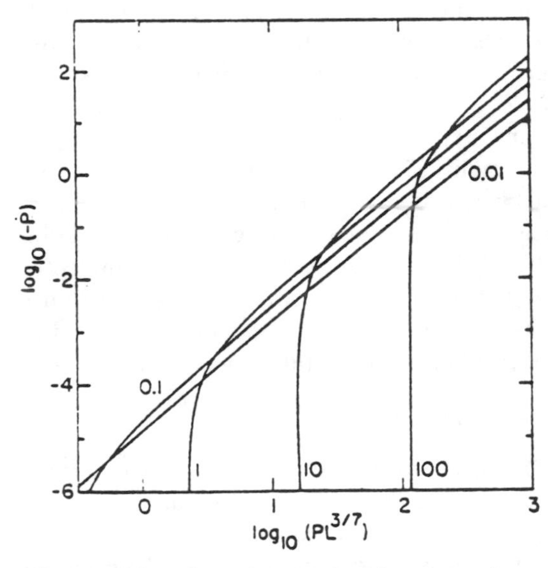


FIGURE 11.3.2 Theoretical evolutionary tracks of AXPs in the spin-up phase, plotting the derivative of the pulse period $\dot{P}$ against the product $PL_x^{3/7}$. The various curves are labeled with the value of the magnetic moment $\mu = BR^3$ in units of $10^{30}$ G cm$^3$, while the units of $\dot{P}$, $P$, and $L$ are s yr$^{-1}$, s, and $10^{27}$ ergs s$^{-1}$, respectively. From P. Ghosh and F. K. Lamb, 1979, *Ap. J.*, 234, 296.

AXPs, there is some indication that for those AXPs which show QPO oscillations, e.g. Cen X-3, either the torque (11.3.4) does not apply or the beat frequency model of QPOs does not apply (Shibazaki, 1989), at least in their straightforward form. It is clear, however, just from dimensional arguments, that the accretion torque for disk systems cannot be too different from (11.3.4), although the details of the function $n(\omega_s)$ may vary. In wind accretion systems, the torque is much more uncertain, the results of numerical hydrodynamic calculations of the amount of accreted angular momentum being still preliminary (see Chap. 7.2*a*). However, because in wind systems the specific angular momentum is much smaller, one expects from small inhomogeneities and perturbations a more random behavior, leading to frequently alternating episodes of spin-up and spin-down (e.g. Taam and Fryxell, 1988).

*d) External Torques and Spin Evolution in RPPs*

For rotation-powered pulsars, the simplest physical picture of the energy losses indicates that the order of magnitude of the external torque (eq. 8.2.14) may be comparable to that due to the radiation losses of a vacuum dipole,

$$K_s = -\left(\kappa/8c^3\right)\left(B_0 R^3\right)^2 \Omega^3, \tag{11.3.5}$$

where $\kappa \sim 1$ and $B_0$ is the surface field strength. Observationally, however, the dependence on the angular velocity, $\propto \Omega^n$, may be different from $n = 3$ (see Chap. 8.1). For instance, the vacuum radiation losses would give $n < 3$ if inertial effects straighten the lines out, the extreme value being $n = 2$ for a wind-type solution with straight lines. On the other hand, other loss mechanisms could give $n > 3$, e.g. gravitational radiation. There is also uncertainty concerning the parameter $\xi$, which can be a function of the inclination angle $\alpha$ between the directions of the magnetic dipole $\vec{\mu}$ and rotation $\vec{\Omega}$ axes, and differs for specific models (e.g. eqs. 8.2.15 and 8.3.13–8.3.14, which predict an alignment or a counteralignment tendency, respectively). Restricting oneself to electromagnetic processes only, Blair and Candy (1988) write a generalized torque equation

$$\left(\frac{3c^3 I}{8\pi^2 R^6}\right) P\dot{P} = \eta B^2 \sin^2\alpha + \xi B^2 \cos\alpha + \lambda B^2 \beta^2 \sin^2\alpha. \tag{11.3.6}$$

Here $\alpha$ is the angle between the magnetic and rotation axes, and $\eta$ and $\xi$ are constants which distinguish between a pure vacuum dipole model ($\eta = 1$, $\xi = \lambda = 0$) and a real pulsar with a dense magnetosphere which generates field-aligned currents ($\xi > 0$), or nondipole radiation ($\eta \neq 1$), or relativistic beams ($\lambda > 0$) with beam width $2\beta$. The older scenarios (e.g. Ostriker and Gunn, 1970) assumed $\alpha = \pi/2$ and $\eta = 1$, $\xi = \lambda = 0$. More generally, (11.3.6) allows for the following possibilities:

1) $\eta = 1$, $\xi = \lambda = 0$, $\alpha =$ constant: the slowdown is dominated by the vacuum magnetic torque, with negligible torque from magnetospheric currents and no geometrical evolution of the field.

2) $\eta = 0$, $\xi = 1$, $\lambda = 0$, $\alpha =$ increasing: the slowdown is dominated by field-aligned currents in the magnetosphere, and there is geometrical evolution, in particular field counteralignment (as in Beskin, Gurevich, and Istomin, 1983).

3) $\eta = 1$, $\xi = \lambda = 0$, $\alpha =$ decreasing: the slowdown is dominated by vacuum dipole torques, with geometric evolution leading to alignment (e.g. Kundt, 1981; Jones, 1976).

The effect of a number of these scenarios on the $P, \dot{P}$ distribution has been discussed by Blair and Candy (1988). Several of the physical scenarios themselves are discussed in Chapter 8, above. Notice that in the above equations $B_0$ was assumed to be constant in time, but if the latter evolves as well (c.f. § 11.4 below), the torque evolution equation will be correspondingly affected (e.g. Shibazaki and Lamb, 1989).

Unlike in the case of AXPs and aside from glitch episodes related to internal torques, the RPP external torques have the common feature that they predict a spin-down as the pulsar loses rotational energy. Assuming that the magnetic field is approximately constant until a time $\sim \tau_D$ (e.g. eq. 11.2.9), the evolutionary tracks of RPPs on a $\dot{P}$, $P$ diagram are $\dot{P} \propto P^{-1}$ followed by a rapid exponential drop at the value where the magnetic field decays (eq. 8.2.16). Alternatively, in the BGI model, the tracks are $\dot{P} \simeq$ constant (eq. 8.3.19) followed by a drop after a time $\tau_D$ (see Fig. 11.3.3). Assuming that the magnetic field is deduced from the $\dot{P}$, $P$ values (e.g. from eq. 8.2.17), the tracks in a $B$, $P$ plane would be in both cases a horizontal track to the right followed by an exponential drop at the $P$ value corresponding to $\tau_D$. As discussed in Chapter 8.1, $\tau_D$ actually represents the torque decay time scale, and this could be due e.g. to alignment. Furthermore, there is now evidence that a simple exponential magnetic field decay process leads to inconsistencies in some cases. Clearly, to accommodate the evidence accumulated in the last four or

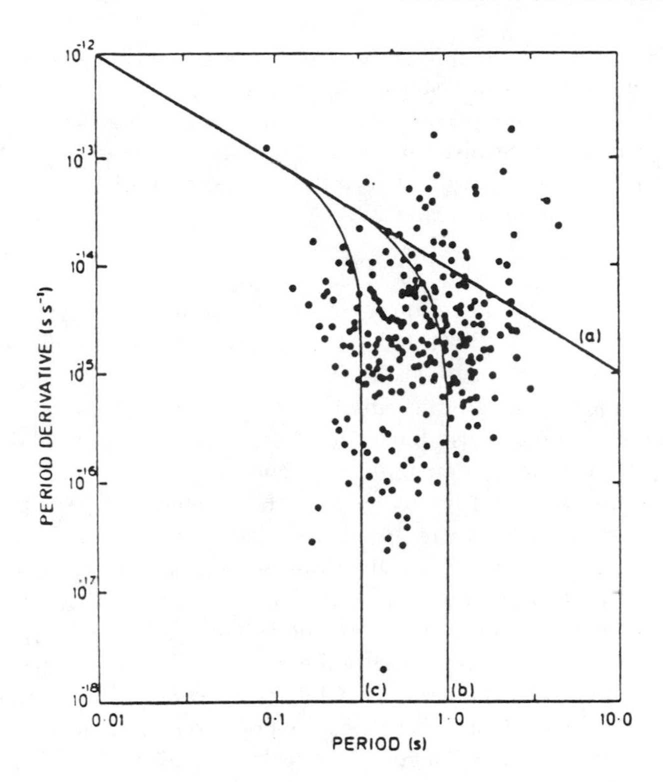


FIGURE 11.3.3 Evolutionary tracks of RPPs in the $\dot{P}$, $P$ plane for the standard model assuming (a) a constant magnetic field, (b) a magnetic field decay time of 10 Myr, and (c) a decay time of 1 Myr. For stronger or weaker initial magnetic fields, the curves would be higher or lower by half a decade in both $P$ and $\dot{P}$ for each decade in $B$. For the BGI model, the field evolution is almost constant in $\dot{P}$ until the field decay time is reached, at which point they drop in the same manner as those shown here. From A. G. Lyne, R. N. Manchester, and J. H. Taylor, 1985, *M.N.R.A.S.*, 213, 613.

five years on millisecond pulsars, many of which are in the lower left of the $\dot{P}$, $P$ diagram, one has to involve considerations which go beyond the simple assumptions used in the figure.

Although MSPs may in principle be born with such low periods (the centrifugal breakup period is just under $\sim 1$ ms), the most common view is that they are or were part of a binary system with a low-mass companion, from which they start accreting after the original pulsar slowed sufficiently to cross the death line. The system becomes a LMXB and begins to spin up due to the accreted angular momentum. After mass transfer stops, the neutron star has a short period and becomes a radio pulsar for a second time (Alpar *et al.*, 1982; Radhakrishnan and Srinivasan, 1982). Since the neutron star will spin up to its equilibrium period, where the magnetospheric radius (7.2.4) equals the corotation radius (7.2.6),

$$P_{eq} \simeq 1.9 R_6^{18/7} M_{1.4}^{-5/7} B_9^{-3/7} \text{ ms}, \tag{11.3.7}$$

the magnetic field must be low, $B \lesssim 10^9$ G, in order to reach periods as low as 1.5 ms for PSR 1937 + 1. In fact, a measurement of $\dot{P}$ indicates from equation (8.2.17) a field $\sim 5 \times 10^8$ G. A single MSP could presumably arise from disruption or dispersal of the binary companion. There are also binary MSP systems, such as PSR 0655 + 64, where the binary companion is a cool WD whose age must be $> 10^9$ yr. In single MSPs, even though no companion is left, the amount of mass accretion required ($\sim 0.1\ M_\odot$) at a critical accretion rate ($\dot{M} \sim 10^{-8}\ M_\odot\ \text{yr}^{-1}$) requires the neutron star to be older than $\sim 10^9$ yr. In this time, the field would have decayed well below $10^8$ G if pure exponential decay were assumed. This led to the hypothesis (see § 11.4*c*) that perhaps it is only the crustal field that decays on the time scale 2–10 Myr suggested by the statistics of pulsar kinetic and characteristics ages (Chap. 8.1*c*), while a core component of the field could decay on a much longer time scale, $\gtrsim 10^9$ yr (see § 11.4*c*). There may therefore be an asymptotic lower value of the field, which from the observations seems to be $\sim 5 \times 10^8$ G. The evolution of a recycled MSP would look schematically as illustrated in Figure 11.3.4. In this picture, in the $B, P$ plane the initially large crustal field remains constant as the period lengthens, until it decays and eventually radio emission stops as it crosses the death line $B \propto P^2$. Eventually mass transfer starts and spins the pulsar up toward shorter period, until $P_{eq} \propto B^{6/7}$ is reached. The crustal magnetic field continues to decay while the LMXB stage continues, so shorter and shorter equilibrium periods are achieved, and the neutron star dribbles down along the equilibrium spin-up line until mass transfer stops. At this point the pulsar begins its second life, as a MSP. It starts to spin down a second time, but the time needed to reach the death line is now longer than a Hubble time, so one finds these MSPs between the spin-up line and the line marking the age of the galaxy. An alternative view of MSPs, proposed by Ruderman (1991) and discussed in § 11.4*d*, involves an AIC origin of a very fast rotating neutron star which actually spins down to reach its present $\gtrsim$ ms period.

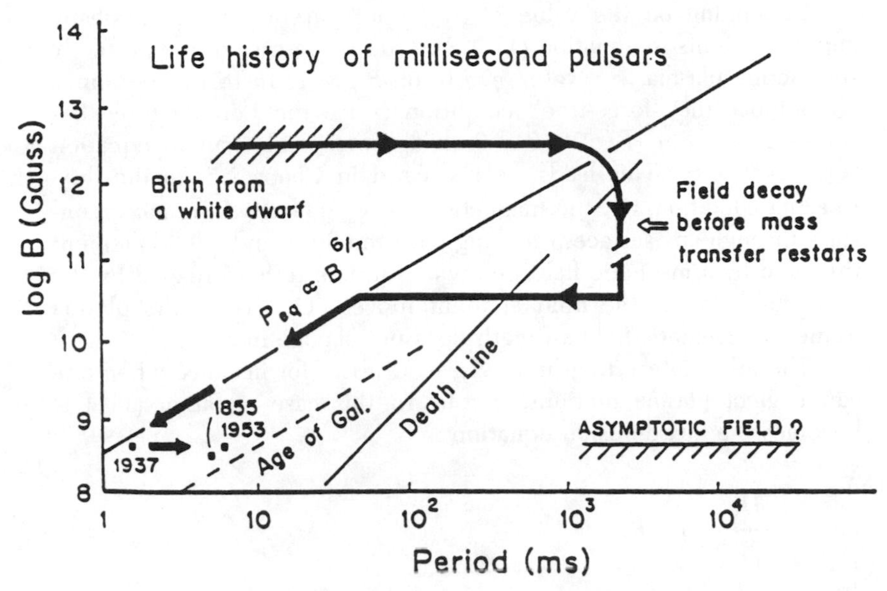


FIGURE 11.3.4 The LMXB accretion spin-up scenario for the formation of recycled millisecond pulsars. Reprinted with permission from G. Srinivasan, 1990, *Adv. Space Res.*, 10, no. 2, 167; © 1990 Pergamon Press PLC.

## 11.4 Magnetic Evolution of Neutron Stars

### *a) The Time Dependence of the Magnetic Flux in a Plasma*

The microscopic Maxwell equations for a plasma (eqs. 2.5.3 and 2.5.5) give an explicit time dependence for the magnetic field, which together with Ohm's law (eq. 2.5.27) for a moving medium yields (eq. 2.5.28) an equation for the time dependence of the magnetic field in a conductive medium,

$$\frac{1}{c}\frac{\partial \vec{B}}{\partial t} = -\vec{\nabla} \times \left(\vec{v} \times \vec{B}\right) + \eta \nabla^2 \vec{B}, \tag{11.4.1}$$

where $\eta = c/4\pi\sigma$ is the resistivity and $\sigma$ is the conductivity of the plasma. For plasma volumes with a characteristic size $l$ and characteristic velocity $v$, the ratio of the second to the first term on the right-hand side of equation (11.4.1) is the magnetic Reynold's number,

$$R_m = lv/\eta. \tag{11.4.2}$$

Depending on the value of $R_m$, there are two astrophysically important limits to equation (11.4.2). One is the case of the perfectly conducting plasma, $\sigma \to \infty$ or $\eta \to 0$, or $R_m \gg 1$. In this case, Ohm's law implies the "force-free" condition (since the Lorentz force becomes zero), $\vec{E} + (\vec{v}/c) \times \vec{B} = 0$, and the second term in equation (11.4.1) is zero. This leads, as discussed in Chap. 2.5, to the flux-freezing condition, i.e. the magnetic flux $\Phi_M = \int_S \vec{B} \cdot d\vec{S}$ remains constant through any surface $S$ moving with the fluid, and a fluid element threaded by a magnetic field line will continue to be threaded by the same field line as the fluid element moves. That is, in the plasma frame the magnetic field strength does not change in time.

The other interesting limit, $R_m \ll 1$, arises for instance when one can neglect plasma motions, $\vec{v} \to 0$. In this case equation (11.4.1) becomes the flux diffusion equation

$$\frac{1}{c}\frac{\partial \vec{B}}{\partial t} = \eta \nabla^2 \vec{B}. \tag{11.4.3}$$

Taking the scalar product of equation (11.4.3) with the element of surface $d\vec{S}$ and integrating over the surface $S$, we have

$$\frac{\partial \Phi_M}{\partial t} = \eta \int_S \nabla^2 \vec{B} \cdot d\vec{S}, \tag{11.4.4}$$

and putting $\nabla^2 \vec{B} \sim \vec{B}/l^2$ one sees that $\Phi_M$ decays exponentially with a time constant $\tau$ given by

$$\tau \simeq \frac{l^2}{\eta} = \frac{l^2 4\pi\sigma}{c^2} \simeq 5 \times 10^9 l_6^2 \sigma_{25} \text{ yr}, \tag{11.4.5}$$

where we normalized to illustrative values $l \sim R_* = 10^6$ cm, $\sigma = 10^{25}$ $s^{-1}$. This is the time scale required for the ohmic losses per unit volume $\eta j^2$ to dissipate an energy comparable to the magnetic energy density $B^2/8\pi$, which is a fraction of the Hubble time. Below we discuss how this time scale could be shorter (see eq. 11.4.9).

In the presence of a finite thermal heat conduction and chemical potential gradients, thermoelectric effects must be included in the induction equation (11.4.1) (Urpin and Yakovlev, 1980a; Blandford, Applegate, and Hernquist, 1983). In the limit where the electron chemical potential $\mu_e \gg m_e c^2$, the expanded induction equation reads

$$\frac{1}{c}\frac{\partial \vec{B}}{\partial t} = -\vec{\nabla} \times \left(\vec{v} \times \vec{B}\right) - \vec{\nabla} Q_0 \times \vec{\nabla} T - \vec{\nabla} \times \left(\eta \vec{\nabla} \times \vec{B}\right), \tag{11.4.6}$$

where the $\nabla^2 B$ term of (11.4.1) has been transformed using the vector identity $\vec{\nabla} \times (\vec{\nabla} \times \vec{B}) = \vec{\nabla}(\vec{\nabla} \cdot \vec{B}) - \nabla^2 \vec{B}$; the velocity $\vec{v}$ is now

$$\vec{v} = \vec{v}_b + \frac{1}{n_e \mu_e} \vec{\kappa} \cdot \vec{\nabla} T \frac{\tau_r d \ln \left(\mu_e / \tau_r\right)}{d \ln \mu_e} - \frac{\vec{j}}{e n_e}; \tag{11.4.7}$$

the thermopower $Q_0$ is

$$Q_0 = -\frac{\pi^2 k_B^2 T c}{3 e \mu_e} \frac{d \ln \kappa_0}{d \ln \mu_e} \tag{11.4.8}$$

(e.g. Urpin and Yakovlev, 1980b); $\kappa_0$ is the unmagnetized thermal conductivity entering the heat flux equation $\vec{F} = -\kappa_0 \vec{\nabla} T_0$; and $\tau_r$ is the relaxation time. The terms in the expanded induction equation (11.4.6) represent, respectively:

i) The convection of magnetic field at a velocity $v$, which is the sum of the hydrodynamical bulk velocity $v_b$, the thermal diffusion velocity, and the electron mobility, which for a relaxation time that is independent of the density can be approximated by $v \sim -F/\mu_e n_e$; this is the drift velocity of the field, which is oppositely directed to the heat flux.

ii) A battery term proportional to $\vec{\nabla} n_e \times \vec{\nabla} T$ describing the creation or destruction of field by thermoelectric currents. If $\vec{\nabla} T \cdot \vec{B} \times \vec{F} > 0$, this term contains a part $\propto (-\vec{\nabla} n_e \cdot \vec{F})\vec{B}$ which leads to exponential growth (decay) of the field if the heat flows down (up) the density gradient.

iii) The ohmic field decay term.

The most significant effect in a neutron star crust is the field convection effect (i), which may be described as follows (Blandford, Applegate, and Hernquist, 1983). The difference in thermal velocities of hotter electrons below and cooler electrons above in a horizontal magnetic field which deflects them sideways leads to a horizontal heat flux $F_\perp \sim (ce\tau/\mu_e)\vec{B} \times \vec{F}$, where $\vec{F}$ is the vertical heat flux. (This is for a cooling neutron star; for an accreting case the horizontal gradient changes sign, since hot is above and cold below.) The Fourier components of $\vec{B}$ with horizontal wavelength comparable to the depth $z$

create horizontal temperature gradients $\vec{\nabla} T \sim F_{\perp}/\kappa$, where $\kappa$ is the thermal conductivity. The pressure of the degenerate relativistic free electrons is $P(n_e, T) = P(n_e, 0) + (\pi^2 k_B^2/6) n_e T^2/\mu_e$, so there will be an additional horizontal pressure gradient $\vec{\nabla} P \sim (k_B^2/\mu_e) T \vec{\nabla} T$. This must be balanced by a thermoelectric field $E \sim (k_B^2/\mu_e e) T \nabla T \sim -(v/c)B \sim (F/\mu_e n_e c)B$. This electric field has a nonvanishing curl so that $-\vec{\nabla} \times \vec{E} = (1/c)\,\partial B/\partial t = \Gamma \vec{B}$. The growth or decay rate $\Gamma \sim (F/\mu_e n_e c z)$ is positive (negative) when the heat flow is down (up) the density gradient. For characteristic (cooling) values $F \sim 10^{20}$ ergs $\mathrm{cm}^{-2}$, $\mu_e \sim 4$ MeV, $n_e \sim 3 \times 10^{31}\ \mathrm{cm}^{-3}$, and $z \sim 50$ cm, the growth rate is $\Gamma_+ \sim 300\ \mathrm{yr}^{-1}$, which is comparable to the ohmic decay rate calculated for the conductivity $\sigma \sim 10^{21}\ \mathrm{s}^{-1}$ appropriate for these conditions. The dominance of growth over decay depends therefore delicately on the estimates of the electron collision time, which determine the transport properties and therefore the heat flux.

### *b) Origin and Decay of Neutron Star Magnetic Fields*

The earliest and simplest view of the magnetic fields of neutron stars is that they are given by the frozen-in value of the primeval magnetic field of the parent star in its presupernova stage (e.g. Woltjer, 1964; Pacini, 1967). In turn, the magnetic fields of the parent stars are thought to have arisen via a self-excited hydromagnetic dynamo mechanism (e.g. Parker, 1979; Zel'dovich, Ruzmaikin, and Sokoloff, 1983) linked to internal mass motions and currents. The flux-freezing condition is very well satisfied in a collapsing supernova, since the conductivity is large and the ohmic diffusion time is much longer than the free-fall time of the collapsing star. Taking an initial stellar magnetic field of $B_i \sim 10^2$ G and an initial stellar radius of $R_i \sim 10^{11}$ cm, the flux-freezing condition $BR^2 =$ const. leads to a final field of $B_f \sim 10^{12}$ G. A similar argument can be made for inferring a typical white dwarf field of $B_{\mathrm{wd}} \sim 10^7$ G, and a neutron star field of $B \sim 10^{12}$ G is also obtained if one uses flux freezing starting from such a white dwarf configuration, as in the Fe core collapse of an evolved star or the accretion-induced collapse of a white dwarf in a binary stellar system. On the other hand, from a low stellar field $\sim 1$ G one would get a low-field WD, $B \sim 10^5$ G, and the neutron star magnetic field would be correspondingly weaker, $B \sim 10^9$–$10^{10}$ G, which is comparable to that observed in millisecond pulsars in globular clusters and elsewhere (see, however, below).

Immediately after its collapse, the neutron star material is hot

($T > 10^{10}$ K) and gaseous. The time needed to cool to a characteristic melt temperature $T_m \sim 9 \times 10^9$ K corresponding to a characteristic crustal density $\rho \sim 3 \times 10^{13}$ g cm$^{-3}$ is about $t_{\rm cool} \lesssim 5 \times 10^3$ s, at which point crystallization of the crust takes place (Flowers and Ruderman, 1977). Before this time, while the star is still fluid, MHD instabilities could lead to a significant reduction of the original fossil dipole magnetic moment on a hydromagnetic time scale, which, however, for initial typical pulsar spin rates is much longer than the crystallization time. After the crust has crystallized, the field is anchored there by the very high conductivity ($\sigma \sim 10^{25}$ s$^{-1}$). This conductivity is provided by degenerate relativistic electrons, which scatter off phonons in the crystal lattice or off impurities in the lattice (Flowers and Itoh, 1976, 1981). If the decay is purely exponential and the magnetic field is assumed to be confined to the crust, which has a typical crustal depth $\Delta r_c \sim 10^5$ cm, the typical decay time scale is

$$\tau_D \sim 4(\Delta r_c)^2 \sigma / \pi c^2 \sim 2 \times 10^7 \text{ yr} \tag{11.4.9}$$

(Flowers and Ruderman, 1977). In the core, the density is of order $\rho \gtrsim 5 \times 10^{14}$ g cm$^{-3}$ and the conductivity is much higher, $\sigma_c \sim 10^{31}$ s$^{-1}$, which for a typical length scale $R \sim 10^6$ cm lead to exponential decay times in excess of $10^{14}$ yr if the field extends into the core. Wang and Eichler (1988) have also argued that, if the field in the core is purely poloidal, the dipole component will be subject to MHD instabilities in the superfluid core, which would rearrange it into a higher multipole configuration on a time scale comparable to the decay time scale of the crustal currents.

The approximate agreement between the crustal exponential decay time scale (11.2.9) and the empirically determined pulsar magnetic torque decay time scale of $\tau_{\rm MT} \sim (5\text{–}9) \times 10^6$ yr based on statistics of $\dot{P}/P$ (see Chap. 8.1*c*) is perhaps suggestive, although intrinsic uncertainties render the interpretation debatable (Bhattacharya and Srinivasan, 1990). Different models of pulsar energy loss predict an evolution leading to an alignment or counteralignment of the magnetic and rotational axes, although the time scale for this is not uniquely determined. In the alignment case this would also lead to a decay of the torque, which is given by the component of $\vec{B}$ perpendicular to the $\vec{\Omega}$ (Lyne, Manchester, and Taylor, 1985; Blair and Candy, 1988), and it is difficult experimentally to decide between these two alternatives. Thus, the empirically determined $\tau_D$ need not be directly connected to a decay of the magnetic field strength by ohmic losses.

A further complication of the ohmic decay picture is that, even if the initial field is confined to the crust, the effective decay time scale can be longer than implied by the exponential decay time scale for the crust alone. The reason for this is that, as ohmic losses progress, the field diffuses from the crust both outward (where the conductivity diminishes) and inward toward the core (where the conductivity has a much higher value). The diffusion equation (11.4.6) can be separated for a poloidal field, leading to an eigenvalue equation. Each eigenvalue decays exponentially with its own decay time scale, but the total field, which is made up of a weighted superposition of the different eigenvalues, does not (Sang and Chanmugam, 1987). These authors find that, for an initially dipolar field confined to the entire crust, the field value decays quasi-exponentially by a factor of order $e$ in about $10^7$ yr, and thereafter decays approximately as a power law by a factor of order $10^2$ in a Hubble time. This would still be compatible with the distribution of $\dot{P}/P$ if the torque decay and the decline of pulsar radio emission were caused by e.g. alignment. According to the above calculation, an initial field of $B \sim 10^{12}$ G could not therefore decay to a value as low as $B \sim 10^8$ G, as inferred in some millisecond pulsars. A sufficiently rapid field decay for this to occur over less than a Hubble time might require the initial field to be confined to a more restricted region of the outer crust.

### c) Thermomagnetic Field Growth

As discussed in § 11.4*a*, it is also possible that initial seed magnetic fields in neutron stars can grow or decay as a result of thermomagnetic effects (Urpin and Yakovlev, 1980a; Blandford, Applegate, and Hernquist, 1984; Blondin and Freese, 1986; Urpin, Levshakov, and Yakovlev, 1986). In order for this to occur on an astrophysically plausible time scale, the original or seed field must initially be confined to the outer crust or in a metallic liquid layer above the solid crust, and a substantial heat flux must be present. For nonaccreting neutron stars (e.g. radio pulsars), the heat flux is outward and goes down the density gradient, across the magnetic field lines which are confined to the crust. This leads to a field growth, as discussed in connection with equation (11.4.6), and the existence of a large enough heat flux implies that the growth phase would occur for a limited amount of time after the birth of the neutron star, while it is cooling. In order for this thermally driven growth to outpace the ohmic diffusion losses, Blandford, Applegate, and Hernquist (1983) require that the electron collision time exceed the existing estimates (e.g. Flowers and Itoh, 1981) by a factor of $\sim 3$.

Under the previous conditions, newly formed neutron stars should be able to develop surface fields of $\sim 10^{12}$ G, above which value the instabilities saturate because the subsurface magnetic stresses would exceed the lattice yield stress of the crust and would maintain these fields as long as heat flows through the crust. Thereafter the field would decay on an ohmic diffusion time scale of order $\sim 10^7$ yr (see, however, the previous discussion concerning the validity of the exponential form of the decay time scale). For fast rotating pulsars such as the Crab, which is only $\sim 10^3$ yr old and has a field of $\sim 10^{12}$ G, it may be that the growth time can be accelerated by tapping the rotational energy, which could provide an additional large heat flux by dissipation between the crust and the core. In accreting X-ray pulsars, some of which are old but strongly magnetized (e.g. $t \gtrsim 10^8$ yr, $B \sim 3 \times 10^{12}$ G in the case of Her X-1), the field growth and maintenance phase can be achieved in connection with the heat flux associated with the accretion. The latter occurs over a small fraction of the surface area, and heat is then conveyed from the interior toward the rest of the nonaccreting surface, where it escapes. This satisfies the requirement of a substantial heat flux directed outward along the density gradient, through a region where the magnetic field is horizontal (again, assuming it is confined to the crust). On the other hand, the very low fields ($\sim 10^8$ G) of some millisecond pulsars may be explained by thermally driven field destruction (Blondin and Freese, 1986), which would be appealing if these objects are young, e.g. $t \lesssim 10^6$ yr, as suggested for PSR 1937 + 214 by its kinetic age, being within 30 pc of the galactic plane. In this case one must assume that the neutron star has an inward-directed large heat flux, which could be associated with thermonuclear reactions occurring over the whole surface following a brief ($\sim 10^4$ yr) phase of accretion at a rate $\sim 10^{-5}$ yr$^{-1}$ from a massive companion star. The base of the thermonuclear envelope will have a temperature $T_b \sim 10^9$ K, while the interior will be kept at $T_i \lesssim 10^8$ K because of neutrino losses, driving an inward heat flux which for this accretion rate is of order $F \sim 10^{22}$ ergs cm$^{-2}$ s$^{-1}$. In this case the convection term dominates the ohmic loss term, and Blondin and Freese (1986) find an exponential field decay time scale of $\tau_D \simeq (n_e \mu_e z/F)_{z=z_b} \simeq 3 \times 10^3 (z_B/200\ \mathrm{m})^5 (F/10^{22}\ \mathrm{ergs\ cm^{-2}\ s^{-1}})^{-1}$ yr. Supporting evidence of a qualitative nature for this inverse thermomagnetic effect is provided by a study by Shibazaki *et al.* (1989), who plot the estimated magnetic field strength against the estimated total mass accreted by the neutron star, finding an inverse correlation. An alternative hypothesis for explaining such an inverse correlation may be that of Bisnovatyi-Kogan and Komberg (1974), who suggested that accreting matter could bury and screen the magnetic field of the star so that the dipole component seen from outside would be diminished even if the interior component had not decayed.

*d) Composite Models of Field Decay and Residual Field*

The present observational evidence concerning field decay or field growth is inconclusive, as discussed above and in Chapter 8.1. The statistics of pulsar radio luminosity versus period $P$, and of $\dot{P}$ versus $P$ (Lyne, Manchester, and Taylor, 1985; Blair and Candy, 1988) are consistent with either a field decay or an alignment of the dipole and rotation axis on a time scale $\tau \sim (5\text{–}9) \times 10^6$ yr. Several authors have suggested that the data may be compatible with no decay of the magnetic field at all (Kundt, 1981; Michel, 1986). A further complication is posed by the existence of a number of millisecond pulsars with magnetic fields $B \sim 5 \times 10^8\text{–}10^9$ G, some of which are in binary systems believed to be old, $t \gtrsim 10^8\text{–}10^9$ yr, on the basis of the evolutionary time scale of the low-mass companions (e.g. Kulkarni, 1986). In this case, however, their field should have decayed far below $B \sim 5 \times 10^8$ G, if exponential ohmic decay is assumed with a time scale $\tau_D \sim 5\text{–}10 \times 10^6$ yr. The birthrate of LMXBs in the galaxy is $t_b^{-1} \sim 10^{-7}$ yr$^{-1}$, which in steady state should equal the birthrate of MSPs, if all LMXBs lead to MSPs. The time spent by these pulsars at a field value $B \sim 10^9\text{–}10^8$ G is at most $t_M \sim 2 \times 10^7$ yr, so there should be only a few low-field MSPs in the galaxy at any time. The ratio of low-field MSPs to the total number $\sim 1\text{–}5 \times 10^5$ of active pulsars in the galaxy (Lyne, Manchester, and Taylor, 1985) is therefore $\sim 10^{-5}$. However, out of the observed $\sim 500$ pulsars in the galaxy, there are $\gtrsim 30$ observed low-field MSPs, a ratio of $\sim 10^{-2}$, despite the fact that the latter are harder to detect. The conclusion based on this evidence (Bhattacharya and Srinivasan, 1986; van den Heuvel, van Paradijs, and Taam, 1986; Kulkarni, 1986) is that magnetic fields, at least after they have reached values $B \lesssim 10^9$ G, must decay at a much slower rate ($\tau_{D,\,\mathrm{MSP}} > 10^9$ yr).

The previous arguments do not necessarily rule out the possibility of understanding the field strengths $10^9$ G $\gtrsim B \gtrsim 10^{13}$ G of most known neutron star systems within the hypothesis of an exponential surface field decay with $\tau_D \sim 5\text{–}10 \times 10^6$ yr. In order to show this, Taam and van den Heuvel (1986) developed a number of stellar evolutionary scenarios based on either Fe core collapse or AIC white dwarf collapse leading to neutron stars (see also § 11.1), the outcome of which are high-field objects that are relatively young, as well as low-field old MSP-type objects. It is possible that the presumed long-lived low magnetic fields of MSPs are associated with the neutron star core, while the faster decaying high fields of pulsars and accreting X-ray pulsars are associated with the lower conductivity crust (e.g. Kulkarni, 1986). This hypothesis is born out in part by the field decay calculations of Sang and Chanmugam (1987, 1990) who, however, find an exponential decay only for the first *e*-folding factor in time and a slower decay afterward so that the time scales do not agree in detail. This is not surprising, given the incomplete understanding of the interior field configuration and of the exact transport properties in the crust and core of a neutron star. Nonetheless, as a result of the increasing constraints on models imposed by the recent explosion of data on MSPs (e.g. Table 3) as well as on the magnetic field values of AXPs and GRBs, the need has emerged for increasingly complex models of field evolution.

A composite model involving field generation, decay, and accretion has been proposed by Romani (1990b). In this model the initial field is created either by thermomagnetic effects in the solid crust or by dynamo effects in the still liquid outer parts of the young neutron star, as in Blandford *et al.* (1983). Both of these growth mechanisms are assumed to be limited by the yield stress of the lattice at the melt surface, determined by the critical strain angle $\varepsilon$ and the shear modulus $\mu$, which gives a maximum field

$$B_m = (8\pi\varepsilon\mu)^{1/2} = 3.7 \times 10^{13} \varepsilon_{-2}^{1/2} ZA^{-2/3} \rho_{11}^{2/3} \text{ G.} \qquad (11.4.10)$$

The melt surface at which this field is evaluated moves out as the star cools, until the magnetic field prevents further motion of the melt surface at $\rho \sim 10^7$ g cm$^{-3}$ at $t \sim 10^6$ yr. Thereafter the field is assumed to dissipate by ohmic diffusion, taking into account the varying conductivity in the different regions of the crust, so that the field of the outer crust also takes $\sim 10^6$ yr to dissipate, which leads to a 30% decrease of the overall dipole component, but thereafter ceases to decay because the star is sufficiently cold for flux freezing to be effective. This evolution takes pulsars from an initial seed field stage straight up to the Crab stage (from which point the field essentially stabilizes) through the average pulsar phase and past the death line to the "graveyard" region, which may provide GRBs via e.g. the dormant pulsar model (Chap. 9.2). Low-field pulsars are assumed to be dormant pulsars that were recycled in binary systems and spun up to short periods (see also § 11.3). The magnetic field is assumed to be decreased in these recycled pulsars because of accretion effects. This includes not only (*a*) the inverse thermomagnetic effect but also (*b*) the increased crustal temperature caused by accretion, which lowers the conductivity and accelerates ohmic dissipation, and (*c*) the reduction of the crustal dipole field, by the deposition and compression of accreted matter bringing opposite current loops into close contact. In addition, (*d*) after reaching a certain density at the base of the accretion column, matter drags the field lines from the cap toward the equator, where they are advected under the crust and screened, diminishing the effective dipole component. The AXPs would just be the early versions of such binary systems, where the field has not yet decayed significantly.

Crustal field motions may also occur in neutron stars as a result of the spin-down or spin-up of the star (Ruderman, 1991). This is because the crustal lattice experiences strong stresses from pinned crustal superfluid neutron vortices on the crustal lattice nuclei (see § 11.3), and it is these neutron vortices that carry the bulk angular momentum of the star. In low-field spinning-down MSPs or spinning-up LMXBs these stresses are strong enough to crack the crustal lattice into plates, which then migrate along the surface carrying the magnetic field components embedded in them. For spinning-down MSPs, the superfluid vortices have to move outward and therefore cause the plates to migrate from the pole toward the equator, where they remain (unless regions of opposite polarity come close enough to annihilate). In a spinning-up LMXB the motion is from the equator toward the pole. As a result of this mechanism one would expect MSPs to achieve perpendicular dipole moments ($\vec{B} \perp \vec{\Omega}$), while LMXBs would become aligned rotators. Noting that three of the five lowest-field MSPs appear to be perpendicular rotators (Lyne and Manchester, 1988; Fruchter *et al.*, 1988), Ruderman concludes that these MSPs are low-field pulsars which were born rotating very fast and spun down to their present millisecond periods. This initial fast rotation may be achieved if $\sim 10^{-1}\ M_\odot$ have been accreted by a white dwarf progenitor which eventually undergoes accretion-induced collapse to a neutron star. The initial field could be as low as $\sim 10^9$ G if the WD had the typical low value of $\sim 10^5$ G, and this initial field may be further diminished by subduction and partial cancellation near the equatorial regions. For the two short-period MSPs which appear to be close to aligned, Ruderman argues that the progenitor WD might have been very close to the Chandrasekhar limit, so it would not have accreted enough mass to produce a neutron star rotating that fast, being instead spun up at a later stage in an LMXB. Assuming that most MSPs are perpendicular rotators, this scenario is contrary to the usual view that MSPs are neutron stars spun up in LMXBs, a fact which would solve the apparent discrepancy between MSP and LMXB birthrates discussed previously. It would also explain why LMXBs (e.g. X-ray bursters, QPOs) are not seen to pulsate, since they would be aligned rotators.

An alternative view is that the neutron star magnetic field, or at least the residual magnetic field, is dominated by a core component (Srinivasan *et al.*, 1990). For a core made up mostly of superfluid neutrons, the magnetic field is expected to be confined to flux tubes in the superconducting proton component, which are subject to buoyancy forces trying the expel them (Muslimov and Tsygan, 1985; Jones, 1988). However, the proton flux tubes are also expected to be subject to pinning forces with the neutron quantized vortex lines, forcing them to move together (see § 11.3). As the neutron star spins down, the neutron vortices move outward carrying with them the flux tubes, and thus decreasing the interior magnetic flux. The magnetic flux evolution is thus determined by the spin evolution of the star, which if given by the usual magnetic dipole braking with $\Omega \propto B_s$ (e.g. § 11.3) leads to a decay $B_s \propto t^{-1/4}$, amounting to one order of magnitude decay over a Hubble time scale. This would be adequate for explaining the slow decay of a residual field. However, if the entire field was initially a core field, a much larger spin-down may be achieved during phases of weak accretion in a stellar wind, since among such sources some very long periods are observed. This would decrease the vortex line and flux line density by a much larger factor, according to Srinivasan *et al.* (1990), so that the minimum field would be determined by the longest period achieved by a neutron star. Subsequently, if the neutron star becomes a strong accretor and starts to spin up to become a recycled pulsar, this minimum field value will be carried inward by the vortex lines newly injected into the core, where the field will remain trapped as long as the pulsar remains a fast rotator.

# Appendix A

## Relativistic Electron Wave Functions and Currents

The relativistic wave functions for an electron in a strong uniform magnetic field have been derived by Bussard (1980). Their form, as used here, is given in Alexander (1990), and they are listed here as a reference along with some useful relations. The wave function is composed of a time and a spatial part,

$$\Psi(x) = e^{-iEt}\psi_{nsqp\lambda}(\vec{x}), \tag{A.1}$$

where $n$ labels the Landau level ($n = 0, 1, 2, \cdots$), $s$ is the spin component parallel to the external field ($s = \pm 1/2$), $q$ labels the annulus of the guiding center of the electron's orbit ($q = 0, 1, 2, \ldots$), $p$ is the component of the momentum parallel to the external field and is continuous, and $\lambda = \pm 1$ gives the sign of the energy in the solution to the Dirac equation (i.e. $\lambda = 1$ for electrons, and $\lambda = -1$ for positrons). The guiding center locates the electron's orbit in the plane perpendicular to the external magnetic field, but this does not have any effect on the energy, which is therefore infinitely degenerate. As shown by Johnson and Lippman (1949) and Canuto and Ventura (1977), the guiding center operator is analogous to the harmonic oscillator and gives discrete eigenvalues labeled by the integer $q = 0$, $1, 2, \ldots$ We will use here the convention $\hbar = c = 1$. Calling $\beta$ the set of quantum numbers in equation (A.1), i.e. $\beta \equiv [nsqp\lambda]$, the spatial wave function is written as

$$\psi_\beta(\vec{x}) = \left[\frac{E+m}{2LE}\right]^{1/2}\left[1 - \frac{\vec{\gamma}\cdot\vec{\pi}}{E+m}\right] f_{nq}(\vec{x}_\perp)e^{ipz}u_s^\lambda. \tag{A.2}$$

In equation (A.2), $L$ is a normalization length, $E$ is the total energy,

$$E = \sqrt{p^2 + 2m\omega_c(n + s + 1/2) + m^2}, \tag{A.3}$$

$\vec{\gamma}$ are the Dirac matrices, and $\vec{\pi}$ is the canonical momentum operator given by

$$\vec{\pi} = \frac{1}{i}\nabla + e\vec{A}, \tag{A.4}$$

where $\vec{A}$ is the vector potential of the external magnetic field in the Landau gauge, where $\vec{A} = (-B_y/2, B_x, 0)$ for $\vec{B} = (0, 0, B)$. The $u_s^\lambda$ in equation (A.2) is a four component spinor with positive energy solutions (1, 0, 0, 0) and (0, 1, 0, 0) for $s = \pm 1/2$, respectively, and negative energy solutions (0, 0, 1, 0) and (0, 0, 0, 1) for $s = \pm 1/2$. The transverse part of the wave function, $f_{nq}(\vec{x}_\perp)$, is defined by the raising and lowering operators:

$$\pi_\pm f_{nq}(\vec{x}_\perp) = \sqrt{m\omega_c(n + 1/2 \pm 1/2)}\, f_{n\pm 1q}(\vec{x}_\perp), \tag{A.5}$$

$$X_\pm f_{nq}(\vec{x}_\perp) = \sqrt{\frac{1}{m\omega_c}(q + 1/2 \mp 1/2)}\, f_{nq\mp 1}(\vec{x}_\perp), \tag{A.6}$$

where the ground state wave function is

$$f_{00}(\vec{x}_\perp) = \sqrt{\frac{m\omega_c}{2\pi}}\,\exp\left(-\frac{m\omega_c}{4}x_\perp^2\right), \tag{A.7}$$

and $\pi_\pm$ and $X_\pm$ are operators for the transverse momentum and guiding center in cyclical coordinates (i.e. $X_\pm = (X + iY)/\sqrt{2}$).

The Fourier transforms of the transition currents are defined as

$$\vec{J}_{\beta\beta'}(-\vec{k}) \equiv \int d^3x \overline{\psi_{\beta'}(\vec{x})}\vec{\gamma}\psi_\beta(\vec{x})e^{i\vec{k}\cdot\vec{x}}. \tag{A.8}$$

Using the above wave functions, this is

$$\vec{J}_{\beta\beta'}(-\vec{k}) = \frac{2\pi}{L}\left[\frac{E+m}{2E}\frac{E'+m}{2E'}\right]^{1/2}\delta(p + k - p') \times i^{j-j'}(-1)^{q-q'}e^{i(j'-j-q'+q)\phi}F_{qq'}(\xi)\vec{G}_{\beta\beta'}(-\vec{k}_\perp), \tag{A.9}$$

where $j$ is total magnetic quantum number, $j = n + s + 1/2$, and the $F$ functions are given by

$$F_{qq'}(\xi) = i^{|q-q'|}\left[e^{\xi}\frac{q_<!}{q_>!}\xi^{|q-q'|}\right]^{1/2} L_{q_<}^{|q-q'|}(\xi), \tag{A.10}$$

where $\xi = k_\perp^2/2m\omega_c$ and $L$ represents an associated Laguerre polynomial. An often used relation for these functions is

$$\sum_{q''=0}^{\infty} e^{iq''\eta} F_{0q''}(\xi') F^{*}_{q'q''}(\xi) = i^{q'}\sqrt{q'!}\left(\sqrt{\xi}\,e^{i\eta} - \sqrt{\xi'}\right)^{q'} \times \exp\left[-\frac{\xi+\xi'}{2} + \sqrt{\xi\xi'}\,e^{i\eta}\right]. \tag{A.11}$$

The components of the vector $\vec{G}_{\beta\beta'}$ depend on $j$, $s$, and $\lambda$. They are given on a spin by spin basis for $\lambda = \lambda'$ (the "pure" case) by

$$\begin{aligned}
\left[\vec{G}^{+}_{\beta\beta'}(-\vec{k}_{\perp})\right]_{-} &= i\sqrt{2}\,e^{-i\phi}F_{j,j'-1}(\xi) \\
&\quad\times\left[\theta_{s,s'}\vec{u}_{\perp} + \theta_{-s,-s'}\vec{u}'_{\perp} + \theta_{-s,s'}(\vec{u}'_{\parallel} - \vec{u}_{\parallel})\right], \\
\left[\vec{G}^{+}_{\beta\beta'}(-\vec{k}_{\perp})\right]_{+} &= -i\sqrt{2}\,e^{i\phi}F_{j-1,j'}(\xi) \\
&\quad\times\left[\theta_{s,s'}\vec{u}'_{\perp} + \theta_{-s,-s'}\vec{u}_{\perp} + \theta_{s,-s'}(\vec{u}_{\parallel} - \vec{u}'_{\parallel})\right], \\
\left[\vec{G}^{+}_{\beta\beta'}(-\vec{k}_{\perp})\right]_{z} &= F_{j-1,j'-1}(\xi) \\
&\quad\times\left[\theta_{s,s'}(\vec{u}_{\parallel} + \vec{u}'_{\parallel}) + \theta_{s,-s'}\vec{u}'_{\perp} + \theta_{-s,s'}\vec{u}_{\perp}\right] \\
&\quad + F_{j,j'}(\xi)\left[\theta_{-s,-s'}(\vec{u}_{\parallel} + \vec{u}'_{\parallel}) - \theta_{s,-s'}\vec{u}_{\perp}\right. \\
&\quad \left. + \theta_{-s,s'}\vec{u}'_{\perp}\right],
\end{aligned} \tag{A.12}$$

and for $\lambda = -\lambda'$ (the mixed case) by

$$\begin{aligned}
\left[\vec{G}^{-}_{\beta\beta'}(-\vec{k}_{\perp})\right]_{-} &= i\sqrt{2}\,e^{-i\phi}F_{j,j'-1}(\xi) \\
&\quad\times\left[\theta_{-s,s'}(1 + \vec{u}_{\parallel}\vec{u}'_{\parallel}) - \theta_{s,-s'}\vec{u}_{\perp}\vec{u}'_{\perp}\right. \\
&\quad \left. - \theta_{s,s'}\vec{u}_{\perp}\vec{u}'_{\parallel} + \theta_{-s,-s'}\vec{u}_{\parallel}\vec{u}'_{\perp}\right], \\
\left[\vec{G}^{-}_{\beta\beta'}(-\vec{k}_{\perp})\right]_{+} &= -i\sqrt{2}\,e^{i\phi}F_{j-1,j'}(\xi) \\
&\quad\times\left[\theta_{s,-s'}(1 + \vec{u}_{\parallel}\vec{u}'_{\parallel}) - \theta_{-s,s'}\vec{u}_{\perp}\vec{u}'_{\perp}\right. \\
&\quad \left. - \theta_{s,s'}\vec{u}_{\parallel}\vec{u}'_{\perp} + \theta_{-s,-s'}\vec{u}_{\perp}\vec{u}'_{\parallel}\right], \\
\left[\vec{G}^{-}_{\beta\beta'}(-\vec{k}_{\perp})\right]_{z} &= F_{j-1,j'-1}(\xi) \\
&\quad\times\left[\theta_{s,s'}(1 - \vec{u}_{\parallel}\vec{u}'_{\parallel}) - \theta_{-s,-s'}\vec{u}_{\perp}\vec{u}'_{\perp}\right. \\
&\quad \left. - \theta_{s,-s'}\vec{u}_{\parallel}\vec{u}'_{\perp} - \theta_{-s,s'}\vec{u}_{\perp}\vec{u}'_{\parallel}\right] \\
&\quad - F_{j,j'}(\xi)\left[\theta_{-s,-s'}(1 - \vec{u}_{\parallel}\vec{u}'_{\parallel}) - \theta_{s,s'}\vec{u}_{\perp}\vec{u}'_{\perp}\right. \\
&\quad \left. + \theta_{s,-s'}\vec{u}_{\perp}\vec{u}'_{\parallel} + \theta_{-s,s'}\vec{u}_{\parallel}\vec{u}'_{\perp}\right],
\end{aligned} \tag{A.13}$$

where $\vec{G}_{\beta\beta'}$ is given in cyclical coordinates,

$$G_{\pm} = \frac{1}{\sqrt{2}}\left(G_x \pm iG_y\right). \tag{A.14}$$

Also, in equations (A.12) and (A.13),

$$\theta_{a,b} = \begin{cases} 1 & \text{if } a > 0, b > 0, \\ 0 & \text{otherwise,} \end{cases} \tag{A.15}$$

and

$$\vec{u}_{\perp} = \frac{\sqrt{2m\omega_c j}}{E + m}, \quad \vec{u}_{||} = \frac{p}{E + m}. \tag{A.16}$$

The susceptibility tensor for an electron-positron plasma can be written in terms of the tensor $\vec{\vec{Q}}$ (eq. 5.3.3), related to the $G$ functions through

$$\vec{\vec{Q}} = \frac{E + m}{2E}\frac{E' + m}{2E'}\vec{G}_{qq'}(\vec{k})\vec{G}^*_{qq'}(\vec{k}), \tag{A.17}$$

where

$$\vec{G}\vec{G}^* = \begin{pmatrix} G_-G_-^* & G_-G_+^* & G_-G_z^* \\ G_+G_-^* & G_+G_+^* & G_+G_z^* \\ G_zG_-^* & G_zG_+^* & G_zG_z^* \end{pmatrix}. \tag{A.18}$$

The six independent components of $\vec{\vec{Q}}$ are

$$\begin{aligned} Q(1) &= Q_{11} = \epsilon\left|\left[\vec{G}_{qq'}^{(\lambda\lambda')}(\vec{k})\right]_-\right|^2, \\ Q(2) &= Q_{22} = \epsilon|G_+|^2, \\ Q(3) &= Q_{33} = \epsilon|G_z|^2, \\ Q(4) &= Q_{12} = Q_{21}^* = \epsilon G_-G_+^*, \\ Q(5) &= Q_{23} = Q_{32}^* = \epsilon G_+G_z^*, \\ Q(6) &= Q_{31} = Q_{13}^* = \epsilon G_zG_-^*, \end{aligned} \tag{A.19}$$

where

$$\epsilon \equiv \left[\frac{E + m}{2E}\right]\left[\frac{E' + m}{2E'}\right]. \tag{A.20}$$

The pure ($\lambda = \lambda'$) components of the tensor $Q$ are

$$Q^{(p)}(1) = 2\epsilon\Big[u_\perp^2\,\theta_{s,s'} + u_\perp'^2\theta_{-s,-s'} + \big(u_{||} - u'_{||}\big)^2\theta_{-s,s'}\Big]\,|\,F_{j,j'-1}\,|^2,$$
$$Q^{(p)}(2) = 2\epsilon\Big[u_\perp^2\,\theta_{-s,-s'} + u_\perp'^2\theta_{s,s'} + \big(u_{||} - u'_{||}\big)^2\theta_{s,-s'}\Big]\,|\,F_{j-1,j'}\,|^2,$$
$$Q^{(p)}(3) = \epsilon\Big\{\Big[\big(u_{||} + u'_{||}\big)^2\theta_{s,s'} + u_\perp^2\,\theta_{-s,s'} + u_\perp'^2\theta_{s,-s'}\Big]$$
$$\times\,|\,F_{j-1,j'-1}\,|^2$$
$$+\Big[\big(u_{||} + u'_{||}\big)^2\theta_{-s,-s'} + u_\perp^2\,\theta_{s,-s'} + u_\perp'^s\theta_{-s,s'}\Big]\,|\,F_{jj'}\,|^2$$
$$-2u_\perp u'_\perp\big(\theta_{-s,s'} + \theta_{s,-s'}\big) + F^*_{j-1,j'-1}F_{j,j'}\Big\},$$
$$Q^{(p)}(4) = 2\epsilon\left(\frac{e^{-i\phi}}{i}\right)^2 u_\perp u'_\perp\big(\theta_{s,s'} + \theta_{-s,-s'}\big)F_{j,j'-1}F^*_{j-1,j'},$$
$$Q^{(p)}(5) = \sqrt{2}\,\epsilon i e^{i\phi}\{\big[u_\perp\big(u_{||} + u'_{||}\big)\theta_{-s,-s'}$$
$$+u_\perp\big(u'_{||} - u_{||}\big)\theta_{s,-s'}\big]F_{j-1,j'}F^*_{jj'}$$
$$+\big[u'_\perp\big(u_{||} + u'_{||}\big)\theta_{s,s'}$$
$$-u'_{||}\big(u'_{||} - u_{||}\big)\theta_{s,-s'}\big]F_{j-1,j'}F^*_{j,j'-1}\},$$
$$Q^{(p)}(6) = \sqrt{2}\,\epsilon i e^{i\phi}\{\big[u'_\perp\big(u'_{||} + u_{||}\big)\theta_{-s,-s'}$$
$$-u'_\perp\big(u'_{||} - u_{||}\big)\theta_{-s,s'}\big]F_{j-1,j'-1}F^*_{j,j'}$$
$$+\big[u_\perp\big(u'_{||} + u_{||}\big)\theta_{s,s'}$$
$$+u_\perp\big(u'_{||} - u_{||}\big)\theta_{-s,s'}\big]F_{j-1,j'-1}F^*_{j,j'-1}\}. \tag{A.21}$$

The mixed parts ($\lambda = -\lambda'$) are

$$Q^{(m)}(1) = 2\epsilon\Big[\theta_{-s,s'}\big(1 + u_{||}u'_{||}\big)^2 + u_\perp^2\,u_\perp'^2\theta_{s,-s'}$$
$$+u_\perp^2\,u_{||}'^2\theta_{s,s'} + u_{||}^2\,u_\perp'^2\theta_{-s,-s'}\Big]\,|\,F_{j,j'-1}\,|^2,$$
$$Q^{(m)}(2) = 2\epsilon\Big[\theta_{s,-s'}\big(1 + u_{||}u'_{||}\big)^2 + u_\perp^2\,u_\perp'^2\theta_{-s,s'}$$
$$+u_{||}^2\,u_\perp'^2\theta_{s,s'} + u_\perp^2\,u_{||}'^2\theta_{-s,-s'}\Big]\,|\,F_{j-1,j'}\,|^2,$$
$$Q^{(m)}(3) = \epsilon\Big\{\Big[\theta_{s,s'}\big(1 - u_{||}u'_{||}\big)^2 + \theta_{-s,-s'}u_\perp^2\,u_\perp'^2$$
$$+\theta_{s,-s'}u_{||}^2\,u_\perp'^2 + \theta_{-s,s'}u_\perp^2\,u_{||}'^2\Big]\,|\,F_{j-1,j'-1}\,|^2$$
$$+\Big[\theta_{-s,-s'}\big(1 - u_{||}u'_{||}\big)^2 + \theta_{s,s'}u_\perp^2\,u_\perp'^2$$
$$+\theta_{s,-s'}u_\perp^2\,u_{||}'^2 + \theta_{-s,s'}u_{||}^2\,u_\perp'^2\Big]\,|\,F_{jj'}\,|^2$$
$$+2\big[\big(\theta_{-s,-s'} - \theta_{s,s'}\big)\big(1 - u_{||}u'_{||}\big)$$
$$+\big(\theta_{s,-s'} + \theta_{-s,s'}\big)u_{||}u'_{||}\big]u_\perp u'_\perp F^*_{j-1,j'-1}F_{jj'}\},$$

$$
\begin{aligned}
Q^{(m)}(4) &= 2\epsilon\left(\frac{e^{-i\phi}}{i}\right)^2\big[(\theta_{s,s'} + \theta_{-s,-s'})u_{||}u'_{||} \\
&\quad -(\theta_{-s,s'} + \theta_{s,-s'})(1 + u_{||}u'_{||})\big]u_{\perp}u'_{\perp}F_{j,j'-1}F^{*}_{j-1,j'}, \\
Q^{(m)}(5) &= \sqrt{2}\,\epsilon i e^{i\phi}\{[-\theta_{s,-s'}u_{\perp}u'_{||}(1 + u_{||}u'_{||}) \\
&\quad -\theta_{s,s'}u_{\perp}u_{||}u'^{2}_{\perp} - \theta_{-s,-s'}u_{\perp}u'_{||}(1 - u'_{||}u'_{\perp}) \\
&\quad +\theta_{-s,s'}u_{||}u_{\perp}u'^{2}_{\perp}]F_{j-1,j'}F^{*}_{jj'} \\
&\quad +[-\theta_{s,-s'}u_{||}u'_{\perp}(1 + u_{||}u'_{||}) \\
&\quad -\theta_{s,s'}u_{||}u'_{\perp}(1 - u_{||}u'_{||}) - \theta_{-s,-s'}u^{2}_{\perp}u'_{||}u'_{\perp} \\
&\quad +\theta_{-s,s'}u^{2}_{\perp}u'_{||}u'_{\perp}]F_{j-1,j'}F^{*}_{j-1,j'-1}\}, \\
Q^{(m)}(6) &= \sqrt{2}\,\epsilon i e^{i\phi}\{[-\theta_{-s,-s'}u_{||}u'_{\perp}(1 - u_{||}u'_{||}) \\
&\quad -\theta_{s,s'}u^{2}_{\perp}u'_{||}u'_{\perp} + \theta_{s,-s'}u^{2}_{\perp}u'_{||}u'_{\perp} \\
&\quad -\theta_{-s,s'}u_{||}u'_{\perp}(1 + u_{||}u'_{||})]F_{jj'}F^{*}_{j,j'-1} \\
&\quad +[-\theta_{s,s'}u_{\perp}u'_{||}(1 - u_{||}u'_{||}) - \theta_{-s,-s'}u_{\perp}u_{||}u'^{2}_{\perp} \\
&\quad +\theta_{s,-s'}u_{\perp}u_{||}u'^{2}_{\perp} \\
&\quad -\theta_{-s,s'}u_{\perp}u'_{||}(1 + u_{||}u'_{||})]F_{j-1,j'-1}F^{*}_{j,j'-1}\}. \qquad \text{(A.22)}
\end{aligned}
$$

These pure and mixed components allow one to define a susceptibility tensor $\overleftrightarrow{\alpha}$ (see eq. 5.3.3) which is completely general, including both electron and positron components.

# Tables

TABLE 1. Binary X-Ray Pulsars (from Nagase, 1989a; Ögelman, 1988)

| NAME/ CATALOG | COORD. | PULSE PER. (s) | ORBITAL PER. (days) | $a_x \sin i$ (lt.-sec) | $f(M)$ | Ecc. | $D$ (kpc) | $Lx$ (ergs $s^{-1}$) | COMP. Sp. | REMARKS |
|---|---|---|---|---|---|---|---|---|---|---|
| 1E | 1024–57 | 0.061 | ... | ... | ... | ... | ... | ... | ... | |
| A | 0535–668 | 0.069 | 16.66 | ... | ... | 0.4 | 50 | $1\times10^{39}$ | B2 III–IVe | Be/LMC |
| SMC X-1 | 0115–737 | 0.717 | 3.892 | 53.46 | 10.8 | <0.007 | 65 | $5\times10^{38}$ | B0 I | HMB/SM |
| Her X-1 | 1656+354 | 1.24 | 1.700 | 13.18 | 0.9 | <0.003 | 5 | $2\times10^{37}$ | A9-B | LMB |
| H | 0850–42 | 1.78 | ... | ... | ... | ... | ... | $1\times10^{37}$ | ... | TR |
| 4U | 0115+634 | 3.61 | 24.31 | 140.13 | 5 | 0.34 | 3.5 | $1\times10^{37}$ | O-Be | Be |
| V | 0331+530 | 4.37 | 34.25 | 48 | 0.1 | 0.31 | 2–4 | $4\times10^{35}$ | Be | Be |
| Cen X-3 | 1119–603 | 4.84 | 2.087 | 39.79 | 15.5 | <0.0008 | 8 | $1\times10^{38}$ | O6-8f | HMB |
| 1E | 1048–593 | 6.44 | ... | ... | ... | ... | 3 | $3\times10^{34}$ | Be | Be |
| GX 109–1 | 2259+586 | 6.98 | <0.08 | <0.2? | ... | >0.1 | 3.6 | $5\times10^{35}$ | ... | LMB/SM |
| 4U | 1627–673 | 7.68 | 0.0228 | ... | ... | ... | 6 | $1\times10^{37}$ | UV | LMB |
| 2S | 1553–542 | 9.26 | 30.7±2.8 | 164 | 5 | 0.09 | ... | ... | Be | Be |
| LMC X-4 | 0532–664 | 13.5 | 1.408 | 26 | 15 | <0.02 | 50 | $6\times10^{38}$ | O7 III–V | HMB/LMC |
| 2S | 1417–624 | 17.6 | >15 | ... | ... | ... | ... | ... | Be? | Be? |
| GS | 1843+00 | 29.5 | ... | ... | ... | ... | ... | ... | ... | TR |
| OAO | 1657–41 | 38.2 | ... | ... | ... | ... | 1–5 | $0.3–6\times10^{37}$ | ... | HMB |
| EXO | 2030+375 | 41.8 | 45.6–47.5 | ... | ... | ... | 5 | $1\times10^{38}$ | Be | Be |
| Cep X-4 | 2137+57 | 66.2 | ... | ... | ... | ... | ... | ... | ... | TR |
| GS | 1843–024 | 94.8 | ... | ... | ... | ... | ... | ... | ... | TR |
| A | 0535+262 | 104 | 111 | 500 | 20 | 0.3 | 2.4 | $2\times10^{37}$ | B0 III–Ve | Be |
| Sct X-1 | 1833–076 | 111 | ... | ... | ... | ... | ... | ... | ... | TR |
| GX 1+4 | 1728–247 | 114 | 304 ? | ... | ... | ... | 9 | $1\times10^{38}$ | M6 III | LMB |
| 4U | 1230–61 | 191 | ... | ... | ... | ... | ... | ... | ... | TR |

TABLE 1. *Continued*

| | | | | | | | | | | |
|---|---|---|---|---|---|---|---|---|---|---|
| GX 304 − 1 | 1258–613 | 272 | 133 | 500 | ··· | ··· | 2.4 | $1\times10^{36}$ | B2 Ve | Be |
| Vela X-1 | 0900–403 | 283 | 8.965 | 112 | 20 | 0.09 | 2.0 | $5\times10^{36}$ | B0.5 Ib | HMB |
| 4U | 1145–619 | 292 | 187.5 | 600 | ··· | ··· | 1.5 | $1\times10^{35}$ | B0-1 Ve | Be |
| 1E | 1145.1–614 | 297 | 5.648? | ··· | ··· | ··· | 8 | $3\times10^{34}$ | B2 I–IIa | HMB |
| A | 1118–616 | 405 | ··· | ··· | ··· | ··· | 5 | $5\times10^{36}$ | O9.5 IV–Ve | Be |
| GPS | 1722–363 | 413 | ··· | ··· | ··· | ··· | ··· | ··· | ··· | TR |
| 4U | 1907+097 | 438 | 8.38 | 80 | 9 | 0.22 | 7 | $4\times10^{37}$ | OB I | HMB |
| 4U | 1538–522 | 529 | 3.730 | 55 | 13 | ··· | 7 | $4\times10^{36}$ | B0 I | HMB |
| GX 301 − 2 | 1223–624 | 696 | 41.50 | 367 | 31 | ··· | 1.8 | $3\times10^{36}$ | B1.5 Ia | HMB |
| X-Per | 0352+309 | 835 | 580? | ··· | ··· | ··· | 0.35 | $4\times10^{33}$ | O9.5 III–Ve | Be |

TABLE 2. Some Nonpulsing LMXBs with Known Periods (from van der Klis, 1989; van den Heuvel, 1987; van Paradijs, 1983)

| NAME/ CATALOG | COORD. | X-RAY TYPE | ORBITAL PER. | COMP. Sp. | REMARKS |
|---|---|---|---|---|---|
| NGC 6624 | 1820–30 | X-QPO | 0.18 h | ⋯ | Episodic reg. X-ray bursts, X-ray mod. |
| | 1916–053 | XRB | 0.33 h | ⋯ | Periodic X-ray dips |
| | 1323–62 | XRB | 3 h | ⋯ | |
| | 1636–53 | X | 3.8 h | ⋯ | Reg. X-ray bursts, optical brightness variations |
| EXO | 0748–676 | X | 3.82 h | ⋯ | |
| GX 9+9 | 1728–169 | X-QPO | 4.2 h | ⋯ | X-ray mod. |
| | 1735–44 | QPO | 4.6 h | ⋯ | Reg. X-ray bursts |
| Cyg X-3 | 2030+407 | X-$\gamma$ | 4.8 h | ⋯ | X-rays vary w. $P_{orb}$; MeV-TeV-PeV $\gamma$-rays; radio flares-jets |
| | 1254–69 | XRB | 4.8 h | ⋯ | |
| | 1822–37 | X | 5.6 h | ⋯ | |
| | 2129+47 | X | 5.6 h | ⋯ | X-ray and optical intensity variations |
| | 1659–29 | X | 7.1 h | ⋯ | Transient |
| | 0620–00 | X | 7.8 h | K | Transient |
| Cen X-4 | 1455–315 | X | 8.2 h? | ⋯ | X-ray variations |
| GX 339–4 | 1659–487 | QPO | 14.8 h? | ⋯ | |
| Sco X-1 | 1617–155 | X-QPO | 19.2 h | ⋯ | QPO, radio source, jets, optical variations |
| GX 17+2 | 1813–140 | X-QPO | 19.8 h? | ⋯ | QPO, occasional X-ray bursts |
| | 0921–63 | X | 9.0 d | F giant | Optical variations |
| Cyg X-2 | 2142+380 | X-QPO | 9.8 d | F giant | Occasional XRB, X-ray and optical variations |
| SS 433 | 1908+05 | X | 13.1 d | ⋯ | Radio jets, X-ray variations, $L_x/L_0=10^{-3}$ |
| Cir X-1 | 1516–56 | XRB | 16.6 d | O-Be | $L_x/L_0=10^{-3}$ |

TABLE 3. Some Binary, Millisecond, and Globular Cluster Pulsars (after Backer and Kulkarni, 1990; Backer, 1990)

| PSR | $P$ (ms) | $\dot{P}$ ($10^{-18}$ ss$^{-1}$) | log $B$ | $P_{orb}$ (d) | Ecc. | M. fcn. | $M_{comp}$ | GC/f | Reference |
|---|---|---|---|---|---|---|---|---|---|
| 0021−72A | 4.479 | <1 | ··· | 0.02 | 0.33 | $1.6\times10^{-8}$ | 0.022 | 47 Tuc | 88 IAU 4602 |
| 0021−72C | 5.757 | ··· | ··· | ··· | ··· | ··· | ··· | 47 Tuc | 89 IAU 4892 |
| 0021−72D | 5.357 | ··· | ··· | ··· | ··· | ··· | ··· | 47 Tuc | 90 workshop |
| 0655+64 | 195.671 | 0.68 | 10 | 1.03 | <0.00005 | $7\times10^{-2}$ | 0.7−1.3 | f | APJ 253 L57 |
| 0820+02 | 864.873 | 103.9 | 11.5 | 1232.47 | 0.0119 | $3\times10^{-3}$ | 0.2−0.4 | f | APJ 236 L25 |
| 1257+12 | 6.218 | ··· | ··· | ··· | ··· | ··· | ··· | f | 90 IAU 5073 |
| 1310+18A | 33.163 | ··· | ··· | 255.84 | ··· | 0.01 | ··· | M53 | 89 IAU 4853 |
| 1516+02A | 5.553 | ··· | ··· | ··· | ··· | ··· | ··· | M5 | 89 IAU 4880 |
| 1516+02B | 7.947 | ··· | ··· | 6.85 | 0.126 | 0.00065 | ··· | M5 | 89 IAU 4880 |
| 1534+12 | 37.904 | ··· | ··· | 0.421 | 0.27 | 0.32 | ··· | f | 90 IAU 5073 |
| 1620−26A | 11.076 | 0.82 | 9.5 | 191.44 | 0.02532 | $8.0\times10^{-3}$ | 0.35 | M4 | NAT 332 45 |
| 1639+36A | 10.378 | <0.05 | ··· | ··· | ··· | ··· | ··· | M13 | 89 IAU 4819 |
| 1745−20 | 288.6 | ··· | ··· | ··· | ··· | ··· | ··· | NGC 6440 | 89 IAU 4905 |
| 1744−24A | 11.563 | ··· | ··· | 0.0708 | 0.003 | 0.00032 | ··· | TER 5 | 90 IAU 4974 |
| 1744−24B | 442.84 | ··· | ··· | ··· | ··· | ··· | ··· | TER 5? | 90 IAU 4974 |
| 1802−07A | 23.10 | ··· | ··· | 2.62 | 0.22 | 0.0097 | ··· | NGC 6539 | 90 IAU 5013 |
| 1820−11 | 279.828 | 1378 | 11.8 | 357.76 | 0.79462 | $6.8\times10^{-2}$ | 1.4 | f | 89 NAT |
| 1820−30A | 5.440 | ··· | ··· | ··· | ··· | ··· | ··· | NGC 6624 | 90 IAU 4988 |
| 1820−30B | 378.59 | ··· | ··· | ··· | ··· | ··· | ··· | NGC 6624 | 90 IAU 4988 |
| 1821−24A | 3.054 | 1.62 | 9.3 | ··· | ··· | ··· | ··· | M28 | 87 IAU 4401 |
| 1831−00 | 520.947 | 14.3 | 10.9 | 1.81 | 0.0001 | $1.2\times10^{-4}$ | 0.06−0.13 | f | NAT 328 399 |
| 1855+09 | 5.362 | 0.017 | 8.5 | 12.33 | 0.000021 | $5.6\times10^{-3}$ | 0.2−0.4 | f | NAT 322 714 |

TABLE 3. *Continued*

| PSR | $P$ (ms) | $\dot{P}$ ($10^{-18}$ ss$^{-1}$) | log $B$ | $P_{orb}$ (d) | Ecc. | M. fcn. | $M_{comp}$ | GC/f | Reference |
|---|---|---|---|---|---|---|---|---|---|
| 1908+00A | 3.6 | ⋯ | ⋯ | ⋯ | ⋯ | ⋯ | ⋯ | NGC 6760 | 90 IAU 5010 |
| 1913+16 | 59.030 | 8.64 | 10.3 | 0.32 | 0.6171 | $1.3\times10^{-1}$ | 1.4 | f | APJ 201 L55 |
| 1937+21 | 1.558 | 0.11 | 8.6 | ⋯ | ⋯ | ⋯ | ⋯ | 4C 21.53 | 82 NAT |
| 1953+29 | 6.133 | 0.03 | 8.6 | 117.35 | 0.00033 | $2.4\times10^{-3}$ | 0.2–0.4 | f | 83 NAT |
| 1957+20 | 1.607 | 0.016 | 8.3 | 0.38 | <0.001 | $5.2\times10^{-6}$ | 0.02 | f | NAT 333 237 |
| 2127+11A | 110.665 | −20. | ⋯ | ⋯ | ⋯ | ⋯ | ⋯ | M15 | NAT 337 531 |
| 2127+11B | 56.133 | >0 | ⋯ | ⋯ | ⋯ | ⋯ | ⋯ | M15 | 89 IAU 4762 |
| 2127+11C | 30.529 | 4.99 | ⋯ | 0.335 | 0.6814 | $1.53\times10^{-1}$ | 1.4 | M15 | 89 IAU 4772 |
| 2127+11D | 4.803 | <0 | ⋯ | ⋯ | ⋯ | ⋯ | ⋯ | M15 | 90 workshop |
| 2127+11E | 4.651 | ⋯ | ⋯ | ⋯ | ⋯ | ⋯ | ⋯ | M15 | 90 workshop |
| 2303+46 | 1066.371 | 569.3 | 11.3 | 12.340 | 0.6584 | $2.5\times10^{-1}$ | 1.4 | f | NAT 317 787 |

TABLE 4. Coefficients $A^i_\alpha$ (after Mészáros and Ventura, 1979)

a) Coefficients $A^i_\alpha$ for the ordinary mode, $i = 1$, as a function of the frequency. The corresponding values for the extraordinary mode are $A^2_\alpha = 1 - A^1_\alpha$. Note also that the relation $\sum_\alpha A^1_\alpha = 1.5$ is satisfied reflecting the normalization of the $e^\wedge$ vector in Eq. (3.4.34).

| $\omega/\omega_B$ | $A^1_z$ | $A^1_+$ | $A^1_-$ |
|---|---|---|---|
| $10^{-3}$ | 1 | 0.2536 | 0.2464 |
| $10^{-2}$ | 0.9997 | 0.2806 | 0.2197 |
| $10^{-1}$ | 0.9894 | 0.3951 | 0.1155 |
| 0.2 | 0.9726 | 0.4609 | 0.0664 |
| 0.5 | 0.9219 | 0.5644 | 0.0136 |
| 1 | 0.8562 | 0.6438 | 0 |
| 2 | 0.7749 | 0.7136 | 0.0115 |
| 5 | 0.6676 | 0.7849 | 0.0474 |
| 10 | 0.5990 | 0.8243 | 0.0767 |
| 100 | 0.5107 | 0.8707 | 0.1185 |

b) Vacuum-modified coefficients $A^1_+$, $A^1_-$, as a function of frequency and density, for the extraordinary mode (1) and $B = 10^{-1}\ B_c$. The remaining coefficients are obtained from these through $A^2_\pm = 1 - A^1_\pm$, $A^i_z = 1.5 - A^i_+ - A^i_-$. The line running through the table separates the vacuum-modified case (*right*) from the plasma-dominated case (*left of the line*). The empty spaces at the upper left corner indicate the regime where $\omega_p \gtrsim \omega$, for which collective effects can become important.

| $\omega/\omega_H \backslash \omega_p^2/\omega_H^2$ | | $10^{-4}$ | $10^{-5}$ | $10^{-6}$ | $10^{-7}$ | $10^{-8}$ | $10^{-9}$ | $10^{-10}$ | $10^{-11}$ |
|---|---|---|---|---|---|---|---|---|---|
| $10^{-3}$ | + | | | 0.75 | 0.75 | 0.75 | 0.75 | 0.75 | 0.75 |
| | − | | | 0.75 | 0.75 | 0.75 | 0.75 | 0.75 | 0.75 |
| $10^{-2}$ | + | | 0.72 | 0.72 | 0.72 | 0.72 | 0.72 | 0.80 | 0.75 |
| | − | | 0.78 | 0.78 | 0.78 | 0.78 | 0.78 | 0.70 | 0.75 |
| 0.1 | + | 0.61 | 0.60 | 0.60 | 0.59 | 0.93 | 0.77 | 0.75 | 0.75 |
| | − | 0.88 | 0.88 | 0.89 | 0.90 | 0.55 | 0.73 | 0.75 | 0.75 |
| 0.2 | + | 0.54 | 0.54 | 0.53 | 0.44 | 0.83 | 0.76 | 0.75 | 0.75 |
| | − | 0.93 | 0.93 | 0.94 | 0.99 | 0.67 | 0.74 | 0.75 | 0.75 |
| 0.5 | + | 0.44 | 0.43 | 0.40 | 0.95 | 0.80 | 0.76 | 0.75 | 0.75 |
| | − | 0.99 | 0.99 | 1.00 | 0.51 | 0.70 | 0.74 | 0.75 | 0.75 |
| 1 | + | 0.36 | 0.36 | 0.36 | 0.36 | 0.36 | 0.36 | 0.36 | 0.36 |
| | − | 1.00 | 1.00 | 1.00 | 1.00 | 1.00 | 1.00 | 1.00 | 1.00 |
| 2 | + | 0.30 | 0.40 | 0.60 | 0.72 | 0.75 | 0.75 | 0.75 | 0.75 |
| | − | 0.99 | 1.00 | 0.89 | 0.78 | 0.75 | 0.75 | 0.75 | 0.75 |
| 5 | + | 0.44 | 0.65 | 0.73 | 0.75 | 0.75 | 0.75 | 0.75 | 0.75 |
| | − | 0.99 | 0.84 | 0.77 | 0.75 | 0.75 | 0.75 | 0.75 | 0.75 |
| 10 | + | 0.64 | 0.73 | 0.75 | 0.75 | 0.75 | 0.75 | 0.75 | 0.75 |
| | − | 0.86 | 0.77 | 0.75 | 0.75 | 0.75 | 0.75 | 0.75 | 0.75 |
| 100 | + | 0.75 | 0.75 | 0.75 | 0.75 | 0.75 | 0.75 | 0.75 | 0.75 |
| | − | 0.75 | 0.75 | 0.75 | 0.75 | 0.75 | 0.75 | 0.75 | 0.75 |

# References


Ables, J. G., *et al.*, 1988, *IAU Circ.*, no. 4602.

Abramowitz, M. and Stegun, I., 1964, *Handbook of Mathematical Functions* (Dover, New York).

Adler, S. L., *Ann. Phys.* (New York), 67, 599.

Adler, S. L., Bahcall, J. N., Callan, C. G. and Rosenbluth, M. N., 1970, *Phys. Rev. Letters*, 25, 1061.

Alcock, C., Farhi, E. and Olinto, A., 1986, *Phys. Rev. Letters*, 57, 2088.

Alcock, J. 1987, in *IAU Symp. 125, Origin and Evolution of Neutron Stars*, ed. D. Helfand and J.-H. Huang (Reidel, Dordrecht), p. 413.

Alexander, S. and Mészáros, P., 1991a, *Ap. J.*, 372, 545.

Alexander, S. and Mészáros, P., 1991b, *Ap. J.*, 372, 554.

Alexander, S., Mészáros, P. and Bussard, R. W., 1989, *Ap. J.*, 342, 928.

Alexander, S. G., 1990, Ph.D. thesis, Pennsylvania State University.

Alexander, S. G. and Mészáros, P., 1989, *Ap. J.*, 344, L1.

Alexeenko, V. V., *et al.*, 1987, in *Proc. 20th ICRC* (Moscow), OG 3.1-12.

Alexeenko, V. V., *et al.*, 1989, paper presented at the 3d VHE Gamma-Ray Astronomy Meeting.

Alpar, M. A., 1977, *Ap. J.*, 213, 527.

Alpar, M. A., 1989, in *Timing Neutron Stars*, ed. H. Ögelman and E. v. d. Heuvel (Kluwer, Dordrecht), p. 431.

Alpar, M. A., Cheng, K. S. and Pines, D., 1989, *Ap. J.*, 346, 823.

Alpar, M. A. and Ögelman, H., 1987, *Astron. Ap.*, 185, 196.

Alpar, M. A., Pines, D. and Cheng, K. S., 1990, *Nature*, 348, 707.

Alpar, M. A. and Sauls, J. A., 1988, *Ap. J.*, 327, 723.

Alpar, M. A., *et al.*, 1982, *Nature*, 300, 728.

Alpar, M. A., *et al.*, 1984a, *Ap. J.*, 276, 325.

Alpar, M. A., *et al.*, 1984b, *Ap. J.*, 278, 791.

Alpar, M. A., *et al.*, 1984c, *Ap. J.*, 282, 533.

Aly, J. J., 1980, *Astron. Ap.*, 86, 192.

Aly, J. J., 1986, in *Plasma Penetration in Magnetospheres*, ed. N. Kylafis *et al.* (Crete University Press, Crete, Greece), p. 125.

Anderson, P. W. and Itoh, N., 1975, *Nature*, 256, 25.

Anzer, U. and Börner, G., 1983, *Astron. Ap.*, 122, 73.

Anzer, U., Börner, G. and Mészáros, P., 1976, *Astron. Ap.*, 50, 305.

Aprile, E., *et al.*, 1985, in *Proc. Int. Europhysics Conf. on High Energy Physics*, Bari.

Arnett, D. and Bowers, R., 1977, *Ap. J.* (*Suppl.*), 33, 415.

Arons, J., 1981, *Ap. J.*, 248, 1099.

Arons, J., 1983a, *Ap. J.*, 266, 215.

Arons, J., 1983b, in *AIP Conf. Proc. 101, Positron-Electron Pairs in Astrophysics*, ed. M. Burns, A. Harding, and R. Ramaty (AIP, New York), p. 163.

Arons, J., 1987, in *IAU symp. 125, The Origin and Evolution of Neutron Stars*, ed. D. Helfand (Reidel, Dordrecht), p. 207.

Arons, J., Klein, R. and Lea, S., 1986, in *Plasma Penetration into Magnetospheres*, ed. N. Kylafis *et al.* (Crete University Press, Crete, Greece), p. 141.

Arons, J., Klein, D. and Lea, S. M., 1987, *Ap. J.*, 312, 666.

Arons, J. and Lea, S., 1976a, *Ap. J.*, 207, 914.

Arons, J. and Lea, S., 1976b, *Ap. J.*, 210, 792.

Arons, J. and Lea, S., 1980, *Ap. J.*, 235, 1016.

Arons, J. and Scharlemann, E. T., 1979, *Ap. J.*, 231, 854.

Arons, J., *et al.*, 1982, in *AIP Conf. Proc. 115, High Energy Transients in Astrophysics*, ed. S. Woosley (AIP, New York), p. 215.

Atteia, J., *et al.*, 1987, *Ap. J. (Supp.)*, 64, 305.

Baade, W. and Zwicky, F., 1934a, *Phys. Rev.*, 45, 138.

Baade, W. and Zwicky, F., 1934b, *Proc. Nat. Acad. Sci.*, 20, 254.

Baan, W. and Treves, A., 1972, *Astron. Ap.*, 22, 421.

Backer, D. C. , 1990, private communication.

Backer, D. C. and Kulkarni, S. R., 1990, *Phys. Today*, 43, no. 3, 26.

Bahcall, J. N. and Wolf, R., 1965, *Phys. Rev.*, B140, 1452.

Baltrusaitis, R. M., *et al.*, 1985a, *Ap. J.* (*Letters*), 297, L145.

Baltrusaitis, R. M., *et al.*, 1985ba, *Ap. J.* (*Letters*), 293, L69.

Baring, M. G., 1988, *M.N.R.A.S.*, 235, 51.

Barnard, J. J., 1988, in *The Origin and Evolution of Neutron Stars*, ed. D. Helfand and J.-H. Huang (Reidel, Dordrecht), p. 56.

Baron, E. and Cooperstein, J., 1991, in *Supernovae*, ed. S. Woosley (Springer, Berlin), p. 342.

Barrow, J. D. and Tipler, F. J., 1986, *The Anthropic Cosmological Principle* (Oxford University Press, Oxford).

Basko, M. M., 1976, *Astrophysics (Aztrofizika)*, 12, 169.

Basko, M. M. and Sunyaev, R. A., 1975a, *Astron, Ap.* , 42, 311.

Basko, M. M. and Sunyaev, R. A., 1975b, *Sov. Phys. JETP*, 41, 52.

Basko, M. M. and Sunyaev, R. A., 1976a, *M.N.R.A.S.*, 175, 395.

Basko, M. M. and Sunyaev, R. A., 1976b, *Sov. Astron.*, 20, 537.

Battistoni, G., *et al.*, 1985a, *Phys. Letters*, 155B, 465.

Battistoni, G., *et al.*, 1985b, in *Proc. 19th ICRC* (La Jolla).

Baylin, C. D. and Grindlay, J. E., 1990, *Ap. J.*, 353, 159.

Baym, G. and Pethick, C. J., 1979, *Ann. Rev. Astron. Ap.*, 17, 415.

Baym, G., Pethick, C. J. and Sutherland, P., 1971, *Ap. J.*, 170, 299.

Baym, G., *et al.*, 1969, *Nature*, 224, 673; 224, 872.

Baym, G., *et al.*, 1985, *Phys. Letters*, 160B, 181.

Bekefi, G. 1966, *Radiation Processes in Plasmas* (Wiley, New York).

Belian, R. D., Conner, J. P. and Evans, W. D., 1976, *Ap. J.* (*Letters*), 207, L33.

Berezinski, V. S., Castagnoli, C. and Galeotti, P., 1986, *Ap. J.*, 301, 235.

Berger, Ch., *et al.*, 1986, *Phys. Letters*, 174, 118.

Beskin, V. S., Gurevich, A. V. and Istomin, Ya. A., 1983a, *Ap. Space Sci.*, 102, 301.

Beskin, V. S., Gurevich, A. V. and Istomin, Ya. A., 1983b, *JETP*, 58, 235.

Beskin, V. S., Gurevich, A. V. and Istomin, Ya. A., 1984, *Ap. Space Sci.*, 102, 301.

Beskin, V. S., Gurevich, A. V. and Istomin, Ya. A., 1986, *Sov. Phys. Usp.*, 29, 946.

Beskin, V. S., Gurevich, A. V. and Istomin, Ya. A., 1988, *Ap. Space Sci.*, 146, 205.

Bethe, H. and Critchfield, C. H., 1938, *Phys. Rev.*, 54, 248.

Bezchastnov, V. G. and Pavlov, G. G., 1988, *Ap. Space Sci.*, 148, 257.

Bezchastnov, V. G. and Pavlov, G. G., 1989, *Sov. Phys. JETP*, 95, 832.

Bezchastnov, V. G. and Pavlov, G. G., 1991, *Ap. Space Sci.*, in press.

Bhat, C. L., *et al.*, 1986a, *Ap. J.*, 306, 587.

Bhat, C. L., *et al.*, 1986b, *Astron. Ap.*, 159, 299.

Bhat, P. N., *et al.*, 1980, *Astron. Ap.*, 81, L3.

Bhat, P. N., *et al.*, 1986c, *Nature*, 319, 127.

Bhat, P. N., *et al.*, 1987, in *Proc. 20th ICRC* (Moscow), OG 3.1–14.

Bhattacharya, D. and Srinivasan, G., 1986, *Curr. Sci.*, 55, 327.

Bhattacharya, D. and Srinivasan, G., 1990, in *Neutron Stars*, NATO ASI, Crete, Greece, ed. J. Ventura and D. Pines (Cambridge University Press), in press.

Bignami, G. F. and Hermsen, W., 1983, *Ann. Rev. Astron. Ap.*, 21, 67.

Bignami, G. F., *et al.*, 1984, *Nature*, 310, 464.

Bionta, R. M., *et al.*, 1987, *Phys. Rev.*, D36, 30.

Bisnovatyi-Kogan, G., 1974, *Ap. Space Sci.*, 26, 25.

Bisnovatyi-Kogan, G., 1975, *Ap. Space Sci.*, 35, 23.

Bisnovatyi-Kogan, G., 1990, in *Los Alamos Workshop on Gamma-Ray Bursts*, ed. R. Epstein *et al.* (Cambridge University Press), in press.

Bisnovatyi-Kogan, G. S. and Komberg, B. V., 1974, *Sov. Astr.*, 18, 217.

Bjorken, J. D. and Drell, S. D., 1964, *Relativistic Quantum Mechanics* (McGraw-Hill, New York).

Blaauw, A., 1961, *Bull. Astron. Inst. Netherlands*, 15, 265.

Blaes, O., Blandford, R., Goldreich, P. and Madau, P., 1989, *Ap. J.*, 343, 839.

Blaes, O., Blandford, R., Madau, P. and Koonin, S., 1990, *Ap. J.*, 363, 612.

Blair, D. G. and Candy, B. N., 1988, in *Timing Neutron Stars*, ed. H. Ögelman and E. v. d. Heuvel (NATO ASI, Kluwer, Dordrecht), p. 609.

Blandford, R., Applegate, J. and Hernquist, L., 1983, *M.N.R.A.S.*, 204, 1025.

Blandford, R. and Eichler, D., 1987, *Phys. Rep.*, 154, no. 1, 1.

Blondin, J. and Freese, K., 1986, *Nature*, 323, 786.

Boer, M., *et al.*, 1991, *Astron. Ap.*, in press.

Bonazzola, S., 1982, *Astron. Ap.*, 108, 19.

Bonazzola, S., Heyvaerts, J. and Puget, J. L., 1979, *Astron. Ap.*, 78, 53.

Bonazzola, S., *et al.*, 1984, *Astron. Ap.*, 136, 89.

Bondi, H. and Hoyle, F., 1944, *M.N.R.A.S.*, 104, 273.

Bonnet-Bidaud, J.-M. and Chardin, G., 1988, *Phys. Rep.*, 170, 6, 325.

Bonsema, P. and van den Heuvel, E., 1985, *Astron. Ap.*, 46, L3.

Boone, J., *et al.*, 1984, *Ap. J.*, 285, 264.

Börner, G. and Mészáros, P., 1979, *Plasma Phys.*, 21, 357.

Boynton, P. E., 1981, in *IAU Symp. 95, Pulsars*, ed. W. Sieber and R. Wielebinski (Reidel, Dordrecht), p. 279.

Boynton, P. E., Crossa, L. and Deeter, J. E., 1980, *Ap. J.*, 237, 169.

Boynton, P. E. and Deeter, J. E., 1979, in *Compact Galactic X-Ray Sources*, ed. F. K. Lamb and D. Pines (University of Illinois, Urbana), p. 168.

Boynton, P. E., *et al.*, 1984, *Ap. J.* (*Letters*), 283, L53.

Boynton, P. E., *et al.*, 1986, *Ap. J.*, 307, 545.

Brainerd, J. J., 1987, *Ap. J.*, 313, 714.

Brainerd, J. J., 1989, *Ap. J. (Letters)*, 341, L67.

Brainerd, J. J. and Lamb, D. Q., 1987, *Ap. J.*, 313, 231.

Brainerd, J. J. and Mészáros, P., 1991, *Ap. J.*, 369, 179.

Brainerd, J. J. and Petrosian, V., 1987, *Ap. J.*, 320, 703.

Braun, A. and Yahel, R. Z., 1984, *Ap. J.*, 78, 349.

Brazier, K. T. S., *et al.*, 1990, *Ap. J.*, 350, 745.

Buccheri, R., 1976, in *The Structure and Content of the Galaxy and Galactic Gamma-Rays*, ed. C. Fichtel, and F. Stecker (NASA Goddard, Greenbelt, MD), p. 52.

Buccheri, R., 1980, in *Adv. Space Exploration* (Pergamon Press, Oxford), vol. 7, p. 17.

Buccheri, R., *et al.*, 1985, *Nature*, 316, 131.

Burnard, D. J., Arons, J. and Klein, R. I., 1991, *Ap. J.*, 367, 575.

Burnard, D. J., Klein, R. I., and Arons, J. 1988, *Ap. J.*, 324, 1001.

Burnard, D. J., Klein, R. I., and Arons, J. 1990, *Ap. J.*, 349, 262.

Burns, M. L. and Harding, A. K., 1984, *Ap. J.*, 285, 747.

Burrows, A., 1980, *Phys. Rev. Letters*, 44, 1640.

Bussard, R. W., 1980, *Ap. J.*, 237, 970.

Bussard, R. W., 1984, *Ap. J.*, 284, 357.

Bussard, R. W., 1986a, in *Gamma-Ray Bursts*, ed. E. Liang and V. Petrosian (AIP, New York), p. 147.

Bussard, R. W., 1986b, private communication.

Bussard, R. W., Alexander, S. G. and Mészáros, P., 1986, *Phys. Rev.*, D34, 440.

Bussard, R. W. and Lamb, F. K., 1982, in *AIP Conf. Proc. 115, High Energy Transients in Astrophysics*, ed. S. Woosley (AIP, New York), p. 189.

Bussard, R. W., Lamb, F. K. and Pakey, D., 1986, SSL preprint 86-130.

Bussard, R. W., Mészáros, P. and Alexander, S. G., 1985, *Ap. J. (Letters)*, 297, L21.

Canal, R., Isern, J. and Labay, J., 1990, *Ann. Rev. Astron. Ap.*, 28, 183.

Canuto, V. and Kelly, D. C., 1972, *Ap. Space Sci.*, 17, 277.

Canuto, V., Lodenquai, J. and Ruderman, M., 1971, *Phys. Rev.*, D3, 2303.

Canuto, V. and Ventura, J., 1972, *Ap. Space Sci.*, 18, 104.

Canuto, V. and Ventura, J., 1977, *Fund. Cosmic Phys.*, 2, 203.

Caraveo, P., Bignami, G. F., Mitrofanov, I. G. and Vacanti, G., 1988, *Ap. J.*, 327, 203.

Carraminana, A., *et al.*, 1988, in *Timing Neutron Stars*, ed. H. Ögelman and E. v. d. Heuvel (NATO-ASI, Kluwer, Dordrecht), p. 369.

Cassiday, *et al.*, 1990, in *Proc. 21st ICRC* (Adelaide), 2, 60.

Castor, J. I., 1972, *Ap. J.*, 178, 779.
Cawley, M. F., *et al.*, 1985, in *Proc. 19th ICRC* (La Jolla), 1, 173.
Chadwick, P. M., *et al.*, 1985a, in *Proc. 19th ICRC* (La Jolla), 1, 251.
Chadwick, P. M., *et al.*, 1985b, *Astron. Ap.*, 151, L1.
Chadwick, P. M., *et al.*, 1985c, *Nature*, 317, 236.
Chadwick, P. M., *et al.*, 1987a, in *Very High Energy Gamma-Ray Astronomy*, ed. K. Turver (Reidel, Dordrecht), p. 115.
Chadwick, P. M., *et al.*, 1987b, in *Very High Energy Gamma-Ray Astronomy*, ed. K. Turver (Reidel, Dordrecht), p. 121.
Chanan, G. A., *et al.*, 1978, *Ap. J.* (*Letters*), 228, L71.
Chandrasekhar, S., 1935, *M.N.R.A.S.*, 95, 207.
Chandrasekhar, S., 1960, *Radiative Transfer* (Dover, New York).
Chanmugam, G. and Brecher, K., 1985, *Nature*, 313, 767.
Chardin, G., 1986, in *Accretion Processes in Astrophysics*, ed. J. Audouze and J. T. T. Van (Frontières, Gif-sur-Yvette), p. 63.
Cheng, A. F., 1989, *Ap. J.*, 339, 291.
Cheng, K. S., Ho, C. and Ruderman, M, 1985a, *Ap. J.*, 300, 500.
Cheng, K. S., Ho, C. and Ruderman, M, 1985b, *Ap. J.*, 300, 522.
Cheng, K. S., and Ruderman, M, 1989, *Ap. J.* (*Letters*), 337, L77.
Cheng, K. S., and Ruderman, M, 1991, *Ap. J.*, 373, 187.
Chiu, H.-Y. and Salpeter, E. E., 1964, *Phys. Rev. Letters*, 12, 413.
Chudakov, A. E., *et al.*, 1965, *Trans. Consultants Bureau*, 26, 99.
Chudakov, V. V., *et al.*, 1985, in *Proc. 19th ICRC* (La Jolla), 9, 441.
Clark, G. W., *et al.*, 1990, *Ap. J.*, 353, 274.
Clayton, D. D., 1983, *Principles of Stellar Evolution and Nucleosynthesis* (University of Chicago Press, Chicago).
Cocke, W. J., Disney, M. J. and Taylor, D. J., 1969, *Nature*, 221, 525.
Cohen, R., Lodenquai, J. and Ruderman, M., 1970, *Phys. Rev. Letters*, 25, 467.
Colgate, S., 1990, in *Los Alamos Workshop on Gamma-Ray Bursts*, ed. R. Epstein *et al.* (Cambridge University Press), in press.
Colgate, S. and Petschek, A., 1981, *Ap. J.*, 248, 771.
Corbet, R. H. D., 1984, *Astron. Ap.*, 141, 91.
Cordes, J. M., 1978, *Ap. J.*, 222, 1006.
Cordes, J. M., 1987, in *The Origin and Evolution of Neutron Stars*, ed. D. Helfand and J.-H. Huang (Reidel, Dordrecht), p. 35.
Cordes, J. M. and Downs, G. S., 1985, *Ap. J.* (*Suppl.*), 59, 343.
Cordes, J. M. and Helfand, D. J., 1980, *Ap. J.*, 239, 640.
Coté, J. and Pylyser, E. H. P., 1989, *Astron. Ap.*, 218, 131.

Daugherty, J. K. and Bussard, R. W., 1980, *Ap. J.*, 238, 296.
Daugherty, J. K. and Harding, A. K., 1982, *Ap. J.*, 252, 337.
Daugherty, J. K. and Harding, A. K., 1983, *Ap. J.*, 273, 761.
Daugherty, J. K. and Harding, A. K., 1986, *Ap. J.*, 309, 362.
Daugherty, J. K. and Ventura, J., 1978, *Phys. Rev.*, D18, no. 4, 1053.
Davidson, K., 1973, *Nature Phys. Sci.*, 246, 1.
Davidson, K. and Ostriker, J. P., 1973, *Ap. J.*, 179, 585.
Davies, R. E. and Pringle, J. E., 1980, *M.N.R.A.S.*, 191, 599.
Davis, L. and Goldstein, M., 1970, *Ap. J. (Letters)*, 159, L81.
Deeter, J. E., *et al.*, 1989, *Ap. J.*, 336, 376.
de Jager, O. C., *et al.*, 1987, Ph.D. thesis, Potchefstroom University.
de Jager, O. C., *et al.*, 1990, *Nucl. Phys. B (Proc. Suppl.)*, in press.
Demianski, M. and Proszynski, M. 1983, *M.N.R.A.S.*, 202, 437.
Dermer, C. D., 1989, *Ap. J. (Letters)*, 347, L13.
Dermer, C. D., 1990, *Ap. J.*, 360, 197.
Dermer, C. D. and Vitello, P., 1990, in *Los Alamos Workshop on Gamma-Ray Bursts*, ed. R. Epstein *et al.* (Cambridge University Press), in press.
Desai, U., 1986, in *AIP Conf. Proc. 141, Gamma-Ray Bursts*, ed. E. Liang and V. Petrosian (AIP, New York), p. 8.
Domokos, G. and Nussinov, S., 1987, *Phys. Letters*, 187B, 372.
Dothwaite, J. C., *et al.*, 1984a, *Astron. Ap.*, 136, L14.
Dothwaite, J. C., *et al.*, 1984b, *Ap. J. (Letters)*, 286, L35.
Dothwaite, J. C., *et al.*, 1984c, *Nature*, 309, 691.
Downs, G. S., 1981, *Ap. J.*, 249, 687.
Drees, M., *et al.*, 1988, preprint, MAD/PH/424.
Drury, L. O., *Rep. Prog. Phys.*, 46, 973.
Dyson, F., 1969, *Nature*, 223, 486.
Dzikowski, T., *et al.*, 1983, *J. Phys. G*, 9, 459.
Eames, P., *et al.*, 1987, in *Very High Energy Gamma-Ray Astronomy*, ed. K. Turver (Reidel, Dordrecht), p. 179.
Edwards, P. G., *et al.*, 1985, *J. Phys. G: Nucl. Phys.*, 11, L101.
Eggleton, P., 1983, *Ap. J.*, 268, 368.
Eichler, D., 1986, in *Accretion Processes in Astrophysics*, ed. J. Audouze and J. T. T. Van (Moriond Meetings, Gif-sur-Yvette, Editions Frontières), p. 27.
Eichler, D. and Vestrand, T., 1985, *Nature*, 318, 345.
Elsner, R. F. and Lamb, F. K., 1976, *Nature*, 262, 356.
Elsner, R. F. and Lamb, F. K., 1977, *Ap. J.*, 215, 897.
Elsner, R. F. and Lamb, F. K., 1984, *Ap. J.*, 278, 326.

Epstein, R. I., 1985, *Ap. J.*, 291, 822.

Epstein, R. I., 1988, *Phys. Rep*, 163, 155.

Epstein, R. I. and Hurley, K., 1988, *Astron. Letters and Comm.*, 27, 229.

Erber, T., 1964, *Rev. Mod. Phys.*, 38, 626.

Ericksson, R. A., *et al.*, 1976, *Ap. J.*, 210, 539.

Euler, H. and Kockel, B., 1934, *Naturwissenschaften*, 23, 246.

Evans, W. D. and Laros, J. G., 1986, in *The Origin and Evolution of Neutron Stars*, ed. D. Helfand (Reidel, Dordrecht), p. 477.

Fabian, A., *et al.*, 1975, *M.N.R.A.S.*, 197, 15P.

Fawley, W., Scharlemann, E. T. and Arons, J., 1977, *Ap. J.*, 217, 227.

Fazio, G. G., *et al.*, 1972, *Ap. J. (Letters)*, 175, L117.

Feautrier, P., 1964, C. R. *Acad. Sci. Paris*, 258, 3189.

Fegan, D. J., 1990, in *Proc. 21st ICRC* (Adelaide), 2, 111.

Fegan, D. J., *et al.*, 1987, in *Very High Energy Gamma-Ray Astronomy*, ed. K. Turver (Reidel, Dordrecht), p. 111.

Felten, J. E. and Rees, M. J., 1972, *Astron. Ap.*, 17, 226.

Fenimore, E. E., *et al.*, 1982, *Nature*, 297, 665.

Fenimore, E. E., *et al.*, 1988, *Ap. J. (Letters)*, 335, L71.

Feynman, R. P., Leighton, R. B. and Sands, M., 1963, *Lectures on Physics*, vol. 1 (Addison-Wesley, Reading, MA).

Fichtel, C. E., Thompson, D. J. and Lamb, R. C., 1987, *Ap. J.*, 319, 362.

Flanagan, C., 1990, *Nature*, 345, 416.

Flowers, E. and Itoh, N., 1976, *Ap. J.*, 206, 218.

Flowers, E. and Itoh, N., 1981, *Ap. J.*, 250, 750.

Flowers, E. and Ruderman, M., 1977, *Ap. J.*, 215, 302.

Flowers, E. G., *et al.*, 1977, *Ap. J.*, 215, 291.

Forrest, D. J., *et al.*, 1980, *Solar Phys.*, 35, 15.

Frank, J., King, A. A. and Raine, D. J., 1985, *Accretion Power in Astrophysics* (Cambridge University Press, Cambridge).

Fried, B. D. and Conte, S. D., 1961, *The Plasma Dispersion Function* (Academic Press, New York).

Friedman, B. and Pandharipande, V. R., 1981, *Nucl. Phys.*, A361, 502.

Friedman, J., Ipser, J. and Parker, L., 1986, *Ap. J.*, 304, 11.

Friman, B. L. and Maxwell, O. V., 1979, *Ap. J.*, 232, 541.

Fritz, G., *et al.*, 1969, *Science*, 164, 709.

Fruchter, A. S., *et al.*, 1988a, *Nature*, 333, 237.

Fruchter, A. S., *et al.*, 1988b, in *Timing Neutron Stars*, ed. H. Ögelman and E. v. d. Heuvel (NATO ASI, Kluwer, Dordrecht), p. 163.

Gaisser, T. and Stanev, T., 1987, *Phys. Rev. Letters*, 54, 2265.

Gasiorowicz, S., 1974, *Quantum Physics* (Wiley, New York).

Geprägs, R., Kaiser, H., Herold, H. and Ruder, H., 1990, in *Neutron Stars*, ed. J. Ventura and D. Pines (NATO ASI, Kluwer, Dordrecht), in press.

Ghosh, P. and Lamb, F. K., 1978, *Ap. J.* (*Letters*), 223, L83.

Ghosh, P. and Lamb, F. K., 1979a, *Ap. J.*, 232, 259.

Ghosh, P. and Lamb, F. K., 1979b, *Ap. J.*, 234, 296.

Giaconni, R., *et al.*, 1971, *Ap. J.* (*Letters*), 167, L67.

Gibson, A. I., *et al.*, 1982, *Nature*, 296, 833.

Gilman, D., *et al.*, 1980, *Ap. J.*, 236, 951.

Ginzburg, V. L., 1970, *Propagation of Electromagnetic Waves in Plasmas* (Pergamon, Oxford).

Gnedin, Yu. N. and Nagel, W., 1984, *Astron. Ap.*, 138, 356.

Gnedin, Yu. N. and Pavlov, G. G., 1974, *Sov. Phys. JETP*, 38, 903.

Gnedin, Yu. N., Pavlov, G. G. and Shibanov, Yu. A., 1978a, *JETP Letters*, 27, 305.

Gnedin, Yu. N., Pavlov, G. G. and Shibanov, Yu. A., 1978b, *Sov. Astron. Letters*, 4, 117.

Gnedin, Yu. N. and Sunyaev, R. A., 1973, *JETP*, 65, 102.

Gold, T., 1968, *Nature*, 218, 731.

Goldreich, P., 1970, *Ap. J.* (*Letters*), 160, L11.

Goldreich, P. and Julian, W. H., 1969, *Ap. J.*, 157, 869.

Goldstein, H., 1950, *Classical Mechanics* (Addison-Wesley, Reading, MA).

Good, M. L. and Ng, K. K., 1985, *Ap. J.*, 299, 706.

Goodman, J., 1986, *Ap. J.* (*Letters*), 308, L47.

Gorham, P. W., *et al.*, 1986, *Ap. J.* (*Letters*), 308, L11; *Ap.J.*, 309, 121.

Gorham, P. W., *et al.*, 1987, in *Very High Energy Gamma-Ray Astronomy*, ed. K. Turver (Reidel, Dordrecht), p. 125.

Grindlay, J., *et al.*, 1976, *Ap. J.* (*Letters*), 205, L127.

Grindlay, J., *et al.*, 1983, *Nature*, 300, 730.

Grindlay, J. E., *et al.*, 1975a, *Ap. J.* (*Letters*), 197, L9.

Grindlay, J. E., *et al.*, 1975b, *Ap. J.*, 201, 82.

Grindlay, J. E., *et al.*, 1976, *Ap. J.*, 209, 592.

Gruber, D. E. and Rothschild, R. E., 1984, *Ap. J.*, 283, 546.

Gruber, D. E., *et al.*, 1980, *Ap. J.* (*Letters*), 240, L127.

Gudmundsson, E. H., Pethick, C. J. and Epstein, R. I., 1983, *Ap. J.*, 272, 286.

Gunn, J. E. and Ostriker, J. P., 1970, *Ap. J.*, 160, 969.

Gupta, S. K., 1983, Ph.D. thesis, University of Bombay.

Gupta, S. K., *et al.*, 1978, *Ap. J.*, 221, 268.

Haines, L. and Roberts, D., 1969, *Am. J. Phys.*, 37, 1145.

Halzen, F., 1990, in *Astrophysical Aspects of the Most Energetic Cosmic Rays*, ed. M. Nagano (Tokyo University Press), in press.

Halzen, F., *et al.*, 1987, *Phys. Letters*, 190B, 211.

Hameury, J. M. and Lasota, J. P., 1986, in *AIP Conf. Proc. 144, Gamma-Ray Bursts*, ed. E. Liang and V. Petrosian (AIP, New York), p. 164.

Hameury, J. M. and Lasota, J. P., 1989, *Astron. Ap.*, 211, L15.

Hameury, J. M., *et al.*, 1980, *Astron. Ap.*, 90, 359.

Hameury, J. M., *et al.*, 1982, *Astron. Ap.*, 111, 242.

Hameury, J. M., *et al.*, 1985, *Ap. J.*, 293, 56.

Hamilton, P. A., King, E. A., McDonell, D. and McCulloch, P. M., 1989, *IAU Circ.*, no. 4708.

Harding, A. K., 1981, *Ap. J.*, 245, 267.

Harding, A. K., 1986, *Ap. J.*, 300, 167 and 462 (corrigendum).

Harding, A. K., 1990, private communication.

Harding, A. K. and Daugherty, J. K., 1991, *Ap. J.*, in press.

Harding, A. K. and Gaisser, T. K., 1990, *Ap. J.*, 358, 561.

Harding, A. K., Mészáros, P., Kirk, J. G. and Galloway, D., 1984, *Ap. J.*, 278, 369.

Harding, A. K., Petrosian, V. and Teegarden, B., 1986, in *Gamma-Ray Bursts*, ed. E. Liang and V. Petrosian (AIP, New York), p. 75.

Harding, A. K. and Preece, R., 1987, *Ap. J.*, 319, 939.

Harding, A. K. and Preece, R., 1989, *Ap. J.* (*Letters*), 338, L21.

Hartle, J. B. and Sabbadini, A. G., 1977, *Ap. J.*, 153, 807.

Hartmann, D., Epstein, R. I. and Woosley, S. E., 1989, *Nucl. Phys. B* (*Proc. Suppl.*), 10B, 27.

Hartmann, D., Epstein, R. I. and Woosley, S. E., 1990, *Ap. J.*, 348, 625.

Hartmann, D. and Woosley, S., 1988, in *Multiwavelength Astrophysics*, ed. F. Córdova (Cambridge University Press), p. 189.

Hartmann, D., Woosley, S. and Arons, J., 1988, *Ap. J.*, 332, 777.

Harwit, M. and Salpeter, E. E., 1973, *Ap. J.* (*Letters*), 186, L37.

Heemskerk, M. and van Paradijs, J., 1989, *Astron. Ap.*, 223, 235.

Heintzmann, H., 1981, *Nature*, 292, 811.

Helfand, D. J. and Tademaru, E. P., 1977, *Ap. J.*, 216, 842.

Helmken, H. F. and Weekes, T. C., 1979, *Ap. J.*, 228, 531.

Henrichs, H. F., 1983, in *Accretion Driven Stellar X-Ray Sources*, ed. W. Lewin and E. v. d. Heuvel (Cambridge University Press, Cambridge), p. 393.

Henriksen, R. N. and Norton, J. A., 1975, *Ap. J.*, 201, 431.

Hermsen, W., 1981, *Phil. Trans. Roy. Soc. London*, A401, 519.

Hermsen, W., *et al.*, 1987, *Astron. Ap.*, 175, 141.

Hernquist, 1984a, *M.N.R.A.S.*, 213, 313.

Hernquist, 1984b, *Ap. J. (Suppl.)*, 56, 325.

Herold, H., 1979, *Phys. Rev.*, D19, 2868.

Herold, H., Ruder, H. and Wunner, G., 1981, *Plasma Phys.*, 23, 775.

Herold, H., Ruder, H. and Wunner, G., 1982, *Astron. Ap.*, 115, 90.

Herold, H., Ruder, H. and Wunner, G., 1985, *Phys. Rev. Letters*, 54, 1452.

Herold, H., Wolf, K. and Ruder, H., 1987, *Ap. Space Sci.*, 131, 591.

Hewish, A., *et al.*, 1968, *Nature*, 217, 709.

Higdon, J. C. and Lingenfelter, R. E., 1990, *Ann. Rev. Astron. Ap.*, 28, 401.

Higdon, J. C. and Schmidt, M., 1990, *Ap. J.*, 355, 13.

Hillas, A. M., 1984, *Nature*, 312, 50.

Hillas, A. M., 1987, in *Very High Energy Gamma-Ray Astronomy*, ed. K. Turver (Reidel, Dordrecht), p. 243.

Hillebrandt, W. and Müller, E., 1976, *Ap. J.*, 207, 589.

Hillier, R. R., *et al.*, 1970, *Ap. J. (Letters)*, 162, L177.

Hinrichs, H. F., 1983, in *Accretion Driven Stellar X-Ray Sources*, ed. W. Lewin and E. v. d. Heuvel (Cambridge University Press, Cambridge), p. 393.

Ho, C., 1988, *M.N.R.A.S.*, 232, 91.

Ho, C., 1989, *Ap. J.*, 342, 396.

Ho, C., 1990, paper at the 175th A.A.S. Meeting, Washington, D.C.

Ho, C. and Arons, J., 1987, *Ap. J.*, 321, 404.

Ho, C. and Epstein, R., 1989a, *Ap. J.*, 343, 277.

Ho, C. and Epstein, R., 1989b, in *Proc. GRO Science Workshop*, ed. W. N. Johnson (NASA Goddard, Greenbelt, MD), pp. 4–436.

Ho, C., Epstein, R. and Fenimore, E., 1990, *Ap. J. (Letters)*, 348, L25.

Hoffman, J. A., *et al.*, 1978, *Nature*, 271, 630.

Holloway, N., 1973, *Nature Phys. Sci.*, 246, 6.

Holt, S. S. and McCray, R., 1982, *Ann. Rev. Astron. Ap.*, 20, 323.

Howard, W. M., Wilson, J. R. and Barton, R. T., 1981, *Ap. J.*, 249, 302.

Hoyle, F., *et al.*, 1953, *Phys. Rev.*, 92, 649.

Huebner, W. F., *et al.*, 1977, Los Alamos Report, LA-6760-M.

Hueter, G. J., 1984, in *AIP Conf. Proc. 115, High Energy Transients in Astrophysics*, ed. S. Woosley (AIP, New York), p. 373.

Hueter, G. J. and Gruber, D., 1982, in *Accreting Neutron Stars*, ed. W. Brinkmann and W. Truemper, MPE Rept. 177, p. 213.

Hui, A. K., *et al.*, 1978, *J.Q.S.R.T.*, 19, 509.

Hulse, R. A. and Taylor, J. H., 1974, *Ap. J. (Letters)*, 191, L59.

Humlicek, J., 1979, *J.Q.S.R.T.*, 21, 309.

Hurley, K., 1986a, in *AIP Conf. Proc. 141, Gamma-Ray Bursts*, ed. E. Liang and V. Petrosian (AIP, New York), p. 3.

Hurley, K., 1986b, in *The Origin and Evolution of Neutron Stars*, ed. D. Helfand (Reidel, Dordrecht), p. 489.

Hurley, K., 1990, *Adv. Space Res.*, 10, 179.

Hurley, K., Cline, T. and Epstein, R., 1986, in *AIP Conf. Proc. 141, Gamma-Ray Bursts*, ed. E. Liang and V. Petrosian (AIP, New York), p. 33.

Iben, I. and Renzini, A., 1984, *Phys. Rep.*, 105, 330.

Iben, I. and Tutukov, A. V., 1984, *Appl. J. (Suppl.)*, 54, 335.

Illarionov, A. F. and Sunyaev, R. A., 1975, *Astron. Ap.*, 39, 185.

Imamura, J. N. and Epstein, R. I., 1987, *Ap. J.*, 313, 711.

Inoue, H., 1975, *Pub. Astron. Soc. Japan*, 27, 311.

Iwamoto, N., 1980, *Phys. Rev. Letters*, 44, 1637.

Jackson, J. D., 1975, *Classical Electrodynamics* (Wiley, New York).

Jauch, J. M. and Rohrlich, F., 1976, *The Theory of Photons and Electrons* (Springer, New York).

Jennings, C., 1982, *Ap. J.*, 258, 110.

Jennings, D. M., *et al.*, 1974, *Nuovo Cim.*, 20, 71.

Johnson, M. H. and Lippman, B. A., 1949, *Phys. Rev.*, 241, 1153.

Jones, P. B., 1976, *Ap. J.*, 209, 602.

Jones, P. B., 1986, *M.N.R.A.S.*, 218, 477.

Jones, P. B., 1988, *M.N.R.A.S.*, 233, 875.

Joss, P. C. and Rappaport, S., 1983, *Nature*, 304, 419.

Joss, P. C. and Rappaport, S., 1984, *Ann. Rev. Astron. Ap.*, 22, 537.

Kadomtsev, B. and Kudryavtsev, V., 1971a, *Sov. Phys. JETP Letters*, 13, 9.

Kadomtsev, B. and Kudryavtsev, V., 1971b, *Sov. Phys. JETP Letters*, 13, 42.

Kafka, P. and Mayer, F., 1984, in *High Energy Transients in Astrophysics*, ed. S. Woosley (AIP, New York), p. 578.

Kahabka, P., 1987, Ph.D. thesis, University of Munich.

Kalkofen, W., 1987, in *Numerical Radiative Transfer*, ed. W. Kalkofen (Cambridge University Press, Cambridge).

Kaminker, A. D., Pavlov, G. G. and Mamradze, P. G., 1986, *Proc. Varenna-Abastumani Int. Workshop on Plasma Astrophysics* (ESA SP-251, Paris).

Kaminker, A. D., Pavlov, G. G. and Mamradze, P. G., 1990, *Ap. Space Sci.*, 174, 241.

Kaminker, A. D., Pavlov, G. G. and Shibanov, Yu. A., 1982, *Ap. Space Sci.*, 86, 249.

Kanbach, G., *et al.*, 1980, *Astron. Ap.*, 90, 163.

Kanno, S., 1975, *Pub. Astron. Soc. Japan*, 27, 287.

Kanno, S., 1980, *Pub. Astron. Soc. Japan*, 32, 117.

Karzas, W. J. and Latter, R., 1961, *Ap. J.* (*Suppl.*), 6, 167.

Katz, J. I. and Smith, I. A., 1988, *Ap. J.*, 326, 733.

Kaul, *et al.*, 1985, in *Proc. 19th ICRC* (La Jolla), 1, 165.

Kazanas, D. and Ellison, D., 1986, *Nature*, 319, 380.

Kelley, R., *et al.*, 1983, *Ap. J.*, 268, 790.

Kifune, T., *et al.*, 1987, in *Very High Energy Gamma-Ray Astronomy*, ed. K. Turver (Reidel, Dordrecht), p. 173.

Kii, T., *et al.*, 1986, *Pub. Astron. Soc. Japan*, 38, 751.

Kippenhahn, R. and Weigart, A., 1967, *Z. f. Astrophysik*, 65, 251.

Király, P. and Mészáros, P., 1988, *Ap. J.*, 333, 719.

Kirk, J. G., 1980, *Plasma Phys.*, 22, 639.

Kirk, J. G., 1984, Max-Planck Rep. MPA 158 (invited talk at the 1984 meeting of the Australian Astron. Soc.).

Kirk, J. G., 1985, *Astron. Ap.*, 142, 430.

Kirk, J. G., 1986, *Astron. Ap.*, 158, 305.

Kirk, J. G. and Cramer, N. F., 1985, *Australian J. Phys.*, 38, 715.

Kirk, J. G. and Galloway, D. J., 1982, *Plasma Phys.*, 24, 339 and 1025.

Kirk, J. G. and Melrose, D. B., 1986, *Astron. Ap.*, 156, 277.

Kirk, J. G. and Mészáros, P., 1980, *Ap. J.*, 241, 1153.

Klebesadel, R. W., Fenimore, E. and Laros, J., 1984, in *AIP Conf. Proc. 115, High Energy Transients in Astrophysics*, ed. S. Woosley (AIP, New York), p. 429.

Klebesadel, R. W., Strong, I. B. and Olson, R. A., 1973, *Ap. J.* (*Letters*), 182, L85.

Klein, R. I. and Arons, J., 1989, in *Proc. 23rd ESLAB Symp. on Two Topics in X-Ray Astronomy*, ed. J. Hunt and B. Battrick (ESA SP-296), p. 89.

Komesaroff, M. M., 1970, *Nature*, 225, 612.

Kössl, D., *et al.*, 1988, *Astron. Ap.*, 205, 347.

Kouveliotou, C., *et al.*, 1988, *Ap. J.* (*Letters*), 330, L101.

Koyama, K., *et al.*, 1989, *Pub. Astron. Soc. Japan*, 41, 461.

Kozlenkov, A. A. and Mitrofanov, I. G., 1985, *Sov. Astron.*, 29, 591.

Kozlenkov, A. K. and Mitrofanov, I. G., 1987, *Sov. Phys. JETP*, 64, 1173.

Krall, N. A. and Trivelpiece, A. W., 1973, *Principles of Plasma Physics* (McGraw-Hill, New York).

Kraus, U., *et al.*, 1989, *Astron. Ap.*, 223, 246.

Krausse-Polstorff, J. and Michel, C. F., 1985, *M.N.R.A.S.*, 213, 43P.

Kristian, J., *et al.*, 1989, *Nature*, 338, 234.

Kulkarni, S. R., 1986, *Ap. J.* (*Letters*), 306, L85.

Kulkarni, S. R., and Narayan, R., 1988, *Ap. J.*, 335, 755.

Kundt, W., 1981, *Astron. Ap.*, 98, 207.

Kuznetsov, A. V., *et al.*, *Sov. Astron. Letters*, 12, 755.

Lamb, D. Q., 1982, in *AIP Conf. Proc. 77, Gamma-Ray Transients and Related Astrophysical Phenomena*, ed. R. Lingenfelter *et al.* (AIP, New York), p. 249.

Lamb, D. Q., 1988, in *AIP Conf. Proc., Nuclear Spectroscopy of Astrophysical Plasmas*, ed. N. Gehrels and G. Share (AIP, New York), p. 265.

Lamb, D. Q. and Lamb, F. K., 1977, *Ann. N.Y. Acad. Sci.*, 302, 261.

Lamb, D. Q., Lamb, F. K. and Pines, D., 1973, *Nature Phys. Sci.*, 246, 52.

Lamb, D. Q., Wang, J. C. and Wasserman, I., 1990, *Ap. J.*, 363, 370.

Lamb, D. Q. *et al.*, 1989, *Ann. N.Y. Acad. Sci.*, 571, 460.

Lamb, F. K., 1985, in *Galactic and Extragalactic Compact X-Ray Sources*, ed. Y. Tanaka and W. Lewin (ISAS, Tokyo), p. 19.

Lamb, F. K., 1988, in *Timing Neutron Stars*, ed. H. Ögelman and E. v. d. Heuvel (NATO ASI vol. 262, Kluwer, Dordrecht), p. 649.

Lamb, F. K. and Pethick, C. J., 1974, in *Astrophysics and Gravitation, 16th Int. Solvay Congress* (Univ. Bruxelles, Brussels, Belgium), p. 135.

Lamb, F. K., Pethick, C. J. and Pines, D., 1973, *Ap. J.*, 184, 271.

Lamb, F. K., *et al.*, 1978, *Ap. J.*, 225, 582.

Lamb, R. C., *et al.*, 1977, *Ap. J.* (*Letters*), 212, L63.

Lamb, R. C., *et al.*, 1987, in *Very High Energy Gamma-Ray Astronomy*, ed. K. Turver (Reidel, Dordrecht), p. 139.

Lamb, R. C., *et al.*, 1988, *Ap. J.* (*Letters*), 328, L13.

Landau, L. D., 1932, *Phys. Abh. Sov. Union*, 1, 285.

Landau, L. D. and Lifshitz, E. M., 1965a, *Mechanics* (Addison-Wesley, Reading, MA).

Landau, L. D. and Lifshitz, E. M., 1965b, *Quantum Mechanics* (Addison-Wesley, Reading, MA).

Landau, L. D. and Lifshitz, E. M., 1977, *Statistical Physics* (Pergamon, Oxford).

Lang, K. R., 1980, *Astrophysical Formulae* (Springer-Verlag, Berlin), p. 453.

Lang, P., *et al.*, 1990, in *Proc. 21st ICRC* (Adelaide), 2, 139.

Langer, S. H., 1981, *Phys. Rev.*, D25, 1157.

Langer, S. H., McCray, R. and Baan, W. A., 1980, *Ap. J.*, 238, 731.

Langer, S. H. and Rappaport, S., 1982, *Ap. J.*, 257, 733.

Laros, J. G., *et al.*, 1982, *Ap. Space Sci.*, 88, 243.

Laros, J. G., *et al.*, 1984, in *AIP Conf. Proc. 115, High Energy Transients in Astropohysics*, ed. S. Woosley (AIP, New York), p. 378.

Latal, H. G., 1986, *Ap. J.*, 309, 372.

Lewin, W., 1980, in *Globular Clusters*, ed. D. Hanes and B. Madore (Cambridge University Press), p. 315.

Lewin W., van Paradijs, J. and van der Klis, M., 1988, *Space Sci. Rev.*, 46, 273.

Liang, E. P., 1987, *Comm. Ap.*, 12, 35.

Liang, E. P., Jernigan, T. and Rodrigues, R., 1983, *Ap. J.*, 271, 766.

Lichti, G. G., *et al.*, 1980, in *Adv. Space Exploration* (Pergamon, Oxford), vol. 7, p. 49.

Lieu, R., 1983, *M.N.R.A.S.*, 205, 973.

Lloyd-Evans, J., *et al.*, 1983, *Nature*, 305, 784.

Loudon, R., 1959, *Am. J. Phys.*, 27, 649.

Lovelace, R., 1976, *Nature*, 262, 649.

Lund, N., 1990, in *Los Alamos Workshop on Gamma-Ray Bursts*, ed. R. Epstein *et al.* (Cambridge University Press), in press.

Lyne, A. G., 1987a, in *High Energy Phenomena around Collapsed Stars*, ed. F. Pacini (NATO ASI, Kluwer, Dordrecht), p. 121.

Lyne, A. G., 1987b, in *The Origin and Evolution of Neutron Stars*, ed. D. Helfand and J.-H. Huang (Reidel, Dordrecht), p. 63.

Lyne, A. G. and Manchester, R. N., 1988, *M.N.R.A.S.*, 234, 447.

Lyne, A. G., Manchester, R. N. and Taylor, J. H., 1985, *M.N.R.A.S.*, 213, 613.

Lyne, A. G. and Smith, F. G., 1990, *Pulsar Astronomy* (Cambridge University Press).

Lyne, A. G., *et al.*, 1988, *IAU Circ.*, no. 4537.

Lyubarsky, Yu. E., 1986, *Astrophysics (Astrofizika)*, 25, 383.

Maile, T., *et al.*, 1989, *Astron. Ap.*, 223, 251.

Makino, F., and the *Ginga* Team, 1989, *IAU Circ.*, no. 4872.

Makishima, K., *et al.*, 1990, paper delivered at the Astr. Soc. Japan meeting.

Makishima, K., *et al.*, 1991, *Ap. J.* (*Letters*), 365, L59.

Manchester, R. N., 1987, in *The Origin and Evolution of Neutron Stars*, ed. D. Helfand (Reidel, Dordrecht), p. 3.

Manchester, R. N. and Taylor, J. H., 1977, *Pulsars* (Freeman, San Francisco).

Manchester, R. N. and Taylor, J. H., 1981, *Astron. J.*, 86, no. 12, 1953.

Maraschi, L. and Cavaliere, A., 1977, *Highlights of Astronomy*, 4, 127.

Maraschi, L., Reina, C. and Treves, A., 1974, *Astron. Ap.*, 35, 389.

Marshak, M. L., *et al.*, 1985a, *Phys. Rev. Letters*, 54, 2079.

Marshak, M. L., *et al.*, 1985b, *Phys. Rev. Letters*, 55, 1965.
Matsuoka, M., *et al.*, 1985, in *Galactic and Extragalactic Compact X-Ray Sources*, ed. Y. Tanaka and W. Lewin (I.S.A.S., Tokyo), p. 45.
Matz, S. M., *et al.*, 1985, *Ap. J.* (*Letters*), 288, L37.
Maxwell, O. V., 1979, *Ap. J.*, 231, 201.
Maxwell, O. V., *et al.*, 1977, *Ap. J.*, 216, 77.
Mazets, E. P., *et al.*, 1981, *Nature*, 290, 378.
Mazets, E. P., *et al.*, 1982a, *Ap. Space Sci.*, 82, 261.
Mazets, E. P., *et al.*, 1982b, *Ap. Space Sci.*, 84, 173.
Mazets, E. P., *et al.*, 1988, *Adv. Space Res.*, 8, no. 2, 669.
McCray, R. and Lamb, F. K., 1976, *Nature*, 262, 356.
McCulloch, P. M. and Hamilton, P. A., 1990, *Proc. IAU Coll. 128*, Lagow, Poland, in press.
McKenna, J., 1988, in *Timing Neutron Stars*, ed. H. Ögelman and E. v. d. Heuvel (NATO ASI, Kluwer, Dordrecht), p. 143.
Meegan, C. A., Fishman, G. J. and Wilson, R. B., 1985, *Ap. J.*, 291, 479.
Melia, F., 1988, *Nature*, 336, 658.
Melia, F., 1990a, *Ap. J.*, 351, 601.
Melia, F., 1990b, *Ap. J.*, 357, 161.
Melia, F. and Fatuzzo, M., 1989, *Ap. J.*, 346, 378.
Melrose, D. B., 1974, *Plasma Phys.*, 16, 845.
Melrose, D. B. and Kirk, J. G., 1986, *Astron. Ap.*, 156, 268.
Melrose, D. B. and Padden, W. E. P., 1986, *Australian J. Phys.*, 39, 961.
Melrose, D. B. and Parle, A. J., 1983, *Australian J. Phys.*, 36, 799.
Melrose, D. B. and Stoneham, R. J., 1976, *Nuovo Cim.*, A32, 435.
Melrose, D. B. and Zheleznyakov, V. V., 1981, *Astron. Ap.*, 95, 86.
Mestel, L. and Wang, Y.-M., 1979, *M.N.R.A.S.*, 188, 799.
Mestel, L. and Wang, Y.-M., 1982, *M.N.R.A.S.*, 198, 405.
Mészáros, P., Harding, A. K., Kirk, J. G. and Galloway, D., 1983, *Ap. J.* (*Letters*), 266, L33.
Mészáros, P. and Nagel, W., 1985a, *Ap. J.*, 298, 147.
Mészáros, P. and Nagel, W., 1985b, *Ap. J.*, 299, 138.
Mészáros, P., Nagel, W. and Ventura, J., 1980, *Ap. J.*, 238, 1066.
Mészáros, P., Pavlov, G. G. and Shibanov, Yu. N., 1989, *Ap. J.*, 337, 426.
Mészáros, P. and Riffert, H., 1988, *Ap. J.*, 327, 712.
Mészáros, P., Riffert, H. and Bagoly, Z., 1989, *Ap. J.* (*Letters*), 337, L23.
Mészáros, P. and Ventura, J., 1978, *Phys. Rev. Letters*, 41, 1544.
Mészáros, P. and Ventura, J., 1979, *Phy. Rev.*, D19, 3565.
Mészáros, P., *et al.*, 1988, *Ap. J.*, 324, 1056.

Michel, F. C., 1974, *Ap. J.*, 192, 713.

Michel, F. C., 1979, *Ap. J.*, 227, 579.

Michel, F. C., 1982, *Rev. Mod. Phys.*, 54, 1.

Michel, F. C., 1985, *Ap. J.*, 290, 721.

Michel, F. C., 1986, *Phys. Today*, 39, no. 10, p. 9.

Michel, F. C., 1987, *Nature*, 309, 311.

Michel, F. C., 1991, *Theory of Neutron Star Magnetospheres* (University of Chicago Press, Chicago).

Michel, F. C. and Goldwire, H. C., 1970, *Ap. Letters*, 5, 21.

Migdal, A. B., Chernoustan, A. J. and Mishustin, I. N., 1979, *Phys. Letters*, 83B, 158.

Mihalas, D., 1978, *Stellar Atmospheres*, 2d ed. (Freeman, San Francisco).

Mihara, T., *et al.*, 1990, *Nature*, 346, 250.

Miller, G. S., Epstein, R. I., Nolta, J. P. and Fenimore, E. E., 1990, in *Los Alamos Workshop on Gamma-Ray Bursts*, ed. R. Epstein *et al.* (Cambridge University Press), in press.

Miller, G. S., Salpeter, E. E. and Wasserman, I., 1987, *Ap. J.*, 314, 215.

Miller, G. S. and Wasserman, I., 1985, *Phys. Rev.*, A31, 120.

Miller, G. S., Wasserman, I. and Salpeter, E. E., 1989, *Ap. J.*, 346, 405.

Mitrofanov, I. G., 1984, *Ap. Space Sci.*, 105, 245.

Mitrofanov, I. G. and Ostryakov, V. M., 1981, *Ap. Space Sci.*, 77, 469.

Mitrofanov, I. G., *et al.*, 1986, *Sov. Astron.*, 30, 659.

Mitsui, F., *et al.*, 1987, *Proc. 20th ICRC* (Moscow).

Molnar, L., 1985, Ph.D. thesis, Harvard University.

Morello, C., *et al.*, 1983, *Proc. 18th ICRC* (Bangalore), 1, 127.

Morse, R., 1987, in *Very High Energy Gamma-Ray Astronomy*, ed. K. Turver (Reidel, Dordrecht), p. 197.

Motch, C., *et al.*, 1985, *Astron. Ap.*, 145, 201.

Müller, E., 1984, *Astron. Ap.*, 130, 415.

Murakami, T., 1989, in *Proc. 23rd ESLAB Symposium on Two Topics in X-Ray Astronomy*, ed. J. Hunt and B. Battrick (ESA SP-296), p. 173.

Murakami, T., 1990, *Adv. Space Res.*, 10, no. 2, 63.

Murakami, T., *et al.*, 1988, *Nature*, 335, 234.

Murakami, T., *et al.*, 1990, in *Los Alamos Workshop on Gamma-Ray Bursts*, ed. R. Epstein *et al.* (Cambridge University Press), in press.

Muslimov, A. G. and Tsygan, A. I., 1985, *Sov. Astron. Letters*, 11, 80.

Nagase, F., 1989a, *Pub. Astron. Soc. Japan*, 41, 1.

Nagase, F., 1989b, in *Proc. 23rd ESLAB Symposium on Two Topics in X-Ray Astronomy*, ed. J. Hunt and B. Battrick (ESA SP-296), p. 45.

Nagase, F., 1990, private communication.

Nagase, F., *et al.*, 1990, *Ap. J.* (*Letters*), 351, L13.

Nagel, W., 1980, *Ap. J.*, 236, 904.

Nagel, W., 1981a, *Ap. J.*, 251, 278.

Nagel, W., 1981b, *Ap. J.*, 251, 288.

Nagel, W., 1982, Ph.D. thesis, University of Munich.

Nagel, W., 1983, in *Accreting Neutron Stars*, ed. W. Brinkmann and J. Trümper, MPE Rep. 177, p. 302.

Nagel, W. and Ventura, J., 1983, *Astron. Ap.*, 148, 66.

Narayan, R. and Vivekanand, M., 1983, *Astron. Ap.*, 122, 45.

Neshpor, Yu. A., *et al.*, 1979, *Ap. Space Sci.*, 61, 349.

Nomoto, K., 1986, in *IAU Symp. 125, The Origin and Evolution of Neutron Stars*, ed. D. Helfand and J. H. Huang (Reidel, Dordrecht), p. 281.

Nomoto, K. and Iben, I., 1985, *Ap. J.*, 297, 531.

Nomoto, K. and Kondo, J., 1991, *Ap. J.* (*Letters*), 367, L19.

Nomoto, K. and Tsuruta, S., 1986, *Ap. J.* (*Letters*), 305, L19.

Nomoto, K. and Tsuruta, S., 1987, *Ap. J.*, 312, 711.

Norris, J. P., *et al.*, 1986, *Ap. J.*, 301, 213.

North, A. R., *et al.*, 1987, *Nature*, 326, 567.

Novick, R., *et al.*, 1977, *Ap. J.* (*Letters*), 215, L117.

Ögelman, H., 1988, in *Timing Neutron Stars*, ed. H. Ögelman and E. v. d. Heuvel (NATO ASI vol. 262, Kluwer, Dordrecht), p. 193.

Ohashi, T., *et al.*, 1984, *Pub. Astron. Soc. Japan*, 36, 699.

Oppenheimer, J. R. and Volkoff, G. M., 1939, *Phys. Rev.*, 55, 374.

Oster, L., 1960, *Phys. Rev.*, 119, 1444.

Ostriker, J. P. and Gunn, J. E., 1969, *Ap. J.*, 157, 1395.

Ostriker, J. P. and Gunn, J. E., 1970, *Nature*, 223, 813.

Oyama, Y., *et al.*, 1986, *Phys. Rev. Letters*, 56, no. 9, 991.

Pacini, F., 1967, *Nature*, 216, 567.

Pacini, F. and Ruderman, M., 1974, *Nature*, 251, 399.

Paczyński, B., 1970, in *Mass Loss and Evolution of Close Binaries*, ed. K. Gyldenkerne and R. West (Kopenhagen Univ. Publ. Funds), p. 142.

Paczyński, B., 1986, *Ap. J.* (*Letters*), 308, L43.

Paczyński, B., 1990a, in *Los Alamos Workshop on Gamma-Ray Bursts*, ed. R. Epstein *et al.* (Cambridge University Press), in press.

Paczyński, B., 1990b, *Ap. J.*, 348, 485.

Pakey, D. D., 1990, Ph.D. thesis, University of Illinois.

Pakey, D. D., Bussard, R. W. and Lamp, F. K. 1989, preprint.

Pandharipande, V. R., Pines, D. and Smith, R. A., 1976, *Ap. J.*, 208, 550.

Parker, E. N., 1979, *Cosmical Magnetic Fields* (Oxford University Press, Oxford).

Parmar, A. N., *et al.*, 1991, *Ap. J.*, 366, 253.

Pavlov, G. G., 1986, *Proc. Varenna-Abastumani Workshop on Plasma Astrophysics* (ESA SP-251, Paris), p. 383.

Pavlov, G. G. and Bezchastnov, V. G., 1988, *Adv. Space Res.*, 8, nos. 2–3, p. 563.

Pavlov, G. G. and Gnedin, Yu. N., 1984, in *Ap. Space Phys. Rev.* [*Sov. Sci. Rev.*, sect. E], ed. R. Sunyaev, vol. 3, p. 197.

Pavlov, G. G. and Golenetskii, S. V., 1986, *Ap. Space Sci.*, 73, 33.

Pavlov, G. G. and Panov, A. N., 1976, *JETP*, 44, 300.

Pavlov, G. G. and Shibanov, Yu. A., 1979, *Sov. Phys. JETP*, 49, 741.

Pavlov, G. G., Shibanov, Yu. A. and Yakovlev, D. G., 1980, *Ap. Space Sci*, 73, 33 (PSY).

Pavlov, G. G. and Yakovlev, D. G., 1976, *Sov. Phys. JETP*, 43, 389.

Pederson, H., *et al.*, 1983, *Ap. J.* (*Letters*), 270, L43.

Phinney, E. S., *et al.*, 1988, *Nature*, 333, 832.

Pineault, S., 1990, in *Los Alamos Workshop on Gamma-Ray Bursts*, ed. R. Epstein *et al.* (Cambridge University Press), in press.

Pineault, S. and Poisson, E., 1989, *Ap. J.*, 347, 1141.

Pines, D., 1980, *Science*, 207, 597.

Pines, D., 1985, in *High Energy Phenomena around Collapsed Stars*, vol. 195, ed. F. Pacini (NATO ASI, Kluwer, Dordrecht), p. 193.

Pines, D., 1987, in *High Energy Phenomena around Collapsed Stars*, ed. F. Pacini (NATO ASI, Kluwer, Dordrecht), p. 193.

Pines, D. and Alpar, M. A., 1985, *Nature*, 316, 27.

Pizzichini, G., 1986, *Ap. J.*, 301, 641.

Pollock, E. and Hansen, J., 1973, *Phys. Rev.*, A8, 311.

Pomraning, G. C., 1978, *Radiation Hydrodynamics* (Pergamon, Oxford).

Pravdo, S., *et al.*, 1977, *Ap. J.* (*Letters*), 216, L23.

Pravdo, S. H. and Bussard, R. W., 1981, *Ap. J.* (*Letters*), 246, L120.

Pringle, 1981, *Ann. Rev. Astron. Ap.*, 19, 137.

Pringle, J. E. and Rees, M. J., 1972, *Astron. Ap.*, 21, 1.

Pröschl, P., *et al.*, 1982, *J. Phys.*, B15, 1959.

Proszinski, M. and Przybcien, D., 1984, in *Green Bank Workshop on Millisecond Pulsars*, ed. S. Reynolds and D. Stinebring (NRAO, Green Bank), p. 151.

Protheroe, R. J., 1987, *Proc. 20th ICRC* (Moscow).

Protheroe, R. J. and Clay, R. W., 1985, *Nature*, 315, 205.

Protheroe, R. J., *et al.*, 1984, *Ap. J.* (*Letters*), 280, L47.

Quenby, J. and Lieu, R., 1987, in *Proc. 20th Int. Cosmic. Ray Conf.* (Moscow), 2, 252.

Radhakrishnan, V. and Cooke, D. J., 1969, *Ap. Letters*, 3, 225.

Radhakrishnan, V. and Rankin, J. M., 1990, *Ap. J.*, 352, 258.

Radhakrishnan, V. and Srinivasan, G., 1983, *Curr. Sci.*, 51, 1096.

Ramana Murthy, P. V. and Wolfendale, A. W., 1986, *Gamma-Ray Astronomy* (Cambridge University Press, Cambridge).

Ramaty, R., Lingenfelter, R. and Bussard, R., 1981, *Ap. Space Sci.*, 75, 193.

Ramaty, R. and Mészáros, P., 1981, *Ap. J.*, 250, 384.

Ramaty, R., *et al.*, 1980, *Nature*, 287, 122.

Rankin, J. M., 1983, *Ap. J.*, 274, 333 and 359.

Rankin, J. M., 1990, *Ap. J.*, 352, 247.

Rankin, J. M., *et al.*, 1970, *Ap. J.*, 162, 707.

Rao, M. V. S., *et al.*, 1989, *IAU Circ.*, no. 4883.

Rappaport, S., Verbunt, F. and Joss, P., 1983, *Ap. J.*, 275, 713.

Raubenheimer, B. C., *et al.*, 1986, *Ap. J.* (*Letters*), 307, L43.

Raubenheimer, B. C., *et al.*, 1989, *Ap. J.*, 336, 394.

Ray, A. and Datta, B., 1984, *Ap. J.*, 282, 542.

Ray, A. and Kluzniak, W., 1990, *Nature*, 344, 415.

Rebetzky, A., *et al.*, 1988, *Astron. Ap.*, 205, 215.

Rebetzky, A., *et al.*, 1989, *Astron. Ap.*, 225, 137.

Rees, M. J., 1982, in *Accreting Neutron Stars*, ed. W. Brinkmann and W. Truemper, MPE Rep. 177, p. 179.

Resvanis, L., *et al.*, 1987a, in *Very High Energy Gamma-Ray Astronomy*, ed. K. Turver (Reidel, Dordrecht), p. 105.

Resvanis, L., *et al.*, 1987b, in *Very High Energy Gamma-Ray Astronomy*, ed. K. Turver (Reidel, Dordrecht), p. 131.

Resvanis, L., *et al.*, 1987c, in *Very High Energy Gamma-Ray Astronomy*, ed. K. Turver (Reidel, Dordrecht), p. 135.

Resvanis, L., *et al.*, 1988, *Ap. J.* (*Letters*), 328, L9.

Rhoades, C. E. and Ruffini, R., 1974, *Phys. Rev. Letters*, 32, 324.

Ricker, G., 1990, in *Los Alamos Workshop on Gamma-Ray Bursts*, ed. R. Epstein *et al.* (Cambridge University Press), in press.

Riffert, H., 1980, *Ap. Space Sci.*, 71, 195.

Riffert, H., 1983, Ph.D. thesis, University of Munich.

Riffert, H., 1987, *Astron. Ap.*, 172, 241.

Riffert, H., Mészáros, P. and Bagoly, Z., 1989, *Ap. J.*, 340, 443.

Romani, R. W., 1990a, in *Supernova and Stellar Evolution*, ed. A. Ray and T. Velusamy (World Scientific, Singapore), in press.

Romani, R. W., 1990b, *Nature*, 347, 741.

Romani, R. W., Kulkarni, S. and Blandford, R., 1987, *Nature*, 329, 309.

Rösner, W., *et al.*, 1984, *J. Phys.*, B17, 29.

Rudak, B. and Mészáros, P., 1991, *Ap. J.*, in press.

Ruder, H., *et al.*, 1981, *Phys. Rev. Letters*, 46, 1700.

Ruderman, M., 1971, *Phys. Rev. Letters*, 27, 1306.

Ruderman, M., 1974, in *IAU Symp. 53, Physics of Dense Matter*, ed. C. Hansen (Reidel, Dordrecht), p. 117.

Ruderman, M., 1975, *Ann. N.Y. Acad. Sci.*, 262, 164.

Ruderman, M., 1976, *Ap. J.*, 203, 213.

Ruderman, M., 1987, in *High Energy Phenomena around Collapsed Stars*, ed. F. Pacini (NATO ASI, Reidel, Dordrecht), p. 145.

Ruderman, M., 1991, *Ap. J.*, 366, 261.

Ruderman, M. and Cheng, K. S., 1988, *Ap. J.*, 335, 306.

Ruderman, M. and Shaham, J., 1983, *Nature*, 304, 425.

Ruderman, M., Shaham, J. and Tavani, M., 1989, *Ap. J.*, 336, 507.

Ruderman, M. and Sutherland, P. G., 1975, *Ap. J.*, 196, 51.

Ruderman, M., *et al.*, 1989, *Ap. J.*, 343, 292.

Rybicki, G. B. and Lightman, A. P., 1979, *Radiative Processes in Astrophysics* (Wiley, New York).

Sakurai, J. J., 1967, *Advanced Quantum Mechanics* (Addison-Wesley, Reading, MA).

Salpeter, E. E., 1952, *Ap. J.*, 115, 326.

Salpeter, E. E., 1967, in *Lectures in Applied Mathematics*, vol. 10 (American Mathematical Society).

Samorski, M. and Stamm, W., 1983, *Ap. J.* (*Letters*), 268, L17.

Sang, Y. and Chanmugam, G., 1987, *Ap. J.* (*Letters*), 323, L61.

Sang, Y. and Chanmugam, G., 1990, *Ap. J.*, 363, 597.

Sauls, J. A., 1988, in *Timing Neutron Stars*, ed. H. Ögelman and E. v. d. Heuvel (NATO ASI, vol. C-262, Kluwer, Dordrecht), p. 457.

Savonije, G. J., 1983, in *Accretion Driven Stellar X-Ray Sources*, ed. W. Lewin and E. v. d. Heuvel (Cambridge University Press), p. 343.

Savonije, G. J. and Papaloizu, J. C., 1983, *M.N.R.A.S.*, 203.

Schaeffer, B. E., *et al.*, 1983, *Ap. J.* (*Letters*), 270, L49.

Scharlemann, E. T., Arons, J. and Fawley, W. M., 1978, *Ap. J.*, 222, 297.

Scharlemann, E. T. and Wagoner, R. V., 1973, *Ap. J.*, 182, 951.

Scheepmaker, T., *et al.*, 1981, *Space Sci. Rev.*, 30, 325.

Schreier, E., *et al.*, 1972, *Ap. J.* (*Letters*), 172, L79.

Schwarzschild, M., 1965, *Structure and Evolution of Stars* (Dover, New York).

Seward, F. D. and Harnden, F. R., 1982, *Ap. J. (Letters)*, 256, L45.

Seward, F. D., *et al.*, 1984, *Ap. J. (Letters)*, 287, L19.

Shabad, A. E. and Usov, V. V., 1982, *Nature*, 295, 215.

Shabad, A. E. and Usov, V. V., 1984, *Ap. Space Sci.*, 102, 327.

Shabad, A. E. and Usov, V. V., 1985, *Ap. Space Sci.*, 117, 309.

Shafranov, V. D., 1967, *Reviews of Plasma Physics*, vol. 3, ed. M. A. Leontovich (Consultants Bureau, New York).

Shakura, N. I. and Sunyaev, R. A., 1973, *Astron. Ap.*, 24, 337.

Shapiro, S. L. and Lightman, A. P., 1976, *Ap. J.*, 204, 555.

Shapiro, S. L. and Salpeter, E. E., 1975, *Ap. J.*, 198, 671.

Shapiro, S. L. and Teukolsky, S. A., 1983, *Black Holes, White Dwarfs and Neutron Stars* (Wiley, New York).

Share, G. H., *et al.*, 1986, *Adv. Space Res.*, 6, no. 4, 15.

Shibazaki, N., 1989, in *Proc. 23rd ESLAP Symposium on Two Topics in X-Ray Astronomy*, ed. J. Hunt and B. Battrick (ESA SP-296).

Shibazaki, N. and Lamb, F. K., 1989, *Ap. J.*, 346, 808.

Shibazaki, N., *et al.*, 1989, *Nature*, 342, 656.

Shinoda, K., *et al.*, 1988, in *Physics of Neutron Stars and Black Holes*, ed. Y. Tanaka (Acad. Press, Tokyo), p. 67.

Silant'ev, N. A., 1981, *Ap. Space Sci.*, 82, 363.

Simola, J. and Virtamo, J., 1978, *J. Phys.*, B11, 3309.

Sitenko, A. G., 1964, *Electromagnetic Fluctuations in Plasmas* (Academic Press, New York).

Skinner, G. K., *et al.*, 1982, *Nature*, 297, 568.

Slane, P. and Fry, W. F., 1989, *Ap. J.*, 342, 1129.

Smith, F. G., 1986, *M.N.R.A.S.*, 219, 729.

Smolukowski, R. and Welch, D., 1970, *Phys. Rev. Letters*, 24, 1191.

Sobolev, V. V., 1963, *A Treatise on Radiative Transfer* (Van Nostrand, Princeton).

Soffel, M., *et al.*, 1983, *Astron. Ap.*, 126, 251.

Soker, N. and Livio, M., 1984, *M.N.R.A.S.*, 211, 927.

Sokolov, A. A. and Ternov, I. M., 1968, *Synchrotron Radiation* (Akademie Verlag, Berlin).

Sokolov, A. A. and Ternov, I. M., 1986, *Radiation from Relativistic Electrons* (AIP, New York).

Spitzer, L., Jr., 1962, *Physics of Fully Ionized Gases* (Wiley, Interscience, New York).

Srinivasan, G., 1990, *Adv. Space Res.*, 10, no. 2, 167.

Srinivasan, G., Bhattacharya, D., Muslimov, A. G. and Tsygan, A. I., 1990, *Curr. Sci.*, 59, 31.

Stanev, T., *et al.*, 1985, *Phys. Rev.*, D32, 1244.
Stecker, F., 1968, *Phys. Rev. Letters*, 21, 1016.
Stecker, F., 1971, *Cosmic Gamma-Rays* (NASA SP-249).
Stecker, F., 1973, *Ap. J.*, 185, 499.
Stella, L. and Rosner, R., 1984, *Ap. J.*, 277, 312.
Stella, L., White, N. and Priedhorsky, W., 1987, *Ap. J.* (*Letters*), 315, L49.
Stepanian, A. A., 1982, *Ap. Space Sci.*, 84, 347.
Stinebring, D. R., *et al.*, 1984, *Ap. J.* (*Suppl.*), 55, 279.
Stix, M., 1962, *The Theory of Plasma Waves* (McGraw-Hill, New York).
Storey, M. and Melrose, D. B., 1987, *Australian J. Phys.*, 40, 1989.
Sturrock, P. A., 1971, *Ap. J.*, 164, 529.
Sturrock, P. A., 1986, *Nature*, 321, 47.
Sturrock, P. A., Harding, A. K. and Daugherty, J. K., 1989, *Ap. J.*, 346, 950.
Suga, K., *et al.*, 1985, in *Techniques in UHE Gamma-Ray Astronomy*, ed. R. J. Protheroe and S. A. Stephens (University of Adelaide), p. 48.
Sunyaev, R. A., 1976, *Sov. Astron. Letters*, 2, 111.
Sunyaev, R. A. and Titarchuk, L. G., 1980, *Astron. Ap.*, 86, 121.
Svetozarova, G. I. and Tsytovich, V. N., 1962, *Radiofizika*, 5, 658.
Taam, R. E., 1983, *Ap. J.*, 270, 694.
Taam, R. E. and Fryxell, B. A., 1988, *Ap. J.* (*Letters*), 327, L73.
Taam, R. E. and van den Heuvel, E. P. J., 1986, *Ap. J.*, 305, 235.
Taam, R. E., *et al.*, 1978, *Ap. J.*, 222, 269.
Tademaru, E., 1974, *Ap. Space Sci.*, 30, 179.
Tananbaum, H., *et al.*, 1972, *Ap. J.* (*Letters*), 174, L143.
Taylor, J. H., 1987, in *IAU Symp. 125, The Origin and Evolution of Neutron Stars*, ed. D. Helfand and J.-H. Huang (Reidel, Dordrecht), p. 383.
Taylor, J. H., Manchester, R. N. and Hugenin, G. R., 1975, *Ap. J.*, 195, 513.
Taylor, J. H. and Stinebring, D. R., 1986, *Ann. Rev. Astron. Ap.*, 24, 285.
Taylor, J. H. and Weisberg, J. H., 1989, *Ap. J.*, 345, 434.
Teegarden, B. and Cline, T., 1980, *Ap. J.* (*Letters*), 236, L67.
Teegarden, B. and Cline, T., 1981, *Ap. Space Sci.*, 75, 181.
Thompson, D. J., Fichtel, C. E., Kniffen, D. A. and Ögelman, H. B., 1975, *Ap. J.* (*Letters*), 200, L79.
Toll, J. S., 1952, Ph.D. thesis, Princeton University.
Tremaine, S. and Zitkow, A., 1986, *Ap. J.*, 301, 155.
Trimble, V., 1982, *Rev. Mod. Phys.*, 54, 1183, and 55, 511.
Trümper, J., 1987, in *Very High Energy Astronomy*, ed. K. Turver (NATO ASI vol. 199, Reidel, Dordrecht), p. 7.
Trümper, J., *et al.*, 1978, *Ap. J.* (*Letters*), 219, L105.

Trümper, J., *et al.*, 1982, in *Accreting Neutron Stars*, ed. W. Brinkmann and J. Trümper (MPE rep. 177, Max Planck Inst. Extr. Phys.), p. 134.

Trümper, J., *et al.*, 1986, *Ap. J.* (*Letters*), 300, L63.

Tsai, W. and Erber, T., 1974, *Phys. Rev.*, D10, 492.

Tsygan, A., 1975, *Astron. Ap.*, 44, 21.

Tueller, J., *et al.*, 1984, *Ap. J.*, 279, 177.

Tutukov, A. V. and Yungelson, L. R., 1979, in *Mass Loss and Evolution of O-Type Stars*, ed P. Conti and C. de Loore (Reidel, Dordrecht), p. 401.

Urpin, V. A., Levshakov, S. A. and Yakovlev, D. G., 1986, *M.N.R.A.S.*, 219, 703.

Urpin, V. A. and Yakovlev, D. G., 1980a, *Soviet Astron.*, 24, 126.

Urpin, V. A. and Yakovlev, D. G., 1980b, *Soviet Astron.*, 24, 425.

Usov, V. V. and Shabad, A. E., 1983, *Sov. Astron. Letters*, 9, 212.

Usov, V. V. and Shabad, A. E., 1985, *Sov. Phys. JETP Letters*, 42, 19.

van Beveren, D., 1980, Ph.D. thesis, Free University of Brussels.

van den Heuvel, E. P. J., 1983, in *Accretion Driven Stellar X-Ray Sources*, ed. W. Lewin and E. v. d. Heuvel (Cambridge University Press), 303.

van den Heuvel, E. P. J., 1988, in *Timing Neutron Stars*, ed. H. Ögelman and E. v. d. Heuvel (NATO ASI vol. 262, Kluwer, Dordrecht), p. 523.

van den Heuvel, E. P. J., van Paradijs, J. and Taam, R., 1986, *Nature*, 322, 153.

van der Klis, M., 1983, Ph.D. thesis, Univ. of Amsterdam, the Netherlands.

van der Klis, M., 1989, *Ann. Rev. Astron. Ap.*, 27, 517.

van der Klis, M., *et al.*, 1985, *Nature*, 316, 225.

van Paradijs, J., 1983, in *Accretion Driven Stellar X-Ray Sources*, ed. W. Lewin and E. v. d. Heuvel (Cambridge University Press).

van Paradijs, J., 1988, in *Timing Neutron Stars*, ed. H. Ögelman and E. v. d. Heuvel (Kluwer, Dordrecht), p. 191.

Van Riper, K. and Lamb, D. Q., 1981, *Ap. J.* (*Letters*), 244, L13.

Ventura, J., 1973, *Phys. Rev.*, A8, 3021.

Ventura, J., 1979, *Phys. Rev.*, D19, 1684.

Ventura, J., 1988, in *Timing Neutron Stars*, ed. H. Ögelman and E. v. d. Heuvel (NATO ASI, C-262, Kluwer, Dordrecht), p. 491.

Ventura, J., Nagel, W. and Mészáros, P., 1979, *Ap. J.* (*Letters*), 233, L125.

Verbunt, F., 1988, in *The Physics of Neutron Stars and Black Holes*, ed. Y. Tanaka (Inst. Space and Astronaut. Sci., Tokyo).

Verbunt, F. and Rappaport, S., 1989, *Ap. J.*, 345, 210.

Verbunt, F., *et al.*, 1987, *Nature*, 309, 311.

Vestrand, T. and Eichler, D., 1979, in *AIP Conf. Proc. 56, Particle Acceleration Mechanisms in Astrophysics*, ed. J. Arons *et al.* (AIP, New York), p. 281.

Virtamo, J. and Jauho, P., 1975, *Nuovo Cim.*, 26B, 537.

Vishwanath, P. R., 1982, in *Proc. Int. Workshop on Very High Energy Gamma-Ray Astronomy*, ed. P. V. Ramana Murthy and T. C. Weekes (Ootacamund), p. 21.

Vitello, P. and Dermer, C., 1991, *Ap. J.*, in press.

Vladimirsky, B. M., *et al.*, 1973, *Proc. 13th ICRC* (Denver), 1, 456.

Voges, W., *et al.*, 1982, *Ap. J.*, 263, 803.

von Ballmoos, P., *et al.*, 1989, in *Proc. GRO Science Workshop*, ed. W. N. Johnson (NRL, Washington, DC), pp. 4–182.

von Weizsäcker, C. F., 1937, *Phys. Zeits.*, 38, 176.

Wallace, P. T., *et al.*, 1977, *Nature*, 266, 692.

Wallace, R. K., Woosley, S. E. and Weaver, T. A., 1982, *Ap. J.*, 258, 696.

Wang, J. C., Wasserman, I. M. and Salpeter, E. E., 1988, *Ap. J.* (*Suppl.*), 68, 735.

Wang, J. C., Wasserman, I. M. and Salpeter, E. E., 1989, *Ap. J.*, 338, 343.

Wang, J. C., *et al.*, 1989, *Phys. Rev. Letters*, 63, 1550.

Wang, Y.-M., 1981, *Astron. Ap.*, 102, 36.

Wang, Y.-M. and Frank, J., 1981, *Astron. Ap.*, 93, 255.

Wang, Y.-M., Nepveu, M. and Robertson, J. A., 1984, *Astron. Ap.*, 139, 93.

Wang, Y.-M. and Welter, G. L., 1981, *Astron. Ap.*, 102, 97.

Wang, Z. and Eichler, D., 1988, *Ap. J.*, 324, 966.

Wasserman, I. M. and Salpeter, E. E., 1980, *Ap. J.*, 241, 1107.

Webbink, R., Rappaport, S. and Savonije, G. J., 1983, *Ap. J.*, 270, 678.

Weekes, T. C., 1986, in *Accretion Processes in Astrophysics*, ed. J. Audouze and J. T. T. Van (Frontières, Gif-sur-Yvette), p. 57.

Weekes, T. C., 1988, *Phys. Rep.*, 160, 1.

Weekes, T. C., 1990, in *Proc. Snowmass High Energy Physics Meeting*, in press.

Weekes, T. C., *et al.*, 1989, *Ap. J.*, 342, 379.

Weisberg, J. M. and Taylor, J. H., 1984, *Phys. Rev. Letters*, 52, 1348.

Wheaton, W. A., *et al.*, 1979, *Nature*, 282, 240.

White, N., Swank, J. and Holt, S. S., 1983, *Ap. J.*, 270, 711.

Wills, R. D., *et al.*, 1982, *Nature*, 296, 723.

Wolszcan, A., *et al.*, 1988, *IAU Circ.*, no. 4552.

Woltjer, L., 1964, *Ap. J.*, 140, 1309.

Wood, K., Desai, U. and Schaefer, B., 1986, in *AIP Conf. Proc. 141, Gamma-Ray Bursts*, ed. E. Liang and V. Petrosian (AIP, New York), p. 4.

Woosley, S. E., 1982, in *Accreting Neutron Stars*, MPE Rep. 177, ed. W. Brinkmann and J. Trümper (Max Planck Inst., Garching), p. 189.

Woosley, S. E. and Taam, R., 1976, *Nature*, 263, 101.

Woosley, S. E. and Wallace, R. K., 1982, *Ap. J.*, 258, 716.

Woosley, S. E. and Weaver, T. A., 1985, in *Nucleosynthesis and Its Implications on Nuclear and Particle Physics*, 5th Moriond Conf., ed. J. Audouze and T. V. Than (Reidel, Dordrecht).

Woosley, S. E. and Weaver, T. A., 1986, *Ann. Rev. Astron. Ap.*, 24, 205.

Wunner, G., 1979, *Phys. Rev. Letters*, 42, 79.

Wunner, G. and Herold, H., 1979, *Ap. Space Sci.*, 63, 503.

Wunner, G., Herold, H. and Ruder, H., 1983, in *Positron-Electron Pairs in Astrophysics*, ed. M. Burns *et al.* (AIP, New York), p. 411.

Wunner, G. and Ruder, H., 1980, *Ap. J.*, 242, 828.

Wunner, G., Ruder, H. and Herold, H., 1981, *J. Phys.*, B14, 765.

Wunner, G., *et al.*, 1986, *Astron. Ap.*, 170, 179.

Yahel, R. Z., 1980, *Astron. Ap.*, 90, 26.

Yakovlev, D. G., 1982, *Sov. Astron.*, 26,416.

Yakovlev, D. G., 1984, *Ap. Space Sci.*, 98, 37.

Yoshida, A., *et al.*, 1989, *Pub. Astron. Soc. Japan*, 41, 509.

Yoshida, A., *et al.*, 1990, in *Los Alamos Workshop on Gamma-Ray Bursts*, ed. R. Epstein *et al.* (Cambridge University Press), in press.

Zdziarski, A., 1984, *Astron. AP.*, 134, 301.

Zdziarski, A., 1987, in *13th Texas Symp. Relativistic Astrophysics*, ed. M. Ulmer (World Scientific, Singapore), p. 553.

Zdziarski, A., Coppi, P. S. and Lamb, D. Q., 1990, *Ap. J.*, 357, 149.

Zdziarski, A. and Lamb, D. Q., 1987, *Ap. J.* (*Letters*), 309, L79.

Zel'dovich, Ya. B. and Raizer, Yu. P., 1972, *Physics of Shock Waves and High Temperature Hydrodynamic Phenomena* (Academic Press, New York).

Zel'dovich, Ya. B., Ruzmaikin, A. A. and Sokoloff, D. D., 1983, *Magnetic Fields in Astrophysics* (Gordon and Breach, New York).

Zel'dovich, Ya. B. and Shakura, S. S., 1969, *Sov. Astron.*, 13, 175.

Zheleznyakov, V. V., 1977, *Electromagnetic Waves in Cosmic Plasmas* (Nauka, Moscow).

Zheleznyakov, V. V. and Litvinchuk, A. A., 1987, *Sov. Astron.*, 31, 159.

Zyskin, Y. L. and Mukhanov, D. B., 1985, *Proc. 19th ICRC* (La Jolla), 1, 165.

Zyskin, Y. L., *et al.*, 1987, *Astron. Tsirk.*, no. 1508, p. 11.

# Index